FRANCIS GALTON'S GENIUS: 1865–1911

Francis Galton's Genius: 1865–1911

Nature Untangled from Nurture

Gavan Tredoux

galton.org

ISBN: 978-1-963225-05-1
United Kingdom Edition.

Francis Galton: a Lifetime of Exploration.

1. *Francis Galton's Nature and Nurture:* 1822–1865.
2. *Francis Galton's Genius:* 1865–1911.

Supplements:

1. *Sir Francis Sacheverel Darwin:* Charles Darwin's Adventurous Uncle.
2. *Francis Galton at Cambridge:* Letters and Diaries, 1840–1844.
3. *The Diary of Charles John Andersson:* 1850–1851.
4. *Francis Galton on Mars:* the Discontinuous Variation Notebook.
5. *Francis Galton and Alphonse de Candolle:* Notes and Correspondence.
6. *Francis Galton and Alfred Binet:* Correspondence.
7. *Francis Galton:* Selected Papers, Notebooks and Diaries.

I have no patience with the hypothesis occasionally expressed, and often implied, especially in tales written to teach children to be good, that babies are born pretty much alike, and that the sole agencies in creating differences between boy and boy, and man and man, are steady application and moral effort. It is in the most unqualified manner that I object to pretensions of natural equality. The experiences of the nursery, the school, the University, and of professional careers, are a chain of proofs to the contrary. ... A collection of living magnates in various branches of intellectual achievement is always a feast to my eyes; being, as they are, such massive, vigorous, capable-looking animals.

— Francis Galton, F. Galton 1869a.

———oOo———

Mr. F. Galton would refer first to the purely ethnological part of the memoir, which dwelt upon the difficulty of defining the Bantu race. He thought that ethnologists were apt to look upon race as something more definite than it really was. He presumed it meant no more than the average of the characteristics of all the persons who were supposed to belong to the race, and this average was continually varying. The popular notion seemed based upon some idea like that of a common descent of the different races, from a parent Noachian stock, whence the aborigines of each county were derived, and where they lived in unchanged conditions till the white man came. Nothing can be further from the truth. We know how in South Africa the Bantu population has been in constant seethe and change; how, in much less than a single century, Chaka and his tribe, Mosilekatse and his tribe, and others, have in turn become prominent nations, and the average of the whole Bantu population must thereby have differed at different times. This same fluctuation of the average qualities of the population must, for anything we can see to the contrary, have gone on for many thousands of years. He therefore thought the phrase of Bantu race, as signifying some invariable and definite type, to be a mere chimera. In the earlier part of his memoir, Sir Bartle Frere had compared our mode of treating uncivilized races to that of the Romans. He heartily wished that the resemblance held in certain essential points. Our military hold was as firm, our tolerance of local customs was as great, our dealings were as just, and more just than theirs. But we did not amalgamate with them as the Romans did, we did not intermarry; by means of our missionaries we pressed upon them a form of religion which was not the most congenial. Our civilization was stiff. This, and much more, was pointed out in a very able and most pathetic memoir by Mr. [Edward Wilmot] Blyden, the present Minister of Liberia to England, who is a full-blooded negro. The article appeared in "Fraser's Magazine" some years ago, and it showed the repressive effect of White civilization upon the Negroes, as contrasted with that of the Mohammedans. It was a shame to us as an Imperial nation, that representatives of the many people whom we governed, did not find themselves more at home among us. They seldom appeared in such meetings as the present one; they did not come to England. We did not see them in the streets. It was very different in ancient Rome, where the presence of foreigners from all parts of the then known world was a characteristic feature of every crowd. He did not now suggest any action, but merely wished to lay stress on this serious drawback to our national character as rulers of a great Empire.

— Francis Galton, F. Galton 1881c.

Contents

Preface

Like his better-known cousin Charles Darwin, Francis Galton was an acute observer. He grasped the nature of variation in a way that few of his contemporaries could match, absent revelation. Whereas they thought of variation as mere error—a regrettable nuisance—he understood it as the key to understanding mankind and the natural world.

This second volume concerns itself with the mature phase of Galton's enquiries. For the first phase of his career, see *Francis Galton's Nature and Nurture: 1822–1865.* Once he settled into his stride he would conduct the kind of path-finding forays that suited his temperament best, continually moving on to new ground while remaining focused on the same central questions, leaving the details, once the lie of the land was established, to the elaborations and footnotes of his successors.

As a result, between 1865, when he was nearly 43, and his death in 1911 when he was just short of 89, Galton constructed the research programme we now recognize as *behaviour genetics.* It is an attempt to understand the forces that shape differences in the way that humans actually turn out to be, rooted within the framework of natural selection. Though he got his start in African exploration, the technology of travel and the institutional world of Victorian science, it was this late phase that produced his enduring legacy.

To get there, Galton had to invent measures of the 'way that we are', or rather of the ways in which we differ. Then he had to find a means to collect large numbers of those measures, no easy thing to do in a loosely connected world without databases, governed by prudery and stifled by religion. Machines of his own invention simplified the task. Once that data was in hand he required techniques for representing and *analysing* it. Therefore he invented the core statistical techniques that underlie scientific reasoning today. Along the way this most unusual man unleashed a 'space program' of spin-offs: composite photographs, stereoscopic maps, fingerprints, branching processes, eugenics, mental testing. An entire research programme made from scratch.

To an extraordinary extent, many of the questions that Galton grappled with and contributed most to—the 'nature and nurture' synthesis that he named—are still very much alive today.

Figure 1: Francis Galton in his thirties, National Portrait Gallery.

Chapter 1

Frank's Ordinary Day

'Your bath is ready, Mr. Galton'. Gifi withdrew his head from the doorway to let his master perform his ablutions in private. *Things* had been happening, but the manservant pretended not to notice. A slim balding man, tall for his day at nearly six foot, rose and slipped on his dressing gown. On the way to the bathtub he picked up a pair of diving spectacles and a copy of the *Times*. Would those new adjustments work? Sitting comfortably in the warm water, he pressed the spectacles firmly on his nose, sealing them. Grasping the newspaper in both hands, he kept it safely in the air as he steadily sank down, until his head was covered with water. He stared up through the water at the newsprint, aided by the optically-correcting goggles.

Though he was well below the water, Galton found that he could now read the headlines clearly. Flipping to the scheduled events of the day, he verified that Dr. Livingstone's letters would be discussed at the Royal Geographical Society that evening. He had promised to attend, for the first time in years. Turning to other news, he glared at the usual distortions of fact. Holding the paper fully at arms length, the smallest print could still be read. By now a minute had passed, and an alarming feeling developed. After considering this for a while, the cause was apparent. He had forgotten to breathe. Bursting up into the air, some deep breaths restored his equilibrium. But now the newspaper was damp.

After his bath, Frank proceeded in his bathrobe to the water closet, tucked away at the top of a steep flight of stairs. When the door was latched from the inside, a semaphore deployed on the outside. Visitors on the ground floor could see that the room was engaged before bothering up the steps. Aged sisters contemplating the ascent from the ground floor were grateful.

Breakfast at 42 Rutland Gate was sparse when visiting explorers were not in town. Louisa, thin and pinched, had perennial gastric complaints and was not much of an eater. Her husband conformed without complaint. The teapot attracted most of his interest. Taking a notebook out of his pocket, he carefully noted the steeping time and took the temperature of the brew, deploying a large thermometer. There were a dozen entries from prior experiments, converging on a result.

The instrument was a Negretti & Zambra, recently tested at the Kew Laboratory, to which Frank had more or less retreated. The device worked well, provided it was not bumped, making it too delicate for use in African exploration. A sip suggested the emulsion was on the strong side, which suited his own taste. Duly noted. The mystery of the teapot would soon be cleared up. Louisa grimaced. These domestic experiments had amused her at first, but one of her week-long headaches was coming on.

For once it was not raining, suggesting a brisk walk to nearby Hyde Park while 'Lou' retired to her room. Perhaps there would be a procession to observe. Into a coat compartment dropped an object resembling a parcel wrapped in brown paper and string. Into a trouser pocket, a piece of paper resembling a cross with asymmetric arms joined a thimble with a needle protruding discreetly from the end, just enough to penetrate the paper but no more. A thumb could assume the thimble inside the pocket without attracting the least public attention. Hat and walking stick completed the ensemble, and Frank stepped out into the clear autumn day.

As he walked up through Knightsbridge, heading toward Kensington Road and over to Hyde Park, the emerging scientist surveyed the other pedestrians drawn out by the mild weather. Before he passed the fenced copse of grass and trees in the middle of Rutland Gate, two women bustled past in the opposite direction. Galton was ready for them. The thimble grasped the paper cross and inserted two marks next to each other on the short arm. They were both hideous—trout in crinoline. Possibly visitors from Aberdeen. Before he got to Kensington Road, a Brougham trundled past. Were the horses spying on him? It was hard to completely shake the feeling of paranoia he had worked for weeks to establish, until he was satisfied that he could reflexively suspect a blinkered horse of surveillance.

On the outskirts of Hyde Park a throng of people pressed behind an advance guard of young women, each of whom wore a green piece of clothing: a hatband or feather, perhaps a glove, scarf or rustic stole. Undecided about these specimens, marks were placed near the middle of the paper cross. The throng surged into the park, sticking as closely to each other as Damara cattle worried by lions. Street horrors were clambering into the trees to see events from the top branches. On a slight promontory, the lead-ox was already warming to his theme. Amnesty for Fenians! Galton paused at a distance to survey the reaction of the crowd as the speaker took command of them. A pinkish flush of cheeks intensified with the excitement, then slowly dissipated. Taking out a notebook and pencil, Galton noted down 'Tints for measuring aggregate mood'. But what colour scale would do?

In order to get a closer look at the Fenian speaker, the geographer walked up to the edge of the crowd. Hats and the odd tall person got in the way. People were getting progressively taller, ever since his days as an undergraduate at Cambridge, when he was able to comfortably see over the heads of a crowd of his fellow students. Now as he watched impassively—Amnesty for Fenians! Unconditional release!—the scientist slowly slipped his hand into his inner coat pocket and took out the brown parcel. Keeping it concealed, he let the attached string unravel. The parcel descended unseen within his coat, until it came to rest in front of his shoes. Then he slowly stepped forward onto the wooden block disguised by the wrapping paper, rising comfortably above the hats.

A mixture of types. The thimble made a succession of marks on the trousered paper cross. The ox began to grow a stalactite of saliva on his beard, which would occasionally twinkle in the mild sunlight, swelling and threatening to disburse its contents onto his green scarf. Satisfied, Frank stepped back and deftly reeled the parcel back up into his coat compartment. For some time he had been working on a less troublesome shift to deal with such situations: his periscopic stovepipe hat. But that still needed a lot of refinement to make it less conspicuous. The best gazing was anonymous. Over in the distance another crowd was gathering.

Mr. Punch was suffering a welter of blows to his prominent nose from Judy, a muscular figure with a jaw like a pickle jar. The modest audience roared. Urchins ran past. Snatching a purse, they provoked a hue and cry. Blighters! Galton kept his focus on Punch. Could he come to worship the hideous figure if he deliberately trained himself to do so? After the performance ended, he walked over to the performers and asked a small unshaven man in a dirty coat how much he wanted for the puppet. Leaving a note for Gifi with Mr. Punch's minder, he arranged for payment on delivery to Rutland Gate.

Finding a park bench, the inquisitive scientist sat back to survey passers-by. After a desultory number of surreptitious pin pricks—was London turning into an outpost of Aberdeen?—he turned instead to the long avenue of trees leading away from the bench past the lawns. They reminded him of Bushey Park, beyond Richmond, where he had amused himself a few weeks ago by counting the flower stems of the famous blooming chestnuts. Now he tried the same technique on the leaves.

By fixing on some points of reference—a lamp post, that deformed branch, a distant monument—he progressively divided one tree into twenty portions. Then he carefully counted the leaves in one portion. Then another. A few more. A narrow notebook and pencil kept score. Repeating the process on more portions, chosen at random, he averaged the results—wonderful things, averages!—and multiplied by 20 to produce an estimate for one tree. Carefully totalling the number of trees, he multiplied to get the leaf count for the avenue. No more than 60,000! A million leaves would require miles of trees. Numbers are just like old friends.

Walking on, a church showed briefly through the houses at the edge of the park. In Venice he had counted the pigeons at the basilica of St Mark with Louisa. Thunder concentrated his inspection. The church doorway offered itself, and he hurried over. Just in time. While the rain hammered down within inches of his face, he pondered—dogs, Sir, do a good deal of *pondering*—over a conundrum that had been bothering him for some time. If churches are protected by divine favour, why do they all have lightning conductors? How many times are they struck, really? And what about clergymen, do they live longer, in consequence of their devotions? Perhaps prayer for oneself is not effective. But what about prayer for others? Without data, a hopeless question. Sovereigns receive endless prayers on their behalf, but do *they* live longer? Pencil to diary: *lives of the sovereigns, compare lifespan to national averages of the same class* (who receive few prayers).

As the rain softened, the sheltering observer, disappointed with the dearth of female passers-by, decided to walk over to Regents Park Zoo, there to continue some experiments that he had been conducting on more dependable subjects. Pedestrianism, that Regency obsession, was a tradition in the family. His great-uncle, the strongman Captain Robert Barclay-Allardice, had once walked a thousand miles in a thousand hours for a wager, then shipped off to fight in the Napoleonic wars the next day. Today's route was less complicated, since Captain Barclay's nephew did not have to worry about paid thugs waylaying him to swing the outcome of the swollen side-wagers. Specially installed gas lights had deterred them then, and had allowed Barclay's manservant to aim the welter of blows that he launched to keep the Captain awake when he appeared to fall asleep on his feet, in between draughts of port wine. But the street lights were not lit now, and Frank, who was not at all sleepy, was diverted by the idea of Gifi emerging from an alley to set about his shoulders with a stick.

Preferring to grapple with ideas rather than servants, Frank busied himself along the way with some mental exercises that were proving difficult to work out, but might lead somewhere. As ordinary street objects succeeded one after another, he forced himself to tackle the very first mental association that came to mind for each one, silently keeping score as he walked. Extreme self-discipline was required to grab and inspect an association immediately. Faint and fickle, they could still be distinguished if seized quickly enough, like pedestrians surprised by a Regency footpad. Developing immediate revulsion to the onset of an idea helped. Today's ideas were more or less the same writhing knot he had bagged some weeks earlier when first trying the exercise. A puzzle. They were mostly scenes from earlier days, as if he was a guest in Earl Stanley's menagerie at Knowsley, wandering off among exhibits of images but somehow always landing back at exactly the same spectacle.

The Zoo was thronged with curious visitors, including regulars long known to Galton. They too had grown used to the sight of the scientist pausing in front of the enclosures, peering intently at the animals while gently squeezing the knob at the head of his walking stick. An India-rubber pipe forced air through a whistle hidden inside the staff, emitting sounds audible only to keen-eared animals. The monkeys paid no attention at all, but lions started, elephants shuffled uneasily, and zebra panicked. In Berne, where dogs of all descriptions overran the city, he had noticed that lapdogs were far more acute than the St Bernards who booby-trapped the pavements. Insects did not seem to hear the notes at all (or so he thought). Several visitors to the Zoo shot looks in his direction—there was obviously dramatic variation between individuals in hearing ability. Where could the sounds be coming from? Faintly embarrassed, the shy scientist left the Zoo and proceeded through the park to its welcome lawns.

Resting his stick on a grassy patch, well clear of the crowds, Galton was reminded of a dodge suggested to him some time ago at the Athenaeum—was it his philosophical friend Herbert Spencer?–on the subject of his recurrent breakdowns. Giddiness and nausea invariably accompanied these, which had persecuted him since his days at Cambridge nearly thirty years earlier. Firmly grasping the stick, he leaned forward, placing his forehead on his hands. Shutting his eyes, he walked round in a circle three times, keeping to his right-angled position. As he straightened up to walk off, giddiness overwhelmed him. Tumbling onto the grass, he took several minutes to collect himself again, relieved that nobody in the park seemed to have noticed. Would the trick work for everyone? Perhaps he could persuade Tyndall to try it.

Working his way back to Pall Mall, the still-giddy scientist was forced more than once to stop and lean against doorways, to regain his balance. Regiments of brilliant sunbeams—never common in England—brought forth from an auxiliary coat pocket the heliostat which always travelled with him. It was his own design. By placing it to the eye one could aim flashes of light by fixing on the mock-sun projected by a system of internal mirrors, showing where the signals were visible. Distant office workers were intrigued by the repeated flashes they saw coming from a long way down their street. They could not decipher the rules of arithmetic that the flashes were encoding; that would have required an audience of inquisitive and exceptionally well-born Martians. And so Galton's unsteady figure made its way by stages to the Athenaeum Club. A quiet afternoon in its well-stocked library tempted.

A few months earlier, Captain Richard Francis Burton had been ensconced every day in the neoclassical building designed by his namesake Decimus. Battlements of folio volumes in front of him were rapidly reduced. The *Hadji* had stopped over briefly in the capital after a long absence in South America, en route to the Ottoman domains in Syria. Wife Isabel had begged a consulship in Damascus for her vexed husband, who had been menacing swathes of the sub-continent in the company of the Tichborne Claimant. A slight nod of heads on those awkward afternoons acknowledged a friendship ended by the ambiguous death of Burton's Nile foe, John Hanning Speke. Once, they had discussed the characters of Africans and Indians for hours within these walls. Today the *Hadji* was beyond reach in the Levant, having run out of leave to stretch.

After puzzling over that quarrel, while catching his breath, Galton approached Tedder, the Athenaeum librarian, to ask for the latest number of *Notes and Queries on China*. Over the last year he had ransacked biographical dictionaries, compiling tables of relatives of outstanding judges, scientists, writers, painters, oarsmen and wrestlers. Eminence appeared to breed eminence, but international evidence was wanted. The long-standing examination system in China for civil service mandarins promised rich data. Unfortunately there were still no replies to the plea he had placed in *Notes and Queries*, so his attention returned again to calculating the number of relatives a person could be expected to have on average. Galton already had a good estimate for how many eminent relatives his eminent subjects had, but were they overrepresented in the crowd? A difficult problem to be sure, but his friend the Rev. H. W. Watson, a high Wrangler at Cambridge, would soon send him some ideas.

Over in the easy chairs, Herbert Spencer was glaring at a new periodical calling itself *Nature*. 'Norman Lockyer! I gave him the idea after the collapse of the *Reader*'. Galton was resigned to Spencer's insistence on priority. News of the Regent's Park giddiness and subsequent tumble drew dour satisfaction from the philosopher, followed by a minute account of recent ailments. Galton eventually wearied, leaving behind the comforts of the club and his deductive friend—in so many ways his opposite—for a stroll over more solid ground. He would go to the evening meeting of the Royal Geographical Society.

In Hyde Park, the Fenians had long since dispersed. A rabbit flashed past, pursued by a spaniel towing an elderly gentleman in obvious distress. An explorer would have unleashed the dog by now. Galton had been slowly gathering the nerve to ask his cousin Charles Darwin to procure a warren of rabbits for some experiments to test the naturalist's theory of inheritance by Pangenesis. The idea was to transfuse blood between does with different coat colours. If all went as expected, the colours would be transferred when they bred. Then, as the veteran of Damaraland paused at an anastomosis of paths, ruminations of rabbits and blood transfusions were supplanted by memories of Speke and the memorial to Nile explorers, still unbuilt. The subject simply had to be taken up again that evening.

At the Geographical, Livingstone's letters and dispatches from Africa were being read aloud, while most of the attendees dozed or shifted uncomfortably. The Nile was old hat. The missionary was determined to get as lost in the wilds as possible. Now he was far beyond the reach of geographical sponsors not under divine guidance. In the middle of a long communication to Sir Bartle Frere, Galton lost all interest in the coy Scotsman and scanned the elderly audience instead. At times the bored gentlemen dropped their guard and swayed from side to side—some rather more than others. Randomly selecting an audience section of about 50 people, Galton began to count the number of fidgets per minute, using his heartbeat as a metronome. About 45. Intermissions of heightened interest reduced the number of movements by half, muting their amplitude and period, smartening their slow and sluggish yawing. The slim pocket journal now recorded additional readings. A more-than-usually-boring performance, but there were still only spotty records to compare with.

The rain had been goaded by news of Livingstone and the lakes into a tropical downpour. Buttonholing Sir Roderick Murchison about building a memorial to Speke in Hyde Park produced more of the same vague sympathy and studied inaction. 'Sir Roddy' had not forgotten the blonde explorer's paranoid concoctions and threats to reclaim the trackless interior for the glory of the French. Relief came when the dark arts of city clubmen conjured a cab. The driver was nearly Nilotic himself, barely visible in the dusk. Stepping into the vehicle, Galton took care to request 'Rutland Gate, 42', an address that he often alternated with '42 Rutland Gate'. Through the hansom's window the moon showed a rich yellow. During the day it was always white. At twilight it was purple. Why had nobody bothered to explain that before? On arrival, the time taken was discreetly noted: '42 Rutland Gate' was well ahead of 'Rutland Gate, 42'.

Oblivious to the colours of the moon, 'Lou' had already retired for the night to nurse her head and delicate digestive tract. Some absorbing hours of leisure time could be invested in the workshop then. Herbert Spencer had recently suggested identifying ideal types by superimposing transparent images, but Frank's new photographic equipment offered better alternatives. What if one repeatedly exposed the same plate for brief moments, capturing successive images and forming an aggregate type of face? He set about it.

Chapter 2

Hereditary Talent and Character

The first detailed description of Galton's emerging research programme appeared in 1865 as a two-part series of articles in *Macmillan's Magazine*, in the June and August issues (see Appendix C, p. 399). This appears to be an odd forum nowadays, but Victorian journals often carried a comprehensive mixture of serious research, miscellaneous news, editorial commentary, book reviews and rarely-memorable serialized novels and stories. Galton's two pieces contained a rough outline of many of the themes that would dominate in years to come, and would eventually form the discipline of differential psychology in the next century; they also foreshadowed the development, much later, of sociobiology and evolutionary psychology.[1]

Two propositions form the basis of Galton's thinking: that human behaviour is inherently variable, differing between individuals and groups; and that human behaviour has been shaped by natural selection. Though at first it may seem odd, this progression from travel and exploration to social biology, by way of comparative ethnography, made perfect sense in his era. Reports from observant travellers abroad seeded the just-emerging discipline of anthropology, and few were better placed than Galton was—by his personal experience, by his network of contacts among explorers, and by his command of the travel literature—to synthesize the two.

Darwin's *On the origin of species by means of natural selection, or the preservation of favoured races in the struggle for life* had been published only a few years earlier, on Thursday the 24th of November, 1859.[2] It dealt almost exclusively with animal examples, and notoriously made no more than a cryptic reference to man near the very end. 'Light will be thrown on the origin of man and his history'.[3] The implications that the *Origin* held, once man was rightly considered an animal too, were obvious to those prepared to make the effort. Man had appeared in Darwin's notebooks from the very beginning, and he would later write a full-length consideration of *The Descent of Man* in 1871.[4] But throughout the 1860s Darwin held his fire and left discussion about natural selection of humans to proxies.

Galton had read the *Origin* immediately. He was sent a copy directly on its sold-out publication by Darwin himself.[5] Less than two weeks later he must have finished absorbing it, since he sent his cousin an appreciative note on the 9th of December. 'Pray let me add a word of congratulation on the completion of your wonderful volume, to those which I am sure you will have received from every side. I have laid it down in the full enjoyment of a feeling that one rarely experiences after boyish days, of having been initiated into an entirely new province of knowledge which, nevertheless, connects itself with other things in a thousand ways'.[6] He was able to add a correction: whereas Darwin had claimed that a young rhinoceros had no predators, Galton knew from his African experiences that it did: the wild dogs which

1 Barkow, Cosmides, and Tooby 1992; Wilson 2014.

2 C. Darwin 1859, Freeman 1977.

3 C. Darwin 1859, 488. This was expanded to 'Much light' in the sixth edition of 1872.

4 C. Darwin 1871a.

5 His copy survives and contains a note stating that it was presented to him by Darwin, GALTON/1/5/2.

6 FG to Darwin, 1859-12-09, DARWIN, DCP-LETT-2573.

continually harass it.[7]

A few years later, in an exchange of letters over Christmas 1869, prompted by the publication of his *Hereditary Genius*, Galton half-mocked this reclamation from benightedness. 'I always think of you in the same way as converts from barbarism think of the teacher who first relieved them from the intollerable[8] burden of their ~~old~~ superstition. I used to be wretched under the weight of the old fashioned "arguments from design", of which I felt though I was unable to prove to myself, the worthlessness. Consequently the appearance of your "Origin of Species" formed a real crisis in my life; your book drove away the constraint of my old superstition as if it had been a nightmare and was the first to give me freedom of thought'.[9] Four years after Darwin's death, when Galton received the Gold Medal of the Royal Society, at Burlington House on Saint Andrew's day of 1886, he paid public tribute to the naturalist's influence on him. 'Few can have been more profoundly influenced than I was by his publications. They enlarged the horizon of my ideas. I drew from them the breath of a fuller scientific life, and I owe more of my later scientific impulses to the influences of Charles Darwin than I can easily express. I rarely approached his genial presence without an almost overwhelming sense of devotion and reverence, and I valued his encouragement and approbation more, perhaps, than that of the whole world besides'.[10] Again, when writing his *Memories* two decades after the death of his reclaimer, Galton contrasted his own ready conversion to the less-than-enthusiastic reactions of many among his contemporaries.

> The publication in 1859 of the *Origin of Species* by Charles Darwin made a marked epoch in my own mental development, as it did in that of human thought generally. Its effect was to demolish a multitude of dogmatic barriers by a single stroke, and to arouse a spirit of rebellion against all ancient authorities whose positive and unauthenticated statements were contradicted by modern science.
>
> I doubt, however, whether any instance has occurred in which the perversity of the educated classes in misunderstanding what they attempted to discuss was, more painfully conspicuous. The meaning of the simple phrase "Natural Selection" was distorted in curiously ingenious ways, and Darwinism was attacked, both in the press and pulpit, by persons who were manifestly ignorant of what they talked about. This is a striking instance of the obstructions through which new ideas have to force their way. Plain facts are apprehended in a moment, but the introduction of a new idea is quite another matter, for it requires an alteration in the attitude and balance of the mind which may be a very repugnant and even painful process. On my part, however, I felt little difficulty in connection with the *Origin of Species*, but devoured its contents and assimilated them as fast as they were devoured, a fact which perhaps may be ascribed to an hereditary bent of mind that both its illustrious author and myself have inherited from our common grandfather, Dr. Erasmus Darwin.

In earlier times, Charles had regular contact with his Galton cousins, and especially with Bessy, an older sister of Francis.[11] When he was studying medicine in London, between the summers of 1839 and 1840, Francis consulted Charles several times for career advice, and the older man gave considerable attention to the student. Afterward, the cousins seem to have lost contact, until the publication of Galton's *Narrative of an Explorer in Tropical South Africa* in 1853, describing his expedition to South Africa of 1850–1852.[12] This drew an admiring letter from

[7] *Lycaon pictus*.
[8] An archaic spelling.
[9] FG to Darwin, 1869-12-24, DARWIN, DCP-LETT-7034.
[10] F. Galton 1886a.
[11] For a full treatment of the Darwin-Galton connection, see Tredoux 2023c.
[12] F. Galton 1853.

the reclusive naturalist—by then settled in the country at Down House, some six miles south of Bromley, east-south-east of central London and anywhere between 3 and 5 hours travel from South Kensington in those days—which was forwarded through the publisher they shared, John Murray. 'You will probably be surprised, after the long intermission of our acquaintance, at receiving a note from me; but I last night finished your volume with such lively interest, that I cannot resist the temptation of expressing my admiration at your expedition, & at the capital account you have published of it'.[13]

Galton's reply has not survived, but he was soon mining Darwin's South American experiences for the *Art of Travel*,[14] which first appeared in December 1854 and established his name as the leading technician of travel for many years to come. From his cousin there was the nugget that bones make good firewood, or at least they did on wood-starved Tierra del Fuego. In May of 1854, Galton had asked about shifts specifically for tinder and fire-starting, but the naturalist could only apologetically refer him to some remarks by FitzRoy about the Fuegians striking iron pyrites to make sparks.[15] Darwin warmly appreciated a presentation copy of the book from the author, 'I always thought your idea of your Book a very good one, & that you would do it capitally, & from what I have seen my forethought is, I am sure, quite justified. I hope that your volume will have a large sale, but what I fully expect is that it will have a long sale, & if you save from some disasters half-a-dozen explorers, I feel sure that you will think yourself well rewarded for all the trouble your volume must have cost you'.[16]

The naturalist reciprocated by requesting information from Galton related to the domestication of animals, an important element of his long-developing but still-unpublished analogy between artificial and natural selection. Galton was able to put Darwin in touch with Charles John Andersson, his old travelling companion from the South-West African expedition, for details about the peculiar Damara cattle.[17] Andersson had stayed on in the region, but was then in London to fight for the publication of his lavishly illustrated travel narrative, *Lake Ngami* (1856). Darwin would continue to harvest information about domestication from foreign travellers through Galton, whose position within the RGS, and ongoing correspondence for new editions of the *Art of Travel*, kept him in direct touch with the leading explorers of the day. And as we have seen previously, the naturalist signed Galton's 1857 application for membership of the Royal Society, from both 'personal' and 'general' knowledge.

Galton's eager 'devouring' of the *Origin* suggests that he really was prepared for it. He had long left the more narrow commercial Quaker background of the Galton family behind. After the death of his father he had wandered adventurously in Egypt, the Sudan and the Levant, and settled down for six months in a village retreat in the hills outside Damascus. There he lived as a Muslim. For many years after, he thought his left hand unclean. Mild Islamophilia also persisted, at least where Africa was concerned. The South African expedition that followed had also exposed Galton to a strikingly different world than he had seen in the Northern hemisphere.

Galton was never a naturalist or a specimen collector like his cousin, but displayed the same intensely keen powers of observation. In Damaraland he had been especially struck by differences in behaviour displayed by the oxen that his expedition depended so critically on. These had to be carefully selected and trained for many months, mostly from scratch, before the expedition could even set off—which was one reason why the expedition was so much harder to launch from there than it would have been in South Africa itself, where a ready supply of trained oxen could be bought. Only a very small fraction of highly-prized independently-minded oxen were suitable for training as team leaders.

[13] Darwin to FG, 1853-07-24, DARWIN, DCP-LETT-1525.

[14] F. Galton 1855.

[15] Darwin to FG, 1854-05-28, DARWIN, DCP-LETT-1881.

[16] Darwin to FG, 1855-01-01, DARWIN, DCP-LETT-1627.

[17] See volume 1, Tredoux 2023c, for full details.

The 'savage races' which he had encountered, and which gave him 'material to think about all the rest of my life', consisted of modest numbers of diminutive Bushmen, large numbers of yellowish Namaqua Hottentots, presided over by Oerlams (Hottentots with some European admixture); and many negroid Hereros (Damaras) and Ovambos. Interspersed were pockets of an archaic population of *Ghou Damup* (Hill Damaras), who were of negroid origin but spoke only the Nama language. In stark contrast to these were the small number of European settlers in the area.

By this time, Galton had already become a freethinker of sorts, prompting the stern Rhenish missionary Carl Hugo Hahn to note with exasperation in his diary that the clever young Englishman was a 'miserable infidel goblin (*wicht*)'.[18] Hahn would later add that he had judged the traveller, who became a firm friend, a little harshly. Whereas his travelling companion Andersson often made commonplace biblical and religious references in his diaries, Galton never did. But, as we have noted before, when the sensational Baptist preacher Charles Haddon Spurgeon (1834–1892) worked London in late 1856, Galton went to see him not once but twice.[19]

There is no direct evidence that Darwin shared his theory of natural selection with Galton prior to the publication of the *Origin*, but the fact that Galton had become extensively involved in Darwin's inquiries into domestication suggests that he had some idea of its broad outline by 1856 or so. However, Galton's annotations to his copy of the *Origin* were sparse. Darwin's section on the Malthusian rate of increase of a pair of breeding elephants attracted some marginal jottings; Galton questioned his cousin's conclusion that the pair would, if all survived, produce roughly 15 million offspring within five centuries.[20] Darwin's speculation 'can absolute perfection everywhere be found' was more skeptically marked up 'anywhere'.[21] Beside these, some phrases were underlined, and paragraphs marked 'X'. No doubt Galton took far more detailed notes elsewhere, as he did from other sources, but these have not survived. It took several years for the new theory to produce an effect on Galton's own research, which was then turning from travel and exploration to meteorology.

It is not certain how Galton was introduced to the Ethnological Society. His friend John Crawfurd was active in both the Ethnological and the Geographical, as was his engineer cousin Douglas Galton, so it was probably inevitable that Galton, a rich source in his own right of comparative ethnology derived from his African explorations, would be drawn to it. He was attending meetings by February 1862 at the latest, when from the floor he offered comments doubting a link between innumeracy and civilization, in response to one of Crawfurd's innumerable papers, 'Numerals as evidence of civilization'. The following year, Lubbock presided over the Ethnological, and Galton was on its council, serving with Thomas Wright as a secretary to the society. By then the Anthropological Society of Hunt and Burton had established itself as a rival forum, but Galton remained loyal to the Ethnological. His involvement continued to grow: Louisa's Diary noted throughout 1864 that her husband's evenings were now often devoted to the Ethnological—she was taken there herself by one of the Gurneys, on at least one occasion—with frequent visits to and from the Lubbocks.[22]

Galton read his first paper to the Ethnological on December 22nd, 1863, under the title 'First Steps Toward the Domestication of Animals'. John Crawfurd presided. The paper only appeared in the *Transactions* of the society in 1865. Slow and intermittent publication was one of the longstanding defects of the Ethnological. In the meantime, Galton brought out the piece as a pamphlet and distributed it to friends and learned societies. Another version was read at the Leeds Philosophical and Literary Society on March 1st, 1864, as 'On the early

[18] Hahn 1984.
[19] Louisa's diary entries, 1856-11-12 and 1857-01. See Tredoux 2023c.
[20] Annotations to C. Darwin 1859, 64, GALTON/1/5/2.
[21] Annotation to C. Darwin 1859, 206, GALTON/1/5/2.
[22] Louisa's Diary entries for 1863, GALTON/1/2/1/7, GALTON/1/2/1/10/1.

Domestication of Animals'. It was also presented to 'Section D–Zoology and Botany' at the meeting of the British Association at Bath on Thursday September 22nd. He took his first full foray into the field seriously.

Scholars have traditionally made little sense of this paper, treating it as a curious diversion.[23] In fact it was a crucial beginning for Galton, who later incorporated it into his *Inquiries into human faculty* (1883).[24] The *Origin* had based its argument on a sophisticated analogy between artificial and natural selection. Now Galton argued explicitly and at length that animal *behaviour* can also be controlled by artificial selection. The same analogy would suggest that animal behaviour had been shaped by *natural* selection, with obvious applications to human behaviour. But first it had to be established that behaviour was responsive to selection at all, as a heritable property.

Galton framed the question differently in the paper itself, not yet fully aware of where this line of inquiry would lead in the future. 'Have extraordinary geniuses arisen who severally taught their contemporaries to tame and domesticate the dog, the ox, the sheep, the hog, the fowl, the camel, the llama, the reindeer, and the rest? Or again, is it possible that the instincts of savages, combined with the qualities of the animals in question, may have sufficed to originate every instance of established domestication?' To answer this, Galton turned to his network of travellers to supplement his own experience, which suggested that 'savages' everywhere regularly tamed wild animals. 'I have been chiefly indebted in writing this essay to general recollections, which I have not had time to verify, to the conversations of recent travellers, and to the memoranda which many of them have been so kind as to favour me with'.

Charles John Andersson, then trading in South-West Africa, received a query to help Galton verify his own information from the region, but the gloomy Swede's response was too late to include. The Abyssinian experiences of the ex-Trinity man Mansfield Parkyns, whom Galton had first chanced on at Khartoum in 1846—practically naked, bronzed by the sun and greased with butter—were easier to incorporate now that Parkyns lived in England.[25] Darwin's co-evolutionist Alfred Russel Wallace was able to supply examples from his years in the Malay Archipelago. The embarrassed 'widow' and travelling companion of Thomas Witlam Atkinson, the Siberian traveller and bigamist, had been able to supply information about the Kirghis. Sir Charles Nicholson testified that Aboriginal women in Australia were passionate about cats. T. H. Huxley informed Galton that in New Guinea, where he had spent some time on the *Rattlesnake*, women sometimes nursed suckling pigs at their own breasts. The influential botanist Dr. Berthold Seeman (1825–1871) mentioned that the Fijians tamed flying foxes, lizards and the 'parroquet', and so on. In this way, Galton easily multiplied examples from contemporary travellers and ancient authorities.

Having satisfied himself that contemporary 'savages' are generally known to try taming wild animals, Galton proposed three characteristics of an animal that would render it a suitable candidate for domestication by repeated selection *in that way*. '*Hardiness.*—First, it must be able to shift for itself and to thrive, although it is neglected; since, if it wanted much care, it would never be worth its keep. *Fondness for Man.*—Secondly, it must cling to man, notwithstanding occasional hard usage and frequent neglect'. A third quality required is 'usefulness to man', as food or for labour. Critically, animals vary in those qualities, the first two of which are really behavioural characteristics. 'It is certain that some animals have naturally a greater fondness for man than others'. Moreover it is 'notorious that attachments and aversions exist in nature'.

Thus some animals are better candidates than others for the 'vast number of half unconscious attempts' that 'have been made throughout the course of ages' so that 'ultimately, by slow degrees, after many relapses, and continued selection, our several domestic breeds became

[23] See for example Bulmer 2003.
[24] F. Galton 1883d.
[25] Tredoux 2023c.

firmly established'. Galton supposed that all the animals which met his conditions for unconscious domestication by barbarians had already been domesticated by the modern era. Their instinctive behaviour must have been modified by continued selection over many generations.

The attendance at the Ethnological Society when Galton's paper was read, that Tuesday the 22nd at 8 p.m., was distinguished, even though Christmas was imminent. Richard Owen, the leading but petulantly anti-Darwinian anatomist, offered praise and remarks on the domestication of the American turkey, perhaps without realizing the Darwinian framework of Galton's argument. Alfred Russel Wallace recollected tame fish and squirrels in the Malay Archipelago. Clements Markham, from the Geographical, noted the taming of Gulls in the Andes. Henry Walter Bates, who had identified the evolutionary logic of animal mimicry, raised the case of domesticated tapirs and tamed birds in the Amazon. John Crawfurd doubted whether sub-Saharan Africans were even capable of taming elephants, which might explain, he thought, why they had been domesticated only in India and Northern Africa.[26]

'First steps' accurately foretells Galton's own immersion into ethnographic waters during the year that follows. In July 1864 he started a notebook recording his rapidly evolving ideas, which had moved squarely to human character and behaviour, but now ranged freely over all facets of the latter. 'The ethnological aspects of geography now began to attract me more than the physical ones. It was about this time that the fact dawned on scientific men that the key to the origin of society among civilized nations and to many of their unexplained customs was to be found in the habits of contemporary barbarians'.[27] He continued working in this notebook until sometime early in the new year of 1865. By 'character' he meant roughly the same thing now referred to as personality. Galton suspected that this was also subject to selection, and that evidence for that could be found in the way that character varied geographically among relatively isolated sub-populations and breeding groups, i.e. races. His references show that he was reading widely in the field, delving into the domestic and continental literature.[28]

Within the notebook, a passage from Lucretius appears which crisply states Galton's personal attitude to the nature of scientific inquiry, where it deals with taboo subjects. 'For even as children are flurried and dread all things in the thick darkness, thus we in the daylight fear at times things not a whit more to be dreaded than those at which children shudder in the dark & fancy sure to be. This terror therefore & darkness of mind must be dispelled not by the rays of the sun & flittering shafts of day, but by the aspect and law of nature'.[29] To this Galton appended an observation of his own. 'Sin—injustice to society or untruth to one's own nature. (F.G.)'. Taboos have been inverted since Galton's time. Then, the sections of the notebook which deal with religion might have been foresworn, whereas today those dealing with national character might jar.

Galton's sources for this investigation were, as before, mixed accounts by contemporary and past explorers (like himself), and those drawn from the classical literature. He was especially interested in early contacts with the American Indians by the Spanish. He jotted down a reference to a report in the *Reader*, about a paper 'On Fixity of Type'. This was read at the British Association at Bath that September by the Reverend Frederick William Farrar, a Cambridge man and future Dean of Canterbury. This reveals the crux of Galton's thinking: Farrar laid the groundwork for a theory of *innate* character. 'There was at one time a universal impression that the diversities of type and complexion observable in the human race might easily be accounted for from the effects of climate, custom, food, and manner of life. The opinion is now entirely abandoned by the majority of scientific men'.[30]

[26] There is no clear distinction being made here between merely taming animals, without altering their breed, and continuous selection resulting in a different breed, which is the modern usage.

[27] F. Galton 1886a.

[28] See Appendix A, p. 381, for a full transcription of this important notebook.

[29] Lucretius II, 55.

[30] *Reader* Oct. 29, 1864, 550–551. For Farrar, Frederic William, see page 452.

From the Scottish historian William Robertson's influential *History of America* (1777) and several other sources, Galton noted that the Native Americans were reported to have 'Smallness of appetite', 'Absence of love', 'Love of vengeance', 'Cruelty to prisoners', 'Indifference to death', 'fortitude under torture', 'cold phlegmatic temper', 'utter indifference to melody' and 'insensibility to beauty'. It was reported that 'the natives treat their women with coldness & indifference, neither tender attachment nor ardent desire'. 'The missionaries express astonishment at their coldness'. 'No sympathy—the Spanish sometimes found it necessary to enforce the common duties of humanity by positive laws', 'never care for their gods', 'is a serious melancholy taciturn animal—will sit whole days in one posture without opening the lips marching Indian file—rowing canoe all in silence', 'Secrecy—dissembling—An insurrection at [Fera] was hatching for 30 years unsuspected', 'Perseverance & dignity', 'Partisanship (to community) zeal for honour of their tribe'.

Similar descriptions were noted in sources such as: Edwin Bryant in *What I Saw in California* (1848), who reported the Indians' 'utter indifference to melody' and 'chivalrous, vivid sense of honour'; the first Spanish governor of the Louisiana region, Antonio de Ulloa (1716–1795) (from Pinkerton's compendium of travels, which Galton had ransacked for the *Art of Travel*);[31] and in writers such as Jean-Baptiste Du Tertre (1610–1687), author of *Histoire naturelle et morale des îles Antilles de l'Amérique avec un vocabulaire caraïbe* (1658), who draws a direct contrast with the Negro character.[32] Perhaps, Galton wondered, the African or the Hindu is instead 'disinterested' in the sense that he 'forgets himself too readily & throws himself into others people's ideas & schemes'.

In Australia, Sir George Grey found that 'The savage is naturally of a thoughtful restless disposition' with 'labor neither constant nor steady', while the Chinese—who themselves show 'stability of character'—find savages in general to be 'unstable and unpersevering'. Galton juxtaposed Edwin Chadwick's characterization of the Anglo Saxons. 'The primary qualities of those whom we are accustomed to call Anglo Saxons including the lowland Scotch are great bodily strength, applied under the command of a steady persevering will, mental self-contentedness, impassability to external irrelevant impressions which carries them through the continued [repetition] of toilsome labour "steady as time"'.[33] Galton went so far as to identify a 'strong natural hope' among Anglo Saxons 'that there is a purpose & that we are more than mere sports of nature', apparently in contrast to the fatalism prevalent in some other groups.

Galton deduced that the peculiar temperament of American Indians was inbred, given its invariant distribution. 'The many independent communities in America having common qualities shows that these are due to nature not tradition'. This was not an unusual line of thinking—Darwin had made similar observations in his unpublished 'M Notebook' on man: 'must grant that conscience varies in different races. No more wonderful than dogs should have different instincts'.[34] 'He who doubts about National Character let him compare the American whether in the cold regions of the North,—the elevated table land of Peru, the hot plains of the Amazons & Brazil, with the Negroes of Africa (or again the black man of Van Diemen's land & the energetic copper coloured natives of New Zealand)—the American in Brazil is under the same conditions as Negro on the other side of the Atlantic. Why then is he so different in organization—same cause as colour & shape & ideosyncracy'.[35]

The notion of selectable animal behaviour and evidence of domestication swam alongside the national characters of humans in Galton's notebook. Henry Walter Bates informed Galton

31 Volume XIV, 522.

32 Volume II, 490.

33 Copied by Galton from *Address of Edwin Chadwick, Esq., C.B., as vice president of the Public Health Section to the General Meeting of the National Association for the Promotion of Social Science*, Glasgow, September 29th, 1860.

34 Gruber and Barrett 1974, 279: 75.

35 Gruber and Barrett 1974, 281: 86.

that the King of the Barotse once kept a menagerie, some inmates of which may have been pets. In Sir George Simpson's *Voyage Round the World* (1847), Galton found mention of half-domesticated fur seals. In Armand Quatrefages' *Metamorphoses of Man and the Lower Animals* (1864) he found 'Coquettishness even in Butterflies' [p. 53] and superstition and vanity among gregarious animals. This suggested that 'Dull individuals have their thought reinforced by another's activities'. Selection operates on character. 'A white feather in a pigeon punished with death from a hawk as much as indolence or any vice'. Galton mused that even though English horses were generally unsound, one could 'breed what tempers you like', leading him to mark the *Turf Annals* for 1844 published by Longman. And he anticipated in an offhand way the later development of factor analysis applied to personality. 'Character of the animal generally does not consist of number of independent factors any of which may be separately eliminated', which is to say that dimensions of personalities *tend to go together.* The social insects pop up as analogues for human behaviour. 'Bees ants patriotism loyalty'.

Religion constantly cropped up among these notes. There was the tantalizing question of the possible concentration of higher than usual levels of insanity among members of inbred religious sects. Suspecting that religious seers of visions—he gives the example of the resurrection of Christ and apparitions in the streets of Jerusalem—have an innate tendency to do so, he suggested that they 'might be bred' for that capacity. They could also be viewed as codified repositories of the moral systems emanating from the shared characters of their adherents. He would return to this trait a decade later in his researches into visionaries.[36] There are also glimpses in these pages of another later theme, the extent to which a religious vocation protects the health and longevity of its practitioners: 'Dr Guy's papers Statistical Journal xvii 20. life of clergy & medical men'. He noticed the recently published treatise by the Reverend John Hodgson, *Observations on the Mortality of the Clergy of England and Wales* (1865). William Spottiswoode provided him with a reading list for getting up to speed on world religions: Eugène Burnouf on Buddhism, Spiegel on Zoroaster, Wilson and Rammohan Roy on the Rig Veda, and Henry Thomas Colebrook's *Essays.* Notes quickly appear from these sources.[37]

A tangential line of inquiry was suggested by the way in which groups of men seemed to cohere, challenging the notion of individuality, as combinations broke up and reformed, as if they were perhaps organic or holistic things-in-themselves with life-cycles like organisms. 'Identity—what does the Llama of Thibet think of his identity'. Galton found some commonplace examples from cooperation. 'Council meetings, forget all other matters for the time', 'temporary individuality of the meeting', 'ditto of states, families, schools regiments'. Sacrifices are related to this dissolving, as in the crucifixion of Jesus. 'That great suicide of J to save mankind. We (encourage) allow suicide to save others or for the good of others not for one's own relief'. There is also the suicide and practice of 'suttee' by widows: 'Sawyer's opinion of s. of widows'. In Prosper Lucas (see below) Galton found the idea that 'Hindoos think son is a perpetuation of father's life'. Unrelated ideas were inserted with the others as they happened to catch Galton's interest. For example, the whimsical assertion by Thomas Taylor Meadows, in his *The Chinese and their rebellions* (1856), that one might construct a vocabulary for communicating with dogs and other animals by omitting hard consonants and only using vowels with soft consonants: 'woomah', 'celoo', 'wuh!'.

A less indirect line of thinking seems to have been sparked by Richard Chenevix, an Irish chemist of Huguenot descent, whose *An Essay Upon National Character* (1832) attracted attention. 'Chenevix on national character 39 in library of Athenaeum quote in my appendix his passage [on] the influence of individual geniuses in moulding the destiny of nations vol. 2 p 41–43 & after read his chapter on morality of different nations vol 1, 137'. This led on to a long consideration of the extent to which exceptional talent runs in families, starting with

[36] See chapter 9, p. 159.

[37] For Galton's researches on prayer, see chapter 6, p. 109. For Spottiswoode, William, see page 458.

the Lord Chancellors of England, and several pages of preliminary genealogies, derived from biographical dictionaries, appear. Academic merit seemed similar: 'Mitchell of Oxford is said by Brodrick[38] to have made out that all the 1st classmen of Oxford belong to a very limited number of families'. All this made up an avenue of inquiry that would consume most of Galton's energy for the rest of the decade.

The nature of heredity itself cropped up naturally in connection with the themes of the notebook. Here Galton's starting point seems to have been a long thematic review by his fellow *Reader* editor, the philosopher and critic George Henry Lewes. Titled 'Hereditary Influence, Animal and Human', this was published anonymously in *The Westminster Review* in 1856,[39] and covered three books on heredity. Since Galton noted that Lewes (who is now best-known as George Eliot's lover) was the author, it was probably recommended to him—perhaps by Darwin, who had noticed it at the time, or even by Lewes himself at the *Reader*. The central puzzle for Lewes was the way in which heredity assured both constancy, with like begetting like, while at the very same time producing considerable variation. He also realized, in a piece of unusually clear thinking, that 'species' are really arbitrary classifications invented for convenience, 'an abstract idea, not a concrete fact', attachment to which represented 'lingering remains of old Metaphysics'. Remarkably, Lewes anticipated Darwin in key points. 'Differences of degree become differences of kind, when the gap is widened'.

The first of the three books reviewed by Lewes was a three-volume tome by the French medical doctor Prosper Lucas, *Traité philosophique et physiologique de l'hérédité naturelle dans les états de santé et de maladie du système nerveux*, which had appeared between 1847 and 1850. Lewes maintained that Lucas favoured quantity over quality when collecting his enormous store of 'facts' about heredity. Galton nevertheless acquired a copy and read it closely, taking extensive notes on hereditary longevity, drunkenness, gambling, sexual passion, crime, eloquence, artistic ability, immunity, melancholy, sensations, scent, colour blindness, hearing, and sentiments. From Lucas, Galton was led to Pierre Paul Broc (1782–1848) *Essai sur les races humaines considérées sous les rapports anatomique et philosophique* (1836); Charles George (1723-1789) *Lettres philosophiques sur l'intelligence et la perfectibilité des animaux* (1802); Louis François Charles Girou de Buzareingues (1773–1856) *De la génération* (Paris, 1828), which was also reviewed by Lewes; and other continental sources. The other book reviewed by Lewes, and also noticed by Darwin, was by the Sunderland ophthalmologist and chicken-breeder Reginald Orton (1810–1862), titled *On the physiology of breeding: Two lectures, delivered to the Newcastle Farmers' Club* (1855). Darwin owned and annotated a copy and corresponded with the author. Galton must have read Orton's compendium of breeding lore and 'curious facts', after taking down the title from Lewes, but left no detailed notes.

Though Lewes' review makes interesting reading today for the way in which it anticipated some elements of later thinking, the idea of acquired characters was at its core, which owed much to his friend Herbert Spencer's dogmatic influence. Thus human civilization supposedly ingrained itself in its subjects to become heritable in offspring, even increasing brain volume (to receive the habits) as it did so! Darwin would be increasingly drawn back to acquired characters in successive editions of the *Origin*, as he struggled to reconcile his blending theory of inheritance with persistence of adaptive traits.[40] However Galton—who seems from the start to have been a 'hard hereditarian', in the sense of selection acting on blind variation—was never attracted to the idea. 'Acquired characters' are completely absent from his notes.

The ethnography notebook led Galton to search more broadly for evidence bearing on the selectability of human cognitive ability. He was unable to make direct analogies from artificial selection on animals since he could not find instances where domesticated stock had been

[38] George Charles Brodrick (1831–1903), Warden of Merton.
[39] Lewes 1856.
[40] Darlington 1959.

selected specifically for something like general ability. Dogs, for instance, were usually bred for more specific traits, such as the ability to point, to herd or to guard. Instead he extended his inquiries into the genealogies of people of high ability, like the Lord Chancellors whose family relationships seemed so suggestive. In the absence of a well-defined standard for ability in any particular field, he chose to make *relative* comparisons instead, in such a way that the standard used did not matter much. This was a brilliant insight, as it circumvented inconclusive debates about the meaning of ability, objectivity of standards and a thicket of problems that would have been impossible to resolve to everyone's satisfaction. This was exactly how he dealt with the Lord Chancellors: by choosing an arbitrary standard of eminence, judged by sources independent of himself, and then calculating the proportion of family relationships among those judged eminent. As long as the standard of eminence had no connection itself with the likelihood of qualifiers having family relationships, the proportion would be more or less the same regardless of how it was varied, more or less strictly.

Independent sources of classification by eminence in all fields, to supplement Campbell's *Lives of the Lord Chancellors*, were needed. There were plenty to hand. The eccentric publisher and bookseller Sir Richard Phillips—whose many quirks included vegetarianism—had compiled the compendium *A Million of Facts* in 1832. The book had gone through many editions, and was carried on after his death in 1840. Phillips and his successors included a listing of 605 'original minds' in all fields who lived between 1453 and 1853. Contemporary people of eminence were covered by Edward Walford's *Men of the Time: A Biographical Dictionary of Eminent Living Characters (Including Women)*, which had been published in 1862 and received an update in 1865.[41] An old Trinity connection, the Rev. Charles Hole—then a curate at Shanklin in the Isle of Wight, where Galton likely visited him on one of many walking trips to the island—was just about to publish *A Brief Biographical Dictionary* covering over 19,000 persons. Hole was happy to share the proof sheets in advance. More specialized sources were available. For painters, Michael Bryan's standard *Dictionary of painters and engravers*, a new (posthumous) edition of which had appeared in 1849. For musicians, there was François-Joseph Fétis (1784–1871) *Biographie universelle des musiciens et bibliographie générale de la musique* (Paris, 1837), which went through many editions. He also considered the Cambridge list of Senior Classics since its introduction in 1824.

Each of these sources used its own essentially arbitrary standard for inclusion, given that each had different numbers of entries, but this did not matter. Galton calculated the proportion within each of these lists of *males* who had relations judged eminent by the same standard, conceding that female eminence would be underreported. To save labour, since he was just doing a preliminary exploration, he sampled the lists of Hole, choosing the 1141 names beginning with M, and Walford, examining 85 names starting with A there. Likewise, from Bryan he chose the 391 artists starting with A, and the 515 of the same letter in Fétis' list of musicians. Then he hunted through these to identify family relationships: fathers and sons, brothers, uncles and nephews, and more distant connections.

In each case Galton found a large proportion with family connections. In Phillips' list he found $^1/_6$ who were near male relations, in Hole's list of the first-letter M, $^1/_{11}$. Walford's letter A had fully $^1/_{3.5}$, while the artists and musicians under A revealed $^1/_6$ and $^1/_{10}$. For the Lord Chancellors it was $^1/_3$, for the Cambridge senior classics $^1/_6$. The estimated proportions were bound to differ by the list. Identifying relatives was a procedure whose uncertainties varied by compendium used, and Galton realized this would lead to an *underestimate* of the true proportions, since some family connections would be overlooked. He was more concerned about a possible confounding effect: eminent people might improve the chances that their relatives would become eminent merely by exerting social influence, an objection he found potentially compelling for the military and for statesmen, where open competition was less prevalent and

[41] Walford appears to have taken this project over from earlier incarnations, which omitted women.

rapid promotion was crucial.

To check this, Galton restricted attention to only the literary and scientific men, reasoning that these fields were more closely meritocratic, and that it was unlikely that influence would make a literary or scientific reputation *last* over time. Now he found that in the reduced list of 330 names, about 51 or $^1/_{6.5}$ men had near male relations distinguished for literary ability. Since by his estimate Europe had produced at least 1,000,000 candidate scribblers over that period, the chance of anyone being eminent, assuming all have an equal chance regardless of family, was no more than $^{330}/_{1,000,000} = {}^1/_{3,000}$. It is easy to show using the binomial distribution that the probability then of 51 or more successes is practically zero, even if every family network of candidates is large.[42]

Galton now had a strong *prima facie* case for hereditary influence on human ability, even though he had no precise measures of ability. He was also able to address some popular misconceptions about heredity, well represented by the G. H. Lewes review mentioned above.[43] In particular, the idea that ability primarily descended from the female line, and the belief that people of exceptional ability are, beside being weakly and sickly, disinclined to breed anyway. The notion of maternal descent was refuted by the figures above, which deal only in male lines of relationship, and Galton could see no compelling reason why both sexes would not contribute to their descendants. His own experiences at Cambridge, and in Switzerland with the many scientific and literary members of the Alpine Club—including John Tyndall, and Leslie Stephen—directly refuted the idea that there was any necessary physical enfeeblement brought on by high ability. Scanning the lists he had found plenty of great men with numerous descendants, so the idea that they were sterile could be discarded.

In June of 1865, Galton published a preliminary summary of his results in *Macmillan's Magazine*.[44] The journal was still edited by his friend from *The Reader*, David Masson, and sustained many of the same contributors. Running to nearly ten pages, under the title 'Hereditary Talent and Character: Part I', Galton's first installment appeared alongside no less than three pieces dealing with Abraham Lincoln and his assassination, one of them by Edward Dicey; new installments of the serialized novels 'The Dove in the Eagle's Nest' by Charlotte Yonge, and 'Cradock Nowell' by R. D. Blackmore; a sympathetic article on 'Women and the Fine Arts'; a review by Henry Kingsley of a treatise on salmon; a review of a life of Josiah Wedgwood, a Lunar Society notable; some fresh translations from Horace; and an article on the New Zealand Maori war.

The focus of the first article was on human ability but Galton specified that his larger concern was really with human behaviour and its shaping by selection. The statistical results he presented for ability were just a preliminary report on a much more ambitious inquiry, but were already 'amply sufficient to command conviction', even though, as Galton freely conceded, little was then known about the precise laws of hereditary transmission of ability. 'I can only say that the general resemblances in mental qualities between parents and offspring, in man and brute, are every whit as near as the resemblance of their physical features; and I must leave the existence of actual laws in the former case to be a matter of inference from the analogy of the latter'. Galton's later statistical investigations of heredity at the level of the population, 'biometry', would prove tremendously influential right up till the present day.

Since ability appeared to be heritable, it could be shaped by deliberate selection, leading Galton to his first published formulation of directed human evolution, which he later dubbed *eugenics*. 'It would seem as though the physical structure of future generations was almost as plastic as clay, under the control of the breeder's will'. He concluded whimsically, imagining a

42 First one needs the probability of having at least one eminent relation, again using the binomial theorem and an estimate of the average family size.

43 Lewes 1856.

44 F. Galton 1865b. See Tredoux 2023a.

solemn socialist utopia in which the state shaped the malleable nature of man by encouraging breeding of the most able through targeted marriage subsidies: 'what a galaxy of genius might we not create! We might introduce prophets and high priests of civilization into the world, as surely as we can propagate idiots by mating *crétins*'. This improvement of the human stock was necessary because of the growing complexity of the modern world. 'We are living in a sort of intellectual anarchy, for the want of master minds. The general intellectual capacity of our leaders requires to be raised, and also to be differentiated. We want abler commanders, statesmen, thinkers, inventors, and artists'.

The Victorian world had the delightful habit of widespread critical commentary even on magazine articles. Galton's first installment was covered very thoughtfully, and for the most part favourably. The *Royal Cornwall Gazette* reported that the piece had 'caused some sensation in literary circles' and quoted generously from its concluding proposals for selective breeding of humans.[45] The *Newcastle Courant* referred to his 'very interesting speculation', with an accurate summary of his conclusions.[46] Closer to home, the *Leamington Spa Courier* drily commented that the investigation was 'opportune, for it is a subject on which diverse opinion exist [sic], only to be removed by the accumulation of facts'.[47] The *Cambridge Independent Press* described it only as 'cleverly written'.[48]

Even the *Irish Ecclesiastical Gazette* gamely conceded that Galton 'controverts with much ability the common impression that great power of intellect seems to have been inherited through the mother's side, and that one son often runs away with the talent of the whole family'.[49] But the *Coleraine Chronicle*, after stating some of the principal statistics, hoped that they would not 'discourage anyone from following the bent of original thought' merely on the basis of inadequate genealogy.[50]

Over the Scottish border, the *Fife Herald* noticed 'the first of a series of papers of curious interest' with 'a number of very plausible statistics', quoting the wellborn conclusion.[51] The *Dundee Courier* suggested that the author, who 'argues strongly', would 'not be easily controverted'.[52] The pattern continued in the *Paisley Herald*, which described it as 'an excellent paper', 'Mr. Galton will continue his subject, and we will gladly follow him'.[53] Several papers, such as the *London Evening Standard* and the *Berkshire Chronicle*, ran extracts as a *precis*.[54]

Corresponding with his friend the Harrow schoolmaster (and former Cambridge Apostle) Frederic William Farrar, declining a weekend invitation, Galton revealed the difficulties involved in choosing the right term for the intellectual abilities he was concerned with.[55]

June 12, 1865

Dear Farrar,

I should have much liked to come, but, what with engagements that may keep me late on that day & what with the difficulty of getting back, except at an early or very late hour by train, I must ask you to excuse me.

Thank you for your criticism on the word "Talent". I had not known it was of questionable mark, and had simply taken it as a useful current coin, without more thought. But I see the force of what you say, & have observed that the word is avoided

45 Friday 16 June 1865, 6.
46 Friday 09 June 1865, 5.
47 Saturday 03 June 1865, 3.
48 Saturday 03 June 1865, 6.
49 Saturday 17 June 1865, 6–17.
50 Saturday 10 June 1865, 10.
51 Thursday 01 June 1865, 2.
52 Monday 12 June 1865, 2.
53 Saturday 03 June 1865, 1.
54 Wednesday 07 June 1865, 4; Saturday 03 June 1865, 7.
55 Wellcome Library, London, MS 8376/2.

in the short notices that have appeared about my article in the different papers, & hearing it is not generally approved of, I must try to find a substitute, when I write my book: but it is hard to do so. Genius is inappropriate; ability includes the aid that might be afforded by circumstances. Intellectual gifts is longer & not much better than talents nous verrons.[56]

My next part—the concluding one—is very heterodox & goes into theory.

very faithfully yours
Francis Galton

The second part of the series appeared in *Macmillan's* in August, alongside new installments of 'The Dove in the Eagle's Nest' and 'Cradock Nowell'; J. M. Ludlow's 'Gallery of American Presidents'; John Stuart Mill's friend John Cairnes on 'The Negro Suffrage' in the South (he was in favour of it); light poetry; travel narratives; and ruminations on 'Scotch Deerhounds' (apparently far superior to greyhounds). Now Galton moved on from statistics to analogies and supporting instances, from ability to personality or 'character'.[57]

This was exactly where Galton had started in his notebook; he had published the papers in reverse sequence. However, he first addressed some criticisms he had received about the feasibility of selective breeding as an effective mechanism for changing the nature of a population. His arguments anticipated later developments in population genetics, as he pointed out with some simple calculations that a group of the able would inexorably crowd out the remainder (i.e. the less able) if they maintained a higher effective birth rate over a large number of generations. It would take more or less time to replace the remainder, depending on their relative disadvantage in effective reproduction, which could be set at will.

When addressing character, Galton had to choose a different approach than he had for ability. There were no handy or respectable lists to mine for the family relationships of people independently classified by character. Instead he considered groups of men, specifically *races*, which he believed could be reliably differentiated on the basis of their innate characters.

The precise concept of race used by Galton here requires some explanation. For him it was simply a breeding group, invariably of mixed origin, which could be delineated at the resolution required by the needs of the context. In particular it had nothing to do with ideal platonic forms or types. There is an important statement of his thinking in some remarks he made, nearly twenty years later, about a paper on 'civilized and savage life', delivered at the Anthropological Society by Henry Bartle Frere—who had recently been recalled as High Commissioner for Southern Africa. 'Mr. F. Galton would refer first to the purely ethnological part of the memoir, which dwelt upon the difficulty of defining the Bantu race. He thought that ethnologists were apt to look upon race as something more definite than it really was. He presumed it meant no more than the average of the characteristics of all the persons who were supposed to belong to the race, and this average was continually varying. The popular notion seemed based upon some idea like that of a common descent of the different races, from a parent Noachian stock, whence the aborigines of each county were derived, and where they lived in unchanged conditions till the white man came. Nothing can be further from the truth. We know how in South Africa the Bantu population has been in constant seethe and change; how, in much less than a single century, Chaka and his tribe, Mosilekatse and his tribe, and others, have in turn become prominent nations, and the average of the whole Bantu population must thereby have differed at different times. This same fluctuation of the average qualities of the population must, for anything we can see to the contrary, have gone on for many thousands of years. He therefore thought the phrase of Bantu race, as signifying some invariable and definite type, to be a mere chimera'.[58]

56 'We will see'.
57 F. Galton 1865c.
58 F. Galton 1881c.

A few years later, Galton was also eager to point out internal variation within races. 'The social history of alien peoples is full of problems. Thus it is a frequent observation that men may thrive and be happy under the most cruel governments, but the explanation of the several cases is imperfect. We have yet to learn how far it depends on hidden compensations, and how far on the genius of the people. ... The world of human mind and instinct is richly variegated, persons even of the same sex and race differing sometimes more widely than ordinary men differ from ordinary women, though of course in other ways, and this amount of difference is indeed large'.[59]

Galton argued that differences in racial characters had developed as a result of extended natural selection in past times, but, as the preceding quote shows, and the following confirms, he was well-aware that there were confounding effects. 'In founding argument on the innate character of different races, it is necessary to bear in mind the exceeding docility of man. His mental habits in mature life are the creatures of social discipline, as well as of inborn aptitudes'. The 'social discipline' he refers to might today be called 'culture' or (somewhat more misleadingly) 'environment'. He proposed that the effects of 'inborn aptitudes' and 'social discipline' could only be distinguished by 'observing several individuals of the same race, reared under various influences, and noting the peculiarities of character that invariably assert themselves'. In modern language, he was proposing that identifiable social influences first be 'controlled for' and then comparisons made to see if racial differences persisted anyway. He was confident that they did: 'even when we have imposed these restrictions to check a hasty and imaginative conclusion, we find there remain abundant data to prove an astonishing diversity in the natural characteristics of different races'.

The principal example in this paper involved the American Indians, who had featured so prominently in his ethnology notebook from the beginning. Though these were found in very diverse environments all over the Americas, north and south, and had been colonized by different European empires, they shared, he believed, a common character at their core. 'The men, and in a less degree the women, are naturally cold, melancholic, patient, and taciturn. A father, mother, and their children, are said to live together in a hut, like persons assembled by accident, not tied by affection. The youths treat their parents with neglect, and often with such harshness and insolence as to horrify Europeans who have witnessed their conduct. The mothers have been seen to commit infanticide without the slightest discomposure, and numerous savage tribes have died out in consequence of this practice. The American Indians are eminently non-gregarious. They nourish a sullen reserve, and show little sympathy with each other, even when in great distress. The Spaniards had to enforce the common duties of humanities by positive laws. They are strangely taciturn. When not engaged in action they will sit whole days in one posture without opening their lips, and wrapped up in their narrow thoughts. They usually march in Indian file, that is to say, in a long line, at some distance from each other, without exchanging a word. They keep the same profound silence in rowing a canoe, unless they happen to be excited by some extraneous cause. On the other hand, their patriotism and local attachments are strong, and they have an astonishing sense of personal dignity. The nature of the American Indians appears to contain the minimum of affectionate and social qualities compatible with the continuance of their race'.

As a contrast Galton cited the common character, as he understood it from the reports of travellers and explorers, of the West African Negro. 'He is more unlike the Red man in his mind than in his body. Their characters are almost opposite, one to the other. The Red man has great patience, great reticence, great dignity, and no passion; the Negro has strong impulsive passions, and neither patience, reticence, nor dignity. He is warm-hearted, loving towards his master's children, and idolised by the children in return. He is eminently gregarious, for he is always jabbering, quarrelling, tom-tom-ing, or dancing. He is remarkably domestic, and he is

[59] F. Galton 1885d.

endowed with such constitutional vigour, and is so prolific, that his race is irrepressible'.

The character of the American people made up a third example, but with some crucial differences. The country was initially colonized by European rebels and non-conformists, generally disorderly people who had been victims of persecution. The population was therefore quickly selected by emigration and conflict from a larger European pool in which these heritable tendencies already existed in pockets or scattered throughout, and could be drawn on. As a result, the partially-mixed descendants were 'enterprising, defiant, and touchy; impatient of authority; furious politicians; very tolerant of fraud and violence; possessing much high and generous spirit, and some true religious feeling, but strongly addicted to cant'. This example, like the others, owes much of its force not so much to what evidence existed for it at the time, but rather to its usefulness as a *thought experiment*—in this case for working out how behavioural traits could be 'selected for' rapidly from already existing variation.

These observations about races led Galton to conclude that character must be heritable in the same way as ability, and thus selectable. 'If we were to select persons who were born with a type of character that we desired to intensify ... and if we compelled marriage within the limits of the society so selected, generation after generation; there can be no doubt that the offspring would ultimately be born with the qualities we sought, as surely as if we had been breeding for physical features, and not for intellect or disposition'.

But Galton was quick to reject the idea that acquired characteristics could be inherited, dismissing in turn many common examples often given—pointing in dogs, fondness for man, tameness—as really examples of (usually unconscious) repeated selection on pre-existing innate characteristics, leading an increase in their frequency.

Nor was he a fatalist, pointing instead toward directed evolution. 'Never let it be forgotten that man, as a reasonable being, has the privilege of not being helpless under the tyranny of uncongenial requirements, but that he can, and that he does, modify the subjects in which nature examines him, and that he has considerable power in settling beforehand the relative importance in the examination that shall be assigned to each separate subject'.

Behaviours selected for in the past included those which give rise to religion, which Galton identified as at root a mechanism for group formation and cohesion. This is continuous with general animal behaviour. 'All evidence tends to show that man is directed to the contemplation and love of God by instincts that he shares with the whole animal world, and that primarily appeal to the love of his neighbour'. The institution of celibacy provided a direct example of the conversion of reproductive energy into religious commitment. This is a special form, he argued, of 'disinterested affection', now known as altruism, and was especially important given the long childhood of humans and their dependence on social support for survival. 'Extreme selfishness is not a common vice'. By this disinterest he really comprehended a sort of identity dissolution, in which the actor is subsumed into a larger body, which had been a spin-off from his notebook musings.

Galton did not state explicitly whether he thought the unit of selection here was the set of heritable elements, or the group, but since race and relatedness were both essential to his outlook, they amount to much the same thing in his usage. Given the centrality of religion to the organization of society, he was arguing for natural selection of heritable elements which produced not just a physical body but rather an 'extended phenotype' encompassing ideas and the full range of the mental products. A universal Darwinism.[60]

Another theme, only touched on when closing the first paper, was developed more fully in the second. He had been reaching for this in his ethnology notebook, where it appears as a Darwinian version of 'original sin': not as a *fall* from grace, but instead a *rise* from barbarism. It is analogous to a founding principle of the field that would eventually be known as 'evolutionary psychology': humans became adapted to a very different environment for most

[60] Dawkins 1982.

of their evolutionary history, and are still adapting to a rapidly altered one.[61] 'The natural qualifications of our race are no greater than they used to be in semi-barbarous times, though the conditions amid which we are born are vastly more complex than of old'.

Although traits beneficial to civilization—'endurance of steady labour, tameness of disposition, and prolonged development'— had been selected for by the success and efficacy of that civilization, they had only been *weakly* and superficially selected, in proportion to the length of time spent under civilization: 'we are still barbarians in our nature, and we show it in a thousand ways'. This substrate of base nature predestines man to fall short of complex moral ideals: 'the development of our nature, under Darwin's law of natural selection, has not yet overtaken the development of our religious civilization'. Thus 'the members of old civilizations are far less sensible than those newly converted from barbarism of their nature being inadequate to their moral needs. The conscience of a Negro is aghast at his own wild, impulsive nature, and is easily stirred by a preacher, but it is scarcely possible to ruffle the self-complacency of a steady-going Chinaman'.

To support the idea that 'wild, untameable restlessness is innate with savages', Galton claimed to know of 'numerous instances' of people 'of a low race' who had been adopted young and raised in settler families, but had reverted to ancestral ways after running away to rejoin their biological relatives. This was data he had sought directly the previous year (see Appendix B, p. 397) by distributing a comprehensive and carefully thought-out circular of questions about such 'individuals who have been reared since childhood in European or American schools, families, asylums or missionary establishments'. Transracial adoption studies would be pursued nearly a century after this was written to try and determine whether innate differences persist between races regardless of upbringing.[62]

Here Galton formulated another argument which anticipated a later discovery—the possibility that natural selection can enhance the fitness of some individuals for a while but paradoxically decrease the overall fitness of their race or species at the same time, perhaps even driving it into a downward spiral to extinction. It concerned early maturity, or precocity, which he believed varied by race. He proposed that precocity could be selected for in the working classes of the Industrial Revolution, but weaken the broader breed—'in the great mortality that besets the children of our poor, those who are members of precocious families, and who are therefore able to help in earning wages at a very early age, have a marked advantage over their competitors. They, on the whole, live, and breed their like, while the others die. But, if this sort of precocity be unfavourable to a race—if it be generally followed up by an early arrest of development, and by a premature old age—then modern industrial civilization, in encouraging precocious varieties of men, deteriorates the breed'. Which is to say that locally optimal adaptations might not be globally optimal.

There are also glimmers in-between of later concerns that would be famously elaborated on. Consider this enigmatic remark about development, heritability and twins. 'The interval that separates the full-grown animal from its embryo is too important to be disregarded. It is in this interval that Darwin's law of natural selection comes into play; and those conditions are entered into, which affect, we know not how, the "individual variation" of the offspring. I mean those that cause dissimilarity among brothers and sisters who are born successively, while twins, produced simultaneously, are often almost identical'. Ten years later Galton would test a more specific idea—that innate similarities prevail over external differentiators—by gathering extensive data for the first twin study, comparing identical twins with fraternal twins.[63]

This second installment by Galton drew mixed reviews, with more heat than the first, due to its forthright grounding of religion in natural selection, and its inversion of original sin. The

[61] Barkow, Cosmides, and Tooby 1992.

[62] Scarr and Weinberg 1983.

[63] See chapter 7, p. 127.

Illustrated London News was enthusiastic, offering some evidence of its own and a corollary. 'Mr. Galton's second paper on the hereditary transmission of talent and character is as able and suggestive as the first. His views derive considerable support from the ascertained degeneracy of classes like the Spanish aristocracy, where prejudice has so narrowed the field of matrimonial selection as to virtually almost forbid the possibility of improvement by alliances contracted on physiological grounds. ... He does not, however, remark that ... the counterpoise to the progress of education occasioned by the more rapid comparative multiplication of the inferior orders of intellect is one great cause of the slow advance of mankind as a whole'.[64] But the *Illustrated Times* affected a yawn. 'Mr. Francis Galton's papers on "Hereditary Talent and Character" are unquestionably interesting; but I fail to discover anything new in them, and I fail also to discover that the author recognises the extreme limits of his own principles'.[65]

Over in the West Country, the *Exeter and Plymouth Gazette* was more demanding. 'His theory of breeding clever people from clever people does very well for a magazine article, is peculiarly interesting, and one could wish to live two hundred years to see the experiment fairly tried. But how is the thing to be accomplished?'[66]

In the practical north, the *Durham County Advertiser* did not like the idea of man rising from barbarism, rather than falling from grace, and suggested that if Galton believed that then he 'should number among his intimate acquaintance creatures whom ordinary people despise'.[67] But the *Yorkshire Gazette* saw sound sense in the 'well written' paper, and, perhaps on the basis of first-hand experience, 'thoroughly endorsed' Galton's comments about the need for a steady disposition in barren northern climes.[68]

Over the border in Clackmannanshire the *Alloa Advertiser*[69] offered muted approval—'an able treatise, and contains much logical truth'—but the *Stirling Observer*, while admitting that the argument was elaborate, baulked at the 'somewhat dry and dull style'.[70] The *Buchan Observer* was appropriately measured. 'The writer is evidently master of his ground, and has treated the subject in a thorough lucid and exhaustive manner, indeed so thoroughly, that we cannot pretend in a few words to convey any correct impression of the argument brought forward'.[71]

Across the sea in Ireland, reviews were bad in parts. The *Irish Ecclesiastical Gazette* cautioned that though it was 'an able paper', 'there is a semi-sceptical tone which is much to be deprecated'.[72] The publication especially objected to the notion of mankind experiencing '*myriads* of years of barbarism' (their italics). Likewise, the *Dublin Evening Post* found it 'striking and clever; but his theories are unsatisfactory'.[73] In the north, the *Londonderry Sentinel* solemnly pronounced that the 'views expressed at the conclusion of the paper on "Natural Selection" and "Original Sin," will not be generally adopted'.[74]

Through the remainder of 1865, Galton continued combing through mountains of biographical dictionaries and memoirs to find family relationships between men of ability, expanding to military commanders, sport, theology and beyond. His goal was to publish a full-length study—something like the *Origin* for human intelligence and personality. His statistical methods needed to expand too. Inspired by a William Spottiswoode foray into statistical geography,

[64] Saturday 05 August 1865, 9.
[65] Saturday 05 August 1865, 7.
[66] Friday 04 August 1865, 3.
[67] Friday 11 August 1865, 6.
[68] Saturday 05 August 1865, 8.
[69] Saturday 19 August 1865, 2.
[70] Thursday 03 August 1865, 8.
[71] Friday 04 August 1865, 4.
[72] Thursday 17 August 1865, 11.
[73] Wednesday 30 August 1865, 3.
[74] Tuesday 15 August 1865, 1.

which applied the Gaussian distribution to mountain ranges,[75] he became absorbed in applications of that 'Law of Errors' to the problem. By ranking individuals by their *relative* standing in this distribution, which he called 'statistics by intercomparison', Galton was able to reason about the population-level properties of human intelligence without needing an absolute measure of it, which we still do not have today. The direction he would have to take quickly became clearer. The techniques were emerging. The data was highly suggestive and raw mounds of loose notes were accumulating rapidly. But by February of the new year he would collapse under their weight.

75 Spottiswoode 1861.

Figure 2.1: Francis Galton at age 42 in 1864, K. Pearson 1924. Restored.

Figure 2.2: Louisa, Mrs Francis Galton, in 1862. National Portrait Gallery, UK.

Figure 2.3: The Athenaeum Club, designed by Decimus Burton. Taken in 2016 while it was undergoing exterior renovations.

Figure 2.4: Herbert Spencer (1820–1903) by John Bagnold Burgess. Galton often enjoyed his conversation at the Athenaeum Club.

Murray, John, publisher, Londo

THE

KNAPSACK GUIDE

FOR

TRAVELLERS

IN

SWITZERLAND.

NEW EDITION, REVISED.

WITH CLUE MAPS, PLANS, AND MOUNTAIN OUTLINES.

LONDON:
JOHN MURRAY, ALBEMARLE STREET.
PARIS: A. & W. GALIGNANI & CO.; AND A. XAVIER.
1867.

The right of Translation is reserved.

Figure 2.5: Galton completely revised the 1865 and 1867 editions of Murray's Knapsack Guide for Travellers in Switzerland. *Not credited.*

Chapter 3

Breakdown

Shortly before he turned 44, Francis Galton experienced a catastrophic mental breakdown. It was February of 1866 and he had only recently found the subject matter and general techniques that would become his research programme for the next forty years and make his lasting reputation for brilliant originality. But now he could do nothing but obsess and grind endlessly over inconsequential, unsolvable details. Dizziness. Heart palpitations. They were back.

On the 7th, his wife Louisa, seldom entirely well herself, noted in her diary that Frank was 'Poorly and giddy'. Three days later they saw their regular family physician, Henri Guéneau De Mussy. Though the Frenchman had followed his Royal patron Louis Philippe from Paris into London exile in 1848, and had built up a thriving and sophisticated practice there, De Mussy's advice now was as simple as a country doctor's. He ordered them abroad at once for a long rinse through the European spas and haunts for invalids that they had already been touring hopefully for years.

Scientific circles in London also set in motion. The following day, Thomas Henry Farrer and Justice William Robert Grove called on the Galtons at 42 Rutland Gate, Knightsbridge. Farrer, from the Board of Trade, had been working actively with Galton on a committee investigating the controversial weather forecasting service that Vice-Admiral Robert FitzRoy—Darwin's commander on the *Beagle*—had run before his suicide the previous year. Grove, an influential figure in the Royal Society who had made an early reputation in chemistry and electricity, likely played a role in appointing Galton as the society's representative on the Committee. Their visit now did not help. Louisa's Diary recorded that Frank was 'worse by talking to Farrer & Grove' on the 11th.[1]

Galton must have written to his family to state the situation. They immediately recognized its seriousness. His mother Violetta wrote back to underline Dr. De Mussy's common-sense advice.

> 5 Bertie Terrace
> Leamington
> 14th Feby 1866
>
> My dearest Francis,
>
> I am quite unhappy about your health, do take your mother's advice, & give up all writing, & all head work for a year, remember you have not a strong constitution & overworking your mind falls upon your bodily health. Do my love, be persuaded to lay aside all ambition, & give yourself a complete years rest, & perhaps it may be the saving of your life, come to me whenever you like, & you shall have perfect quiet, my mind is with you by day & by night, I long to have you with me—with love to you & Louisa believe me ever your affectionate mother V. Galton

[1] The committee is covered in Tredoux 2023c. Galton had been elected a member of the Royal Society Council for 1866 in the previous year, 1865. See *Proceedings of the Royal Society of London* 14 (1865): 513. For Grove, Justice William Robert, see page 455. For Farrer, Thomas Henry, see page 452.

Let me my dear, wish you many many happy returns of your coming Birthday,[2] & I pray you may be blest with good health & every happiness this world can give you—God bless you.

His affectionate sister Emma Sophia added more urging, and pointed to an obvious suspect: his steadily expanding research into 'Hereditary Talent and Character'.

Feby 14th

My very dearest Frank

It makes me sad to hear that you are so unwell.

Your birthday is on Friday, & it is a day for making good resolutions. Now do make & keep this one—that is, to take a good long holyday & you will fully reap the benefit in the Autumn & be brisk & well for the B. Association.

Your "Hereditary" will also be better, by returning to it with a fresh eye & refreshed head. Do ponder on this, & make us all happy. It will do Louisa good also. With my best love

Ever dearest Frank

Your very affect. sister

E. S. Galton

How often we shall think of you on the 16th & wish you many happy returns of the day.[3]

Her reference to the British Association meeting—to be held in Nottingham in late August, and for which Galton was a General Secretary—must have added material for rumination. Then within the week more distinguished colleagues had weighed in.

General Sir Edward Sabine, the President of the Royal Society, who had made his scientific reputation through his research into magnetism while on voyages to the Arctic, had already formed a close relationship with Galton. Sir Roderick Impey Murchison, who had turned to geology after serving in the Peninsular Wars, was a frequent president of the Royal Geographical Society. He had been an early supporter of Galton when he was first setting out on his exploration of South-West Africa in 1850. Sir Roderick had seen him rise rapidly in the RGS after winning its Gold Medal in 1853, joining its Council in 1854 and then serving as one of its Secretaries from 1857 onward. After a sharp disagreement in 1863 with the other Secretary of the RGS, Norton Shaw, as a result of which both men resigned their positions, Galton had become the Foreign Secretary of the RGS in 1865.[4] We don't know exactly what they discussed now, but Louisa's Diary for the 20th was gloomy. Her husband was 'very poorly after talk with Sabine & Murchison'. Murchison wrote to Galton a few days later to wish him well. The letter gives a good idea of the political arm-wrestling that Galton had been involved in behind the scenes at the British Association.[5]

16 Belgrave Square
Feb. 23
1866

My dear Galton,

I was very sorry indeed to find that your health prevented your attending the Council of the B Ass^{n}.

[2] The 16th of February.
[3] His birthday.
[4] For this controversy see Tredoux 2023c.
[5] Murchison to Galton, GALTON/3/2/1/95.

We all regretted your absence. Every thing went off well & the two chairs in which I was interested were given C. t[o] Hamilton & E t[o] Sir Charles Nicholson.[6]

I ascertained by feeling pulses beginning with Sabine that Sir H. James[7] would not be acceptable to them & it was quite out of the question to have had Max Muller [as he] has never been inside of the Geographical Society.

Hoping that you will soon be yourself again we shall have enough to talk about as respects your [...] & the new law about the composition of the General Committee.

Owing to the [tepid] way in which Mr. Griffith talked of applications from Plymouth (where we had a very poor meeting) for 1867 & also of Exeter!! a place which could not hold a tithe of the Society. I was rather vexed that they seemed to ignore Dundee.

I said a few words about old [...] & spoke of the Duke of Buccleuch[8] as a very proper President in case we had should go to Scotland.

Yours sincerely
Rod. Murchison.

In early March the Galtons had taken Dr. de Mussy's advice and were in Cannes, by way of Boulogne. They had borrowed 'Villa Allegria' there from Louisa's wealthy friend Mrs. Robert Hollond, a bluestocking and society beauty who had married a wealthy Whig M.P., but kept a Parisian salon attended by the liberal Royalists known as Orleanists.[9] The report of the committee investigating FitzRoy's weather forecasting was submitted to Parliament on April 13. It must have been finalized in Galton's absence, but the scientific meat in it was largely due to him—Louisa's diary recorded that he had finished his part of it that past December.

The Galtons had moved along the coast to Italy by late April, touring into early June, taking in Bordighera, Alassio, Genoa, Milan, Cadenabbia, Lugano, and Stresa. Frank joined his friend Vaughan Hawkins from the Alpine Club for some diverting climbing in the limestone peaks around Lecco and Orta. Some years earlier Hawkins had made a notable attempt on the Matterhorn in the company of another of Galton's close friends, the experimental physicist Professor John Tyndall (see Figure 3.1, p. 48), and the three of them had gone on walking tours together in Cornwall.[10]

Tyndall had suffered his own mental breakdown between 1859 and 1860, with some symptoms similar to those Galton suffered from now. He even described them in a collection of travels edited by Galton, *Vacation Tourists and Notes of Travel in 1860*, to which Hawkins had also contributed. 'Mental exertion, unwisely persisted in, had brought on a curious kind of giddiness, last August, the state which became more and more easily excited, until, finally, the writing of a letter, or the reading of a newspaper, sufficed to convert my head into a kind of electric battery, from which thrills were sent to my fingers' ends. I had more than once been compelled to pause in directing a note; fearful lest the effort required to complete the address should produce some terrible catastrophe in my brain'.[11] According to T.H. Huxley, 'From the first, Tyndall suffered from sleeplessness, with the nervous irritability which is frequently cause and consequence of that distressing malady'. 'Tyndall "consumed his own smoke" better than most people, and though that faculty is worthy of the highest admiration, I suspect that the exercise of it tells a good deal upon the furnace. When things got bad with him, his one remedy was to rush off to the nearest hills and walk himself into quietude'.[12]

6 Sections of the B.A. meetings were numbered alphabetically. Nicholson (1808–1903), M.D. Edinburgh, had been Chancellor of Sydney University.

7 Colonel Sir Henry James, FRS (1803–1877).

8 Walter Montagu Douglas Scott (1806–1884).

9 For Hollond, Mrs. Robert, see page 455.

10 For Hawkins, Francis Vaughan, see page 455. For Tyndall, John, see page 458.

11 F. Galton 1861, 305.

12 T. H. Huxley 1894, 7.

A trip to Killarney—he had been born and raised in County Carlow—had only given Tyndall some temporary relief, so in 1860 he had returned to the Alps, where he showed that he had become a formidable climber since some earlier visits there. Huxley called him 'caprine'. By the end of his trip, 'the tide of health set steadily in the right direction' and soon there was 'no remembrance of any further exhibition of the symptoms which had driven me to Switzerland. Each day's subsequent exercise made both brain and muscles firmer'—though he strongly recommended that the innkeepers there could ease cases like his by providing 'wholesome, tender mutton', rather than the 'wicked tissue' they served under that name.

Searching for a taste of Tyndall's relief, rather than that wicked tissue, the Galtons crossed to Switzerland, over the Simplon to Brieg, Glion and Neuchâtel. Frank was no doubt already taking notes for a second edition of his *Knapsack Guide to Switzerland*, trying out new walking routes and revisiting long-familiar ones. The *Guide*, which had appeared anonymously in 1864, had sold well, and Galton's extensive familiarity with the region and travel literature had also improved the 1865 edition of Murray's *Handbook for travellers in Switzerland, and The Alps of Savoy and Piedmont*; another edition would accompany the revised *Guide* in 1867.[13]

They had returned to Boulogne by the 20th of June, staying at the Hôtel des Bains. At 11 o'clock in the morning Galton heard 'a violent explosion, followed by a low and long-continued rumbling'. Seeing crowds gathering, the Englishman went out and noted 'the long, narrow, smoke-like train of a meteor hanging in the sky', with a 'final puff that indicated the place of explosion with perfect distinctness'. Characteristically, he had a travelling altazimuth on hand, and recorded the bearings of the smoke 'train', informing the *Times* that a comparison of his measurements with those of others would 'indicate the path of the meteor, and will accurately fix its height above the earth at the time of explosion'.[14] Rest and diversion in Europe included exact instrumental observations! But it had clearly not been enough, as they established themselves in the Lake District at Grasmere and Ambleside, from July to mid-August. 'Rain constant', Louisa glumly diarised.

The British Association, that annual scientific pilgrimage of which Galton was inordinately fond, was meeting later that month for the 36th time, as his sister Emma had reminded him. This year it was at Nottingham, with Justice Grove presiding. As one of the General Secretaries, Galton was expected to be present, and beside that he had as usual prepared some papers, this time flowing out of his work on weather forecasting for the Board of Trade: 'On an error in the usual method of obtaining meteorological statistics' and 'On the conversion of wind into passage charts'. He also contributed some detailed observations of the 'aerolitic meteor' seen over Boulogne.[15] These were more mundane topics than his submission the previous year, which concerned underwater vision, or 'Spectacles for Divers'. He had tested those extensively in the bath at 42 Rutland Gate, successfully reading the newspaper while submerged underwater, until an odd feeling brought on by lack of oxygen eventually roused him to the surface.[16] This year's bulging programme included Alfred Russel Wallace on adaptive mimicry, Huxley on human crania, Edward Tylor on the origins of high civilization, Thomas Baines on the Limpopo and Zambezi, Samuel White Baker on the Abyssinian tributaries of the Nile, and Paul du Chaillu on Western Equatorial Africa.

But shortly after Galton got to the meeting at Nottingham, he broke down once more, due to 'the sustained racket of it', which 'too long continued for my own nerves'. He was later convinced that 'it would have been playing with death had I continued'.[17] Louisa noted that he was 'giddy' again and 'done up' on the opening day, the 22nd of August—the day of Justice

[13] F. Galton 1864b, 1865a, 1867b.
[14] F. Galton 1866b. F. Galton 1868.
[15] F. Galton 1866a,c,e.
[16] F. Galton 1865d.
[17] F. Galton 1908, 215.

Grove's Presidential address. They took the next train to Leamington, staying there with his mother. His British Association papers would be read in his absence. At his mother's house Frank continued to be 'poorly' (Sep. 3) so the help of Leamington's internationally-known physician Dr. Henry Jephson was enlisted.

Jephson had tended to the Galton family since the early 1830s, when his phenomenally successful practice at Leamington first took off. He had started out modestly in 1823, as an apothecary's apprentice, succeeding in blowing two of his fingers off in a freelance chemistry experiment.[18] He had taken time off to get a medical degree from Glasgow between 1826 and 1828, when it became clear that his manner with patients was exceptionally winning. In some years he earned over £20,000 in fees, as the well-heeled, including Samuel Tertius Galton, flocked to Leamington to take the sulphurous waters, internally and externally, and submit to Jephson's rule.

Two decades of unstoppable success had enabled Jephson to build a mansion at Leamington, 'Beech Lawn', which is where he would have received Galton. This was a special family favour, since Jephson had already retired from practice in 1848 after going completely blind, and now spent most of his time, and some part of his fortune, on philanthropy. His treatments had remained more or less unaltered since the beginning, involving changes in diet to go with the Leamington waters, and inspiration. The doctor favoured what he called 'plain' eating, but was broad-minded enough to allow his patients necessities like sherry and game, while administering a concoction of his own invention, known ominously as the 'broom', to cleanse their innards. Louisa recorded that, on September 15th, Jephson put Galton on an especially sparse diet, and that her husband grew 'very weak under the treatment'. It was not the first time that Jephson had treated his patient for this malady.[19]

By the end of the month they were back in London, but Galton's situation does not appear to have improved. Early in October the Assistant Secretary to the British Association, George Griffith, called on official business.[20] Two days later, on the 11th, Louisa noted that Frank reported being 'quite tired'. Then it was back to Dr. de Mussy in early November, and the stringencies of another visit by General Sabine. So the Galtons fled once more, this time to lodgings in Breed Place, Hastings, and the welcome company of Louisa's friends Frederick North and his accomplished daughter Marianne, a widely-travelled botanical illustrator—her paintings remain on exhibit today at Kew Gardens in the North Gallery, which she erected at her own expense. The Galtons would stay with the Norths into the New Year, while Galton desperately sought distraction through four-hour horse rides.

The year of 1866 had turned into a blank. Galton had got no more scientific work done. Aside from the two papers at the British Association, and secondary reports about them, his only other appearance in print had been that brief letter to the *Times* about the meteor over Boulogne. His fledgling research programme in hereditary abilities was stalled. Yet his condition appeared to have remained unchanged, so more desperate measures were considered. The Galtons were still at Hastings on the 21st of January 1867 when they heard that the spendthrift Lord Ashley (Anthony Ashley Cooper, 1831–1886) had agreed to take a lease on their house at 42 Rutland Gate for the remainder of the year. That enabled them to set off on an exceptionally long tour of Europe, perhaps in the hope that something would turn up, somewhere, eventually. At Hastings they looked up Miss Eleonora Anne Brandram, whom they had first met at Aigle in Switzerland, in September 1862, where she had often walked with Frank. Miss Brandram agreed to accompany them now to Italy, as a companion to Louisa.[21] On the 26th of January, Galton felt robust enough at least for a high-powered lunch at the

[18] E. A. Galton 2003, 95.
[19] See Tredoux 2023c. For Jephson, Dr. Henry, see page 456.
[20] For Griffith, George, see page ??.
[21] For Brandram, Eleonora Anne, see page 451.

Athenaeum Club with T. H. Huxley and William Spottiswoode. Then, after an inventory of 42 Rutland Gate had been compiled, the Galtons were off again to Italy on the 31st, via Paris.,

From Paris they went through Dijon to Hyères in Provence, and on to Cannes and Mrs. Hollond for a few days. Working along the coast through Nice, Mentone, Oneglia and Alassio, they were in La Spezzia and Genoa on the 10th, and saw the English-built ironclad ram *Affondatore*, the 'Sinker'. It was an unfortunate name, since the ship had itself sunk in a storm the previous year. Now it was being rehabilitated, after being raised by the Italians, perhaps in hope of another go at the Austrians. The remainder of February was spent ambling towards Rome through Pisa and Perugia. In Rome on the 26th, they indulged one of Galton's peculiar observational fancies, the start of a horse race. 'I especially enjoy the start of the horses, for their coats shine so brightly in the sunshine, the jockeys are so sharp and ready, and the delays due to false starts give opportunities of seeing them well. I don't care much for its conclusion'.[22] Eight years later, Galton recalled that he found some respite in the ancient capital, and attributed it to a lack of advertising![23]

> A great deal of energy is wasted in attempting to seize more than can be grasped. There is a feverish tendency, fostered by the daily press, to interest oneself in all that goes on, which leads to perpetual distraction, and curtails the time available for serious and sustained effort. It may be worth while to mention a curious little morbid experience of my own, as suggestive of much more mischief; it is this:—A few years ago, I had foolishly overworked myself, as many others have done, misled by a perverted instinct which goaded to increased exertion, instead of dictating rest. The consequence was, that I fairly broke down, and could not, for some days, even look at a book or any sort of writing. I went abroad; and though I grew much better and could amuse myself with books, the first town where I experienced real repose was Rome. There was no doubt of the influence of the place—it was strongly marked; and for a long time I sought in vain for the reason of it. At last, what I accept as a full and adequate explanation, occurred to me; simply that there were no advertisements on the walls. There was a picturesqueness and grandeur in its streets which sufficed to fill the mind, and there were no petty distractions to fret a weakened eye and brain. When we are in health we take little count of the racket of English life, which may keep apathetic minds from stagnation, but which causes needless wear and tear to active ones, suggesting nothing useful, and teasing, distracting and wearying. I have heard German professors speak with wonder at our waste of energy in mere fidget, and in so-called amusements, which are mostly very dull, and ascribe the successful laboriousness of their own countrymen to the greater simplicity of the lives they lead; and they are a happier people than we are.

Also in Rome, searching for Aristophanes manuscripts, was William George Clark, an old friend from Galton's Trinity College days, when they had whiled away hours in the 'Caseo-Tostic' and Epigram Societies. Clark was a gifted classicist, second only in 1844 to Henry Sumner Maine, who had often joined them in Galton's rooms. After taking up a fellowship at Trinity he had settled down to a bachelor's life of vacation tourism and unfulfilled literary projects, frustrated by theological constraints and, Galton believed, by his pining for a doomed male cousin from the dim past. Now Frank 'saw much' of him. Together they took a trip to Naples to see the mosaics at Pompeii, which Clark had visited before in 1860.[24] Louisa noted the 'flavour of sulphur' from Vesuvius in her diary on the 4th of March. They remained in the vicinity until early May, encountering 'Mrs. Skinner the American rich lady' and exploring the surrounds, including Sorrento, Amalfi, Ischia and Capri.

By now Galton had taken up some work while on the road, and on the 12th or 13th of April had sent off a paper to Kew on an instrument for coordinating meteorological curves. This was not, as far as we know, published—but it may have been an early contribution towards his compound pantagraph for compressing the tracings of self-recording meteorological instruments

22 F. Galton 1908, 179.
23 F. Galton 1874c, 229-230.
24 See F. Galton 1861.

for compact display. The peculiar pleasures of instruments and measurement.[25]

In May the Galtons were on the move again, through Ancona and Loretto to Venice, where on the 27th the pigeons covering the plaza outside the Basilica of St. Marks caught Frank's imagination. These would usually be described as 'innumerable' but Galton set about proving otherwise by counting them. This sort of activity was, again, curiously therapeutic for Galton, recurring throughout his life in different contexts. Consider his fond recollection of those mile-long flowering chestnuts in Bushey Park.[26]

> I fixed upon a tree of average bulk and flower, and drew imaginary lines—first halving the tree, then quartering, and so on, until I arrived at a subdivision that was not too large to allow of my counting the spikes of flowers it included. I did this with three different trees, and arrived at pretty much the same result: as well as I recollect, the three estimates were as nine, ten, and eleven. Then I counted the trees in the avenue, and, multiplying all together, I found the spikes to be just about 100,000 in number. Ever since then, whenever a million is mentioned, I recall the long perspective of the avenue of Bushey Park, with its stately chestnuts clothed from top to bottom with spikes of flowers, bright in the sunshine, and I imagine a similarly continuous floral band, of ten miles in length.

Similar methods would count the pigeons of St Mark's. As May closed, Frank and Louisa worked their way up toward Switzerland, through Padua and Verona, up Lago di Garda to Riva. They were in Botzen, Meran and the South Tyrols in early June. From there they struck up to Nauders in Austria, and over to St. Moritz, where they were in lodgings on the 19th. Galton had a definite object in mind in Switzerland: John Murray had asked him to update his *Knapsack Guide* for a second edition, which would be issued later that year. But before he could complete that, he had to finalize another project he appears to have taken with him on the road, the fourth edition of his popular *Art of Travel.*

Galton's collection of 'shifts and contrivances available in wild countries' had first been published in 1855, but a second edition followed the next year, with a third in 1860, as the work grew in scope and length. The fourth edition was to be 'entirely recast' and 'much enlarged'. Miss Brandram[27] had been helping Galton with the work, the novel parts of which he had doubtless been trying out himself in the rugged landscape. But now he had involved himself in perhaps more work than he initially intended. 'Before I could interpolate the new matter prepared for this Edition I found it necessary to recast the last one, by cutting it in to pieces, sorting it in to fresh paragraphs and thoroughly revising the writing—disentangling here and consolidating there'. The greatly enlarged index must have benefited from Miss Brandram's attentions, but he felt compelled to address previous claims that he had plagiarized portions from Randolph Marcy's *Prairie Traveller.* As he pointed out, his second edition of 1856 predated Marcy's 1859 publication and that the opposite was true—the American appropriated those passages from the *Art of Travel* without attribution.[28]

The effort, and perhaps the echo of controversy, told on Galton's mental state. As the book was being put to bed in St Moritz towards the close of June, he was once again 'giddy' and 'poorly', or 'giddy but about', and it was not until mid-July that the manuscript of the fourth edition could finally be sent off to John Murray in London.

Louisa reported that, at St. Moritz, 'the cure did wonders for me, but did not suit Frank'. Instead, endless exercise was available. Miss Brandram was a useful walking companion—it seems that Louisa was not a hiker—as investigations for the *Knapsack Guide* proceeded. To the south-east of St. Moritz lay the spa of Le Prese on the Bernina Pass, via the carriage road built in 1863; Galton was able to confirm that it was 'beautifully sit. N. of the lake (*Inn*, very

[25] F. Galton 1870b. See Tredoux 2023c.

[26] F. Galton 1869a, 11-12. The park is near Richmond, past Kew, and is now called Bushy Park.

[27] For Brandram, Eleonora Anne, see page 451.

[28] Marcy 1859; F. Galton 1867a.

comfortable), is becoming much frequented by the English. Baths well kept; Sulphur-waters. Many Lombard visitors. Numerous pleasant walks'.[29] They met up with a Colonel Weston, who went with them to see how the limestone had been weathered into pyramid-like formations.

From Le Prese they turned back to the north, over the Albula Pass, and lodged in the ancient Roman town of Curia (Coire) on the 29th of July, a mile from the Rhine. They found 'curious domestic architecture', and a 'considerable collection' of books on Natural History in the Library of the Cantonal Schools. Zurich was just a day away, and from there the diggings at Wetzikon, which they inspected on August 1st: 'remarkable for the lacustrine remains in the neighbourhood, especially at Robenhausen, by the small *Pfäffikersee.* Rude structures have been found 6 ft. below a layer of peat, and with them bones of 3 species of urochs also chamois, bisons, beavers, boars, &c. with remains of apples, pears, and cherries'. Their next stop, on the 2nd, was Seeligsberg—on the shores of Lake Lucerne, and home to yet another sanitarium.

In Seeligsberg the Galtons had the company of the Lushingtons and William Spottiswoode. Franklin Lushington, whose elder brothers were intimates of Tennyson, had known Galton at Trinity College. With Henry Maine, he had helped to bring out the literary remains of their friend Harry Hallam, who in 1850 had died of malaria in Italy, while Galton was in South-Western Africa.[30] Despite forming an intense friendship with Edward Lear—painter, traveller and nonsense-poet—on Corfu in the late 1850s, Lushington had married the daughter of a Somersetshire vicar in 1862; the couple seem to have been on tour.

Galton had probably been in more frequent contact with his colleague at the Ethnological, Geographical and Royal Society, Spottiswoode. While he was not rescuing his family's printing firm from bankruptcy, Spottiswoode had active interests in applied mathematics, mountaineering and travel writing. He had been a member of the Alpine Club since 1858, and so knew Switzerland well. And as if all this was not enough, he was also an accomplished Oriental linguist. As we have seen, it was Spottiswoode's paper on the most probable direction of the forces shaping mountain ranges, delivered at the Geographical Society in April 1860, that first directed Galton's attention to the power of the Gaussian statistical distribution, and inspired him to use it in his studies of hereditary talent and character in 1865. Before that, they had proposed exploring another shared interest together in 1859, the exploration of El Tih in the Sinai desert, but were forced to abandon the plan after Galton developed a life-threatening ulcer at the base of his tongue.[31]

It is possible that Spottiswoode and Lushington joined Galton now on an expedition to climb the Rigi, on the shores of Lake Lucerne—which 'owes its celebrity not to its height, because it is only 5910 ft. above the sea, but to its isolated position, which converts it into a natural observatory'. They were unlikely to have needed the boys who 'may be hired for ½ or 2 fr. to show the way, and carry a light knapsack' for the '3 to 4 hour climb'.[32]

In the meantime the explorer Samuel White Baker had written to Galton to offer some advice. Baker had discovered Lake Albert in 1864, after being tipped-off at Gondokoro by Speke and Grant, who were expecting to be resupplied there.[33] He knew Galton well: from the Monday-evening soirées of explorers they had regularly attended back in the 1850s, at Rear-Admiral Henry Murray's bachelor-apartment in the Albany;[34] from their interaction at the Geographical Society, where Galton planned expedition logistics; and as a source of practical shifts for *The Art of Travel.* It was for more advice of the last kind that Galton had written to Baker, hoping to include his information in the new revision, but he had also kept him updated

29 F. Galton 1867a, 386.
30 For Lushington, Sir Franklin, see page 457. For Spottiswoode, William, see page 458. For Hallam, see Hallam, Henry Fitzmaurice, see page 455.
31 See Tredoux 2023c.
32 F. Galton 1867a, 56.
33 See Tredoux 2023c for a full description of the Nile controversy.
34 For Murray, Rear-Admiral Henry Anthony, see page 457.

about his breakdown.

Tortworth Court
Wotton under Edge
Glostershire[35]
20 / Aug / 67

Dear Mr. Galton

Your note has followed me about & reached me here. I am glad to hear that you are preparing another edition of your charming work, "The Art of Travel", and in a day or two I will send you a few "dodges" that may be worthy of a place in your useful treasury of hints.

I am exceedingly sorry to hear that your health is still a cause of uncertainty. Why not try Egypt? I think it would do you more good than Europe.

Very sincerely yours
Sam. W. Baker

On the day this letter was being written, the Galtons were already in Baden-Baden, by way of Basle, and not remotely in an Egypt frame of mind. They would not have known that Fyodor Dostoevsky had just spent over forty days there in July and August, enriching the pawnbrokers while pursuing his 'unbeatable' roulette system, accumulating the kind of mad ruin already depicted in *The Gambler*, before heading south through Basle to Geneva. Nor would they have known that Ivan Turgenev was living there. Nor that the author of *Oblomov*, Goncharov—a man who had stayed in Mrs. Welch's hotel in Cape Town not long after Galton had fetched up in it—had just visited the spa town.

They stayed less than a week at the gaming resort, moving on to Carlsruhe on the 26th, and Heidelberg on 28th. Galton was probably brooding over his failure to attend the British Association, to be held that year at Dundee, starting on 9th of September. He was after all one of their General Secretaries; but the disaster of the previous year dominated these misgivings. Instead, the Galtons ambled their way around southern and central Germany through September to early October, stopping off at Stuttgart, Ulm, Augsburg, Munich, Ratisbon, Würtzburg, Frankfort and Bonn. Though Frank was kept away from the Dundee meeting, he seems to have been active mentally. Aside from finishing the reconstructed index for the *Art of Travel*, Louisa's diary recorded that, when they were in Munich, 'FG coins tinfoil': but if she meant the material we bake potatoes in, the name and the foil itself go back to the 15th century at least.[36] They were finally back in England by way of Brussels on the 12th of October, after an absence of over nine months. Their first month back was spent on the usual round of family visits, to the Butler clan in Harrow, and the Galtons in Leamington and Claverdon. Still, on another front, Frank was active once more.

In 1865, the talented mathematician H. W. Watson, another member of the Alpine Club, had gratefully swapped his position as a teacher at Harrow for a Rectorship at Berkswell, a village near the boundary between Warwickshire and Coventry.[37] On the 12th of November, Galton paid Watson a visit, probably in search of information about the spread of examination marks in the Cambridge Tripos, which Watson had moderated. This was part of Galton's stalled research into hereditary abilities. When he was at Cambridge himself, he had noted how large the gaps were at the very top of the mathematics Tripos, where a Senior Wrangler might be separated from his nearest competitor by as many as a thousand marks. The examinations stretched over five fatiguing days, so that even a modest edge could be amassed into a large absolute numerical advantage, but Galton suspected that at the top, natural ability produced

[35] So spelled in the letter-head.
[36] *Oxford English Dictionary*.
[37] For Watson, Rev. Henry William, see page 459.

extreme differences between the outlying candidates. A return to these topics suggested hope that he was finally on the mend. Or a desperate gamble.

There was a cab strike on in London shortly after Frank and Louisa returned on the 28th of November, to reclaim 42 Rutland Gate from the high-living Lord Ashley. They seldom went out that winter anyway, but a few days before Christmas, the notable Alpinist and illustrator Edward Whymper was over to lunch. In 1861, on one of his many attempts to climb the Matterhorn, Whymper had tested a canvas tent made according to Galton's suggestions. The structure was supported by *alpenstocks.* 'It looked very pretty when set up in London, but it proved thoroughly useless in the Alps'.[38] When it threatened to blow away, they had been forced to take it down and sit on it, but it was also cold even when they wrapped themselves in its folds. Substituting lancewood poles shod with conical iron points for the *alpenstocks*, bolting them together, and adding an integrated mackintosh floor, Whymper made a tent that was more stable and to his liking. He used this tent when, in 1865, he eventually succeeded in reaching the summit of the Matterhorn—even the loss of four lives seemed worth it. Galton mentioned Whymper's modified 'Alpine tent' in the fourth edition of the *Art of Travel*, but found that the integrated mackintosh floor retained the damp and too easily disturbed the walls.[39]

Galton plainly did not get a full year of rest in 1867, whatever his initial intentions may have been. He had taken work with him on the road, including the revised editions of the *Art of Travel*, and the *Knapsack Guide to Switzerland.* He had also worked on the paper for Kew; and he had obviously been working again at his inquiries into hereditary ability, since he had a stock of questions on hand for H. W. Watson when he got back to England. Despite these attempts, 1867 was one of his least productive years scientifically.

Charles Darwin was also in touch in January of the new year, after being congratulated by Galton on the publication of his latest book *Variation of Animals and Plants under Domestication* (1868), a subject dear to them both. 'I sh^{d} have been still better pleased, if you had told me how you were. I have enquired whenever I have had a chance of hearing, but did not hear some two months ago a very good account. I trust & hope by your note that you are well again, but I know that over mental work takes long in recovery'.[40] Darwin spoke from experience; the parallels between his case and that of Galton will be explored further below.

Later that January, Galton had another setback when he was unable to accept an invitation to dinner with John Crawfurd, the President of the Ethnological Society. Once again he was too 'poorly' to do so. Crawfurd was over 84 and only had a few more months to live, but was still sharp-witted, having contributed some 38 memoirs to the Ethnological since the age of 78. Galton had served as a secretary to the Society, but knew Crawfurd well from the Geographical, where the older man had been a council member since 1857; from the Athenaeum Club, where Crawfurd was invariably established in the vestibule in the afternoon; and from his extensive travel writing. Crawfurd had served as a Colonial administrator, governing Singapore for three years in the 1820s, though he had originally trained as surgeon at Edinburgh. He had travelled widely in the Orient on diplomatic and military missions, compiling histories, dictionaries and an important fossil collection along the way. When he returned to Britain in 1828 he ran several times for parliament as a radical and free-trader, but without success, so he channelled

38 Whymper 1871, 92. Note that Whymper's description in this book is very confusing—instead see Whymper 1865.

39 F. Galton 1867a, 160. Whymper 1865 mistakenly claimed that Galton had advocated waterproofing sheepskin sleeping bags, used by French customs officers in the Pyrenees, in F. Galton 1861. Waterproof sleeping bags are described in F. Galton 1860, 45-6, where Galton discusses his experiments with waterproofed *calico* in the Westmorland mountains (near the Lake District). Galton only discovered the sheepskin bags later that year, after his expedition to see the eclipse in Spain. This error by Whymper—who found waterproofed sleeping bags hot and sweaty—is amplified with distortion into 'heavy criticism' in Forrest 1974, 75, 322.

40 Darwin to FG, circa 1868-01, DARWIN, DCP-LETT-5800.

his energy into scientific directions instead. Galton's warm relationship with Crawfurd suggests that he must have been in very bad condition to miss this opportunity. The old man had apparently nicknamed Galton and Thomas Wright[41] his 'two sons' at the Ethnological. Galton recalled him as 'a vigorous old man of considerable moral weight and of great experience, with not a few amusing peculiarities'. On the 12th of May that year he would attend old Mr. Crawfurd's funeral.

By now, Galton must have realized that endless immersion in the sulphurous waters and classical evocations of the continent, even the countless hours of exercise in the Alps, could do no more for him, however much it suited Louisa. Most of 1868, though not all, would be spent in England. After his visit to H. W. Watson a few months prior, he had been picking up the threads of his research into the inheritance of natural ability and personality, and was re-engaging with at least some of the scientific societies he had previously frequented. Here there were still some swings to be negotiated. On February the 17th, according to Louisa's diary, Frank had a 'small row' with General Sabine at the Meteorological Society, the details of which have now vanished. Throughout his long career, Galton was involved in remarkably few personal clashes. He had an aversion to controversy, and was inclined to avoid it if he could. Sabine had been an important promoter of his career, and whatever the difference was about, it seems to have been repaired quickly.

The Royal Society held its annual soirée at Burlington house on Saturday March the 7th, and Galton felt well enough to attend, for the first time in several years. The Prince of Wales was in attendance, with a horde of foreign dignitaries. John Tyndall demonstrated gas extraction and magnetisation of light. The Kew Committee were able to show off their new weather semaphores and instruments for automatically recording observations, with which Galton was closely connected. Imaginative depictions of Abyssinia by Thomas Baines were highly topical, given the military expedition there the previous year. Galton knew the artist and explorer from Kings Lynn well, as it was Baines' interception on the Mooi River by the Boers that had diverted his own expedition of 1850 to Lake Ngami through the West Coast. Alongside Baines, the original drawings by Cornwallis Harris of African wildlife, dating back to 1842, were on display. Conversations with Harris in early 1850 had helped to inspire Galton to go to Southern Africa, rather than the Nile. New electrical methods for railway signalling were demonstrated, and must have reminded Galton of his own early work on printing telegraphs, which involved mechanisms for remote (digital) control of machinery.[42] Sir William Thomson's rotatory device for inducing static electricity, and James Clerk Maxwell's instruments for measuring capillarity, were also on display. They, like Galton, had been pupils of William Hopkins at Cambridge.

It is possible that this immersion back into the institutional life of science was too much for Galton at such an early stage of his cure, as he attended no further events for the rest of the year. He proceeded to simplify matters further by regretfully resigning his position at the British Association on March 14th, citing 'considerations of health'. His friends and fellow council members, William Spottiswoode and John Lubbock, hoped that he would be able to return when his health improved, but he must have known himself that there would be no return. Even so, he was a member of a deputation from the British Association, including T.H. Huxley, J.D. Hooker, William Spottiswoode and P.L. Sclater, to visit the prime minister Benjamin Disraeli at 10 Downing Street on the 23rd.[43]

After the deputation to Disraeli was concluded, Galton was able to visit the mathematical Reverend H. W. Watson at Berkswell once again, during the Easter break, riding over on horseback from his mother's house at Leamington (roughly an hour and a quarter's ride today

41 Galton mistakenly wrote *Laurence Oliphant* here in his memoirs—see F. Galton 1908, 172—but the other secretary of the ethnological under Crawfurd was the prolific antiquarian Thomas Wright (1810–1878).

42 See Tredoux 2023c for a full description of Galton's Southern African expedition and *Telotype*.

43 *London Daily News*, 24th March, 1868: 5.

by bicycle, or a 4 hour hike). The fact that he went on horseback suggests that he still felt the need for distraction, since this had been a common recourse for him in the past. Galton must have wanted more information from Watson about the Cambridge Tripos examination marks.

In August, a letter from Rutland Gate, dated June 24, appeared in the Hongkong journal *Notes and Queries on China and Japan.* It requested 'trustworthy information as to whether the history of public competition in China, tends to show that certain families are eminently endowed with scholastic activity', hoping for lengthy replies. In October, a reply by William Frederick Mayers to Galton's query appeared, recording several specific instances but regretting that the author could not supply more general evidence.[44] The whole research programme into heredity had steadily picked up again, as can be seen from the remains of Galton's correspondence. Two years previously, in 1866, his queries about pedigrees seem to have petered out, as his condition persisted. Then he had a response from the poet John Addington Symonds, and one from Julia Lennard *nee* Hallam, the sister of his doomed friend Harry, whose family was packed with coming men extinguished young. Aside from his visits to the Rev. Watson, in March of 1868 he had already been in touch with the literary critic Francis Turner Palgrave about the pedigree of that distinguished family. Gifford Palgrave, the critic's brother, was an Arabian explorer, while his father Sir Francis was a noted historian and antiquary. They were also closely connected with the eminent Turner and Hooker families.[45] Come December, Galton was sorting through detailed pedigree information from the philologist Rev. Isaac Taylor (1829–1901), the eminent classicist Benjamin Hall Kennedy (1804–1889, son of Rann Kennedy whose tenure as the second master at the Birmingham Free School overlapped with Galton's unhappy education there), the jurist James Fitzjames Stephen (1829–1894, brother of Galton's Alpining companion Leslie Stephen) and the engineer Charles Hutton Gregory (1817–1898).

It seems that Galton concentrated from 1868 onward squarely on the research that would produce, in March of the following year, a lengthy paper in *Macmillan's Magazine* on the 'Judges of England'. This was followed by a magnum opus at the end of that year, *Hereditary Genius.*[46] He published no other scientific papers in either 1868 or 1869, and kept away for a full year even from the Royal Geographical Society, where in previous years he had made regular appearances in the give-and-take after papers were read. He reappeared at the RGS only in November 1869, for discussions concerning Dr. Livingstone, who, resenting the interference of secular entities in London, had once again 'lost' himself in the African wilds. Moreover, Galton's work with Kew and the Meteorological Committee was now chiefly about instruments, their calibration and comparison, and the invention of gadgets to simplify their use—relaxing work.

A brief but rainy two-month holiday in France and Switzerland, in September and October of 1868, followed long-established patterns from more normal years, as did Louisa's digestive complaints. Christmas was spent once again at Hastings with the Norths, where they had passed time during the hot summer too. All records of Galton's breakdown are lost at this point. When he came to write his memoirs in 1908, he had to remind himself from old diary entries, often copied from Louisa's journals, about the details.

Galton's crisis produced two lost years in his prime. We have no reason to expect that he understood his own life history any better than most individuals, who are notoriously prone to rationalization and retrospective story-telling. He did know that his breakdown of 1866 to 1868 was not an isolated incident. A similar crisis at Cambridge in 1840-3 had stunted his studies there, forcing him to withdraw from the intensely competitive examinations for Mathematical honours, and side-lining him for years afterward. Moreover, when he was copying out his wife's diaries—now lost or destroyed—Galton annotated her entry for February 1866, describing how he had broken down after Sabine and Grove's visit, 'my regular breakdown', and circled the

---

44 *Notes and Queries on China and Japan* (1868), 154-5.
45 GALTON/2/4/1/2/5.
46 F. Galton 1869a,b.

remark in pencil, underscoring earlier diary entries about related incidents in blue (see ??, p. ??).

The first difficulty in understanding this intermittent debilitating syndrome is the vagueness with which Galton and his wife describe it. His symptoms had some purely mental components: fixation on trivial problems, and the sensations of a mental grist-mill (implying auditory hallucinations), resulting in an inability to think about more important things. There were also 'other maladies prejudicial to mental effort', but they were left undescribed. Though he obsessed *day and night*, he claimed that he did not suffer from insomnia, which is inconsistent. But there were also physical symptoms: dizziness (arguably a mental symptom too), heart palpitations, an intermittent pulse, and whatever else is covered by feeling 'poorly', 'done up' and 'knocked up' (most likely severe fatigue). The physical symptoms were so serious that Galton, always thoroughly sober and practical, thought them *life-threatening*, halting his research at a crucial stage for fully two years, trying to quieten them. We must assume then that there were more symptoms which went undescribed. Some oblique hints may exist in the description in *The Art of Travel* of altitude sickness, called *puna* in the Andes: 'giddiness, dimness of sight and hearing, headache, fainting-fits, blood from mouth, eyes, nose, lips, and a feeling like sea-sickness. Nothing but time cures it'. 'The symptoms of overwork are not wholly unlike those of the puna, and many young travellers who have felt the first, have ascribed them to the second'.[47] That 'overwork' need not only be physical.

A year before Galton's major breakdown, William Hopkins (1793–1866), the prodigious mathematical coach at Cambridge, had quickly begun to subside into madness. Hopkins had been Seventh Wrangler in 1827, entering Cambridge at the late age of nigh-on 30 after the death of his first wife, frustrated by farming life in Suffolk. Though he was not a creative mathematician himself, dozens of greats were drilled by him into high positions on the competitive examination lists, including a clutch of Senior Wranglers. Stokes, Cayley, Maxwell, Sylvester, Tait, Todhunter, H. W. Watson and Lord Kelvin were among his many pupils. Galton had also been enrolled under him at Trinity, but had dropped out of contention for honours after his decisive breakdown in late 1842.

In 1863, Hopkins had asked Galton to join him as a General Secretary of the British Association—an immensely flattering gesture, since Galton was given to hero-worship. 'I never had a tutor whom I reverenced and loved so entirely as Hopkins'.[48] However, Hopkins' 'health had by then declined and his fine intellect begun to fail'. By February of 1865 Galton had received letters detailing this 'failure of intellect'. In October, his old tutor would die of 'chronic mania' and 'exhaustion', in the Northumberland House Lunatic Asylum near Stoke Newington.[49]

When Karl Pearson was writing his mentor's biography, he was moved to write to his subject's relatives, inquiring if they had seen any suggestion that Galton himself had suffered from some form of madness. They demurred, and in retrospect the idea appears misdirected.[50]

Some twenty years after his 1866 crisis, Galton conducted his own statistical inquiry into 'mental fatigue', using survey results from 116 teachers about themselves and their pupils.[51] On the 16th of May 1887, he had been invited to chair a meeting of the Teacher's Guild, where a paper 'On Thinking in Shape and Pictorial Teaching', by the Rev. E. Thring, was read. Galton, always on the lookout for empirical data, was quick to spot an opportunity for gathering information for his own use. 'Teachers were urged to use their opportunities for making observations in mental science, and to send him definite instances of the following:

[47] F. Galton 1872c, 19–20.
[48] F. Galton 1908, 65. For Hopkins, William, see page 455.
[49] Craik 2008, 107.
[50] See K. Pearson 1914, 1924, 1930a,b and the Karl Pearson Papers at University College London.
[51] F. Galton 1888g. See Tredoux 2023a.

What can be done by a person whose mind is fresh compared with what he does when somewhat fagged? and, What are the first signs, physical and mental, of approaching fatigue?'[52] After collecting responses to a written circular to teachers, he presented his findings a year later to the Anthropological Institute, though he was disappointed that he got no more than 116 replies.

More specifically, the questions circulated by Galton, assisted by the Board of the Teacher's Guild, had included: 'Has illness, due solely to mental overwork, independent of domestic anxiety and worry ever incapacitated you for more than a month at a time from ordinary school work?'; and 'Have you known cases of more or less serious prostration from mental overwork, as distinguished from the effects of domestic or other anxiety?' Irritability and headaches were common symptoms, along with faintness and actual fainting, sleeplessness, and sleep-walking. One teacher reported that she went colour-blind after too much mental exertion, another respondent drew pictures while sleep-walking. 'He awoke in the night, and found himself in his nightgown, sitting at his table with the gas burning and with this grotesque sketch of an elephant's head and of some other animals just completed. The ink was still wet. He had not the slightest recollection'. Some went deaf, others developed over-sensitive hearing: 'the tired brain becomes almost maddened by an organ grinder'.

Most simply found that their minds 'ceased to bite' once fatigued, with faltering memory and vision (some could not even count cups of tea). It was hard to distinguish overwork from anxiety. There were many debilitating breakdowns. 'Taking all the replies together, I find that out of my 116 correspondents no less than 23 of them have at some period of their lives broken down, and that 21 of these have never wholly recovered the effects. There are six other cases of a less serious kind, some of them slight. In other words one out of five teachers has, so far as the evidence before me goes, been severely stricken'. Major break-downs tended to single out the more able, working alone. 'Dull persons protect their own health of brain by refusing to overwork. It is among those who are zealous and eager, who have high aims and ideas, who know themselves to be mentally gifted, and are too generous to think much of their own health, that the most frequent victims of overwork are chiefly found'.[53]

Surely Galton spoke here from personal experience. 'As soon as the bodily exertion has closed for the day, the man lies down and his muscles have rest; but when the mentally fatigued man lies down, his enemy continues to harass him during his weary hours of sleeplessness. He cannot quiet his thoughts and he wastes himself in a futile way'. 'In many other cases the writers express the difficulty they feel in distinguishing between worry and overwork. The latter is a consequence of the former, while the former often results from the gloom, anxiety, and sense of incapacity caused by the latter. It is a self regenerating circle of evil'.

Where did this manifold of Galton's debilities come from? Was it, to use Galtons' own application of Shakespeare's phrase, nature, nurture, or both—or is that question misconceived to begin with? In his case, immediate precipitating factors are far easier to identify than general causes of his condition; but careful consideration of his personal and family history reveals a great deal of information about both.

Given the history of psychogenic trouble described above, a closer examination of Frank's siblings—all older than himself—and other relatives is called for; an inquiry into natural experiments that is suggested by the Galtonian research programme itself. As we have seen, Frank was convinced that his peculiar physical and mental makeup was inherited.[54]

There is no direct evidence that Frank's father or mother suffered from any of the anxiety symptoms that plagued their son. Among Frank's siblings, relevant information about mental states is scarce. They were evidently very unlike the youngest member of the family, in most things not connected to longevity, which most (though not all) of them enjoyed, living to an

[52] 'Teacher's Guild', *Athenaeum* 1887 (3109, May 28): 707-8.

[53] F. Galton 1888g, 166-167.

[54] Tredoux 2023c.

average age of over 82 years, in an era where the national average was below 50.[55] Still, there is some useful information about his brothers.

Galton noted that his oldest brother Darwin, eight years his senior, had little in common with him. They shared a mature aversion to music, but their 'ideals of life differed to an almost absurd degree: he had not the slightest care for literature or science, and I had no taste for country pursuits. Our differences of temperament became more marked the older we grew'.[56] Frank tried country pursuits in-between things in 1847, but, aside from a temporary enthusiasm for hunting, was not satisfied; his brother seems to have done little else, completing the usual transition of industrial money into landed gentry. Darwin's Primrose League politics were immovably antediluvian and rural, in contrast to his brother's urbane but quiet liberalism and circle of freethinking intellectuals. Surely Frank would have been amused by his oldest brother's penchant for rushing to the bottom of his property, even at night, to salute the trains on the Great Western Railroad as they passed. Yet, in middle-age Darwin also appears to have suffered from mysterious attacks. His sister Emma noted in her diary: '1861. Darwin's attack Thursday April 11th to May 30th; 7 weeks; epileptic attack April 17th'.

The more interesting case is Frank's second-oldest brother, Erasmus Galton, about whom some tantalizing detail is known (see **??**, p. **??**). Erasmus was seven years older than Frank and had a reputation for being 'particularly sensitive and touchy'.[57] He had entered the navy as a midshipman in 1828, at age 13, but had left in 1836, perhaps due to severe injuries sustained earlier, in 1834, when he was assaulted while out walking in Malta with a fellow officer. The story ran that Erasmus had been 'dashed' on the head with a heavy stone, by robbers, losing consciousness and missing his ship. Afterwards, his personality was altered, and he quit the service. This is interesting not as a theory, but rather as a signal that a theory was required. His sister Bessy preferred the explanation that her father did not have enough influence to get her brother a preferment.[58] After this episode, Erasmus farmed and volunteered in the militia, as did his older brother Darwin. He also made a notable ascent of Mont Blanc in September 1850, on a cloudless morning, accompanied by six guides, a volunteer and seven porters (see Figure 3.2, p. 49). A fall on the way down led to a narrow escape. The *Illustrated London News* carried his own report, with woodcuts based on his sketches, of the feat.[59] An exhibition was put on in Edinburgh in 1853, describing Erasmus as the 'the Continental Traveller', with a moving panorama depicting the ascent, and 'various novel and striking effects'. However he wrote to the *Times* disavowing any connection with it.[60] Frank went through similar manias for climbing.

Erasmus became notably more eccentric as the years passed. He insisted on living on his property in a Spartan stone hut, without a kitchen, modelled on militia accommodation at Aldershot. In general he lived a 'very original life', without female company or comfort, according to a peculiarly 'strict sense of duty'. After a visit, Frank remarked that 'what an uncomfortable life it would be to most; but he takes real pleasure in it and it suits him to a "T"', 'he keeps minute accounts and reads books and does kind things, and so, although he sleeps little, the day is full'.[61] It is not known for certain whether Erasmus experienced crises comparable to those experienced by his younger brother, but the sharp changes in his career are noticeable, and he may well have experienced such an episode on Malta, shortly before he left the navy. He appears to have kept up a steady stream of inventions, like his grandfather and younger brother, but with strictly agricultural applications, including an appliance for taming

55 They lived to see 39, 98, 73, 93, 89, 94 and 89 years.
56 F. Galton 1908, 84.
57 For Galton, Erasmus, see page 452.
58 Solomon 2023, 158, E. A. Galton 2003.
59 Saturday 1851-02-08, 6.
60 1853-03-28.
61 Galton to Millicent Lethbridge, 1904-09-06, K. Pearson 1930b, 528.

savage bulls, and a method for improving a 'shaking bog'. However, his youngest brother noticed pointedly, after Erasmus' death from the after-effects of a fall in 1909, that he had not achieved his full potential. 'He somehow seems to me to have failed to get as much interest and "go" in life as his circumstances might have given him'. And 'I think frequently of Erasmus and feel that somehow he had not a fair chance in life. Circumstance and *his own temperament* were often much against him; and all that was visible ending in a small shovelful of ashes, scattered over the flower-beds of a crematorium!'.[62]

All three brothers were childless, though Darwin had a son by his first marriage who died in infancy, and remarried twice (but to women past childbearing age). The remaining siblings were older sisters, who had apparently doted on the youngest member of the family. Like his brothers, they are remarkable for their *differences* from their youngest sibling, despite sharing all the same family background and the advantages of a prosperous upbringing. Violetta and Samuel Tertius responded to the other siblings rather differently than they responded to Frank, whose special abilities were apparent from the beginning and a wondrous topic to them, as reflected in their contemporary diaries, letters and notes.[63]

Among Frank's more distant relatives, the most interesting comparison is Charles Darwin. Apart from sharing a grandfather, Frank resembled Charles in many other ways. As young men, they went through phases of fanatical hunting and shooting. They enrolled to study medicine, but threw it over. At university they showed great promise, and were thought highly of by their tutors, but left without competing for honours, only to achieve the highest levels of intellectual distinction later in life. They abhorred controversy and made a point of entering into public disputes as little as possible. Both went on voyages of exploration to remote and dangerous places. They were also freethinkers, and privately broke with the church. Most intriguingly of all, they shared a cluster of psychological disturbances, an anxiety disorder closely linked to their intellectual activities.

For most of his adult life, Charles Darwin was ill with a mysterious syndrome, which came and went but never disappeared completely. Psychological disturbances marched ahead of physical expressions like trembling, weakness, heart palpitations, eczema, boils, flatulence, retching and vomiting. Many episodes were severe enough to prevent him from working, and could last for many months at a time, or as little as a few hours. They drove him from London to the country, where he cloistered himself away at Down, trying to strictly control the excitement and provocation he was exposed to. Various 'cures' had initial success but inexorably faded, and were discarded in turn: cold-water drenching, ice, dubious medicines, peculiar diets and strenuous exercise all failed to make a lasting difference.

The first signs of this condition predate Darwin's voyage on the *Beagle*. Like many, Handel's *Messiah* made his spine shiver, but, like very few, successive exposures to opera performances could affect him so severely that he became *knocked up*—that is, physically exhausted—and had to resist exposure to them. Lip lesions plagued him, and may have been related to periodic outbreaks of atopic (allergenic) facial eczema. Family letters from this period suggest a labile propensity for translating psychological stimuli into physical expressions. In the prelude to the *Beagle* voyage, he sank into depression and fits of anxiety, with heart palpitations, dizziness and mental discomfort. Once he was settled on the ship—it finally left port on 27 December 1831 after several false starts—and had found a routine, these alarms went away. But after his return in early October of 1836, he started to experience headaches and heart palpitations again, with what seem to be new phenomena for him: flatulence and vomiting.

After his marriage to his first cousin Emma Wedgwood at the end of January 1839, Darwin's head symptoms worsened. Most of the rest of the year was lost, and he withdrew from society. It

[62] Letters from Galton to Millicent Lethbridge, 1909-01-31 and 1909-03-21, K. Pearson 1930b, 591, 594. Emphasis added.

[63] For a full discussion see Tredoux 2023c.

was at this time that Galton sought out his cousin's advice about Cambridge and mathematics, reporting to his family that Charles had been unwell. Through his wife's first two pregnancies, Darwin's vomiting and flatulence intensified, and he had episodes of shivering and chills. Here 'flatulence' referred to air expelled from his stomach spasmodically and painfully, through his oesophagus, leading at times to retching. Serious conversation seemed to provoke him further, severely 'knocking' him up.

Eczema recurred and his face was sometimes badly swollen by it. In the 1860s, he grew his famous beard to hide the signs. Curiously, he came to believe that his eczema actively energized him. Boils often erupted. In his case, it may have been psychosomatic.[64] Headaches and dizziness persisted through the 1840s and after, adding variety to his flatulence and retching. At night, episodes of obsessional thoughts sometimes took hold, and he could not sleep. His son Francis recalled 'He was troubled at night by the activity of his thoughts, and would become exhausted by his mind working at some problem which he would willingly have dismissed. At night, too, anything which had vexed him in the day would haunt him'.[65] At other times he complained of a 'swimming head' and showed signs of depression and listlessness. There were stretches of extensive vomiting, in 1848-9, 1863-4 and 1865.

To some extent these symptoms moderated with age, especially from the mid-1860s down to Darwin's death in 1882. But they never wholly cleared up, and he may simply have become more adept at managing them. Over the years he had tried many remedies. The most promising seemed to be cold-water hydropathy, which he started in 1849 and initially thought very effective, but eventually gave up in 1863 as more than likely harmful. Various diets, such as a regime of small meat portions on toast, were also initially effective, but inevitably lost their potency. Mercuric medicines like 'blue-pill' were carefully tried in small doses, but then left off. More effective than all these were active measures he took to avoid long periods of stimulating company or conversation, which he had found disconcerted him and worsened his symptoms. Sticking to a rigid working and relaxation routine seemed beneficial. But even with the most careful management, his condition was never completely controlled.

Determining the causes of Darwin's condition necessarily involves speculation. There seems to be little doubt that it was psychologically modulated, correlating strongly with stress and excitement, suggesting that it was an anxiety disorder. Insofar as Darwin managed to control it, he did so by limiting anxiety and mental stimulation as far as he could. Physical symptoms resulted from his labile nature, which tended to convert mental upsets into physical signs.

Organic explanations for Darwin's condition have also been suggested. In South America he may have come into contact with any number of tropical diseases, including Chagas' disease, supposedly contracted from an insect bite. If he had Chagas' disease, it was not a typical case, as it must have attacked his stomach and not his heart. Cases like that are known, and all diseases are to some extent asymptomatic, simultaneously complicating present diagnosis and easing it in retrospect. A wide range of diseases can be fitted to observed symptoms by emphasizing some and neglecting others, as needed. It has been postulated that the disease destroyed the 'parasympathetic' muscles used to counteract the 'sympathetic' muscles in the stomach, leading to flatulence, retching and vomiting under stress. In response, it has been argued that it is unlikely Darwin could have suffered from Chagas for so long without showing more definite physical destruction.[66]

Many other possible causes have also been floated to 'explain' Darwin's illness, usually as supposedly conclusive solutions. Candidates include a mitochondrial disorder said to be inherited from the Wedgwoods, systemic lupus, lactose intolerance, *helicobacter pylori* infection, diabetes, Crohn's disease, irritable bowel syndrome, immune system dysfunction, adrenal

---

64 Perhaps it evolved among humans as an 'honest signal'? See Smith and Harper 2003.

65 Francis Darwin, *Life and Letters of Charles Darwin*, I, 102.

66 Hayward 2013.

diminution, gout, Victorian pollution, allergies to pigeon droppings, the mercuric medicines he took from time to time, arsenic or lead poisoning from tin cans on the *Beagle*. There are also several 'maternal deprivation' and 'repressed rage at the father' theories of a kind no longer in vogue, along with explanations positing agoraphobia (an especially unlikely idea for a world-traveller), depression and psychosomatic effects of anxiety about his evolutionary heresy. In the absence of hard evidence from pathology, the possibilities are endless. Recall Galton's observation that, given 'the right of occasional suppression and slight modification', 'it is truly absurd to see how plastic a limited number of observations become, in the hands of men with preconceived ideas'.[67]

Among close family members, Darwin's elder brother Erasmus Alvey Darwin (1804–1881) seems to have shared at least some of his symptoms, including chronic stomach trouble and episodes of persistent lethargy. Always 'delicate' as a child, he whiled away his adult years without accomplishing much of anything, even though he had been an Apostle at Cambridge. However, Erasmus Alvey's symptoms are ambiguous and alternative interpretations are easy to offer.

Caveats about 'occasional suppression' aside, the extent to which many of Darwin's symptoms match those of his half-cousin Galton, is striking. The underlying theme of a core group of the symptoms is anxiety, and should probably be separated from the gastrointestinal group, which Galton does not seem to have suffered from. Headaches, dizziness, shivering attacks, auditory disturbances, heart palpitations and prolonged obsessional thoughts were common to both, occurring at times in intensified and lasting episodes, where they became severely debilitating. They suffered from similar attacks of boils. Both men benefited from prolonged periods of rest and employed devices to take their minds off the present focus of their obsessions. Each carefully avoided personal involvement in controversy as much as possible. Given this, their condition may well have been a common family inheritance, as some have suggested.[68]

Curiously, despite a healthy friendship and correspondence, and some occasional scientific collaborations between the two men, no reference by Galton to Darwin's illness has survived. His recollections of Darwin downplayed their connection. 'I made occasional excursions to visit Charles Darwin at Down, usually at luncheon-time, always with a sense of the utmost veneration as well as of the warmest affection, which his invariably hearty greeting greatly encouraged. I think his intellectual characteristic that struck me most forcibly was the aptness of his questionings; he got thereby very quickly to the bottom of what was in the mind of the person he conversed with, and to the value of it'.[69] There is only an oblique reference to Darwin's seclusion—without noting its causes—when he is compared to Mendel: 'how much can be performed by those who, like him and Charles Darwin, never or hardly ever leave their homes'.[70] But Galton was doubtless fully aware of Darwin's debilitating condition. Conversely, Darwin was aware of his cousin's afflictions. Recall his letter of 1868, at the height of Galton's crisis, in which he confided: 'I know that over mental work takes long in recovery'.[71]

The cousins shared a propensity to work single-mindedly, for many years, on hard and novel problems, and a tendency to injure themselves in the process though a combination of anxiety and obsession. They overcame this in similar but not identical ways, by eventually learning to control the circumstances that triggered the underlying susceptibilities, and discovering through experience when to rest. And, eventually, both men triumphed over their conditions.

As we will see in the next chapter, on the 13th of November 1869, nearly four years after

[67] F. Galton 1863a, 5 col. 2.

[68] See Tobias 1991. Of course, if both Galton and Darwin had similar conditions, that argues strongly against the maternal/mitochondrial theory advanced in Hayward 2013, since they did not share a recent female ancestor.

[69] F. Galton 1908, 169.

[70] F. Galton 1908, 169.

[71] Darwin to FG, circa 1868-01, DARWIN, DCP-LETT-5800.

his breakdown, Macmillan finally brought out Galton's *Hereditary Genius.*[72] Although he was already approaching fifty, the bulk of his scientific work actually lay some way ahead of him. The torment that had pursued him before 1869 now went away.

Galton's own interpretation of his personal evolution was that he had received a 'warning' in 1866, which had 'made a revision of my mode of life a matter of primary importance'. By implication, this 'revision', whatever it was, had been effective. But it is not at all clear what it was, exactly. When closely examined, the general form of Galton's life from, say, 1870 onward, was much the same as before, if anything amplified still further. He resumed activities in multiple scientific organizations, including the Anthropological, the Meteorological, the Royal Geographical, the Royal, and even the Indexing Society. His general research efforts were now more intense than ever, on many new fronts. Since most of the results for which he is now remembered came out of this later period, he cannot have been working less. The opposite is true. Aside from hundreds of scientific papers, he produced over a dozen books in this period. His periods of rest abroad after 1870 seem no different than they had been back in the 1850s. The sporadic controversies in which he reluctantly featured were as infrequent as they had always been.

Something had engaged and fallen into place, in a way that Galton did not understand himself, and is no less inscrutable today. Without that improvement in mental outlook, it is unlikely that the rich store of ideas and results that this exceptional Victorian bequeathed to posterity would have come to full fruition.

[72] *The Examiner*, Saturday 13 November 1869, 15.

Figure 3.1: John Tyndall FRS, as seen by Vanity Fair.

WATERLOO ROOMS.

FOR TWO WEEKS ONLY.

MR ERASMUS GALTON'S
(the Continental Traveller)
GRAND MOVING PANORAMA
OF HIS CELEBRATED
ASCENT OF MONT BLANC
IS NOW OPEN.

To be followed by G. A. SALA'S Comico-Grotesqueo-Pictorio-Musico-Eccentressemo-Symposiorama PANOPTICAN ENTERTAINMENT, introducing a GRAND COMIC MOVING PANORAMA of a Procession of all Nations, on an Overland Journey to the Crystal Palace.

Doors Open at Half-past Seven, to commence at a Quarter-past Eight o'clock precisely.

Reserved Seats, 3s.; Second do., 2s.; Back do., 1s.; Gallery, 6d. Juveniles under Ten Years of Age, Half-price to Reserved, Second, and Back Seats.

A GRAND MORNING REPRESENTATION,
On SATURDAY, March 12 and 19,
at Two o'clock.

Figure 3.2: Erasmus Galton, the 'Continental Traveller'. Panorama of Mont Blanc, Edinburgh, March 1853.

Figure 3.3: In the corner stands 42 Rutland Gate, Kensington, where Galton lived for over half a century. The building to the right was bombed in WWII and rebuilt incongruously.

Figure 3.4: Plaque at 42 Rutland Gate, Kensington.

Chapter 4

Genius

Two days before Christmas of 1869, a letter arrived at 42 Rutland Gate from Down House in Kent. Charles Darwin had been listening to his wife Emma reading his cousin's new book on *Hereditary Genius* out aloud and could not contain himself. Galton had sent him a presentation copy, repaying the compliment paid to him by Darwin on the publication of *On the Origin of Species*. 'I have only read about 50 pages of your Book (to the Judges) but I must exhale myself, else something will go wrong in my inside'. This was not unrealistic; the naturalist could emerge *done-up* for days from musical performances which sent thrills down his spine. 'I do not think I ever in all my life read anything more interesting & original'.[1]

The statistical arguments marshalled by Galton—that genius ran in families and was often inherited at birth—had been effective. 'You have made a convert of an opponent in one sense, for I have always maintained that, excepting fools, men did not differ much in intellect, only in zeal & hard work; & I still think there is an eminently important difference. I congratulate you on producing what I am convinced will prove a memorable work'. But Darwin was not himself of a naturally statistical disposition. 'I look forward with intense interest to each reading, but it sets me thinking so much that I find it very hard work; but that is wholly the fault of my brain & not of your beautifully clear style'.

A month earlier, even though it was only ten days after the publication of the book, Jeremy Bentham's botanizing nephew George had cautiously sounded out the eminent naturalist. 'What do you think of Galton's Hereditary Genius?'[2] Bentham may have been asking for a friend. Ten years earlier, the shy geographer had been converted—not too strong a word—by the *Origin of Species*, and was probably too bashful to ask directly. He revered, almost worshipped, his famous older cousin, who had sponsored his membership of the Royal Society. There was no need to worry, since six years later Darwin confirmed his favourable reception of the book in the second edition of his *Variation of Plants and Animals under Domestication*. 'Some writers have doubted whether these complex mental attributes, on which genius and talent depend, are inherited, even when both parents are thus endowed. But he who will study Mr. Galton's able work on "Hereditary Genius" will have his doubts allayed'.[3]

Hereditary Genius was the culmination of a glacially slow recovery from the mental meltdown of four years earlier. Somehow the former explorer had groped his way back to what he would later call the narrow 'tableland of sanity', with its 'unfenced precipices on every side, over any one of which we may fall'.[4] Labour following up on 'Hereditary Talent and Character' had only been resumed in late 1868. Early in 1869 a foretaste of the big work in preparation appeared in *Macmillan's Magazine*, as 'The Judges of England'.[5] The larger work now extended that

---

1 Charles Darwin to FG, 1869-12-23, DARWIN, DCP-LETT-7032.

2 Bentham to Darwin, 1869-11-23, DARWIN, DCP-LETT-7005. For Bentham, George, see page 451.

3 C. Darwin 1875, Vol. 1, 451. The first edition, C. Darwin 1868, Vol. 2, 7, praised the earlier papers F. Galton 1865b and F. Galton 1865c.

4 F. Galton 1908, 38.

5 F. Galton 1869b.

analysis to a mass of evidence derived from the family relationships of poets, statesmen, men of letters, scientists, military commanders, musicians, painters, divines, classicists, and even wrestlers and oarsmen.

The basis of Galton's argument is an application of Darwin's methods to human mental characteristics, which his cousin had barely touched on. Though this was ambitious for the time, it had the virtue of explanatory simplicity. Galton reasoned that mental abilities ought to vary and be subject to inheritance and natural selection in more or less the same way as physical characteristics like height, strength, appearance and resistance to disease. This saves supposing a completely separate means for their transmission and a slew of special theories for the mental domain only. One of Darwin's key insights was that humans are animals and subject to the same laws, even if that was only implicit in the *Origin of Species*. Galton's insight was to recognize that natural selection does not stop at the neck. Here he anticipated an argument made famous over a century later by evolutionary anthropologists: self-domestication through the repeated interaction of humans in ever more complex societies. As Galton put it, 'civilization is the necessary fruit of high intelligence when found in a social animal, and there is no plainer lesson to be read off the face of Nature than that the result of the operation of her laws is to evoke intelligence in connexion with sociability'.[6]

It is inherently more difficult to reason about the mental abilities of humans than it is to treat their physical characteristics. There is far less to quibble about when measuring height or even susceptibility to disease. To make things harder, Galton was especially interested in what we would call intelligence today, a topic which remains (undeservedly) controversial even after more than a century of empirical research.[7] Harder still, Galton believed firmly in a version of Lord Kelvin's dictum, well before that became popularly known: 'when you can measure what you are speaking about, and express it in numbers, you know something about it; but when you cannot measure it, when you cannot express it in numbers, your knowledge is of a meagre and unsatisfactory kind'.[8] This entailed finding some *quantitative* way to treat ability and its inheritance. No such measures or techniques existed then, so he would have to invent them.

Galton needed some way of gauging ability that did not involve examinations, since few people in that era took the same tests. Directly assigning scores or grades himself would have invited personal bias and endless haggling over precision. The solution that he came up with amounted to pragmatic genius in itself. Rather than finding a direct yardstick for ability, he used its prevalence in the population. People could be ranked or grouped relative to each other, allowing an indirect measure. The proportions involved could be reasoned about quantitatively, and reduced to absolute numbers when the population size was known. Later Galton would generalize this technique under the name 'statistics by inter-comparison', though he had to overcome general incredulity at the idea. Eminence was judged on the basis of consistent opinion. He preferred to use independent assessments of lasting *reputation.* Most people have no widely-known reputation, so this meant limiting attention to the small fraction who were distinguished by the settled judgment of posterity, or at least of many contemporaries, as exceptionally able. The exact level of eminence could be fixed by choosing the proportion of the distribution which ought to satisfy it. Later we will consider his arguments about the distribution of abilities in the population as a whole. For now it is enough to understand that he *defined* a level of eminence as a (small) proportion of the population, at the tail.

The standard of eminence that Galton fixed on was inherently arbitrary. The higher the level was set, the less dispute there would be over who actually met it, but it could not be set so high that there would be too few examples to reason about. Pragmatically, he based the desired level on more than one source. Edward Walford (1823–1897) had recently updated a

[6] F. Galton 1869a, 336. See Wrangham 2019.
[7] Haier 2017.
[8] Lord Kelvin, 'Electrical Units of Measurement' in Thomson 1889, Vol. 1, 73.

hand-list of *Men of the Time: a biographical dictionary of eminent living characters, (including women)* in 1865.[9] This listed 2,500 men, of whom 500 were British, 'decidedly well-known' and past the age of 50, enough time to achieve eminence. Since two million men were over the age of 50 in Britain, the proportion from Walford was 250 per million. Using obituary lists from the *Times* newspaper and other sources, Galton estimated their proportion to be about 238 per million. He settled on 250 per million.

There was hard evidence that people really do differ in ability. Here Galton was able use the common experience of educated readers subject to examination systems, which clearly distinguished candidates. An extreme version of this existed at Cambridge, where the Tripos system produced very large differences at the extreme limits of measured ability. Originally an oral examination in mathematics conducted from a three-legged chair (the tripos), it had been joined by a classics Tripos in the 1820s. Lucrative college fellowships with few duties were on offer for high ranking examinees in either track. The ordeal had been progressively extended to eight full days by 1869.

Many students broke under the strain of the Tripos. Galton himself had dropped out of contention for the mathematical track when he was a student at Trinity in the early 1840s, after developing a progressively alarming and debilitating mental condition at the university. It set in shortly after he arrived there and led to a complete breakdown. For those who got through, the Senior Wrangler outscored the Second Wrangler by several thousand marks, with the gap steadily decreasing with the rank of the examinee, until little separated the mediocre. In the most demanding examination in the realm, differences in ability were very large at the extremes. Galton had a confidential source at the university, an examiner who let him see the raw marks for some recent years. We have already met that source: the Rev. Henry William Watson, a Second Wrangler in his own right who had been in touch with Galton since Nov. 1865, and became an important technical collaborator in later years.[10]

An obvious objection to this line of reasoning is that there is no *single* ability, but rather there are very many special abilities, and individuals may excel in one but not another. This was a long-standing point of contention. Even Samuel Johnson had lent his considerable weight to the question a century earlier. In its original sense, 'genius' referred to a special inspiring or guiding spirit, so that people might have a genius for *some particular thing*. In Johnson's increasingly rational age, usage was already in transition, so that someone might *be* a genius rather than *have* one.

Johnson emphatically denied that genius is a specialization. Rather it is superior general ability, which may have fastened onto something by chance, but might just as well have adhered to something else. In the Hebrides at Moray, when the antiquarian Dr. William Robertson suggested that Edmund Burke may not have been suited by nature to practice law, just as some people had more judgment than imagination, Johnson replied firmly: 'No, sir; it is only, one man has more mind than another. He may direct it differently; he may, by accident, see the success of one kind of study, and take a desire to excel in it. I am persuaded that, had Sir Isaac Newton applied to poetry, he would have made a very fine epick poem. I could as easily apply to law as to tragick poetry'. When Boswell protested 'Yet, sir, you did apply to tragick poetry, not to law', Johnson replied 'Because, sir, I had not money to study law. Sir, the man who has vigour, may walk to the east, just as well as to the west, if he happens to turn his head that way'.[11]

Johnson made a similar argument in his *Life* of the poet Abraham Cowley. 'In the window of his mother's apartment lay Spenser's Fairy Queen; in which he very early took delight to read, till, by feeling the charms of verse, he became, as he relates, irrecoverably a poet. Such are

9 Walford 1865a.

10 For Watson, Rev. Henry William, see page 459.

11 Boswell 1887, Vol. 5, 34-5. Johnson repeated this argument in conversation with Mrs. Thrale.

the accidents, which, sometimes remembered, and perhaps sometimes forgotten, produce that particular designation of mind, and propensity for some certain science or employment, which is commonly called Genius. The true Genius is a mind of large general powers, accidentally determined to some particular direction'.[12] (The idea that genius was due to environmental influences also amused Johnson, who wrote a satirical *Adventurer* on the theme of garrets and the abilities of writers. Since top-floor garrets were proverbially necessary to cloister and inspire the starved hack, writers might be ranked conveniently by the *height* of their garrets, rather than reading their opuses.)

Galton himself anticipated important developments in the study of human ability by noticing, as Johnson did, that there is a dominant *general* mental ability. He argued that those who are outstanding at some tasks tend to be good at other tasks too, or at least would be if they chose to take them up. Specialization and training masked this underlying fact. He did not have the technical language needed to express this idea precisely in 1869. In later years he would develop the machinery that makes precise statements possible, in terms of correlations and latent factors. Later work has vindicated his position—starting with Charles Spearman, who was strongly influenced by Galton's lead. A less 'firm' view of the matter might permit special talents to swim alongside general ability as subsidiary factors, as Burt proposed and Spearman duly conceded.[13] However that may be, accommodating this dispute about general and special abilities would complicate Galton's discussion a lot, but it would make no eventual difference to his conclusions about heritability, if carefully stated.

Galton argues that the estimates of eminence he is forced to use actually correspond closely to the native ability that he really wants to reason about. Some might protest that many of those thought to be eminent are just recipients of nepotism and largesse, or conversely, that many of those gifted with natural ability might fall by the wayside due to random misfortune or prejudice. Here Galton has to appeal to more general arguments. The attributions of eminence that he used were not his own opinion but rather the settled opinion of a large number of men, often over many generations, representing the judgment of entire disciplines. While undeserving exceptions might exist, a lasting reputation cannot be sustained by simulated accomplishment. His sense of ability extends to 'three qualities that are separate and independent of one another', 'capacity, zeal, and vigour; for unless these three, or, at the very least, two of them, are combined', people 'cannot hope to make a figure in the world'.[14]

Just as the eminent are able, Galton finds it implausible that the able will not become eminent under reasonable conditions, given that ability is always at a premium. In support, he cites several examples from his detailed case studies later in the book, which show that many of the eminent actually came from extremely poor circumstances. Take for instance the mathematician Jean-Baptiste le Rond d'Alembert (1717–1783) who was raised as a foundling by the wife of a glazier. This was a special case of a more general argument: that even if the cases he dealt with had been *changelings* at birth, that would not have held them down, just diverted their initial development. Conversely, D'Alembert's biological family was distinguished in a way that his adoptive one was not. Galton was reaching here for what has become a standard tool in the behaviour-genetic field: adoption studies.

One could easily argue, as many later critics did, that either reputation or ability might be confounded with, or affected by, the advantages of influential connections and wealth, especially as an artefact of being born into the right family. Galton is aware of this, and outlines a useful way to test the hypothesis. The Roman Catholic Popes were noted for adopting 'sons', who were typically nephews, and could be expected to receive as much favour as possible. Yet these

---

[12] S. Johnson 2006, Vol. 1.

[13] C. Burt 1909; Jensen 1998; Spearman 1927. For Spearman's connection to Galton, and Burt too, see chapter 12, p. 267.

[14] F. Galton 1869a, 84.

showed no especial tendency to be eminent according to his preliminary tabulation, though (frustratingly) he did not present his data. Nevertheless, he identified the correct research design and his conclusions about adoptees have been broadly confirmed by subsequent work, as we will see.

Galton's first major object is to show that eminence, and therefore ability, runs within families. This does not really depend on the very strong links that Galton proposes with ability. If eminence is overrepresented in certain families, it is natural to suppose that ability is overrepresented in those families too. His argument is disarmingly simple: by tracing many genealogies, he is able to find numerous families in which eminence is vastly overrepresented, far more than one would expect given its prevalence (by definition) in the population as a whole. Many fields are covered: law, politics, the military, literature, science, poetry, music, painting, theology, and classics. Physical activities are not left out. Wrestlers and rowers also have over-performing families.

Each field receives its own chapter in the book. Relationships are sorted by degree of remoteness, from brothers and sisters to sons and grandfathers, uncles and nieces. Biographies and biographical dictionaries are scoured to find family connections. Often Galton has to use his extensive network of personal contacts, through the Royal Society, the Royal Geographical Society and the Ethnological, to find family connections not generally known. He freely concedes to readers in the preface that he has probably made errors. Names can easily be confused—critics and general readers would later point out several instances in the text. Information about family connections is often lacking, especially in female lines. Early deaths mean that ability might not be translated into eminence in time to be noticed. But he is confident that he loses rather more cases than he finds by error.

Galton's most detailed evidence comes from the English judges. He worked out their family relationships in extensive detail, covering 286 judges between the years 1660 and 1865. His article in *Macmillan's Magazine* that April had already shown most of the data and tabulations to the public. His principal source was the nine-volume series *The Judges of England: With Sketches of Their Lives* by Edward Foss (1787–1870), a retired London solicitor and antiquarian who was a member of a legal family himself. The 109 judges who had eminent relations—38% of the total—could be organized into just 85 families, which is to say that 24 of the eminent relations were other judges also listed among the 109. Among these were the Coleridges, including the poet, the Finches, Hydes, Montagus, Norths and Vaughans, the Wyndhams, and many others loaded down with legal talent.

As with all of the other categories in *Hereditary Genius*, the argument that the cases considered really meet the standard of eminence—at most 250 per million people—is informal. For judges it rests on the plausible argument that the pool of barristers from which judges are drawn is already highly selected, making it a double distillation process. But selection alone is not enough, the offices must inherently require some high level of ability thought to be rare. Nothing in Galton's own data establishes that fact, which must appeal to the common knowledge of the reader. The general framework for measuring and comparing ability, which we now largely take for granted, would follow much later and can largely be credited to Galton himself. Precision in the standard of eminence does not matter so much. What matters is whether or not the ability to function as a judge is rare enough to rule out pure chance as an explanation for the fact that they cling to family lines. Of course, other non-chance explanations are possible, but as we will see Galton does not ignore them.

Starting from the most eminent member of each family, Galton is able to show that the number of eminent relations drops off sharply as the degree of relationship declines. Though the most eminent judges tend to have eminent relations at one degree, there are fewer: 26% of fathers, 36% of sons, and 23% of brothers are also eminent. At two degrees this drops off further: 9.5% of grandsons, 7.5% of grandfathers, 4.5% of uncles, and 4.75% of nephews are also eminent.

At three degrees, it approaches zero: 1.5% of great-grandsons, 0.5% of great-grandfathers, 2% of great-nephews, 0.5% of great-uncles, and 1.5% of first-cousins are distinguished. With each remove, the proportion of eminent relations roughly quarters.

The 30 Lord Chancellors formed a special sub-group of the most eminent legal men in Galton's list. Fully 80% (24) of these had at least one eminent relative, more than double the rate of the other judges (36%). He found it implausible that this could be explained by sheer nepotism or 'jobbery', the reach of which might be amplified by greater judicial rank. Not chance then but rather corruption. Galton conceded that 'jobbery' might happen from time to time, perhaps most of the time in other countries, but he argued that the English system is dominated in the long run by merit. Regular changes of government moderate party prejudice, allowing the exceptionally able to eventually win promotion to the bench. And most of the relatives of the eminent judges do *not* achieve eminence themselves, suggesting that thrusting 'jobbery' is ineffective.

The data suggests much more than the simple conclusion that talent clusters in families. There appears to be a mechanism for *transmitting* eminence among lawyers, subject to considerable regularity. This mechanism is capable of growing legal ability through intermarriage until it reaches a peak, but can also steadily diminish it by intermarriage. The strength of either effect depends directly on the degree of relatedness. Choices made in marriages appear to influence the outcomes, without completely determining them—there is still a lot of unexplained variation. Eminence is overrepresented in the families he locates, but most of the members of those families do *not* become eminent.

Similar patterns emerge in the other fields considered by Galton, but he concedes that the data sets he compiles for those, while still substantial and adequate enough for his argument, are not as comprehensive as that for the judges. Each of these fields receives similar treatment in separate chapters, with a descriptive and tabulated summary of family relationships, followed by an appendix with a biographical dictionary of the cases he considers.

Statesmen are harder to rank conclusively by ability. There is less general agreement about their merits, and fewer family connections can be traced. There are also far more cases subject to the effects of nepotism and party. Nevertheless Galton selects 35, grouped into 30 families, working from a list compiled by Henry Brougham (1778–1868) in *Historical Sketches of Statesmen who Flourished in the Time of George III* (1856). Galton, who had met Brougham at Cannes the previous year, shortly before the liberal lion of the *Edinburgh Review* died, stipulates that he only included those clear cases he thinks will meet with general consent. This is supplemented by the complete set of premiers since then, and a few other international cases collected by Galton himself. The abler statesmen appear to have noticeably more eminent relatives.

Military commanders seem to be less problematic when limited to those with outstanding success in the field, whom Galton thinks 'giants in ability', usually coming to the fore when still young. The 32 individuals he compiles are grouped into 27 families. For instance, the family of the Roman general Scipio Africanus is loaded with talent. Likewise the military clan connected to Sir Charles Napier, commander of Galton's friend Richard Francis Burton in Scinde, is brimming with notable generals and admirals, and several accomplished governors and diplomats. The Napiers are also connected through Pitt's rival Charles James Fox with outstanding statesmen. At least 14 of the commanders have 4 or more eminent relations.

It is a lot harder to select the novelists, historians, dramatists and critics Galton treats together as 'literary men'. He thinks it less contentious to stick to modern authors, doing his own selection from biographical dictionaries. In this way he finds 52 people, grouped into 33 families, including the Sydneys, Macaulays, Brontës, Taylors, Palgraves, Hallams and Stephens. He knew the latter three families personally, having been at Cambridge with H. F. 'Harry' Hallam, the second son of the eminent historian (both sons died young). The Palgrave historians have also produced outstanding Orientalists and an explorer, Gifford, well-known to Galton through the RGS. He could not know that a daughter of his Alpine Club friend Leslie Stephen would win fame much later as the 20th century novelist Virginia Woolf. Her father went on to edit the *Dictionary of National Biography*, and her uncle was the distinguished jurist James Fitzjames Stephen, mainly remembered now as the author of *Liberty, Equality and Fraternity* (1874).

Men of science pose distinct difficulties, for instance priority disputes and simultaneous discoveries of the same things unfairly attaching to single names. But Galton readily finds many examples of eminent families. The Swiss Bernoulli family contains an astonishing number of talented and long-lived mathematicians and other scientists, starting with the brothers Jacques (Jacob) and Johan (Jean) Bernoulli, only to dwindle away eventually and blend into the general population. The Napiers reappear through the inventor of logarithms and contemporary of Shakespeare, John Napier (1550–1617). The Herschels, already illustrious in astronomy, would provide Galton over twenty years later with a collaborator in the invention of fingerprinting for personal identification. Galton's own family appears in the list through the Darwins, but now includes himself as one of the most original scientists of any age. He could easily have extended that entry to include many other eminent relations beyond his grandfather Erasmus and cousin Charles the naturalist. The many illustrious sons and grandsons of his naturalist cousin—including Galton's own mathematical collaborator in later years, George Howard Darwin (1845–1912)—swell the list considerably now.

Poets, 'a sensuous, erotic race, exceedingly irregular in their way of life', prove easier for Galton to deal with than musicians. Whereas he consumed large amounts of poetry all through his life, he was more or less indifferent to the charms of music and had to rely on an unidentified friend to provide a list of outstanding talents. The musical friend may have been Richard Buckley Litchfield, who had been the long-standing musical critic of the *Reader* magazine, a publication they had collaborated on in the mid-1860s. Litchfield would soon marry Henrietta Emma Darwin, a daughter of Charles, and lead a life quite unlike the musicians in Galton's list—'the irregularity of their lives is commonly extreme'. Galton finds that about 40% of the poets he consider have eminent relatives, and is able to organize 24 leading poets into 20 families, including most notably the Coleridge and Wordsworth clans, who have many distinguished members. (When Galton arrived at Cambridge in 1840, Christopher Wordsworth (1774–1846), a younger brother of the poet William, was Master of Trinity College, where he had helped to introduce the Classical Tripos examinations.) The musicians yield 26 with distinguished relatives, falling into 14 families, most notably the Bachs, who boast more than 20 outstanding musicians and composers. As many as 57 Bachs were good enough to feature in the standard *Dictionary of Music and Musicians* compiled by François Fétis. Galton reported that in 1750 at least 120 Bachs attended the yearly musical reunion of the family.

When considering the painters, Galton limits himself to the Italian and Dutch masters, again working from a list compiled for him by an unnamed expert. Of 42 names considered, 18 of the most illustrious have eminent relations. Adding in 10 cases of less illustrious painters, another 8 with eminent relations can be found, producing a total of 26 eminent painters with eminent relatives, organized into 14 families. The Bellinis, the brothers Jan and Hubert van Eyck, and the 'Titian' (Vecelli) family of at least eight accomplished painters stand out here, but the list is comparatively sparse and unsatisfactory

Much better information is available for the theologians and religious thinkers Galton grouped as 'Divines', thanks to the four volume *Biographia Evangelica: or, an historical account of the lives and deaths of the most eminent and evangelical authors or preachers* compiled by the Reverend Erasmus Middleton in 1786. Middleton's concentration on the Protestant persuasion is acceptable given the far smaller chance that celibate Catholic divines will leave descendants. Galton is satisfied that the author treats rival denominations fairly, and after working through it is left feeling 'much greater respect for the body of Divines than I had before'. Of the 196 names listed by Middleton, Galton finds that at least 33 have eminent relatives, and organizes them into 25 families. Few modern readers will recognize most of these names, long since sunk into obscurity. One exception to this is perhaps John Donne (1572–1631), now remembered for his poetry but in his day the Dean of St Paul's, and credited by Galton with a holy life and an exemplary death. Others may recognize James Ussher (1580–1656), Archbishop of Armagh and compiler of a calendar from the bible—taught to him by blind aunts who knew it by heart—fixing the creation of the earth at 4004 BC. His brother Ambrose Ussher (c1582–1629) was a noted Arabist, while his uncle Henry was also Archbishop of Armagh, his father Arnold was a clerk of Chancery, his maternal Grandfather James Stanihurst (or Stanyhurst) was a speaker of the House of Commons in Ireland and Master of Chancery, and his maternal uncle, Richard Stanihurst (1547–1618),[15] was a noted philosopher, historian and poet in his day.

Academic achievements get far less attention in Galton's analysis than they might today, when they have become vastly more important. He has trouble tracing the achievements of relatives of high wranglers in the demanding Mathematical Tripos at Cambridge, a frustration he puts down to a very high cut-off point for recognizing those abstract abilities. Anything less than that leads, he speculated a little weakly, to no more than 'days in puzzling over problems too insignificant to be of practical or theoretical import', perhaps over insoluble problems. The Classics Tripos is more fruitful, even though it had only been held since 1824, a total of 46 times.[16] Less time for development before the time of writing surely leads to a number of talents not being recognized, but several families show up strongly anyway. Of the 46 sole Senior Classics, the only achievement considered by Galton, 14 have eminent relatives. The Kennedys in particular had already produced 4 sole Senior Classics, and Galton generously counts at least 9 other eminent members. The Hawkins family included F. Vaughan Hawkins, a Senior Classic in 1854 and a walking companion of Galton on a trip to Cornwall in 1859. The family has at least 13 eminent relatives, including the Rev. Charles J. Vaughan, who was a joint Senior Classic in 1838.

When Galton had joined a Cambridge Reading Party at Keswick in the summer of 1841, he had been impressed by the local wrestlers, taking some lessons with a member of the noted Ivison family.[17] Both rowers and wrestlers showed a distinct tendency to pass their talents on. Here Galton has to rely on another collaborator for lists and details: Robert Spence Watson (1837–1911) a Quaker solicitor and fellow-Alpinist from Newcastle-on-Tyne, and a man 'whose sympathies with athletic amusements are strong'. Spence Watson noticed Galton's earlier articles on 'Hereditary Talent' and sent him details about rowers and wrestlers in Cumberland and Westmorland.[18] The solicitor compiled a list of 172 wrestlers, 46 of whom have eminent relatives. The Ivisons are represented, as are the Tinians, who include the titanic all-round athlete John Tinian—once celebrated in the 1820s but now wholly obscure. However the list of oarsmen, mostly from the Tyne where matches would almost close the city of Newcastle down, is modest. Still, of 21 names at least 18 have relatives who are outstanding oarsmen. The Clasper family, with 5 leading oarsmen and many lesser talents, stand out. Galton would later nominate Spence Watson to the Royal Geographical Society. It was too soon to know that Maud Edith Eleanor Watson, daughter of Galton's collaborator the Rev. H. W. Watson, a Second Wrangler, would become the first women's tennis champion at the 1884 Wimbledon tournament.

Galton's results may be summarized over all the fields he covers in Table 4.1, which is a condensed version of his own summary, concentrating on degrees of relationship rather than on the finer detail he included. It estimates the percentage of eminent kin of the most eminent

[15] Mistakenly called James by Galton.

[16] Galton gets this date right at F. Galton 1869a, 299, but wrong twice on the very next page! The second edition, F. Galton 1892b, corrects one of these errors and repeats the second.

[17] See F. Galton 1908, 62, where the wrestling family is misremembered as 'Ivens'.

[18] F. Galton 1865b,c.

family member, at each degree. In arriving at these percentages he has to use rough guesses for the number of relatives in each degree, a vexing problem that he will return to many times in later years.

Table 4.1: Eminence by distance.

Distance	% Eminent.	Details.
0	100.00	Most eminent member
1	35.33	(31+27+48)/3
2	6.25	(8+5+5+7)/4
3	1.20	(1+1+2+1+1)/5

Any explanation of all this data has to account for at least the following features, where we may simply identify 'ability' with 'ability to become recognized as eminent'.

1. Ability can *grow* within family lines, strengthening through successive generations from negligible levels, until it reaches a peak.

2. After the peak level, ability *declines* regularly with each intermarriage, as relatedness decreases, until it is negligible.

3. Most members of families do not become eminent, in other words there is substantial *within-family variation.*

4. The super-eminent have far more eminent relatives than the merely eminent.

5. Abilities sometimes *skip generations*, where they show no evidence of presence.

6. Advantaged circumstances are not *necessary* for eminence, given that many of the eminent are known to have come from disadvantaged backgrounds.

7. Advantaged circumstances are not *sufficient* to grant eminence, given that most family members do not become eminent *even in the most advantaged cases.*

Galton argues that the most natural explanation for the patterns he finds is that they are the product of hereditary influences, just as other human characteristics like height and disease are commonly supposed to be inherited. Rather then proposing an entirely separate mechanism to explain how human abilities are passed on, he prefers the simpler explanation: more of the same sort of inheritance already familiar to us. Although the mechanisms for this heredity were not well-understood in his time, and he would do substantial work on the question himself, the premise of his approach is that whatever explains heredity in general will explain heredity of mental abilities in particular.

Critics of *Hereditary Genius* over the years have seized on alternative explanations, usually involving some sort of learned or cultural transmission of either ability or its purported causes. That ability runs in families may be attributed to learning or development somehow tied to those families and not others. Significantly, none of these explanations attempt to meet the full force of Galton's evidence. Alternatives are suggested in passing, their mere possibility is taken to undermine Galton's argument, and then the matter is dropped. But it is not enough to suggest that alternatives are *possible.* The case has to be made that those alternatives are more likely to be true, otherwise one could accept Galton's proposal anyway, as at least the best among competing explanations. Any such explanation has a lot of work to do if it is to be taken seriously.

Many of the alternative explanations can be summarized non-pejoratively as *secret sauce* theories. Here the secret sauce, which causes or facilitates ability, is supposed to be passed down within families by some learning process. However this happens, it must ultimately do so through the brains of the family members, who learn (or acquire) whatever it is that must be learned, over successive generations. This need not be overtly conscious, let alone a written-down recipe, and might be something like a set of reinforced habits. Whatever it is, some non-biological mechanism for propagating it is required, and whatever that is, it has to be handed down through family contact, specifically from parents to children.

The fact that ability decreases within families after it has reached a peak might be explained by a mixing process, which would dilute the secret sauce. Marriage to a less distinguished family would bring less potent secret sauce to the match. If the two influences blend in an additive way, ignoring for now whatever the mechanism for this is supposed to be, then the process may be analogous to diluting liquids. Successive dilutions lead to the secret sauce steadily vanishing, and a return to mediocrity. If the super-eminent have very strong doses of it, then it could take longer to dissipate, possibly explaining their greater proportion of eminent relatives, though ingenuity is surely required to argue this for their *ancestors* and socially isolated relatives, since the proportion seems to increase in all directions.

It is rather harder to explain how such a process could produce an *increase* in ability. A blend can be no stronger than the strongest component. Thus those families will have at least as much, likely more, of the ability in question, which can therefore never grow by the process. When the sauces combine, they must, one assumes, sometimes discard less effective elements—otherwise the sauce would, loosely speaking, increase in volume to infinity—or interact in special ways to end up being stronger than either sauce taken alone. A lot of explanation is needed to show how this would work over generations while maintaining the additive effects supposed before.

At the level of the siblings within a given family, another challenge is posed by the clear variation between them. Whatever the secret sauce is, siblings apparently do not always get equal doses. Supposing that there is no *anti-secret sauce* also transmitted in variable amounts, the effective sauce must not only be sufficient but also necessary, so that getting none at all would be a real disadvantage. This also rules out obvious suspects for the secret sauce like nutrition and other health enhancing practices and attributes. One expects all family members to get those in more or less equal degree. Moreover Galton collected numerous cases in which ability skipped a generation, being handed down through a less distinguished family member. If the secret sauce is transmitted through superior development producing continued superior development, that is hard to account for. The numerous cases listed of genius emerging from conditions of extreme deprivation, and genius almost always failing to emerge even in the most advantageous circumstances, suggest that the active ingredients of the secret sauce must be something else.

Today we can easily see that the 'secret sauce' must have at least all the characteristics of biological inheritance, and constructing a satisfactory theory of it amounts to nothing less than a duplicate theory of inheritance, located vaguely in the cultural or learned sphere. At the time that Galton wrote, no satisfactory theory of biological inheritance existed. Darwin had been forced to fall back on the blending theories of 'descent by blood', which, like the hypothetical secret sauce above, would eliminate all distinctive characters through vanishing homeopathic doses. The previous year Darwin had proposed a more sophisticated theory, which he called 'pangenesis', which involved transmission of heredity through 'gemmules' in bodily fluids. The 'gemmules' were in some proportion either passed down unchanged from ancestors or unique to the individual, and somehow 'attached' themselves to cells to construct specific characteristics. Characters not expressed in this way could skip generations, latent in the passed-down gemmules, only to re-emerge later. Anticipating later developments, Galton speculated that expression of the gemmules was likely to require a cumulation of self-organizing small effects. Enough change in quantity leads to an emergent shift in quality. He drew an analogy to populations of people and the distinctive character of the types of towns they inhabit. This is exactly how complex characters affected by many genes are understood today: the modern synthesis, due to R.A. Fisher, Sewall Wright and J.B.S. Haldane in the 1920s, established the link between discrete Mendelian inheritance and continuous characters through an accumulation of small effects (see chapter 13, p. 287).

Galton cheerfully concedes that many of the details of the fancied mechanism of pangenesis might be completely wrong. And just a few years later he will show so himself by conducting a definitive experiment (see chapter 5, p. 77). But he argues that the theory is still a useful basis for a quantitative mathematical model of inheritance. He ventures a very first attempt at this, which will be followed in subsequent years by increasingly sophisticated models. The key assumption of his approach is that there are discrete carriers of heredity (the gemmules) which can be quantified by effect (there are finitely many and the particular collection you have matters).[19] So if some gemmules are passed down unchanged, yet others are unique (mutated, in modern language) then after many generations the unchanged proportion will vanish.[20] So practically all of the stuff carrying hereditary information in any individual would have, using modern terminology, mutated at some stage in one or more of that person's ancestors. More interestingly, if expressed characters such as grades of ability are linked to quantities of gemmules, then it might be possible to predict the effect of mixing those sets of gemmules through mating, giving proportions of progeny expected to display the characters. Galton did not know then that an obscure monk in Moravia, Gregor Mendel, had already published a formula for discrete characters which would languish in obscurity until 1900. At this early stage, Galton's grasp of the nature of heredity is a triumph of inspired higher-level thinking over a best-available but definitely faulty lower-level theory about cellular mechanisms.

If abilities really are hereditary, it should be feasible to breed for them, an idea that had preoccupied Galton since at least 1864.[21] Darwin had come to natural selection working by analogy from artificial selection of domestic animals. Galton saw that this process could be inverted. Not only had human characteristics evolved through natural (or at least unconscious selection), it is also possible to explicitly seize control of the process and deliberately breed for specific qualities. After enough selection, distinct populations will arise, with their own distributions of abilities. Natural examples of populations where this has already happened seem to be at hand. For a more rigorous understanding, some method has to be found for comparing the abilities of entire populations to each other. Indeed, Galton's initial interest in the question five years earlier was a consideration of whole populations and their average distinctions in personality and ability.[22]

In the process of his investigation, Galton fell in love with a statistical distribution that he named the 'Normal Distribution' because it is ubiquitous. While it does simplify any discussion of ability, this assumption has no implications for the inheritance of ability per se. Galton's argument for heritability does not rely in any essential way on the particular form in which ability is distributed within the population. It is possible to show that under very modest assumptions the Normal Distribution, also known as the 'Law of Error' or Gaussian Distribution, should be approximately the right choice for *any* human ability. This miraculous result led Galton to declare that 'I know of scarcely anything so apt to impress the imagination as the wonderful

[19] Galton was aware of the problem of gemmules increasing by combination in mating to infinity, requiring some selection mechanism in the process of mating, see F. Galton 1872a.

[20] Suppose the proportion of the gemmules G transmitted unchanged in a single generation is Gr where $r < 1$. Then after n generations only $Gr^n \approx 0$ would be left unchanged, if n is large enough. This is true even if r is very close to 1.

[21] F. Galton 1865b,c.

[22] See Appendix A, p. 381, for Galton's unpublished 'Ethnography' notebook begun in 1864.

form of cosmic order expressed by the "law of error." A savage, if he could understand it, would worship it as a god. It reigns with serenity in complete self effacement amidst the wildest confusion. The huger the mob and the greater the anarchy the more perfect is its sway.'[23]

As already mentioned, the geographer was introduced to the lordly 'Law of Errors' by his friend and fellow Darwinian William Spottiswoode, who worked energetically at applied mathematics when he was not saving the family printing business from bankruptcy. Spottiswoode, who was also active in the RGS, read 'On Typical Mountain Ranges: An Application of the Calculus of Probabilities to Physical Geography' before the society in April 1860, showing off an application of the Gaussian distribution to a geographical problem.[24] It had been speculated that certain collections of mountain ranges were due to a common geological cause, based on their similar directions. Choosing one such collection, Spottiswoode argued that the directions of the mountains in it were not inconsistent with normal deviations from an underlying mean direction. This was, as the businessman stated, just a recapitulation of an argument due to Quetelet, which is where his colleague Galton was led to find out more.[25]

The Belgian statistician Adolphe Quetelet (1796–1874) was already notable to Galton as a meteorologist, and had supplied him with weather data for the project that the geographer published as *Meteorographica* in 1863.[26] Quetelet had pioneered the application of the 'Law of Errors' to social phenomena since the 1830s. By postulating an 'Average Man' he felt able to justify applying the law to a body of measurements taken from a group of men. Prior to that the law was thought of as a description of varying observations of real objects, such as a star. Now the hypothetical Average Man took the place of the underlying object. Quetelet had generalized this argument to the idea that if a set of observations followed the Law of Errors, it was plausible that the observations cohered, and were not just a confusion of measurements. This was the reasoning at the heart of Spottiswoode's geographical application. Following his friend's lead, Galton eagerly absorbed Quetelet's *Letters addressed to H.R.H. the Grand Duke of Saxe Coburg and Gotha, on the theory of probabilities, as applied to the moral and political sciences.* The book had already been translated into English in 1849, but Galton recalled in his *Memories* that he discovered it only in 1863.[27]

Quetelet had demonstrated that chest measurements taken from 5,738 Scottish militia men followed the Law of Error with uncanny precision, even though his tabulation of the data contained many errors.[28] Similar examples were easily found, such as the heights of enlistees in the French army. De Moivre and Laplace had demonstrated in their proof of the Central Limit Theorem in 1810 that this result is inevitable when a measurement is an average effect of many individually small independent causes, as it is reasonable to suppose most human traits are.

The Cambridge Mathematics Tripos marks obtained from the Rev. H. W. Watson were not amenable to this sort of analysis. Examiners used their own standards, which varied from year to year, and it was obviously truncated sharply at the mediocre limits of ability, which were of little concern to the examiners anyway. To gain confidence that the Law of Errors applied to measures of general mental ability, the Woolwich Military Academy's marks for 1868 were conveniently at hand for back-of-the-envelope comparison. These showed reasonable agreement with the expected distribution of 'errors', though there were only 73 scores to work with (see Figure 4.1, p. 74). Galton felt able to proceed under the assumption that ability was 'normally distributed', as he would later put it. It is possible to challenge this argument, since Galton

---

23 F. Galton 1886f.

24 Spottiswoode 1861.

25 For Spottiswoode, William, see page 458.

26 F. Galton 1863b.

27 Translated from the French by Olinthus Gregory Downes, London, Charles & Edwin Layton, 1849.

28 There appear to have actually been 5,732 soldiers, according to Stigler 1986, 206, who gives a recomputed tabulation and argues that Quetelet used a rough approximation of the binomial distribution for the sake of expedience.

produced little further evidence in justification, just as he offered no extensive evidence for the existence of general ability. However, as always he had done far more research on the question than his published account directly disclosed. Over 150 years of subsequent experience have not refuted his conclusion.

The essence of Galton's method is still used. For example, even today there is, strictly speaking, no *scale* of intelligence that can be definitely shown to be normally distributed. In practice the underlying construct, which can only be reached indirectly via IQ tests, is simply *assumed* by practitioners to be approximately normally distributed.[29] Underlying this reasoning lurks the all-powerful Central Limit Theorem, and some common-sense assumptions about many small and independent contributors to intelligence. But as with Galton's own argument, this is merely a convenient device to simplify discussion.[30]

Galton constructed similarly spaced classes of increasingly high ability, which he labelled A, B, C, D, E, F, G, X (corresponding to classes of increasingly low ability a, b, c, d, e, f, g, x). In this scheme, F was the class containing 233 per million, and the tail combination F+G+X contained close to 250 per million. By following this method of relative inter-comparison, Galton enables us to translate his standards into modern equivalents. We express them here in terms of IQ scores. Thus the class F corresponds to what we would call an IQ of between 152 and 163.[31] Table 4.2 gives approximate IQ equivalents for the classes which modern readers may find easier to understand, and shows how the numbers per million in each division dramatically decline. Note that Galton's letter 'grades' correspond to 10 IQ point intervals.

Table 4.2: Grades as IQs.

class	∼ IQ range	Per million
A	100–109	256,791
B	110–119	162,279
C	120–130	63,563
D	131–140	15,696
E	141–151	2,423
F	152–162	233
G	163–172	14
X	173–∞	1

However, in the bulk of his analysis Galton pays little attention to mediocre levels of ability. He is concerned only with eminence as extreme as classes F+G+X (IQ 152 and up), and occasionally with no less than G+X (IQ 163 and up). In some fields his selections are far more exclusive than others. He supposes that his commanders are more able than his statesmen, who are in turn more able than the judges. But he is less certain about the literary men, scientists and divines, who appear to rank equal to the judges. The painters, poets and musicians are again a grade at least above those, consisting mainly of all-time greats. The minimum grades for the wrestlers and oarsmen are unclear.

Armed with the normal distribution, and his evidence for the heritability of ability, Galton

29 See Jensen 1980 for an extensive discussion, also Jensen 1998.

30 Pedants might prefer to use a Sinh-Arcsinh (SHASH, Jones and Pewsey 2009) or similar distribution to capture the longer left tail of intelligence caused by defectives, and other deviations from strict normality. But the gain in precision is usually not worth complicating the discussion.

31 He used a spacing of $k = 1.03 \times e$, where e is the 'probable error'. In modern terminology, e is 0.6745 standard deviations. Class A lies between 0 and 1k, B lies between k and 2k, etc. So F contained those people between 5k and 6k, or between 3.473675 and 4.16841 standard deviations. Note that 1.03 was used to make 250/million fit in $F + G + X$. See Bulmer 2003, 172–175. A standard deviation of 15 IQ points, centred on 100, has been used, and qnorm(250/1000000, 100, 15, lower.tail=F) = 152.2113.

proceeds to examine the question of how distinct populations with markedly higher levels of ability might be deliberately created through selective breeding. In this way the human race as a whole might even be deliberately altered. 'If we could raise the average standard of our race only one grade, what vast changes would be produced! The number of men of natural gifts equal to those of the eminent men of the present day, would be necessarily increased more than tenfold'.[32] But a number of difficulties have to be cleared up first. Here the threads of Galton's argument must be assembled carefully from his diffuse discussion of them.

1. Are the eminent less fertile than the general population, leading inevitably to their disappearance back into the general population?

2. Would the selected be weaker physically and/or more prone to madness and other peculiarities, presenting a downside to selecting for them?

The possibility that the more eminent might be less fertile has already occurred to Galton when considering the extinction of peerages, which is clearly shown by his genealogical research. This is quite distinct from the fact that possession of a title will only preserve distinction if intermarriage over many generations is carefully controlled, which Galton already recognizes, leading him to caution that merely being able to trace *one line of descent* back to a distinguished ancestor is not in the least impressive. This anticipates his later discovery of 'reversion to mediocrity', or regression to the mean, but it will be some years before he is able to obtain a rigorous demonstration of that effect.[33] Peerages also die out, suggesting that peers might be less fertile.

Since initial lines of peerages are usually founded by an eminent man, then that might suggest that *eminence itself* is implicated in lower fertility. Here Galton comes up with an ingenious analysis. It appears that the peers often married heiresses in order to bolster their wealth. The heiresses had money precisely because they came from families without male issue, and were therefore selected for lower fertility. The perverse incentives of capital doomed the families that chose them, especially if the process was repeated. After enough generations, the line died out, a good illustration that the ultimate determinant of long-term evolutionary success is simply leaving more descendants. Wealth only helps if it increases those numbers in the long run, and here the trade-off appears to have been too severe. The wealth of heiresses could not compensate for their lower fertility in the prevailing conditions of English life. For a population of the gifted to be selectively bred, heiress marriage and other harmful practices will have to be avoided. Thus delaying the age of marriage should also be avoided—Galton is easily able to show that the inevitable result of that practice is crowding-out of those who married late from the population.

This does not directly address the possibility that the eminent are on average less fertile, precisely because they are peculiar. That possibility is suggested by analogy from other extreme examples, like exceptionally tall or short people. In practice such people appear to leave fewer descendants, an example of what would come to be called centripetal selection, where extremes are disfavoured *because they are extreme*. This is not the reversion to mediocrity referred to above, which will be dealt with in detail later. It is ultimately due to sexual reproduction, mating practices and related facts. Here the idea is that whatever makes for high ability comes with unavoidable drawbacks—a Faustian bargain. One such drawback could be loss of fertility, which will complicate any explanation of such abilities as a product of natural selection, and perhaps doom any artificial selection for it. Galton points out that, in his data at least, the eminent seem to be at least as fertile as the others. He has many cases of large families being established by his judges, for example.

[32] F. Galton 1869a, 343.

[33] See chapter 10, p. 205.

The idea that high ability is disadvantageous in itself—that the eminent are physically weaker, or more prone to madness and other debilities—is a variant of this sort of reasoning. Galton argues that it is the other way around—greater ability seems to be associated with better physical and mental health overall, though of course there is considerable variation. Many of the distinguished academic performers at Cambridge, for instance, are also athletically gifted. William Thomson (1824–1907), a Second Wrangler at Cambridge in 1845 and more familiar to us as the physicist Lord Kelvin, was a first-rate oarsman, winning the Colquhoun Sculls in 1843. The Alpine Club of Galton's own day was loaded down with eminent scientists, including his friends John Tyndall and William Spottiswoode, who showed great physical powers when clambering all over Switzerland. Literary men are represented by the likes of Leslie Stephen and the judges by members like Alfred Wills (1828–1912). 'A collection of living magnates in various branches of intellectual achievement is always a feast to my eyes; being, as they are, such massive, vigorous, capable-looking animals'.[34] If the effect of eminence on physical well-being is more subtle, those who argued for it will have to produce real data. And indeed, the subject continues to be investigated down to the present day, with the general conclusion that higher ability tends to be associated with better overall mental and physical health, even at the high extremes.[35]

To bolster his argument, Galton looks for examples of natural populations with elevated ability. The most immediate to hand involves a noticeable distinction within the United Kingdom. The lowlands of Scotland seem to contribute disproportionate numbers of talented families. Even the labourers seem to be better equipped mentally. Galton estimates that the difference is worth perhaps a fraction of one of his grades of ability. He gives a similar guess for the industrious midlands, but does not enter into any more detail. Much more dramatic examples are available from classical history.

The Greeks, in particular the Athenians of the era between 530 and 430 BC, boast at least 14 greats such as Socrates, Plato, Thucydides, Pericles, Aristophanes, Aeschylus, Sophocles, Euripides, and Phidias. With an estimated population of 90,000 free-born citizens, Attica produced some 45,000 free-born people in a hundred years, from whom the 14 will have to be selected, so about 1 in 3214. Thus these greats are in roughly class F of the Attica population, since that is a selection ratio of 1 in 4,300 (or 233 per million, corresponding to an IQ of at least 153). Comparing these eminences to the Victorian population, Galton concludes that any one of them will be ranked in *at least* grade X, with many much higher than that. Assuming that the distance between grades in the Attican and Victorian populations (i.e. the variance of the two distributions) is roughly the same, on average the Greeks are two grades superior to the Victorians. Since a grade represents 10 IQ points, that is a 20 point, or 1.33 standard deviation advantage. By reasoning from the upper tails of the two distributions, where we have some data, Galton had made an inference about their differences as a whole, for which we do not have complete data (see Figure 4.2, p. 74). Here his assumption about the normal shape of the distributions is most useful.[36]

Just as some known populations have higher levels of ability, which Galton supposes arose at most semi-consciously through selective breeding, some other populations appear to have lower levels of ability than (say) the English population. Here Galton has little concrete data beyond his own experiences while exploring South-West Africa nearly two decades earlier. The modern era of mass immigration to the developed world had yet to start, with the exception of the slave societies where the effects of servitude confounded all observations. However, Galton surmised that the Negroes he had dealt with were about two grades inferior in average ability to

34 F. Galton 1869a, 332.

35 Brand 1996; Herrnstein and Murray 1994; Jensen 1998.

36 More complicated distributions would just lead to similar arguments, as long as they are tractable, with differences in detail.

Europeans, just as Attica had been two grades superior. Thus, about a 1.33 standard deviation disadvantage. The Aborigines of Australia—here he had no personal experience and relied on reports from his associates at the Ethnographical and Geographical Societies—rated lower than the Negroes, but he freely conceded that he had even less hard evidence for that.

These were differences in degree and not kind. Even with over one standard deviation difference in means, there would be substantial overlap between the distributions. Some years later, when reviewing Henry Morton Stanley's 1875–1877 expedition across Africa and down the Congo, Galton made similar remarks about the human capital to be found there. 'The negro, taken generally, is idle and clumsy, but we must not allow ourselves to speak of him in terms of universal dispraise. The fact is, that while his average pleasure in work and his average manual dexterity are low when measured by a European standard, it is by no means so low as to make it impossible for a few exceptional individuals and even communities to rise to an equality with average Europeans. By picking and choosing the best individuals out of a multitude of negroes, we could obtain a very decent body of labourers and artisans'.[37]

Simply by making this comparison on a shared scale Galton was rejecting the idea that the races were different enough to be classed as separate species, a point that some contemporary anthropologists disagreed about.[38] That the races within the human species differed in mental abilities, as individuals did, was broadly agreed to. Darwin would take the same position two years later, with little further comment, in his chapter 'On the Races of Man' in *The Descent of Man*. 'Their mental characteristics are likewise very distinct; chiefly as it would appear in their emotional, but partly in their intellectual faculties'.[39]

In that Darwinian organ *The Reader*, T. H. Huxley—another member of the Ethnological Society and Galton's collaborator on the magazine, in which John Stuart Mill and Darwin himself were also shareholders—had been more explicit. 'It may be quite true that some negroes are better than some white men; but no rational man, cognisant of the facts, believes that the average negro is the equal, still less the superior, of the average white man. And, if this be true, it is simply incredible that, when all his disabilities are removed, and our prognathous relative has a fair field and no favour, as well as no oppressor, he will be able to compete successfully with his bigger-brained and smaller-jawed rival, in a contest which is to be carried on by thoughts and not by bites. The highest places in the hierarchy of civilisation will assuredly not be within the reach of our dusky cousins, though it is by no means necessary that they should be restricted to the lowest'.[40]

This did not produce any controversy at the time. Darwin, Huxley and Galton were all abolitionists—a simple consequence of their view that all the races shared a common humanity.

It appears to be plausible from these back-of-the envelope estimates, which do not pretend to be rigorous, that viable populations with levels of ability at least as high as Attica, or as low as Aboriginal Australia, can be constructed deliberately. It would be another decade before Galton came up with a catchword to tag his project for improving the human race through selective breeding: 'eugenics', or the 'wellborn' science.[41] His description of it already has shades of the sort of quasi-religious fervour that the environmentalist movement acquired by the 21st century. 'The needs of centralization, communication, and culture, call for more brains and

[37] F. Galton 1878a

[38] The split between the Ethnological and Anthropological societies at that time has often been attributed to this and other ideological disagreements, but that is not accurate. The split was driven more by the personalities of those involved, e.g. Dr. James Hunt, which was why the two societies were easily reunited when the worst offenders like Hunt passed on and wounds had healed.

[39] C. Darwin 1871a, 161.

[40] Originally published in T. H. Huxley 1865, reprinted often e.g. in Thomas Henry Huxley 1896 and Thomas Henry Huxley 1910. For more on the history of the *The Reader*, a direct ancestor of *Nature*, see Tredoux 2023c.

[41] F. Galton 1883d.

mental stamina than the average of our race possess. We are in crying want for a greater fund of ability in all stations of life; for neither the classes of statesmen, philosophers, artisans, nor labourers are up to the modern complexity of their several professions. An extended civilization like ours comprises more interests than the ordinary statesmen or philosophers of our present race are capable of dealing with, and it exacts more intelligent work than our ordinary artisans and labourers are capable of performing. Our race is overweighted, and appears likely to be drudged into degeneracy by demands that exceed its powers'.[42]

Here Galton railed especially against religious impositions like celibacy, arguing that they removed the best and brightest from the breeding population. This guaranteed a hostile reception for his book from a sizeable part of the reading public.

The readers of *Hereditary Genius* had been warned that errors were unavoidable, and after its publication a stream of correspondents pointed out slips. But readers also offered new information about many cases of family ties Galton had overlooked, as he had supposed. Reviews in the newspapers and periodicals appeared quickly. Early in the New Year, Louisa's summary of 1869 was prematurely pessimistic. 'Frank's book "Hereditary Genius" published in November, not well reviewed, but liked by C. Darwin & men of note'. In fact, notices would continue over the next year, and would not show a clear trend. So far only half a dozen or so reviews had appeared in the popular press, most notably in the *Spectator*, *Globe*, *Daily News*, *Saturday Review* and *Scotsman*.[43]

The review in the *Spectator* was anonymous and offhand, confusing Galton with his cousin Douglas. 'We do not see that Captain Galton has contributed very much towards the discussion of the true point at issue in this matter'.[44] The reviewer does not dispute that some mental qualities are transmitted—'evidence for the transmission of hereditary qualities seems to be over-whelming'—but wants to know which ones. Heredity is accepted by all, 'except by a few men who think that if they allow inequality of birth, they disallow equality of rights'. But the writer insists that genius in particular is a discrete quality and thus is hardly ever passed on intact, oblivious to Galton's construction of grades of ability, and evidence of inheritance in attenuated form, from the superior to the merely eminent.

The fairest of the early reviews appeared in early December in the *Globe*.[45] Galton's arguments are expertly summarized, and his 'very startling results' are recommended to the attention of the reader. The perceptive reviewer 'cannot but admit' that Galton 'has beaten down—to use his own expression—many places of refuge for those who insist that each man is an independent creation, and not a mere function, physically, morally, and intellectually, of ancestral qualities and external influences'. But the writer concludes lamely that Galton gives his facts 'a far too universal application' and is 'too apt to ignore the arguments on the other side of the question', without giving specific examples.

Just over a week later the *Daily News* is more enthusiastic, finding Galton's 'remarkable work' a 'careful accumulation of statistics', a technique which the unsigned piece insists is unfairly maligned.[46] Most people concede, the reviewer believes, that physical characters may be inherited, but 'Mr. Galton undertakes to show, and to a large extent undoubtedly succeeds in showing, that genius is equally transmissible, and that ability goes by descent'. But dreams of race improvement face overwhelming obstacles. Here the unknown reviewer is unusually astute. 'No doubt it would be practicable, if the first consideration of marriageable people were the ultimate improvement of mankind. Then fools would give one proof at least of wisdom in refusing to propagate folly, and the feeble and sickly would take a vow of celibacy from sheer

[42] F. Galton 1869a, 345.

[43] Louisa' diary entry has often been quoted, e.g. in Forrest 1974, without this crucial qualification. See Gökyi it 1994 and page 72 for a list of over 25 reviews.

[44] Anonymous 1869a.

[45] Anonymous 1869b.

[46] Anonymous 1869e.

regard to posterity. Unfortunately, young men will fall in love, and girls will marry them without considering the effect of the union upon the race'. Thus 'it is highly improbable marriages will ever be formed on scientific principles'.

On Christmas day the *Saturday Review* weighed in. The anonymous writer finds the railing against heredity by Henry Thomas Buckle, in his *History of Civilization in England* (1857), far more convincing than Galton's refutation of that 'scouting'.[47] At the same time, somewhat paradoxically, the issue is said to be beyond proof, or even scientific investigation, one way or the other. The matter is said to be too complex. Galton's specific evidence for inheritance of mental ability is 'largely mediocre', a compendium of 'disjointed facts', which are 'inert and largely lifeless', and 'logically worth nothing'. Not so for physical characteristics, so that the chapters on Oarsmen and Wrestlers are perfectly acceptable to the reviewer. Little attention is given throughout to the author's specific arguments, data or careful definitions.

Galton was frustrated beyond endurance by the obtuseness of the early reviews. He asked for help from his publisher Alexander Macmillan (1818–1896) to improve the standard of reviewer.

42 Rutland Gate SW
[...] day 1869[48]

My dear Mr. Macmillan

You told me that 'Hereditary Genius' was likely to be reviewed in 'Nature'. If so, the future reviewer would do real service to the cause of the book if he were to preface any remarks he desired to make by a dispassionate & full account of the line of argument adopted in the book. I say so because the reviewer in this week's 'Saturday' and in the 'Spectator' of some 2 or 3 weeks ago have failed to understand it & they hit blows in the air that fall wide of the book and of me.

The *Spectator* is not worth referring to but the *Saturday* states 1st that I fail in neglecting instances against my theory. 2. that where genius is most transcendentally manifest, the theory fails. 3. That I neglect to shew why the great men on my list are above their brothers & sisters at all.

In other words the reviewer writes ignorantly of what I say. 1. I give the list of failures & successes at the head of each appendix, & in the column C of each table I give the percentages of success (the complement is of course the percentage of failure) in each degree of kinship. 2. In my tables [which] I [give] in each of the chapters, I shew that the abler men have on the whole a large number of able relatives. 3. The reviewer seems utterly at sea in respect to the law of deviations from an average on which I found my theories, and as to the latter chapters, he clearly is quite ignorant about them.

Do then use your influence to have my book analyzed as well as criticized. I sincerely want criticism by a master-hand who can grapple with the arguments and not gambol away from them.

Very sincerely yours
Francis Galton

Galton also wrote to request a fellow Darwinian at the Royal Society, the Rev. Frederic William Farrar—then a Harrow schoolmaster but known for philology—to review the book more thoroughly in *Frazer's Magazine*.[49] A vexed letter calls out the *Saturday Review* again and underscores how much the endorsement of his cousin signified.[50]

---

47 Anonymous 1869c.

48 Probably Christmas day, given the reference to the *Saturday Review*, which appeared on that day.

49 Letter from Galton to Farrar, 1869-12-14, Darwin Papers, University of Virginia. For Farrar, Frederic William, see page 452. Farrar would later help Galton to arrange for Darwin's burial in Westminster Abbey.

50 Darwin Papers, University of Virginia.

Dec 24 / 69
42 Rutland Gate SW

My Dear Farrar

I write to say that the present Hooks pedigree is all right in the book. His father (the once famous pamphleteer & Dean of *Worcester*) was the brother of the [first].

I cannot help saying that when I returned home, I found a most enthusiastically eulogistic letter from Charles Darwin (of the book so far as he had read it), which is most gratiful to me, for the reviewers thus far, won't or can't understand the book. Thus the 'Saturday' of this week finds fault with me for not counting failures, whereas I give them at the head of every appendix & (implicitly) for *each degree of kinship* in the column C of each of my tables. Again he says that the argument fails where genius is most transcendently manifest, whereas I shew the contrary—firstly in the Table I of the Chapters & secondly in an analysis of Comte's representative men p. –322–

Cromwell, by the way, is an instance in my favour & Milton not wholly against it. So of his 3 (Shakespeare, Milton, Cromwell) only Shakespeare (of whose family we know nothing) is negatively against me. Take other great English heroes: Bacon; Pitt & Fox; Marlborough & Wellington; Macaulay &c. &c. & the case is quite in favour of me.

Lastly he says I do not show why the illustrious men are better than their brothers and sisters!—I suppose he skipped the 'Deviation from an average' & the latter chapters.

I shall be most grateful for an analysis of my line of argument as prefaratory to criticisms, for the public, as hitherto led by reviewers, simply are misled concerning the scope of the book.

Pardon this my self-justification. I will never trouble you again with more of the same!

Very Sincerely
Francis Galton

The detailed and fair review three days later in the *Scotsman* may have been some relief.[51] Likewise the *Times* was sober and perceptive in early January, noting that the 'Darwinian theories are capable of infinite expansion', that Galton's 'honest' and 'conscientious' arguments are an application of them to human mental phenomena, and that 'all but the extremest bigots will attach some weight to them'. Yet the *Times* is not as sanguine about the triumph of ability over circumstance. 'Mr. Galton is a little too anxious to array all things in the wedding garment of his theory, and will scarcely allow them a stitch of other clothing'.[52] However, later that January the *Athenaeum* reverts to flippancies such as 'anything may be proved by statistics', and the idea that the unpoetic mathematical talent of Byron's daughter Ada is a counter-example.[53]

Counterbalancing this off-handedness, the *Westminster Review* gives the book a crisp summary and pronounces it 'clear' and 'vigorous', noting Galton's 'statical illustrations, his suggestion of a derivation of mathematical formulae from the doctrine of Pangenesis, and the general ability and originality of his treatise' all of which amounted to 'a worthy contribution to the discussion on Natural Selection, which Mr. Darwin so magnificently inaugurated and so intrepidly continues'.[54]

51 Anonymous 1869d.
52 Anonymous 1870a.
53 Doran 1870.
54 Anonymous 1870d.

The best of the early popular reviews, by the astute Anglo-Irish critic John Ormsby (1829—1895), appeared unsigned in the *Pall Mall Gazette* in February.[55] Ormsby would later make a name for his translation of Cervantes' *Don Quixote*, but at this stage his bread and butter was reviewing for the periodicals. His travel writing would also have been familiar to Galton but they had not met, despite a shared interest in the Alpine Club and its circles. Ormsby describes *Hereditary Genius* as 'painstaking' and 'free from dogmatism', helpfully pointing out some obvious errors in the genealogies and several omitted cases of heredity which would have strengthened the argument. But Ormsby questions the two grade advantage calculated for the Athenians, pointing out that it is unclear whether or not 40,000 resident aliens and 400,000 slaves ought to be counted in the population total. Gratified by the fair treatment, Galton quickly replied to the editor of the *Gazette* (then Frederick Greenwood) that doing so would decrease the superiority of the Greeks by about half a grade only. (More precisely, in modern terms, that would place them 12 IQ points, or nearly one standard deviation, higher on average).[56]

42 Rutland Gate SW
Feb 19, 1870

Sir,

I am truly obliged for your just critique on my remarks about the Athenian race, for I had a strong misgiving that my results were too high. If your amendment were accepted pure and simple—but that again w^{d} I think be going too far in the opposite direction, the revised calculation would be as follows:—

- 90, 000 free
- 40, 000 alien
- 400, 000 slaves
- 530, 000 total = 6 x 90,000 about.

I shew on the 90,000 hypothesis that 1 in 4.822 would have become illustrious, of those who survived the age of 22, and 1 in 3.214 of those who survived 50.

Multiplying these figures by 6 we have in the 530,000 population the ratios of 1 in 28,900 & 1 in 19,284.

Now in page 34 of my book the class F is 1 in 4,300 G 1 in 79,000 therefore their (F+G)/2 would rank above our G, equal to a close upon our H, making the difference 1½ & not 2 grades.

Ormsby replied that he was 'very much gratified' by the 'cordial manner' in which Galton responded to his criticism. 'I have read your book with attention and with very great interest, and I sincerely wish it the success which the industry and ability shown in it deserve. I look upon it as a valuable supplement to Darwin and a most useful ally to everyone who has occasion to combat the sham philosophies of these latter days and I hope ere long to see a second edition out, in which your case, strong as it is already, will be made even stronger'.[57]

The very next day after Ormsby's piece in the *Pall Mall Gazette*, an anonymous reviewer in *Chambers' Journal*—whose proprietor Robert Chambers had provoked a proto-evolutionary storm in 1844 through his anonymous *Vestiges of the Natural History of Creation*—delighted in

[55] Anonymous 1869d; Ormsby 1869.

[56] FG to Pall Mall Gazette, 1870-02-19, GALTON/2/4/1/5. Grade X is 1/1000000 and qnorm(1/1000000,mean=100,sd=15,lower.tail=F) = 171.3014. However the Athenians had 1/(6*4822) of that calibre, and qnorm(1/(6*4822),mean=100,sd=15,lower.tail=F) = 159.689. So the shift is 11.61, or about 12 IQ points, which is 1.2 of Galton's grades.

[57] Ormsby to Galton, 1870-02-24, GALTON/2/4/1/5.

baldly asserting precisely what Galton had denied, which is that nepotism explained absolutely all of the extent to which legal eminence clung to families.[58] But that March the review which mattered most to Galton, of all those published, appeared in *Nature*.[59]

The journal had been founded just the previous year by Norman Lockyer, long familiar to Galton from their days at the *Reader*, where they had collaborated on the science coverage in that magazine.[60] As it happened *Nature* was no more financially viable than the *Reader* had been, and was supported for many years by the Macmillans, who published it. Thus Galton's earlier appeal to his publisher may well have improved the calibre of the reviewer, though surely without compromising the independence of his judgement. The piece was signed by the co-discoverer of evolution, Alfred Russel Wallace, and was perhaps even more laudatory than Darwin's letter to his cousin.[61]

Wallace enthusiastically endorses the idea that mental abilities are inherited just like physical abilities are, finding that fact nearly self-evident given Galton's extensive evidence from family pedigrees, crisply summarized in the review. He is most impressed by the comparisons of populations, calling the methods used 'some of the most startling and suggestive ideas to be found in any modern work'. The evolutionist is skeptical that late marriages by the prudent would lead to their eventual replacement by the imprudent but early-marrying, as Galton had suggested. Merely having more children by mating earlier would not necessarily guarantee their survival to adulthood.

The observations Galton makes about the negative effects of religious restrictions, like celibacy rules, are commended and added to. 'The Church acted as if she had aimed at selecting the rudest portion of the community for the parents of future generations; and the rules as to fellowships at our Universities are a relic of this barbarous custom, being bribes to men of exceptional ability not to marry. Religious persecution acted in the same way. The most fearless, truth-seeking, and intelligent were year by year incarcerated in dungeons or burned at the stake; so that, by this twofold selection, human nature was brutalised and demoralised, and we still feel its hateful effects in the long-continued antagonism to the essential requirements of an advancing civilisation'.

Darwin, who did not publish a review himself, wrote enthusiastically to Wallace 'I was excessively pleased at your review of Galton, & I agree to every word of it'.[62]

A few months later the pendulum wobbled once more in the *Edinburgh Review*. Herman Merrivale (1806–1874) was unimpressed by Galton's evidence or arguments, using them as a foil for his own firmly-held opinions on heredity, which he largely subscribed to on other grounds, informed by antediluvian distinctions between general ability, special talents and genius, understood in the pre-Johnsonian sense.[63] Merrivale was a former barrister but by then the permanent under-secretary for India. He reviewed prolifically for the periodicals, with no less than 65 other pieces in the *Edinburgh Review* alone. Much of his review—unsigned as usual—reads like the report of a civil servant tasked with making definite conclusions based on whatever knowledge he happens to have at hand. He energetically denies that specific vocations are hereditary, pointing out that they could be learned by habit, example or instruction within families. This would have puzzled Galton, who was less concerned with choice of vocations than he was with proven ability to excel at them, and at others; indeed, with general ability. Merrivale rules anyway that general ability is mostly hereditary, that special talents for specific things are less so, and that positive genius is hardly ever passed on (that pre-Johnsonian under-

58 Anonymous 1870b; Chambers 1844.

59 Wallace 1870.

60 Tredoux 2023c.

61 For Wallace, Alfred Russel, see page 459. See Tredoux 2023a for a reproduction of the complete review.

62 Darwin to Wallace 1870-03-31, DARWIN, DCP-LETT-7154.

63 Merrivale 1870. The Galton papers at UCL contain a proof copy of this article with hand-written corrections by Merrivale himself, presumably sent to Galton by the reviewer; GALTON/2/4/1/5.

standing of genius at work). Galton's location of all three on the same dimension, differing by degree, does not register with the reviewer. Curiously, the under-secretary denies that ability can ever *increase* through a process of inheritance. 'No combination which we are entitled to conceive as possible of hereditary influences will produce an individual fitted with mental powers beyond a standard, not so definable indeed as that of bodily powers, but quite as certain'. He could not have known about genetic epistasis and mutations.

Swings led to roundabouts. In the new but short-lived *Journal of Anthropology*—an organ of the Anthropological Society, which would soon reunite with the Ethnological Society that Galton had adhered to after their schism in the early 1860s—the retired judge and registrar George Harris prefers to go further than the author.[64] While the work is 'of deep interest' and a 'careful study', even 'correct as far as it goes', it takes a 'limited view of the entire matter'. Harris is convinced that ability is 'not merely directly hereditary', but 'develops' itself in 'various modes and phases during the progress of its descent'. By this he means the then poorly-understood mechanisms of inheritance, which can lead to skipping of generations, discordance between parents and children and so on. Galton would go on to work on this problem himself in later years, with mixed success. The reviewer also points out, as others had done, the possibility that nepotism could explain the transmission of eminence among the judges and in other fields. Nevertheless Harris concludes brightly that 'Mr. Galton has pointed to a new field, where operations promising a rich return may be carried on'.

Frederic William Farrar's piece in *Fraser's Magazine* was the last substantial review to appear, eight months after Galton's appeal to his colleague. The long-delayed response is detailed but not effective.[65] Unsurprisingly Farrar, who would go on to become the Dean of Canterbury, finds the chapter on Divines least satisfactory, discovering in it a 'very ill-concealed dislike' of its subjects, and a 'wholly inadequate' selection of names. He signals clearly that he is not hostile to Darwinism, but he is disappointed that Galton cannot predict exactly *when* high ability will (or will not) be inherited, even if the author establishes that it often is.

Instead of probabilities, Farrar wants laws and assurances. The possibility of mere influence promoting eminence is again suggested. Many of the names Galton considers eminent are not thought so by Farrar. In an echo of things to come—already anticipated by Galton himself, though in the opposite direction—the schoolmaster proposes that adoptions provide important *counterexamples* to heredity. 'The fact that an adopted child has often reflected in a remarkable manner the characteristics of the family into which he has been introduced, is a strong illustration of the effect of influence. To mention but two instances, Marius the Younger, and P. Scipio Africanus Minor, were both of them adopted sons' (current weight of evidence implies the opposite).[66] Farrar also thinks Galton's plans for race improvement unlikely to succeed, arguing that the Athenians, for example, actually paid little attention to their 'nuptial relations'. As a whole the author's conclusions are said to be 'rather theoretical than practical' even though they are 'eminently suggestive'. To Farrar, the chief value of the book lies in its unprecedented compilation of details, not in its main theories.

Louisa's diary entry was overly pessimistic. There is no consistent pattern in this reception. The following is a comprehensive list, with ratings for negativity, neutrality and positivity.

- 1869-03-23. Pall Mall Gazette (Notice of 'Judges'). Ormsby.
- 1869-03-06. Illustrated London News (Notice of 'Judges'). Anon.
- 1869-11-27. Spectator, Anonymous 1869a. Anon. [−−].
- 1869-12-03. Globe, Anonymous 1869b. Anon. [+].
- 1869-12-04. Cambridge Independent Press. Anon.

64 Harris 1870.
65 Farrar 1870.
66 Plomin 2018; Plomin et al. 2012; Polderman et al. 2015.

- 1869-12-16. Daily News. Anon. [+].
- 1869-12-25. Saturday Review, Anonymous 1869c. Anon. [−−].
- 1869-12-28. The Scotsman, Anonymous 1869d. Anon. [+].
- 1870-01—. Westminster Review, Anonymous 1870d. Anon. [++].
- 1870-01—. Journal of Psychological Medicine. Meredith Clymer. [+].
- 1870-01—. British Quarterly Review, Anonymous 1870c. Anon. [−].
- 1870-01-07. The Times. Anon. [++].
- 1870-01-15. Athenaeum, Doran 1870. Dr. John Doran. [−−].
- 1870-01-22. The Graphic. Anon. [++].
- 1870-02-05. The Examiner and London Review. Anon. [+].
- 1870-02-18. Pall Mall Gazette, Ormsby 1869. John Ormsby (unsigned). [++].
- 1870-02-19. Chamber's Journal, Anonymous 1870b. Anon. [−−].
- 1870-03-17. Nature, Wallace 1870. A. R. Wallace. [++].
- 1870-03-23. The Guardian. Anon. [+].
- 1870-04—. Theological Review. Frances Power Cobbe.
- 1870-04-16. Morning Post. Anon. [−].
- 1870-04-19. Evening Standard. Anon. [+].
- 1870-04—. Catholic World.[67] Orestes A. Brownson.
- 1870-06—. Atlantic Monthly. Anon. [++].
- 1870-06-11. Academy. J. R. Thursfield. [−].
- 1870-07—. Journal of Anthropology, Harris 1870. George Harris. [+].
- 1870-07—. Edinburgh Review, Merrivale 1870. H. Merrivale. [−].
- 1870-08—. Fraser's Magazine, Farrar 1870. F. W. Farrar. [−].

Galton found both friends and critics, by parts.[68] His brusque treatment of religion guaranteed opposition from some sections. His application of Darwinian thinking won him converts to his general method at the very least. No doubt the support of both Darwin and Wallace was an important boost and meant more to the author than the animadversions written for general readers, as Louisa implied. However the popular press was as often supportive as it was hostile. In the coming century, opinions would keep fluctuating, even as evidence accumulated from the research programme that Galton had just initiated.

[67] Reprinted in *The Works of Orestes A. Brownson: Scientific theories* (1884).

[68] Gökyi it 1994 finds it positive on balance, but more sources have been considered here.

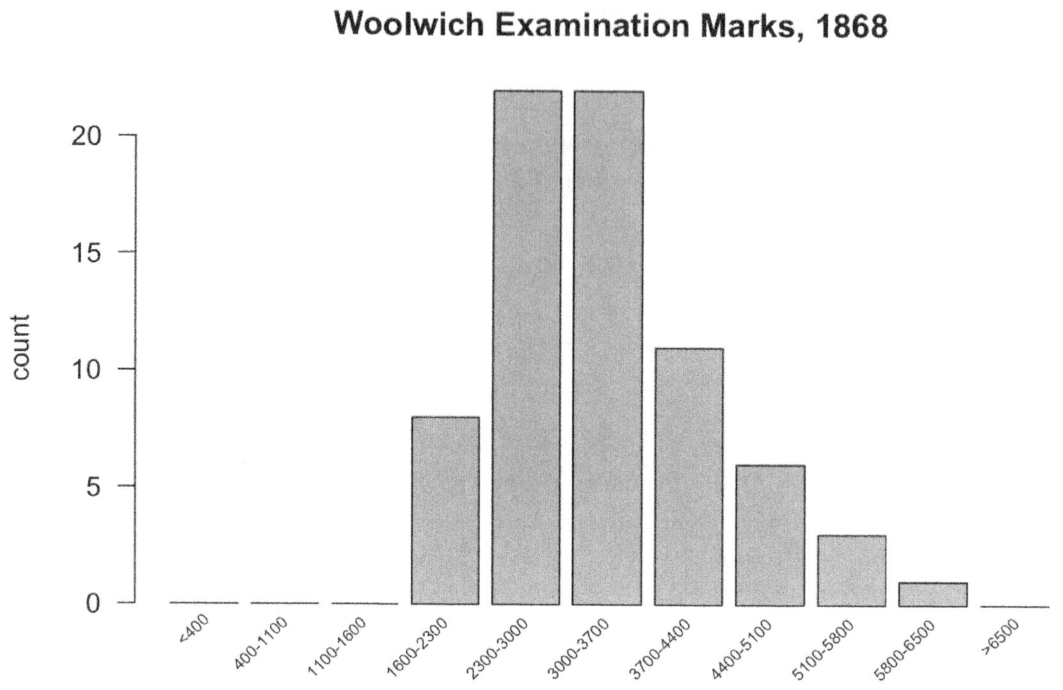

Figure 4.1: Truncated distribution of examination marks.

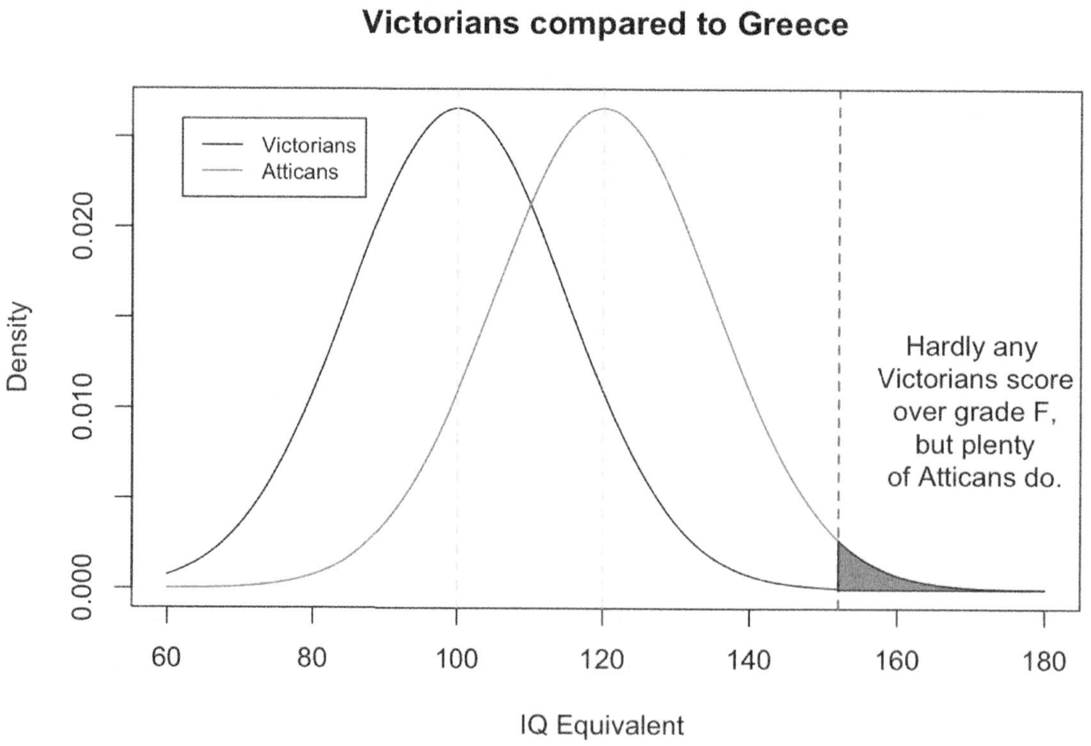

Figure 4.2: High ability populations.

Figure 4.3: Francis Galton in his fifties, taken in the 1870s, K. Pearson 1924. Restored.

Chapter 5

Pangenesis Undone

Late in April of 1871, a testy letter from Charles Darwin appeared in *Nature*.[1] The naturalist—easily excitable, often to the point of dry heaves—seldom involved himself in public disputes. He had been provoked by reports about his cousin Francis Galton's experiments on pangenesis. Galton had, they said, conducted an experiment refuting Darwin's theory of inheritance, 'pangenesis'—that provisional hypothesis he had first advanced three years previously. The reclusive evolutionist claimed that he had been misrepresented, and not refuted at all. But he omitted to mention that he had himself been intimately involved in this experiment at all stages.[2]

> In a paper, read March 30, 1871, before the Royal Society, and just published in the *Proceedings*, Mr. Galton gives the results of his interesting experiments on the inter-transfusion of the blood of distinct varieties of rabbits. These experiments were undertaken to test whether there was any truth in my provisional hypothesis of Pangenesis. Mr. Galton, in recapitulating "the cardinal points," says that the gemmules are supposed "to swarm in the blood." He enlarges on this head, and remarks, "Under Mr. Darwin's theory, the gemmules in each individual must, therefore, be looked upon as entozoa of his blood," &c. Now, in the chapter on Pangenesis in my "Variation of Animals and Plants under Domestication," I have not said one word about the blood, or about any fluid proper to any circulating system. It is, indeed, obvious that the presence of gemmules in the blood can form no necessary part of my hypothesis; for I refer in illustration of it to the lowest animals, such as the Protozoa, which do not possess blood or any vessels; and I refer to plants in which the fluid, when present in the vessels, cannot be considered as true blood. The fundamental laws of growth, reproduction, inheritance, &c. are so closely similar throughout the whole organic kingdom, that the means by which the gemmules (assuming for the moment their existence) are diffused through the body, would probably be the same in all beings; therefore the means can hardly be diffusion through the blood. Nevertheless, when I first heard of Mr. Galton's experiments, I did not sufficiently reflect on the subject, and saw not the difficulty of believing in the presence of gemmules in the blood. I have said (Variation, &c. vol. ii., p. 379) that "the gemmules in each organism must be thoroughly diffused; nor does this seem improbable, considering their minuteness, and the steady circulation of fluids throughout the body." But when I used these latter words and other similar ones, I presume that I was thinking of the diffusion of the gemmules through the tissues, or from cell to cell, independently of the presence of vessels,—as in the remarkable experiments by Dr. Bence Jones, in which chemical elements absorbed by the stomach were detected in the course of some minutes in the

[1] C. Darwin 1871a.
[2] C. Darwin 1868.

crystalline lens of the eye; or again as in the repeated loss of colour and its recovery after a few days by the hair, in the singular case of a neuralgic lady recorded by Mr. Paget. Nor can it be objected that the gemmules could not pass through tissues or cell-walls, for the contents of each pollen-grain have to pass through the coats, both of the pollen-tube and embryonic sack. I may add, with respect to the passage of fluids through membrane, that they pass from cell to cell in the absorbing hairs of the roots of living plants at a rate, as I have myself observed under the microscope, which is truly surprising.

When, therefore, Mr. Galton concludes from the fact that rabbits of one variety, with a large proportion of the blood of another variety in their veins, do not produce mongrelised offspring, that the hypothesis of Pangenesis is false, it seems to me that his conclusion is a little hasty. His words are, "I have now made experiments of transfusion and cross circulation on a large scale in rabbits, and have arrived at definite results, negativing, in my opinion, beyond all doubt the truth of the doctrine of Pangenesis." If Mr. Galton could have proved that the reproductive elements were contained in the blood of the higher animals, and were merely separated or collected by the reproductive glands, he would have made a most important physiological discovery. As it is, I think every one will admit that his experiments are extremely curious, and that he deserves the highest credit for his ingenuity and perseverance. But it does not appear to me that Pangenesis has, as yet, received its death blow; though, from presenting so many vulnerable points, its life is always in jeopardy; and this is my excuse for having said a few words in its defence.

Charles Darwin

It was a surprising and alarming development. There was no more ardent admirer of Darwin than his younger cousin.[3] On the death of Darwin eleven years after this letter to *Nature*, Galton grieved to his sister Emma. 'I feel at times quite sickened at the loss of Charles Darwin. I owed more to him than to any man living or dead; and I never entered his presence without feeling as a man in the presence of a beloved sovereign'. He was actively involved in arranging for the naturalist to be interred in Westminster Abbey, 'laid by the side of Newton as the two greatest Englishmen of Science', and even proposed adding a new stained-glass window to celebrate the discovery of natural selection.[4]

Galton had apparently heard of the *Nature* piece a few days before it was published—probably directly from Darwin himself, but that letter has not survived. A contrite and plaintive note had quickly been dispatched from Rutland Gate to Down House, smoothing over things. The bone of contention was whether or not 'pangenesis' involved transmission of hypothetical hereditary particles, vaguely called 'gemmules', through the blood—his experiment had assumed so—or through some other fluid system.[5]

42 Rutland Gate
April 25 71

My dear Darwin

I am grieved beyond measure, to learn that I have misrepresented your doctrine,—and the only consolation I can feel is that your letter to 'Nature' may place the doctrine in a clearer light and attract more attention to it. I write hurriedly, as time is important, to save the mornings post, in order to point out two·passages which, I hope, in your letter to Nature, you will explain at length, so as to remove the false

3 See chapter 2, p. 8.

4 F. Galton 1908, 169; FG to Emma Sophia Galton, 1882-04-22, in K. Pearson 1924, 198; F. Galton 1882d; F. Galton 1886a.

5 FG to Charles Darwin, 1871-04-25, DARWIN, DCP-LETT-7717.

impression of Pangenesis under which I & probably others labour. In "Domestication Animals &c." p 374 "... throw [off] minute granules or atoms, which circulate freely throughout the system ..." & p. 379 "... the gemmules ... must be thoroughly diffused; nor does this seem improbable considering ... the steady circulation of fluids throughout the body."

(Is there not also, a passage in which the words "circulating fluid" are used?— I cannot hurriedly lay my hand on it, but believe it to exist)

Believe me—necessary in great haste—

very sincerely yours
Francis Galton

The idea of conducting an experiment to test the pangenesis hypothesis had occurred to Galton toward the end of 1869, as he was completing *Hereditary Genius*. As we have seen, there he had seized on the theory as a convenient way to state heredity in mathematical terms. However, the theory had been stated by Darwin in *Variation of Plants and Animals Under Domestication* in very vague language. One reason for this was that it attempted to give a single explanation for all hereditary phenomena and growth. It encompassed both sexual and asexual reproduction, growth of the body and the regrowth of amputated limbs in certain species, as well as supposed inheritance of acquired characters induced by the environment, and atavistic 'reversion' to latent ancestral characters. Darwin called it '*pan*genesis' because he supposed that inheritance was the work not just of dedicated sexual organs but of *every* organ or feature in the body of any organism. Each one of these regularly emitted, he proposed, 'gemmules' which circulated through the body and drove the growth of that organ or feature (in the case of development or regrowth). The gemmules also concentrated in the sexual organs, where the organism reproduced sexually, or in 'buds' where it did not. No essential difference existed between sexual and asexual reproduction on this account. It was suggested that the gemmules could be sub-cellular particles—the existence of cells had only recently been firmly established—but that was left ambiguous.

The inclusion of acquired characters was another step in Darwin's long climb-down from the first edition of his *Origin of Species*. There 'hard heredity', in which existing variation was acted on after the fact by natural selection, had seemed prominent. 'Soft heredity' of acquired characters, purpose-formed by the environment, was barely mentioned—but never entirely absent. Darwin had a theory of selection but no worked-out theory explaining the origin of the variation that selection acted on.

If 'sports' (mutations) arose by chance to provide novel variation, repeated crossing with those who did not have the sport would inexorably blend it out of existence, a problem that Darwin was likely aware of well before Fleeming Jenkin pointed it out in 1867.[6] This is not true if inheritance does not blend hereditary elements, preserving them discretely, but as we will see Darwin never accepted a non-blending theory of heredity. So if sports could not explain anything of long-term evolutionary importance, the alternative was acquisition of characters. Variation was induced by the environment, not merely selected by it. For Darwin, the most important examples were use and disuse of organs, which (unlike his cousin) he accepted without reservation and supposed would eventually become heritable over long periods of time.

The reception of pangenesis, which had been formulated in manuscript as early as 1865, had been mixed. It was easy to point out that versions of the same theory had been proposed many times over the centuries, even by the Greeks. But many, like Galton, seized on it as a working hypothesis. Alfred Russel Wallace wrote to August Weismann (1834–1914) that he found it '*sublime* in its simplicity and the wonderful manner in which it explains the most mysterious of the phenomena of life. To me it is *satisfying* in the extreme. I feel I can never give it up,

[6] Jenkin 1867.

unless it be *positively* disproved, which is impossible, or replaced by one which better explains the facts, which is highly improbable'.[7]

It occurred to Galton that pangenesis could be tested by means of blood transfusions in animals. Since the 'gemmules' of heredity would be found in blood (or so he assumed), pure breeds injected with the blood of other breeds should give birth to mongrels after the operation. He was already familiar with the use of saline transfusions by O'Shaughnessy and Latta during the 'blue epidemic cholera' of 1831.[8] Joseph Lister had recently used antiseptics to safely perform blood transfusions, which had been used on humans with mixed results, perhaps as early as 1795 but certainly by 1818, and on animals as early as the 17th century.[9] Using recently invented anaesthetics, the operations could be done relatively painlessly. The trick in this was to carefully distinguish natural variations in offspring from the variation that mongrelization should induce. Rabbits seemed suitable subjects, since purebred strains were easy to acquire. Galton chose 'silver-grey' rabbits, which usually had (almost) uniformly grey hair (really caused by a mixture of black and white hairs), but occasionally threw a white 'himalaya' variant with black tips, and sometimes a sandy 'himalaya'. Anything other than those would suggest a mongrel effect.

Facilities at Rutland Gate were not suitable, and though his own medical training would be useful, a skilled surgeon with extensive practice using animals would be needed. So Galton sought out the help of the Zoological Society at Regent's Park. From his role in the Royal Geographical Society he already knew the prosector (that is, dissector of animal carcasses), Dr. James Murie (1832–1925). Trained at Glasgow, Murie had been the surgeon on Consul John Petherick's expedition up the Nile, 1861–3 (they were to wait on the river at Gondokoro to relieve John Hanning Speke's overland expedition from Zanzibar, but were famously late, provoking Speke's fury). The Scot had a notoriously bad temper but was highly skilled, completing some 4000 dissections of animals at the Zoo.[10] Galton seemed to get along with him well. 'I called on him to discuss the matter. A dead cobra was lying on his table, and on my remarking that I had never properly seen a poison fang, he coolly opened the creature's mouth, pressed firmly at exactly the right spot, and out started that most delicate and wicked-looking thing, with a drop of venom exuding from it, just in front of his nail. I thought that a man who was so confident of his anatomical knowledge and of his nerve as to dare such an act, must be an especially suitable person to conduct my experiments'.[11] The superintendent of Regent's Park, Abraham Dee Bartlett (1812–1897), an authority on captive animals, would also be useful for advice about the subjects. Meanwhile Galton wrote to Darwin on the 11th of December to ask for help in procuring rabbits for 'peculiar experiments'.[12] Darwin's response has not survived, but before long he would be informed about the minutest details of the blood transfusions. Over the next year, Galton would have three kinds of increasingly complex operations performed on his rabbits.

Throughout, the ears of all the rabbits were tattooed with unmistakable bold Roman numerals. At first Galton tried the relatively simple operation of bleeding as much blood from a silver-grey buck or doe as the animal could stand in one session, up to about an ounce, then replacing it with the same amount of blood procured from a different breed. He used a 5 drachm ebonite syringe, attached to a canula which was inserted into the vein and had a stop-

[7] Wallace 1906, Vol. 2, 422.

[8] Bizarrely, competing treatments at the time included emetics!

[9] Dr. Philip Syng Physick (1768–1837) of the University of Pennsylvania is reported to have performed transfusions as early as 1795 but did not publish or promote the practice, and little is definitely known now about his experiments P. J. Schmidt 1968.

[10] See Tredoux 2023c for Galton's entanglement with Speke and Burton. For Murie, who would in later years become that rare phenomenon, an 'aggressive librarian', see Walker 1997.

[11] F. Galton 1908, 298.

[12] FG to Charles Darwin, 1869-12-11, DARWIN, DCP-LETT-7026.

cock screwed into it. This was tricky since the blood clotted rapidly, but stirring and filtering it removed most of the resulting 'fibrine' protein, so that it could be injected with a syringe. The vein was tied afterward. After doing this for 5 silver-grey does and three bucks, he bred from them. According to his calculations, the animals should have had 1/8 alien 'gemmules' in their systems. While he waited for the rabbits to breed, he kept Darwin up to date. 'As soon as ever I know anything I will write instantly & first to you. For my part, I am quite sick with expectant hope & doubt'.[13] But eventually, out of six litters, almost all of the 30 offspring were true silver-greys. Just one rabbit with a white fore-leg showed faint promise as a mongrel. Reporting the disappointing results to Darwin, Galton announced that he was now going to try more drastic transfusions.[14]

The second type of operation transfused much larger volumes, up to 3 ounces, of wholly 'defibrinized' blood, which had been obtained from multiple rabbits and whisked using a rice straw. Galton calculated that this was the equivalent of transfusing 6 pounds of human blood. This required repeated bleedings and injections. First attempts were made at the close of March 1870. By then Murie had quarrelled decisively with the Zoological Society. His assistant Oscar Louis Fraser, a son of the noted naturalist Louis Fraser (1810–1866), had taken over the surgical operations on the rabbits, though Murie was still able to offer occasional advice. Clotting was now a real challenge, as Darwin was informed in early April. 'I can do nothing with the blood in its natural state it coagulates so quickly, so I defribrinise it.— If I cannot ever succeed in transfusing as much into the rabbits as is necessary to make a fair experiment, I must go to larger animals & try cross circulation with big dogs'.[15] He vowed to 'go on at any waste of rabbit life, until I get at least 1/2 alien blood. The experiment is not fair to Pangenesis until I do'. Darwin responded with a letter of encouragement.[16] However Galton was soon able to obtain one doe and three bucks with larger transfusions, where the bucks had also received the smaller transfusions previously. The results were disappointing since they seemed to be sterile, as Darwin was informed in late June. 'The bucks were as eager as possible for the does; but the latter proving indifferent, I was unable to testify to their union having taken place; so I left them in pairs, in the same hutch, for periods of three days at a time'.[17] Only one produced offspring, pure silver-greys. Speculating that defibrinizing the blood might have filtered out the 'gemmules', or that the particles did not persist for long in the transfused blood, Galton began to devise a more drastic and technically complex method. But due to his summer vacation this had to be postponed.

The Galtons were in Paris on the 15th of July, just in time for the declaration of war with Prussia. They spent two weeks at Grindelwald in Switzerland then retreated back to England via Folkestone, with its bracing sea-bathing, followed by their usual family rounds. The British Association was meeting in Liverpool that September, and Galton delivered a paper on 'Barometric Predictions of Weather' to the Mathematical and Physical Science section. He advised that, for day to day predictions rather than very rare storms, taking the barometer into account was one-third worse than not taking it into account at all.[18] It was October before he was able to resume his experiments in London.

The third operation involved establishing live cross-circulation between the rabbits, by splicing their arteries. He had to use a large canula, 1/13 of an inch, to prevent clotting by a layer of fibrine within it. This made it tricky to insert into the arteries, which were a little too small, but at Dr. Murie's suggestion, delicate curved forceps forced the vessels open just enough. The

[13] Francis Galton to Charles Darwin, 1870-03-15, DARWIN, DCP-LETT-7133.

[14] Francis Galton to Charles Darwin, 1870-03-17 and 1870-03-22, DARWIN, DCP-LETT-7139 and DCP-LETT-7143.

[15] FG to Charles Darwin, 1870-04-08, DARWIN, DCP-LETT-7161.

[16] FG to Charles Darwin, 1870-05-12 and 1870-06-01, DARWIN, DCP-LETT-7185 and DCP-LETT-7212.

[17] F. Galton 1871a; Francis Galton to Charles Darwin, 1870-06-25, DARWIN, DCP-LETT-7245.

[18] F. Galton 1870a.

throats of the rabbits were brought close together on a table, and to prevent entanglement of the doubled equipment, Galton devised a suspension rig of notched bars and counterweights above the table. Every now and then the artery had to be declogged of fibrine, using a improvised corkscrew filed into the end of a piece of wire. In this way prolonged cross-circulation of up to 35 minutes was established between the animals, on October 6, 7, 19, 20 and 27 of 1870.

From the cross-circulated subjects, 88 rabbits were bred in 13 litters. In the new year, Galton wrote to Darwin that 'you must have patience with me & wait yet longer', since 'I have built a good serviceable little house for the rabbits in my own back yard & have all the best of them under my own eye, now'.[19] But by March Galton was forced to conclude, in his write-up of the experiments for the Royal Society, that 'in no single case has there been any evidence of alteration of breed', 'negativing, in my opinion, beyond all doubt, the truth of the doctrine of Pangenesis'.[20] He surely reported this finding directly to Darwin prior to that, but the letter has not been found. That April the rabbits were transferred to Down House for safe-keeping and further breeding. Later that month, as we have seen, Darwin sent his testy letter to *Nature*, after notifying Galton of his disagreement.

Analyzing Darwin's letter, it is apparent that he was relieved to find that he had not actually used the word 'blood' in his formulation of pangenesis. Instead he had used 'fluid'—though he did use the word 'circulate', and which fluids circulate in animals other than blood? He would later change that to 'diffuse'. This allowed him to cling to his theory under the assumption that the gemmules might spread some other way through the body. It was a weak defence and an ad-hoc crutch. He had been, as we have seen, involved in the experiments from the very start in every detail. If he had ever had any doubt about the blood being the carrier of gemmules, he could have raised that at any time over the period of over a year. He admitted this oversight, but clung to the 'fluids' instead. The letter also contains a logical fallacy. While it is true that if pangenesis applied to organisms without blood, as Darwin stated, they would have to circulate some other way, in organisms which *did* have blood, a special explanation is required for how they are kept out of the blood. One of the phenomena he sought to explain was the regeneration of severed organs in certain species like lizards. How else would they be regenerated via gemmules in 'fluids' if not through the blood that flowed to the injured areas?[21]

This is a useful case study for evaluating the popular idea, due to Karl Popper, that science proceeds by falsifying hypotheses.[22] Galton had to sharpen Darwin's hypothesis to even render it testable, since there is no definite meaning to 'fluids'. This sharpened version involving blood had certainly, to any reasonable mind, been falsified. Consider Alfred Russel Wallace, who later recalled that 'Galton's experiments on blood transfusion with rabbits first staggered me, as it seemed to me to be the very disproof I had thought impossible'.[23] In the face of this Darwin himself simply finessed the theory. This supports the contention by the great cytologist C. D. Darlington that Darwin could be *slippery*.[24] However many other defences were available which Darwin did not choose to use. For instance, he might have speculated that the 'gemmules' were not necessarily compatible in their blood form with other animals, and if he had he would have anticipated the discovery forty years later of blood groups.

Galton published a game response to his 'trusted leader' in *Nature*, showing that he was willing to accept any blame for misinterpreting the hypothesis as it was stated.[25] But he cited the language that had supposedly led him astray, minutely. And he ended with a joke based on Darwin's recently published *Descent of Man*. Today that joke reminds the reader of nothing

[19] Francis Galton to Charles Darwin, 1871-01-09, DARWIN, DCP-LETT-7432.
[20] F. Galton 1871a.
[21] Galton made this point in his *Memories*, F. Galton 1908, 296.
[22] Popper 1959.
[23] Wallace 1906, Vol. 2, 422.
[24] Darlington 1959.
[25] F. Galton 1871c.

less than one of the displays of submission that would be described the following year in *The Expression of the Emotions in Man and Animals.*[26]

> It appears from Mr Darwin's letter to you in last week's *Nature*, that the views contradicted by my experiments, published in the recent number of the "Proceedings of the Royal Society," differ from those he entertained. Nevertheless, I think they are what his published account of Pangenesis (Animals, etc., under Domestication, ii, 374, 379) are most likely to convey to the mind of a reader. The ambiguity is due to an inappropriate use of three separate words in the only two sentences which imply (for there are none which tell us anything definite about) the *habitat* of the Pangenetic gemmules; the words are "circulate," "freely," and "diffused." The proper meaning of circulation is evident enough—it is a re-entering movement. Nothing can justly be said to circulate which does not return, after a while, to a former position. In a circulating library, books return and are re-issued. Coin is said to circulate, because it comes back into the same hands in the interchange of business. A story circulates, when a person hears it repeated over and over again in society. Blood has an undoubted claim to be called a circulating fluid, and when that phrase is used, blood is always meant. I understood Mr Darwin to speak of blood when he used the phrases "circulating freely," and "the steady circulation of fluids," especially as the other words "freely" and "diffusion" encouraged the idea. But it now seems that by circulation he meant "dispersion," which is a totally different conception. Probably he used the word with some allusion to the fact of the dispersion having been carried on by eddying, not necessarily circulating, currents. Next, as to the word "freely," Mr. Darwin says in his letter that he supposes the gemmules to pass through the solid walls of the tissues and cells; this is incompatible with the phrase "circulate freely." Freely means "without retardation"; as we might say that small fish can swim freely through the larger meshes of a net; now, it is impossible to suppose gemmules to pass through solid tissue without any retardation. "Freely" would be strictly applicable to gemmules drifting along with the stream of the blood, and it was in that sense I interpreted it. Lastly, I find fault with the use of the word "diffused" which applies to movement in or with fluids, and is inappropriate to the action I have just described of solid boring its way through solid. If Mr Darwin had given in his work an additional paragraph or two to a description of the whereabouts of the gemmules which, I must remark, is a cardinal point of his theory, my misapprehension of his meaning could hardly have occurred without more hesitancy than I experienced, but I certainly felt and endeavoured to express in my memoir some shade of doubt; as in the phrase, p. 404, "that the doctrine of Pangenesis, pure and simple, as I have interpreted it, is incorrect."
>
> As I now understand Mr Darwin's meaning, the first passage (ii, 374), which misled me, and which stands: "... minute granules ... which circulate freely throughout the system" should be understood as "minute granules ... which are dispersed thoroughly and are in continual movement throughout the system"; and the second passage (ii, 379), which now stands: "The gemmules in each organism must be thoroughly diffused; nor does this seem improbable, considering ... the steady circulation of fluids throughout the body," should be understood as follows: "The gemmules in each organism must be dispersed all over it, in thorough intermixture; nor does this seem improbable, considering ... the steady circulation of the blood, the continuous movement, and the ready diffusion of other fluids, and the fact that the contents of each pollen grain have to pass through the coats, both of the pollen tube and of the embryonic sack."

[26] C. Darwin 1871a, 1872.

(I extract these latter *addenda* from Mr Darwin's letter.)

I do not much complain of having been sent on a false quest by ambiguous language, for I know how conscientious Mr. Darwin is in all he writes, how difficult it is to put thoughts into accurate speech, and, again, how words have conveyed false impressions on the simplest matters from the earliest times. Nay, even in that idyllic scene which Mr. Darwin has sketched of the first invention of language, awkward blunders must of necessity have often occurred. I refer to the passage in which he supposes some unusually wise ape-like animal to have first thought of imitating the growl of a beast of prey so as to indicate to his fellow-monkeys the nature of expected danger. For my part, I feel as if I had just been assisting at such a scene. As if, having heard my trusted leader utter a cry, not particularly well articulated, but to my ears more like that of a hyena than any other animal, and seeing none of my companions stir a step, I had, like a loyal member of the flock, dashed down a path of which I had happily caught sight, into the plain below, followed by the approving nods and kindly grunts of my wise and most respected chief. And I now feel, after returning from my hard expedition, full of information that the suspected danger was a mistake, for there was no sign of a hyena anywhere in the neighbourhood. I am given to understand for the first time that my leader's cry had no reference to a hyena down in the plain, but to a leopard somewhere up in the trees; his throat had been a little out of order—that was all. Well, my labour has not been in vain; it is something to have established the fact that there are no hyenas in the plain, and I think I see my way to a good position for a look out for leopards among the branches of the trees. In the meantime, Vive Pangenesis!

Francis Galton.

Louisa recorded in her diary entry for 1871 that 'Frank gave up his rabbit breeding & took to machine inventing', but this was not quite true. It is perplexing to discover that the pangenesis experiments did not stop here but were prolonged for another two years, before they were eventually abandoned sometime in 1873 with no change in result. These remained blood transfusion experiments, regardless of the 'fluids' dodge. Only one outcome could have been accepted: if proof of mongrelization via blood was found, that would vindicate pangenesis. If none was found, that would signify *nothing*—absence of evidence rather than evidence of absence, or 'fluids'. It was not an attempt to falsify the hypothesis, but rather an attempt to confirm it.

Galton gave it his best shot. By September of 1871 he was trying rats as well, 'siamesing' them together to guarantee complete exchange of blood. However the rats caused difficulties in practice, tending to die inconveniently, and by February 1872 he had to abandon them—though one pair were united for as long as 3 months. His assistant Oscar Fraser had departed for Calcutta earlier that year, and the new man—possibly Charles Henry Carter of UCL—was not as skilled mechanically. In the meantime Galton had switched attention back to cross-circulated rabbits, now trying albino Angoras, then silver-greys again.[27] These often went back to Down to be cared for and bred with, as projects for Darwin's servants, then back to London for further cross-circulation. Both Darwin and Galton had already tired of these ongoing experiments by November of 1872, but they kept on, ever hopeful for mongrels. The up-and-coming Trinity-educated biologist Francis Maitland Balfour (1851–1882) got involved early in 1873, trying transplants of tissues on the rabbits. Unsurprisingly Balfour, whose career would be terminated on Mont Blanc less than a decade later, got no further.[28] In September 1875, Galton

[27] Francis Galton to Charles Darwin, 1871-09-13, 1871-11-21, 1871-11-24, 1872-02-01, DARWIN, DCP-LETT-7938, DCP-LETT-8080, DCP-LETT-8085, DCP-LETT-8192.

[28] During his brief career, Balfour was an important influence on the biologists W.F.R. Weldon and William Bateson, who would interact extensively with Galton in coming years over the issue of evolution, see chapter 13, p. 287.

sent Darwin a summary of the rabbit experiments that had been done, but does not appear to have performed any more than those already described above.[29] He recalled in his *Memories* that he transferred his experimental equipment to another up-and-coming Cambridge-educated biologist, George Romanes, who had no more success in producing mongrels than he had.[30] Romanes was actively testing pangenesis by 1874 or 1875, and proposing grafting experiments on plants and animals, including rabbits.[31]

Unlike his cousin, Darwin never accepted that pangenesis had been disproved. It was too useful and comprehensive as a theory to be discarded just like that. A second edition of his *Variation of Plants and Animals under Domestication* appeared in 1875-6. The second volume modified the description of the theory a little, 'clarifying' the language, but stuck to the theory doggedly. 'Mr. F. Galton, after describing his valuable experiments (Proc. Royal Soc. vol. xix. p. 393) on the intertransfusion of the blood of distinct varieties of the rabbit, concludes by saying that in his opinion the results negative beyond all doubt the doctrine of Pangenesis. He informs me that subsequently to the publication of his paper he continued his experiments on a still larger scale for two more generations, without any sign of mongrelism showing itself in the very numerous offspring. I certainly should have expected that gemmules would have been present in the blood, but this is no necessary part of the hypothesis, which manifestly applies to plants and the lowest animals. Mr. Galton, in a letter to "Nature" (April 27, 1871, p. 502), also criticises various incorrect expressions used by me'. Those expressions were now altered. What formerly circulated freely now dispersed throughout the whole system. Darwin appears to have thought that these manoeuvres rescued his theory.[32]

——oOo——

With pangenesis undone, Galton set off in search of his own mechanism for inheritance, regardless of what he told the readers of *Nature*. This resulted in two of his least well-known papers, published in 1872 and 1875. They remained uncollected in his lifetime.[33]

The immediate problem Galton faced was how to explain the origin of differences given a 'strict doctrine of heredity', which suggested similarity. 'It has been thought by some that the fact of children frequently showing marked individual variation in ability from that of their parents is a proof that intellectual and moral gifts are not strictly transmitted by inheritance'. A statistical explanation was proposed. 'I maintain that results derived from large averages are all that can be required, and all we could expect to obtain, to prove that intellectual and moral gifts are as strictly matters of inheritance as any purely physical qualities'.

The theory Galton offered to explain hereditary variability is startlingly modern in many respects, anticipating later developments by many years:

1. It is a *particulate* theory, in which hereditary 'particles' are transmitted entire and not blended *at their level*. Later this would become known as 'continuity of the germ-plasm', a phrase due to Wallace's correspondent August Weismann.
2. It explicitly incorporates *random selection* of the hereditary particles in the process of descent, entailing only a probabilistic relationship between parents and their children. This explains the fact that parents and children resemble each other to a degree, but with considerable variation.
3. Each parent transmits *half* of its hereditary information, which Galton named the 'stirp', to offspring.

[29] Francis Galton to Charles Darwin, 1875-09-24, DARWIN, DCP-LETT-10169.

[30] F. Galton 1908, 298.

[31] Charles Darwin to George Romanes, 1874-12-27, DARWIN, DCP-LETT-9783.

[32] C. Darwin 1875, Vol. 2, 350.

[33] F. Galton 1872a, 1875a. These are reproduced in Tredoux 2023a, and are consistent; they should be read together for a complete description of the theory.

4. *Latent* hereditary particles, unexpressed in their carriers, are distinct from *patent* particles, which are expressed.
5. 'Acquired characters', *if there are any*, must usually affect the expressed particles and therefore are typically not passed on to descendants.
6. Complex organisms require a system of double parentage, most commonly through sexual reproduction, to cope with loss or harmful alteration of hereditary particles over time.
7. There is competition within the 'stirp' between the hereditary particles for representation in offspring.
8. Random influences may alter the expression of hereditary particles, which interact with each other in three-dimensional space.
9. Hereditary particles from *alien species* may in very rare circumstances insinuate themselves into the stirp.

Galton deduces that selection is necessary if a gamete is to contain all the particles of heredity which make up a 'stirp' (partly analogous to our 'genome'), otherwise the storage required would increase geometrically with each generation, given dual parentage, and soon 'would exceed in multitude the wildest flights of rational theory'. Since each parent contributes equally to the traits of offspring, their respective stirps must be halved and the combination passed on, which anticipates our understanding of 'meiosis'.[34] An organism consists of components built by patent or expressed hereditary particles. These are drawn randomly, in a second stage of selection, from the stock of unexpressed latent particles in the stirp. Galton supposes that the latent undeveloped particles transmit via the sexual organs, and that the patent particles reside in what we call the somatic or developed body cells. Thus descendants only receive latent particles. Patent particles are in effect sterile.

Next Galton makes an important mistake. He supposes that children inherit *identical* stirps, halved and combined from their parents, a process he thinks happens once. Yet children vary a lot from each other. Perhaps this is why he introduces that additional level of indirection mentioned above: the particles which are expressed in the children are drawn randomly from the identical stirp they share.[35] (We now know that each stirp, to use his term, is formed afresh in each child, and will on average be half-alike for any two children.) As he also supposes, quite correctly, that random influences will disturb expression of hereditary particles once they are selected, he has plenty of sources of variation between parents and their children.[36] Indeed too many, as he notices himself. How are the siblings remotely alike then? His rather unconvincing resolution invokes adaptation between the stirp and its environment, which reduces variation in the shared stirp itself. Given this, pure strains from good matches will show less variation, he argues, and mongrel strains more.

The motivation behind the crucial *particulate*, or discrete, concept comes from the well-known fact of reversion, which Darwin had dwelt on at length in *Variation*. Characters may skip generations, showing up in ancestors but not immediate descendants, only to appear later in more remote descendants. In Galton's own case, his parents Samuel Tertius and Violetta were much above average but not as exceptional as their son. However his grandfathers and many other ancestors were remarkable men—especially Erasmus Darwin. Galton interpreted this to mean that the hereditary information shaping these characters is preserved.

[34] In F. Galton 1872a this first stage of selection is called, somewhat unhelpfully, 'class representation', by which embryonic elements are drawn.

[35] In F. Galton 1872a this is second stage of selection is called 'family representation' because the stirp is supposed to be shared by all siblings.

[36] See K. Mitchell 2018 for an explanation of this phenomenon as a contributor to non-shared sources of variance between twins.

It is worth lingering over the puzzle that Darwin, well aware of the same facts about reversion, did not reach the same conclusion, opting for blending inheritance anyway. In Darwin's thinking, hybrids are hybridized at the 'gemmule' level, so that the gemmules themselves are blended and are not discrete particles. Exactly how this allows reversion to happen is not explained very clearly, but reflects another strand of his thinking, which operated by analogy with dosage in fluids. Even a small dosage in a blended concoction may still be potent, and lead to a visible threshold effect. The obvious objection is that a pattern must emerge whereby increasingly reduced dosages show attenuated effects. As R.A. Fisher would later point out, it is easy show mathematically that each generation reduces variability under blending inheritance, until it vanishes altogether.[37]

Fisher and others have proposed that awareness of vanishing variability under blending inheritance led Darwin, in successive editions of the *Origin*, to seize on acquired characters as a source for continual reintroduction of variability. Such characters would already be pre-adapted to the environment of the organism. But the relationship between these two ideas may well have been symbiotic. Acquired characters, if they exist, are easier to conceptualize in the continuous realm. Fisher did not help here by referring to them, tongue-in-cheek, as induced 'mutations'. Use and disuse of organs must be slow and progressive, insensibly shading from full functionality to none. If camel-leopards stretch their necks to reach high branches, it is more natural to think of that in terms of continuous expansion like a rubber band, rather than ratchet-notches.[38] Certainly that is how Darwin thought things played out. Consequently, if one believes in acquired characters it is much harder to even think in terms of discrete particulate inheritance.

Galton is extremely skeptical that acquired characters matter at all, though he cannot rule them out entirely. He suggests that they are easily confounded over several generations with selection effects. Or, as in the case of the guinea pigs which supposedly acquired epileptic convulsions from nerve operations on their parents, as claimed by the French physiologist Charles-Édouard Brown-Séquard, are actually learned behaviours.[39] Since environmental influences will usually act on the *patent* (expressed) hereditary particles, they will not easily be passed down to descendants—unless the *latent* particles themselves are directly altered by that environmental condition (many years later this would be demonstrated through use of X rays to induce mutations). Extensive experience shows that mutilations are not inherited. However, since animal breeders retail old wives' tales about instances of acquired characters so often, and sporadic reports are printed from this or that researcher claiming to find them in the laboratory, he simply writes them off as unimportant flukes if they do occur. He supposes that somehow altered hereditary particles might find some path into the progeny. For this he adopts a modification of pangenesis as a hedge against the unlikely.

Darwin had deliberately downplayed the difference between sexual and asexual reproduction in pangenesis, since that did not intrinsically require sex. Galton argues that dual parentage precedes sex, which is just one way of achieving it. A dual system is needed in complex organisms because those require more hereditary particles, and 'there is always a chance of some one or more of the various species of germs in the stirp dying out, or being omitted; and of course when they are gone they are lost for ever, and are irreplaceable by others. From time to time this chance must fall unfavourably, and will cause a deficiency in some of the structural elements'. Dual parentage allows replacement of the missing or damaged elements, engineering redundancy.

Parts of Galton's theory are, we now know, strictly wrong. We have already noted that

[37] Fisher 1930, 1–21. Of course, fluids are actually composed of discrete molecules, and enough homeopathic dilution will eliminate all traces of the needed molecules with probability=1.

[38] We will ignore discrete neck vertebrae here!

[39] This claim of Brown-Séquard (1817–1894), along with some others, was pressed on Galton by Darwin. FG to Charles Darwin, 1873-11-03, DARWIN, DCP-LETT-10238.

siblings do not share identical 'stirps'. His sampling of particles from the stirp is always without replacement, which actually only holds per-locus during meiosis, when the 'stirps' of the parents halve. This led him to construct fallacious explanations depending on a depletion effect at his other sampling stage, so that characters may skip generations because the particles for them have been severely depleted when drawn from the latent stock to form patent elements, requiring the residue to be restocked in more distant generations. While his latent particles correspond intriguingly to epigenetically unexpressed DNA, and his patent ones to expressed DNA, and though it is certainly true that ordinarily it is the totipotent unexpressed DNA that is transmitted to descendants, this does not explain reversion. Atavism is due to recessive genes, epistatic and polygenic effects, and his use of 'latent' only corresponds obliquely with 'recessive'.

Given the state of biological knowledge in Galton's day it is not in the least surprising that the theory is in some key respects wrong. What *is* surprising is how many elements of his thought were not only on-track, but anticipated developments right up to the present day. His disdain for the lore of acquired characters, his embrace of a particle theory and his appreciation of the need for sexual reproduction in complex organisms, showed a sure grasp of the principles involved.

Then consider the notion that hereditary particles compete internally *within* the stirp. This is exactly the idea that there is natural selection within the genome, so that genes may try to promote their own interests and suppress that behavior in other genes—an important phenomenon only properly studied quite recently.[40] Or consider the idea that hereditary particles from other species may find their way into the stirp in rare cases and become fixtures there. We now think that the mitochondrial DNA inside human cells came from a bacterial source and exists there in 'endosymbiosis' with the nuclear DNA, a theory first put forward in 1967 by Lynn Margulis (1938–2011) and only accepted decades later.

There is even a glimpse as early as 1872 of Mendel's then-neglected ratios in this combination of insight, hypothesis and bafflement: 'I should judge, from the impression left by many scattered statistics, that it is perfectly safe to affirm that breeders, when they mate two animals, each having the same unusual characteristic, not through known hereditary transmission, but by supposed variation, would consider themselves fortunate if one quarter of the progeny inherited that quality'.[41]

Galton sent a proof copy of 'On Blood Relationship' to Darwin shortly before it was published. Darwin's comments have not survived, though it is clear from Galton's reply that the naturalist offered suggestions which were incorporated in the final version.[42] A few months later Alfred Russel Wallace enthused to Darwin. 'What a wonderfully ingenious & suggestive paper that is by Galton, on "Blood Relationship". It helps to render intelligible many of the excentricities [sic] of Heredity, Atavism, &c.' The reply from Down House was non-committal. 'As for Galton's paper, I have never yet been able fully to digest it: as far as I have, it has not cleared my ideas, & has only aided in bringing more prominently forwards the large proportion of the latent characters'.[43]

The sequel 'A Theory of Heredity', published at the close of 1875, was received with no more enthusiasm at Down. Galton had been corresponding with his cousin's mathematical son, George Howard Darwin, with whom the hereditarian had developed a sympathetic connection. Early in November, the naturalist wrote to his cousin, mentioning that his son had informed him about a forthcoming paper on heredity. Galton wrote a faintly embarrassed reply the very

[40] A. Burt and Trivers 2006.

[41] F. Galton 1872a.

[42] Francis Galton to Charles Darwin, 1872-05-08, 1872-06-04, DARWIN, DCP-LETT-8357, DCP-LETT-8368.

[43] Alfred Russel Wallace to Charles Darwin, 1872-08-31, 1872-09-02, DARWIN, DCP-LETT-8498, DCP-LETT-8504.

next day, including a useful précis of his theory, freely confessing his 'audacity'. Surely this had not been shared earlier due to its decisive and explicit break from pangenesis. A sore point.

> ... First, I start with the 4 postulates, in favour of which you have so strongly argued, and which may reasonably be now taken for granted:—
>
> 1. Organic units in great number.
> 2. Germs of such units in still greater number & variety (existing somewhere).
> 3. That undeveloped germs do not perish; but multiply and are transmissable.
> 4. Organization wholly depends on mutual affinities.
>
> From these 4 postulates, I logically deduce several results, one of which is the importance and almost the necessity of double parentage in all complex organisations, and consequently, of sex.
>
> Then I argue, that we must not look upon those germs that achieve development as the main sources of fertility; on the contrary, considering the far greater number of germs in the latent state, the influence of the former, i.e. of the personal structure, is relatively insignificant. Nay further, it is comparatively sterile, as the germ once fairly developed, is passive; while that which remains latent continues to multiply. From this follows:— (1) the extremely small transmissibility of acquired modifications, (to which I recur)— 2. the fact that exceptional gifts are sometimes barely transmissable (here the sample was over rich & drained the more fecund residue)— (3.) The fact of some diseases skipping 1 or more generations; (here the supposition is made of the germs of those diseases being peculiarly gregarious, hence the general outbreak of them leaves but a small residue, which has not strength to break out in the next generation, but being husbanded in a latent form, there multiplies & recovers strength to break out in the next or in a succeeding generation).
>
> Next, I go into the question of affinities & repulsions; which I put as necessarily numerous & many sided (while professing entire ignorance of their character) & I argue thence, a long period of restless unsettlement in the newly fertilized ovum accompanied, as we know it to be, with numerous segregations & segmentations, in each of which the dominant germs achieve development, while the residue is segregated to form the sexual elements. But I argue, that as our experience of political and other segregations shews, that they are never perfect, we are justified in expecting that numerous alien germs will be lodged in every structure and that specimens of all of them will be found in almost all parts of the body.
>
> In this way, I account for the reproduction of lost parts, &c. as well as for the inheritance of all peculiarities that had been congenital in an ancestor.
>
> I then consider the cases of inheritance of what had been non-congenital in an ancestor, but acquired by him. I shew, that the deductions usually made, that the structure reacts on the sexual elements is not justified by the evidence of adaptivity of race, when this depends on conditions which act equally on all parts of the body. My reason is, that since the same agents, (viz. the germs,) are concerned both in growth and in reproduction, the conditions that would modify the one, would simultaneously modify the other; hence they would be collaterally affected and the apparent inheritance is not a case of inheritance at all, in the strict sense of the word. Nay the progeny may begin to vary under changed conditions sooner than the parent, (as in the hair or fleece of the young of dogs & sheep, transported to the tropics).
>
> As regards Brown-Séquard's guinea-pigs;— If I rightly understand & am informed of his experiment, it is open to fatal objection. The guinea pigs that were operated on, appear to have been kept separate from the rest. If so, we sh[d] expect the young sometimes to have convulsive attacks from mere imitation, just as we should expect of children brought up in a ward of epileptic patients, or among hysterical people, (revivals, dancing mania, &c.) Besides, there is not the least evidence that the mutilation of the spinal marrow on which the parental epilepsy primarily depended, was inherited. I also disparage much other evidence of the inheritance of acquired modification, leaving but a very small residue to accept.
>
> For this residue, I account by supposing the germs thrown off by the structure during its regular reparation, to frequently find their way into the circulation & some of these occasionally to reach the sexual elements and to become lodged and naturalised there, either by finding an unoccupied place or by dislodging others, like immigrants into an organized society, coming

> from a foreign country. Thus I account both for the fact, and for the great rarity & slowness of the inheritance of acquired modifications
>
> In conclusion, I restate a former definition, that I gave of the character of the relationship between parent and child, which I make out to be, not like that which connects a parent nation and its colonists, but like that which connects the representative government of the parent nation with the representative government of the colonists; with the further supposition, that the govt. of the parent country is empowered to nominate a small proportion of the colonists.
>
> I have now, so far as the limits of a letter admit, made a clean breast of my audacity in theoretically differing from Pangenesis:—
>
> (1) in supposing the sexual elements to be of as early an origin as any part of the body (it was the emphatic declarations of Balbiani on this point that chiefly attracted my interest) and that they are not formed by aggregation of germs floating loose & freely circulating in the system, and (2) in supposing the personal structure to be of very secondary importance in Heredity, being as I take it, a sample of that which is of primary importance, but not the thing itself.—
>
> If I could help, even in accustoming people to the idea that the notion of Organic Germs is certainly that on which the true theory of Heredity must rest, and that the question now is upon details & not on first principles, I should be very happy.

As soon as some proof copies of the forthcoming paper were available, a few days later, they were forwarded to Down for comments. Only a few days passed before comments were returned. The naturalist pleaded complexity. 'I have read your essay with much curiosity & interest, but you probably have no idea how excessively difficult it is to understand. I cannot fully grasp, only here & there conjecture, what are the points on which we differ.—I daresay this is chiefly due to muddy-headiness on my part but I do not think wholly so.—Your many terms not defined, "developed germs"—"fertile" & "sterile germs" (the word "germ" itself from association misleading to me) "stirp"—"sept" "residue" &c.&c. quite confounded me. If I ask myself how you derive & where you place the innumerable gemmules contained within the spermatogen formed by a male animal during its whole life I cannot answer myself.— Unless you can make several parts clearer I believe (though I hope I am altogether wrong) that very few will endeavour or succeed in, fathoming your meaning'.

Darwin noted down specific objections. He clung unapologetically to acquired characters. 'If this implies that many parts are not modified by use & disuse during the life of the individual, I differ widely from you, as every year I come to attribute more & more to such agency'. He would not accept the idea that complex organisms typically require dual parentage, but his counterexample of a *lower* form appears obtuse. 'This seems rather bold, as sexuality has not been detected in some of the lowest forms, though I daresay it may hereafter be'. He did not agree with the argument made about repairing loss of hereditary particles. 'I take a very different view of the meaning or cause of sexuality'. Regarding transmission of injuries to offspring hereditarily, 'I have received new & additional cases, so that I have now not a shadow of doubt'. 'Such cases can hardly be spoken of as very rare, as you w^{d} say if you had received half the number of cases which I have.' In sum, 'I am very sorry to differ so much from you but I have thought that you w^{d} desire my open opinion'.

The naturalist was wrong on all counts, as his adoring cousin certainly realized. Nevertheless, Galton held fire. 'Alas! alas!— and I had taken such pains to express myself clearly, and I see what I mean, so clearly!' he wrote at once. He went ahead anyway. The paper was read before the Anthropological Society on the 9th, as he informed George. 'I read my "Theory of Heredity" at the Anthropological last night, when up got a mad spiritualist who orated, and then offered to address the meeting on the subject as a medium; the spirit speaking through his lips. (This was not accepted.)'.

Darwin must have suspected that he was on less certain ground here. On the same day, the 10th of November, he wrote to apologize. Even a wrong-headed journalist in the *Spectator* 'does not write so odiously as I did in my letter, which you received so beautifully. I sh^{d} be glad to be convinced that the obscurity was all in my head, but I cannot think so, for a clear-headed (clearer than I am) member of my family read the article & was as much puzzled as I was—To this minute I cannot realise what are "developed" "sterile" & fertile germs.—You are a real Christian if you do not hate me for ever & ever'.[44]

A week before Christmas that year the discussion about 'A Theory of Heredity' resumed. Darwin did not follow the discrete theory of inheritance. 'George has been explaining our differences.—I have admitted in new Edit. (before seeing your essay) that perhaps the gemmules are largely multiplied in the reproductive organs;[45] but this does not make me doubt that each unit of the whole system also sends forth its gemmules. You will no doubt have thought of following objection to your view, & I sh^{d} like to hear what your answer is. If 2 plants are crossed, it often or rather generally happens that every part of stems, leaf, & even to the hairs, & flowers of the hybrid are intermediate in character; & this hybrid will produce by buds millions on millions of other buds all exactly reproducing the intermediate character. I cannot doubt that every unit of the hybrid is hybridised & sends forth hybridised gemmules'.

This led to an offhand revelation by Galton which was promptly lost to science for decades: an anticipation of Mendel's ratios. He distinguishes an 'organic unit' (containing hereditary particles) from a 'structural unit', which is the thing built by the hereditary particles.

42 Rutland Gate
Dec 19 / 75

My dear Darwin

The explanation of what you propose, does not seem to me in any way different on my theory, to what it w^{d} be on any theory of organic units. It would be this:—

Let us deal with a single quality, for clearness of explanation, & suppose that in some particular plant or animal & in some particular structure, the hybrid between white & black forms was exactly intermediate, viz. gray,—thenceforward for ever. Then a bit of the tinted structure under the microscope would have a form which might be drawn as in a diagram, as follows:—

(1) 

Whereas in the hybrid, it would be either that some cells were white and others black, & nearly the same proportion of each, thus:—

(2) 

giving on the whole when less highly magnified, a uniform gray tint,—or else this:—

(3) 

in which each cell had a uniform gray tint.

In (1) we see that each cell has been an organic unit (quoad colour).[46] In other words, the structural unit is identical with the organic unit

In (2), the structural unit would not be an organic unit but it would be an organic molecule. It would have been due to the development, not of one gemmule but of a group of gemmules,

[44] Charles Darwin to Francis Galton, 1875-11-02, 1875-11-03, 1875-11-05, 1875-11-07, 1875-11-08, 1875-11-10, DARWIN, DCP-LETT-10237,DCP-LETT-10238, DCP-LETT-10243, DCP-LETT-10245, DCP-LETT-10250. DCP-LETT-10254.

[45] The second edition of *Variation*, C. Darwin 1875.

[46] With respect to colour.

in which the black & white species would, on statistical grounds, be equally numerous (as by the hypothesis, they were equipotent.)

The larger the number of gemmules in each organic molecule, the more uniform will the tint of grayish be in the different units of structure. It has been an old idea of mine, not yet discarded & not yet worked out, that the number of units in each molecule may admit of being discovered by noting the relative number of cases of each grade of deviation from the mean grayness. If there were 2 gemmules only, each of which might be either white or black, then in a large number of cases one quarter would always be quite white, one quarter quite black, & one half would be gray. If there were 3 molecules, we should have 4 grades of colour (1 quite white, 3 light gray, 3 dark gray, 1 quite black) & so on according to the successive lines of "Pascal's triangle". This way of looking at the matter would perhaps shew (a) whether the number in each given species of molecule was constant & (b), if so, what those numbers were.

Ever very faithfully yrs
Francis Galton

This is tantalizing. Darwin believed in hybridization by *blending*, so that in the hybrid hereditary information was somehow fused throughout, which the evolutionist thought explained why the characters (or phenotypes) observed in individuals were intermediate. Galton was proposing that the same effect could happen even if the hereditary information remained particulate and discrete, and was propagated that way entire, never fused, but combined to form an intermediate character only when *expressed* within an individual. This is part-way toward the reconciliation of the biometric (continuous) schools of evolution with the Mendelian (discrete) school which would only emerge much later in 1918 and later (see chapter 13, p. 287 for much more on that controversy, in which both factions were inspired by Galton).

The observed ratios Galton would reason in reverse from are produced by a binomial expansion. For a character controlled by two gemmules a and A, with the resulting combinations aa, aA and AA, there will be a 1:2:1 ratio between the forms in a large population which randomly mates, assuming each parent contributes either a or A to offspring with a 50% chance, through a selection process already comprehended by his theory. (Galton does not anticipate dominance here, in which aA and Aa have the same effect as AA and ratio simplifies to 1:3.) But he is not thinking of a and A as alternative forms of the *same* (what we would call) gene, inherited as a diploid pair. His thinking is more general, and correspondingly vague: these are just any two gemmules affecting a trait. This might explain why his discussion did not lead further, despite this promising start. A real drawback was the particular thought experiment which carried him, in which the coat colour is (in effect) mechanically blended from constituent parts once they have separately developed. From a modern standpoint this is easily generalized to polygenic characters produced by a developmental process, as say an organ is formed, rather than only afterward through the combination of multiple separate organs or parts into a trait (though that certainly works too at an abstract level).

Darwin's reaction to this elementary explanation of a particulate theory of heredity is not known. It does not seem to have made any impression on him. Part of the problem went back to Galton's hedging explanation of whatever cases of acquired characters turned out to be true. The hedge was a modified form of pangenesis, but exactly how it was different from Darwin's own theory remained a puzzle to the naturalist and absorbed his attention—the particle part was at most of secondary interest to him. Shortly before Christmas, Galton struggled to convince George Darwin that his (appended hedging) theory of pangenesis was novel.[47] It would all have been a lot simpler if Galton had not been so ambitious. In 'explaining' all possibilities, just in case, he had tried to have it both ways, which Darwin could not fail to notice.

Alfred Russel Wallace was less encumbered than Darwin. 'Your "Theory of Heredity" seems to me most ingenious and a decided improvement on Darwin's, as it gets over some of the great

[47] Francis Galton to G. H. Darwin, 1875-12-22, DARWIN, DCP-LETT-10317.

difficulties of the cumbrousness of his Pangenesis'.[48] However, nearly fifteen years later he speculated in a letter to the evolutionary biologist Edward Bagnall Poulton (1856–1943) that 'A Theory of Heredity' had simply been written for the wrong audience in the wrong technical language. Even he had promptly forgotten it after he read it.[49]

> Frith Hall,[50] Godalming.
> January 20, 1889.
>
> My dear Mr. Poulton,—My attention has been called by Mr. Herdman, in his Inaugural Address to the Liverpool Biological Society, to Galton's paper on "Heredity," which I read years ago but had forgotten. I have just read it again (in the *Journal of the Anthropological Institute*, Vol. V., p. 329, Jan., 1876), and I find a remarkable anticipation of Weismann's theories which I think should be noticed in a preface to the translation of his book. He argues that it is the undeveloped germs or gemmules of the fertilised ovum that form the sexual elements of the offspring, and thus heredity and atavism are explained. He also argues that, as a corollary, "acquired modifications are barely if at all inherited in the correct sense of the word." He shows the imperfection of the evidence on this point, and admits, just as Weismann does, the heredity of changes in the parent like alcoholism, which, by permeating the whole tissues, may directly affect the reproductive elements. In fact, all the main features of Weismann's views seem to be here anticipated, and I think he ought to have the credit of it.
>
> Being no physiologist, his language is not technical, and for this reason, and the place of publication perhaps, his remarkable paper appears to have been overlooked by physiologists.
>
> I think you will find the paper very suggestive, even supplying some points overlooked by Weismann.—Yours faithfully,
>
> A.R. Wallace.

And so it was that August Weismann eventually received most of the credit for ideas that Galton had already worked out between 1872 and 1875.

[48] A.R. Wallace to Francis Galton, 1876-03-03, K. Pearson 1924, 187.
[49] Marchant 1916, 45–46.
[50] Frith Hill.

Figure 5.1: Down House, residence of Charles Darwin, now a museum. Galton was a regular visitor. Taken in 2022.

Figure 5.2: Charles Darwin, as seen by Vanity Fair.

Figure 5.3: Charles Darwin by John Collier, painted posthumously in 1883.

Figure 5.4: Francis Galton in his fifties, sometime in the 1870s, K. Pearson 1924. Restored.

Chapter 6

Intercomparable Scientific Priests

In the African wilderness, when confronting the robber chieftain Jonker Afrikaner, Galton had primed his self-confidence by dressing up in his full 'riding pink', an outfit he had for no sane reason transported all the way from England. Jackboots, a shotgun, and an old French dragoon's sword acquired in Egypt years before, completed the outfit. Thousands of miles from the Warwickshire hunt, horses were not to be had, so mounted on an ox—'not at all bad leapers'—he had barged his way into the *kraal* of Jonker, there to scold the captain for deeply disappointing conduct: murder, pillage and cattle rustling.

In May of 1873 Galton resorted to dressing up again, when persuading the staid fellows of the Royal Society to complete a circular he had prepared.[1] It asked a great many detailed questions about antecedents, politics, religion, disposition, personal tastes and other qualities, using a novel system of scoring that Galton would later call *inter-comparison.*

> The size of my circular was alarming. Though naturally very shy, I do occasional acts, like other shy persons, of an unusually bold description, and this was one. After an uneasy night, I prepared myself on the following afternoon, and not for the first time before interviews that were likely to be unpleasant, by what is said to have been the usual practice of Buffon before writing anything exceptional, namely, by dressing myself in my best clothes.
>
> I can confidently recommend this plan to shy men as giving a sensible addition to their own self-respect, and as somewhat increasing the respect of others. In this attire I went to a meeting of the Royal Society, prepared to be howled at; but no! my victims, taken as a whole, tolerated the action, and some even approved of it.

Galton was no longer hirsute enough to follow Count Buffon's supplementary practice of having his hair papered and curled the night before, but this deficiency does not appear to have dampened his reception.

By early June, the *London Medical Report* had heard about Galton's project, though they mangled his name and doubted his methods. 'ANTECEDENTS of SCIENTIFIC MEN.—Following out a line of inquiry suggested by the remarkably interesting recent work of M. Decandolle, Mr. Francis, F.A.S. [sic], has issued a schedule of minute and searching questions as to their personal and family antecedents to about 250 of the most eminent scientific men in the United Kingdom. The object is to set forth the influence, through which the dispositions of original workers in science have most commonly been formed, and have afterwards been trained and confirmed. The inquiry is one of much interest, and a considerable amount of curious information is sure to be obtained in this way. It is very doubtful, however, whether in so complex a problem, any certainly useful conclusion can be arrived at by this original method of inquiry'.[2] The *Belfast News-Letter* had more details, listing some extensive samples of Galton's 'rather droll'

[1] Galton wrongly cites 1874 in F. Galton 1908.

[2] *London Medical Record*: June 11, 1873, 367. Reprinted, minus the last skeptical sentence, in *Popular Science Monthly* 438 Volume XX 1873.

questions. They supposed that obtaining answers 'will be rather difficult' but brightened into 'what may not modern science accomplish?'[3]

The printed circular explained that the immediate impetus to the inquiry was a study by the Swiss botanist Alphonse de Candolle titled *Histoire des sciences et des savants depuis deux siècles* (History of science and scholars over two centuries), which had just come out the previous December.[4] The emeritus, already in the middle of a lengthy retirement, had sent Galton a complimentary copy of his book, since he had been motivated to write it by his reading of *Hereditary Genius.* In it, Candolle claimed a 'broader base and reach' than Galton in his coverage of scientific men, with 'a few hundred named instead of 65', all from 'more complete biographical documents' with a wider European scope. The Englishman, whose name he mentions repeatedly in his text, 'highlights heredity', but 'I do not see, however, that he gave the evidence'. The Switzer claimed a more nuanced treatment, taking other important factors into account without, he asserted, denying the role of heredity.

The Candolles were 16th century Huguenot émigrés to Switzerland from Provence. The French-educated Genevan Augustin Pyramus de Candolle (1778–1841), father to Alphonse, had established a botanical tradition in the family that would carry on down to the 20th century over at least 4 generations. The list given by Alphonse of factors favourable to the flowering of science might as well be a description of the Switzerland experienced by refugees of conscience like his own family.[5]

1. A significant number of persons belonging to the rich or independent classes relative to those who have to work for a living and especially to those who have to work by manual labor;
2. A large number of individuals among the rich and independent classes who are contented with their situation, who have a fortune to administer, and who as a result are disposed to concern themselves with intellectual things of little remunerative value;
3. A tradition of intellectual culture which has been directed for several generations towards real things and correct ideas;
4. Immigration of respectable families who have a taste for intellectual things which offer little remuneration;
5. The existence of several families which have traditions favorable to science and to intellectual occupations of all kinds;
6. Well-organized primary, and especially secondary and higher education, which is independent of political parties or religions and which tends to stimulate scientific research and to encourage youths and their professors who are devoted to science;
7. Well-organized facilities for the different kinds of scientific work (libraries, observatories, laboratories, museums);
8. A public interested in truth and reality rather than imaginary or fictitious things;
9. Freedom to state and publish all opinions, at least on scientific topics, without experiencing any serious harm;
10. Public opinion favorable to science and scientists;
11. Freedom to engage in all professions, or not to engage in any, to travel, and to avoid all personal service except that which is undertaken voluntarily;
12. Religion which has little reliance upon authority;
13. Clergy who are friends of education among their own members and among the public;
14. Clergy not tied down or celibate;
15. Common use of one of the three principal languages, English, German, or French. Knowledge of these languages sufficiently extensive among the educated classes;
16. A small independent country or a union of small independent countries;
17. A geographical location with either a northern or temperate climate;
18. Proximity to advanced countries.

[3] *Belfast News-Letter* Friday 27 June 1873, 3.

[4] Candolle 1873. Though it was actually published in December of 1872, it was a common practice at the time to post-date books so that they seemed more current. It has never been translated into English.

[5] Hilts 1975, 8, translating from Candolle 1873, 196-197.

Galton actually found himself broadly in agreement with this sort of reasoning, but took exception to Candolle's criticisms of *Hereditary Genius*. His own private reading notes were sometimes furious: 'most weak', 'did it ever occur', 'stuff', 'stuff & nonsense', 'a concession', 'some crass error here', 'all gleaned from my book', 'concession after concession' and so on. 'I don't say that genius is the sole cause of success—quite the contrary. I do not see that he has [refuted me]. I have taken much higher names, selected with great care, & studied their histories from all kinds of sources. His men as so unknown to fame that he can learn hardly anything about them'.[6]

The Englishman wrote directly to Candolle to complain that the Swiss had misrepresented *Hereditary Genius*. Far from invalidating his theories, Galton scolded, the botanist had 'largely supplemented them, by thoroughly working out a branch of the inquiry into which I never professed to enter, but I literally cannot see that your conclusions, so far as heredity is concerned, differ in any marked way from mine'. He allowed more rancour to show. 'I feel the injustice you have done to me strongly, and one reason that I did not write earlier was that I might first hear the independent verdict of some scientific man who had read both books. This I have now done, having seen Mr Darwin whose opinion confirms mine in every particular'. But he ended the letter emphasizing common ground. 'One of the most striking things to me in your book is the chilling influence on scientific curiosity you prove to result from religious authority'.[7]

Candolle was quick to reciprocate. 'Unless it escapes me, in the 482 pages of my book a sentence or even a word that may cast doubt on my respect for your impartiality, your character and your talent for inquiry, can absolutely only be in error and contrary to my intentions'. 'We are nicely in agreement about the facts. We have the same ideas about races. You have considered a larger number of categories of men. I studied the scholars in a more particular way, employing a different method, but with noticeably close results to yours'. But the Genevan defended the improvements he saw in his own treatment. 'You usually highlight, as the main cause, heredity. When you speak of other causes they are mentioned in passing without trying to unravel their particular importance, individually or in combination'.[8] Writing to Charles Darwin, he again held his ground. 'As a matter of fact, I do not think I have opposed Mr. Galton, but, in succeeding him, I believe I have completed his work by drawing more varied and complete conclusions from similar facts'.[9]

The botanist certainly believed that heredity was important. However he was, unlike Galton, a definite Lamarckian, adding to inheritance of 'acquired characters' his own peculiar belief that state of mind during copulation could be transmitted to offspring. This marvellous idea suggests, if it is true, that Candolle himself thought of flowered vines during the act, since his own progeny became, like their father and grandfather before them, eminent botanists. In a later letter to Candolle, Galton could not resist some satire. 'A strong instance (if accurately recorded) of alcoholism combined both with the evil influences of close interbreeding and of old age on the part of one of the parents, in producing no bad effect on the offspring, is that of Lot and his daughters (Genesis xix. 31)'.[10]

Galton's politely critical review of the *Histoire*, 'On the causes which operate to create scientific men', followed a few months later in the *Fortnightly Review* for March.[11] He complained to the public that in Candolle's work 'my name is frequently referred to and used as a foil to set off his own conclusions'. Still he was confident from repeated statements in the text that the author was 'my ally against his will'. Though he emphasized social causes of scientific eminence

6 F. Galton 1873c. See Tredoux 2023b for a reproduction of the review and for Galton's extensive notes.

7 Galton to Candolle, 1872-12-27. For a full treatment of all the surviving correspondence between Galton and Candolle, with translations, see Tredoux 2023b.

8 Candolle to Galton, 1873-01-02. Tredoux 2023b, 59.

9 Candolle to Darwin, 1873-01-14, DARWIN, DCP-LETT-8737.

10 Francis Galton to Alphonse de Candolle, 1873-05-07, Tredoux 2023b, 79.

11 F. Galton 1873c.

'in a most instructive manner', Candolle nevertheless calls out the primary influence of race on ability. He also recognizes that ability varies hereditarily within races even more than between them, allowing that mental ability is inherited in the same way that physical characteristics are. But Candolle's opinions are often vague, since the work is 'so unfortunately deficient in method, that the author's views on any point have to be sought for in passages variously scattered'.

According to Galton, the most valuable part of Candolle's study is really his investigation of those social influences which are most conducive to science *irrespective of natural ability or any theory of heredity.* The 'blighting effect of dogmatism' identified by Candolle seems undeniable. It is driven by religion, not so much by specific doctrines as by education in narrow habits of enquiry. This appears to be worse in Catholic countries but is also a feature of Protestantism. Innocent of cynicism, Galton reassures readers of the *Fortnightly* that there is nevertheless hope for harmony between science and religion, given the devotion of both spheres to ideas rather than material interests, and the fact that 'both work in a disinterested way for the public good'. He approves of Candolle's argument that classical education in Latin and Greek is stultifying for scientists. 'Nothing is brought out more clearly in the work than that the first desideratum in scientific education is to stimulate curiosity and the observation of real things, and that too much encouragement of the receptive faculty is a serious error'.

Though he finds much to admire in Candolle's *Histoire*, Galton concludes that the Genevan has not taken sufficient care to disentangle all the other influences at work, 'since he has collected next to nothing about the relatives of the people upon whom all his statistics are founded'. Still, even though Galton had forwarded a copy of all this criticism to Geneva, the Swiss did not reply with a broadside. 'If I favoured controversy I would reply to some of your observations, but that is a kind of discourse for which I have nothing but repugnance, and I prefer to rest on the many points on which we are in agreement'. If a second edition of his book is called for, considered responses can be given there.[12]

Now, in early 1873, the Englishman was motivated to take up this sort of 'sociology of science' with more empirical and statistical rigour. He would deal only with contemporary domestic subjects; his circular set the context. 'I have been engaged for some time past in collecting information on the Antecedents of Eminent Men. My present object is to set forth the influences through which the dispositions of Original Workers in Science have most commonly been formed, and have afterwards been trained and confirmed'.[13] He enclosed in the questionnaire a reprint of his review of Candolle's book. 'The result of my past efforts has clearly impressed upon me the fact that a sufficiency of data cannot be obtained from biographies without extreme labour, if at all; therefore, instead of imperfectly analysing the past, it seems far preferable to deal with contemporary instances, and none are more likely to appreciate the inquiry or to give correct information than Men of Science'.

Here Galton was going after the extreme upper echelons of scientific ability. He estimated that there were between 250 and 300 highly eminent scientific men in the United Kingdom, by which he meant a level equivalent to 'Medallists of the chief scientific societies; the Presidents of the same, now and in former years; those who have been elected to serve at various times on the Council of the Royal Society, and similarly, the Presidents of the several sections of the British Association'. If he got responses from half of these, he had half of the entire population with at least the required level of proven ability under study.

The survey came with detailed instructions introducing the method of 'statistics by intercomparison', using the 'standard scale of classes applicable to any quality or group of qualities, mental or physical, founded on the well-known law of deviation from statistical averages'.

> Qualifications to which no strictly defined meaning is attached, such as 'large,'

[12] Alphonse de Candolle to Francis Galton, 1873-04-29, Tredoux 2023b, 73.

[13] 'Antecedents of Scientific Men' circular, enclosed with Galton's proof copy of *English Men of Science.* GALTON/2/11/4.

> 'considerable,' necessarily convey different ideas to different persons, and are unfit to be used statistically. I therefore beg to recommend in their place a notation of Classes which I proposed, justified, and employed in my 'Hereditary Genius,' and which is explained in the following Table. The successive graduations divide the Classes and start from Mediocrity as a zero point, and they proceed both upwards and downwards. They are calculated to show equal increments or decrements of the quality in question. The Classes fill the intervals of, and are bounded by, the successive graduations.
>
> The amount of the quality in question, indicated by each graduation, is that which exists at the limit which divides a select body of individuals from the residue of the population out of whom they have been selected, on the ground of their possessing that quality in an exceptional degree, either of excess or deficiency. The population is understood to be limited to individuals of the same sex, age, and circumstances, and the severity of selection to be as is stated in the first column of the Table. [Table of normal distribution with grades, see chapter 4, p. 63.]

Obtaining more precise estimates of human traits in a form that could be treated quantitatively, was really a *central* aim of the survey. This is shown most clearly in the interim report Galton gave at the close of February, at a Friday evening meeting of the Royal Institution.[14] 'On men of science, their nature and their nurture' describes 'interesting considerations regarding the measurement of qualities', for which he argued there is a generalisable technique 'as applicable to magnitude as they are to intellect and morals, and to every form of animal or vegetable life as they are to men'.

The technique applies equally well, Galton argued, to 'the size of nuts, and peas, and acorns', as to 'the moral and intellectual qualities of mankind'. If a bag of random nuts is arranged in rank order, and the median nut is measured, another bag of random nuts would yield a very similar median. 'Hence, whatever be the length of the series the *relative position* in it of the nut will be a *strict criterion* of its size. This is of course equally true of all groups of qualities or characters whatever, in which the law of statistical constancy prevails, the series, in each case, being arranged according to gradations of the quality in question'. No precise *external measurement scale* is needed, beyond the ability to only rank individuals. Though nuts have obvious external criteria—diameter, volume or weight—other qualities do not.

To illustrate the all-important 'law of errors', Galton brought to the lecture theatre an ingenious mechanical device of his own invention (see Figure 6.2, p. 124). Behind glass, in a box, lead shot was released through a funnel at the top. The shot was interfered with as it fell by an orchard of pins—each pin deflecting the shot left or right with an equal chance—and captured at the bottom in bins. The result was exactly the bell-shaped distribution of shot predicted by the law of errors—strictly speaking, the discrete binomial version of it. Landing up in either tail requires a sustained sequence of deflections in the same direction, which is unlikely. He called this device the 'Quincunx', after the distinctive orchard pattern of the pins, a square of four with a fifth in the middle, as in the 5 on a die, or on Roman bronze coins. In coming years he would make extensive use of elaborations of this device to show how the statistical ideas suggested by his investigations of heredity worked in practice.[15]

---

[14] F. Galton 1874f, reproduced in Tredoux 2023a.

[15] The published version of the lecture in F. Galton 1874f does not mention this wonderful instrument, mentioning only that '[this was illustrated by many experiments]', but the demonstration is recalled in F. Galton 1877d. The first version of the Quincunx was built by the opticians Samuel Charles Tisley & George Spiller of 172 Brompton Road in South Kensington, according to Galton's handwritten notes on the machine which survives in the Galton Collection at UCL today. The instructions on the extant device read 'Instrument to illustrate the principle of the Law of Errors or Dispersion. Francis Galton, FRS. Charge the instrument by reversing it, to send all the shot into the pocket. Then sharply re-reverse and immediately set it upright on a level table. The shot will all drop into the funnel, and running thence through its mouth, will pursue devious courses through the harrow and will accumulate in the vertical compartments at the bottom, these affording

It was an idea important enough to Galton's thinking to justify reiteration in a letter 'On a Proposed Statistical Scale' to *Nature* a week later.[16] Galton asked if readers could think of any universal method of measurement not based on the 'law of errors'. He hoped that by raising the topic he would provoke further discussion. And indeed, it would later become clear that the Normal Distribution is not essential to this method, which works with *any* cumulative distribution.

The reaction by the *Spectator* was not what Galton had in mind. The magazine scorned at the idea, under the heading 'Nuts and Men'.[17] Rhetoric rose to the fore. 'Is it possible to conceive a quainter perversion of a scientific method, really applicable to magnitudes, to the purpose of a description of moral qualities?' Biographers would have to rank their subjects on *all* their qualities, an exhausting exercise. 'Mr. Galton's attempt to give a scientific turn to biography would spoil literature, without enriching science'. Some sharp condemnations of classical education by the leading scientific men, reported approvingly by Galton in his lecture, had cut deep. Thus 'how helplessly men with the physico-scientific bias upon them flounder about in the region of human and moral influences'. Though 'S.S.' might stand for 'Statistical Scale', the critic thought it was more likely to become known as the 'Sham Science' scale. Along similar lines, the *Academy* dismissed Galton's 'mere collection of statistics' since 'inferences based upon statistics must always be misleading, unless the enumeration of instances be so vast as to eliminate minor causes of disturbance'. They hoped that scientists would receive an even more thorough dose of classical education to correct their thinking. They feared that Galton did not understand the principles of induction.[18] Neither publication could foresee that this 'sham' and supposed misunderstanding would be universally adopted in the next century.

A few months later, the anthropologist Edward Burnett Tylor (1832–1917) wrote in to the *Spectator* to deplore its thoughtless braying, arguing that Galton had simply assumed the mantle of the great Belgian statistician Adolphe Quetelet, who had a similar 'mania' for measuring everything. The Belgian had initiated the statistical treatment of moral qualities, even if that neglected poetry. Both men understood that averages were not enough, variation had to be taken into account too.[19] The *Spectator* simply shrugged. Both Tylor and Galton contributed to the handbook *Notes and Queries on Anthropology For the Use of Travellers and Residents in Uncivilized Lands* issued that year. In an appendix, Galton provided his first extended description of his general method of inter-comparison.[20]

Only one F.R.S.—'of high scientific distinction' and already conversant with Galton—actively took offence at his long list of questions.[21] He got enough data out of the other fellows, with 104 replies received from 191 circulars sent, to support a book. *English Men of Science* duly appeared on the 5th of December 1874. But Galton's scheme for scoring qualities by degrees of 'the law of errors' proved too daunting for most respondents. Charles Darwin had replied early in May 1873, 'I have filled up the answers as well as I could; but it is simply impossible for me to estimate the degrees'.[22] George Bentham needed a lengthy consultation on the 1st of June with Galton over the schedule, and recorded in his diary that he had only filled it up 'with some difficulty'.[23] Perhaps this onerous requirement deterred the Zoologist George Busk, who had to be reminded just before Xmas of 1873 that a prompt reply would be 'peculiarly

a representation of the law of dispersion'. Chris Pritchard has argued that the Victorian game of bagatelle, a kind of pinball, may have served as an inspiration for the device, C. B. Pritchard 2006.

16 F. Galton 1874e.

17 *Spectator*, 1874-05-16, 623-624.

18 *Academy* 1874-05-09, 521.

19 Tylor 1874.

20 F. Galton 1874b.

21 F. Galton 1908.

22 Darwin to Galton 1873-05-28, DARWIN, DCP-LETT-8924.

23 B. D. Jackson 1906, 227.

acceptable', since 'I have nearly finished my first & principal batch, & want sadly to test general conclusions—thus far provisional—by wider induction'.[24] The X-Clubman never got around to it. By contrast, the surgeon Sir James Paget had already sent 'full and very instructive replies', encouraging the author to think that 'the circular has been so largely responded to that the fault will soon be with me, if several instructive facts are not established by means of what has been to obligingly communicated to me'.[25] Nonetheless, in the end Galton did not attempt any formal analysis based on graded traits. 'I have by no means abandoned its advocacy, but have learnt the necessity of explaining and exemplifying it in considerable detail before its merits and convenience are likely to become as generally recognised as I believe they deserve to be'.[26]

A total of just over a hundred detailed replies to 'Antecedents of Scientific Men' may seem small. But what a hundred! Nearly all the members of the 'X Club' of Darwinians-without-Darwin were there: the anatomist T.H. Huxley; the botanist Joseph Dalton Hooker; the anthropologist Sir John Lubbock; the mathematician William Spottiswoode; the chemist Edward Frankland; the geometer Thomas Archer Hirst; and the philosopher and sociologist Herbert Spencer. George Busk was the only 'X-Club' fellow who eventually failed to reply. That Darwin-without-the-X-Club, cousin Charles, went beyond the questionnaire and communicated several paragraphs of autobiographical explication. His submission survives but almost all of the others were returned to the subjects. Among those were the geographer George Back, the botanist George Bentham (see Figure 6.3, p. 124), the ophthalmologist William Bowman, the algebraist Arthur Cayley, the geologist John Stevens Henslow (posthumously), the engineer Fleeming Jenkin, the statistician William Stanley Jevons, the physicist James Clerk Maxwell, the anatomist Richard Owen, the chemist Lyon Playfair, the physiologist John Scott Burdon-Sanderson, the botanist Philip Sclater, the mathematician Archibald Smith, the astronomer Piazzi Smyth, and many others now less well-known but then at the forefront of Victorian science.[27]

Galton knew many of the men surveyed extremely well. Most members of the X-Club were personal friends. William Spottiswoode had once planned a joint exploration of the Sinai with Galton, which had to be abandoned at the last minute, and had been a close colleague at the Royal Geographical Society, where they were both honorary secretaries in the early 1860s. Bowman had accompanied young Frank on a 'mediculising' trip to Europe during the summer of 1838, when the future explorer was still an aspiring medical student. Cayley had led a Cambridge reading party including Galton to the Scottish Highlands in 1842. Archie Smith, recently deceased, was a sporadic dinner guest and perhaps a victim of Louisa Galton's frugal housekeeping at 42 Rutland Gate. Frank was also a not-infrequent visitor to Down House and his closeted cousin Charles, who leaned on his guest for statistical advice.

Names of individuals were kept confidential and their disclosures were (sometimes thinly) disguised in the published text. In 1975 the historian of science Victor Hilts was amazed to discover a key to the book nestling in 'Galton's Desk', a venerable piece of furniture which had inhabited the Statistics Department at University College London for decades but had recently been transferred to the university archives. But unknown to Hilts, Galton's annotated copy of the book with attributions and corrections, prepared for a second edition which never materialized, also survives.[28] Together these allow almost all quotes in the book to be identified,

[24] Francis Galton to George Busk, 1873-12-20, photocopy possessed by the author.

[25] Francis Galton to Sir James Paget, 1873-09-27, APS.

[26] F. Galton 1874c, viii, 261.

[27] The complete list of scientists is given in Appendix D, p. 421.

[28] See GALTON/2/11/4/1, GALTON/2/4/2/1 and Hilts 1975. 'Galton's Desk' is today the largest object in the Galton Collection of apparatus at UCL, which includes one of his bookcases and many gadgets. It is described in the catalogue as 'Desk belonging to Galton. Mahogany desk with 4 small drawers on each side and drawer across centre. Red leather top. On wheels. 2 person desk'. For a picture see `http://galtcat.museums.ucl.ac.uk`.

adding tremendous interest for the modern reader. We find out that the physicist James Clerk Maxwell shared Charles Darwin's backbone-shivering nerve for music. 'Strongly affected by music when a child, could not tell whether it was pleasant or painful, but rather the latter; never forget melodies or the words belonging to them and these run through the mind at all times and not merely when the tunes are in fashion; can play on no instrument and never received instruction in music'.[29]

The full questionnaire the scientists responded to was comprehensive and made up seven quarto pages.[30] It asked for information about

- Genealogy—e.g. origin, family size;
- Physical traits—e.g. age, height, build, hat size, hair colour, complexion, eye colour, physical energy, health and robustness;
- Mental traits—e.g. temperament, especially independence of thought and influence of religion, politics, mental energy, business ability;
- Education.

Respondents were asked to self-assess the importance of hereditary or innate properties on their formation, likewise for educational and within-family influences. Unavoidably, many questions relied on personal judgment as to personality and so on. It reached beyond the recipients themselves to include most of the same questions for their parents and the more remarkable of their relatives, as well as their spouses. For these, personal judgment at one or more removes was called for, but it allowed Galton to investigate novel topics. Recipients were also invited to volunteer extra information they thought relevant.

The harvest of survey returns was analyzed in *English Men of Science*, subtitled 'Their Nature *and* Nurture' (emphasis added). The subsequent long-running controversies over nature *versus* nurture easily obscure comprehension of this. Certainly Galton was partly interested here in weighing the relative contributions of nature versus those of nurture, but he was also picking up Candolle's study of the social and cultural conditions favourable to the development of science *irrespective of any theory of heredity*, the theme he had isolated in his review of Candolle.[31] The bulk of the text is given over to this latter theme, the nurture of science. But first Galton carefully describes his understanding of the relative roles and operation of nature and nurture.

One subtlety follows from Galton's usage of 'heredity' as distinct from 'nature'. One can be born with a character, and thus have it by nature, despite it not appearing in a forebear, so that it is not 'hereditary' in Galton's meaning. Modern usage has '*broad-sense* heritability' for his 'nature', and '*narrow-sense* heritability' for his 'hereditary'. His careful treatment of the interplay between nature and nurture shows a clear grasp of issues that over a century of subsequent developments would confirm, despite an avalanche of obscurantism.[32]

> The phrase "nature and nurture" is a convenient jingle of words, for it separates under two distinct heads the innumerable elements of which personality is composed. Nature is all that a man brings with himself into the world; nurture is every influence from without that affects him after his birth. The distinction is clear: the one produces the infant such as it actually is, including its latent faculties of growth of body and mind; the other affords the environment amid which the growth takes place, by which natural tendencies may be strengthened or thwarted, or wholly new ones implanted. Neither of the terms implies any theory; natural gifts may or may not be hereditary;

[29] Hilts 1975, 59, Colp 2008.
[30] Roughly letter-size in modern US terminology, or 11.25 by 8.75 inches.
[31] F. Galton 1873c.
[32] F. Galton 1874c, 12-16, emphasis added.

> nurture does not especially consist of food, clothing, education or tradition, but it includes all these and similar influences whether known or unknown.
>
> When nature and nurture compete for supremacy on equal terms in the sense to be explained, the former proves the stronger. It is needless to insist that neither is self-sufficient; the highest natural endowments may be starved by defective nurture, while no carefulness of nurture can overcome the evil tendencies of an intrinsically bad physique, weak brain, or brutal disposition. Differences of nurture stamp unmistakable marks on the disposition of the soldier, clergyman, or scholar, but are wholly insufficient to efface the deeper marks of individual character. The impress of class distinctions is superficial .... *In the competition between nature and nurture, when the differences in either case do not exceed those which distinguish individuals of the same race living in the same country under no very exceptional conditions, nature certainly proves the stronger of the two.*

The emphasized statement at the close is crucial: the relationship between nature and nurture is contingent, and a property of the world we, or at least most of us, happen to live in. It is easy to construct a thought experiment in which nurture overrides nature decisively. Imagine that at birth all newborns are automatically and equally lobotomized to produce only the sort of ability required for, say, journalism. Whatever each child brings into that world counts for nothing. Given enough ill-will, such a world could certainly be organized, and the required technology has long existed; the difficulties in constructing it are only political, not biological. Most will be grateful that we do not live in such a world. Galton recognized that our task is to find out how the world we *do* inhabit happens to be organized, and not to reason hazily about some other world that we do not inhabit; to avoid confusing *what is* with wishful thinking about what supposedly *might be*.

Galton chose to present his analysis of the returns in terms of (1) 'Antecedents', (2) 'Qualities', (3) 'Origin of Taste for Science' and (4) 'Education'. The last three sections took up the bulk of his pages and may give the impression that the book is mainly a thematic collation of autobiographical opinions. This is not so.

The first section on 'Antecedents' is an extension of the techniques used in *Hereditary Genius*, and the very first statistical investigation of Darwin's new theory of sexual selection, as applied to humans, ever conducted. The returns showed that his scientists had 660 grandfathers or uncles, of which fully 13 were eminent. Since he had previously established from the obituary lists and other sources that the entire kingdom only produced 50 men a year who met a similar standard of eminence, this was an over-representation by many orders of magnitude. From another angle, considering only the graduates of Oxford and Cambridge, Galton estimated that only 3 out of 660 men of similar eminence might be produced. Alternatively, taking only the public schools into account, he estimated that less than 1 in 660 would be at least as eminent. So the cause of the over-representation could not simply be the outcome of university or elite education per se. Similarly, there were 1450 brothers and male cousins of his scientific men. Five of those were first in their year at Cambridge in either classics or mathematics, a frequency of 1 in 300—far in excess of the general population.

Extending another technique from *Hereditary Genius*, Galton calculated that the proportion of eminence, as estimated by the respondents, declined as the 'blood relationship' weakened, from father or brother to grandfathers and uncles (a dilution independent of the personal standards of eminence his respondents were using, assuming that each was at least consistent). He compared this to the relationship he had calculated for 'divines' in 1869, expressed as percentages *of the total number of such relatives* (see Table 6.1).[33]

[33] In the text Galton actually gives the percentages *per family* (column B in *Hereditary Genius*, F. Galton 1869a, 61) so these have been converted using the estimates for numbers of relatives per family (column C there) of

Table 6.1: Eminence by distance and field.

	Scientists	Divines
Fathers	28%	22%
Brothers	24%	28%
Grandfathers	10%	14%
Uncles	10%	10.5%

It is this line of enquiry that Galton must have been 'leisurely engaged' on, as he put it, when Candolle's book gave the impetus to *Men of Science.* He was able to add eight new scientific families and their pedigrees to the list he had already investigated in *Hereditary Genius.* But the most interesting information in his returns concerned something else: sexual selection, which Darwin had brought to prominence in *The Descent of Man* a few years earlier.[34] Since he now had substantial information about the characteristics of the mothers and fathers of 62 of his scientific men, Galton was able to analyze assortative mating statistically. Height was the most straightforward physical trait to investigate. He argued that if the heights of mothers and fathers co-varied—though he did not use that word—then the *sum* of those heights would have a markedly different 'probable error'.

It turned out that the probable error of the sum was roughly equal to the sum of the probable errors. So Galton concluded that there was no substantial association between the heights of mothers and fathers. There was no definite sign in his data that either matching or contrasting heights had been sexually selected for in the generations which produced his scientific men. A modern analysis would directly calculate the covariance from the heights for father and mother pairs and divide that by the standard deviations of those heights, giving a form where a value close to zero means a weak association (and is not an artefact of the scales involved). We do not strictly need the original individual heights. This yields a correlation of roughly 0.19 (see Appendix F, p. 431, for the details). It would be 1888 before Galton himself realized that this formulation is the key, but he was already reaching for that discovery 14 years earlier.[35] In 2017, a meta-analysis of 154 correlation studies of assortative mating for height found 148 positive correlations for partner heights and comparable positive assortative mating overall, $r = 0.23$.[36]

Galton was also able to investigate how well parents matched on other characters, such as

1. Temperament—nervous, sanguine, bilious or lymphatic; 22 cases;
2. Hair color—seven classes from black to fair; 72 cases;[37]
3. Physical build—corpulent, muscular, compact, spare and medium; 71 cases.

Here he simply counted the cases he considered in 'harmony' and compared them to those in 'contrast' and those he classed 'indifferent'. Harmony outweighed contrast for temperament 10 to 2 and for hair colour (44 to 6), but not for figure (24 to 23). His methods are inadequate, but it would be fifty years before better ones were invented (see Appendix F, p. 431, for more analysis).

Most of Galton's scientists were of English origin, by at least five to one. This is less impressive than it may seem, given that the 1801 census counted about 8.3 million residents of England to about 1.6 million in Scotland, and only 0.5 million in Wales. Tracing their birthplaces on the map with smoothing to show concentrations—a recurring technique for

1 father, 1.5 brothers, 2 grandfathers, 4 uncles. Galton was bashful about these estimates in 1869 and had not improved on them, so stuck to the per family estimates here, but that is hard to make sense of.

34 C. Darwin 1871a.

35 F. Galton 1888c.

36 Stulp, Gert and Simons, Mirre J. P. and Grasman, Sara and Pollet, Thomas V. 2017.

37 Galton omitted cases with other colors like 'sandy' and 'auburn' that were hard to compare.

Galton which underlay many of his discoveries—suggested that they came from a narrow strip of the country (see Figure 6.4, p. 125).[38]

Karl Pearson would later argue the area traced in this map corresponds more or less to the parts of the UK where the broader population was concentrated. This appears to be true, taking 1801 as the approximate year of birth, as Figure 6.5, p. 125 suggests for the England and Wales census. However it is certainly possible that areas like East Anglia are under-represented among Galton's men of science.[39]

Several of the questions in the survey were designed to solicit the attitude of the scientists toward religion and its effect on their inquiries. 'To what religious body or bodies have You, your Father, and your Mother respectively belonged'. 'Was either your Father or your Mother descended from persons persecuted for political or religious opinions, or from political or religious refugees? If so state the precise relationship. Mention whether their political or religious opinions became traditional in the family'. They were asked to describe their 'Independence of judgment in social, political or religious matters. Give illustrations'. Again, under 'Strongly marked mental peculiarities bearing on scientific success', Galton suggested that they volunteer their 'religious bias of thought'. These questions try to get at religiosity as an inherent character trait—nature rather than an accident of nurture—and he analyzed his results in the same framework as mental and physical qualities like head size, height, energy, business sense, general health, memory and the like.

As the author expected, and Candolle would have predicted, forebears from non-conformist sects were overrepresented. More than a few of the scientists confided that they had wrestled free from their religious nurture—'from my own reflection' as Darwin put it. But the other results were rather unexpected. About 20% of respondents happily professed *overt religious bias*. The preeminent physicist James Clerk Maxwell said that he was 'Given to theological ideas, and not reticent about them'. The mathematician Robert Main (1808–1878) wrote that he was familiar with the books in his own 'considerable theological library'. The botanist James Bateman (1811–1897) was dogmatic, having 'no more doubt about the plenary inspiration of Scripture' than he did about 'the simplest axiom in mathematics'. Another botanist and geologist, Thomas Cobbold (1828–1886), described himself as 'intensely religious'. Even T.H. Huxley, that professed agnostic and ultra-Darwinian X-Clubber, privately admitted to a 'profound religious tendency, capable of fanaticism' though tempered by skepticism. Many, including Allman, Egerton, Thomson, Williamson, Parker, Gunther, Mivart and Piazzi Smyth, asserted that religious feeling helped rather than hindered them in their science. This especially puzzled Galton. 'I had expected it to have been much more deterrent and hindering than the answers warrant'.[40]

Religion had drawn Galton in an unusual public controversy just two years earlier, when he had published a piece in the *Fortnightly Review* sketching some 'Statistical Inquiries into the Efficacy of Prayer'.[41] It was a piece he had worked up as early as 1866, but the *Fortnightly* had then refused to publish it. G. H. Lewes had only recently been appointed editor. 'I think your paper would [cause] such an outrage as to place me in an awkward position with our proprietor. I am already in the ticklish position of having Xians [Christians] for proprietors being myself avowedly not a Xian'.[42]

The impetus then for the inquiry had come from a controversy sparked by the Anglo-Irish physicist John Tyndall, a close friend and frequent walking companion of Galton, who

---

38 F. Galton 1874c, 20.

39 K. Pearson 1924. Source `https://www.campop.geog.cam.ac.uk/research/projects/occupations/britain19c/population/populationdensity1801.png`.

40 F. Galton 1874c, 135.

41 F. Galton 1872b. Reproduced in full in Tredoux 2023a.

42 G. H. Lewes to Francis Galton, circa 1866, GALTON/2/4/1/4/13, where Lewes is mistakenly catalogued as 'Lewis'.

had written forcefully in the *Pall Mall Gazette* that praying for deliverance from that Indian import, cholera, as encouraged by the *Gazette*, suggested an unscientific belief in miracles. He fell upon the *Fortnightly* next, ruling that those invoking prayer were 'putting the intellect to shame'.[43]

This controversy had flared up again in 1872 when Tyndall applied his bellows, forwarding a letter to the *Contemporary Review* from an anonymous physician (actually Sir Henry Thompson), who suggested that those who believed that prayer worked could try it systematically on hospital patients to find out.[44] Now the *Fortnightly* under John Morley was bold enough to publish Galton's piece.[45]

Several readers of Tyndall and Thompson's animadversions had pointed out that prayer motivated by statistical curiosity hardly counted as pious or sincere. Galton's study improved Thompson's experimental design, using observed data on longevity instead.[46] Since 1844 the statistician Dr. William Augustus Guy had been collecting and publishing the expected age of people who reached at least 30 years for many professions, between the years 1758 and 1843.[47] Included were 97 members of royal houses, who, Galton noted, were prayed for generally and repeatedly, and not as an experiment. Yet Guy's figures showed that their average age upon reaching 30 was only about 64 years, much lower than professions like the law—surely seldom prayed for—where 68 years was expected from a collection of 294 cases. The Gentry, of whom Guy had 1,632 cases, expected 70 years of life (in those naive days, standard deviations for these figures were not given). Galton included many other less-statistical arguments against the worth of prayer, such as the fact that insurers do not give special rates to the pious, that doctors had failed to notice the healing powers of a much-needed intervention, that missionaries to the tropics expired regularly in succession, and that religious corporations appeared to have no business advantages.

Unsurprisingly, letters flew to multiple periodicals. The *Spectator* took aim at Galton, describing his method as 'physicism gone wild, a direct attempt to weigh mental consequences in a pair of brass scales', which did no more than try their patience.[48] The blow was blunted a bit by repeatedly calling the scientist 'Captain Galton'—they were forced to confess that they had 'stupidly' confused him with his cousin Douglas Galton the engineer. In a letter to the editor, Galton thanked them for the civil discussion, but did not buy new-found skepticism about statistics by reformist theologians, who 'complacently ignore that their predecessors have been beaten along their whole line by statistical inquiries, for it is by more or less unconscious use of statistics that the belief in ordeal, duel, augury, and the rest, has disappeared, and now that theologians are summoned on statistical grounds to surrender a belief which I have shown to have much less claim to be considered as intuitive, they start with naive indignation, as at a previously unheard-of and most unreasonable interference'.[49]

Writing from the Athenaeum Club, Sir Henry Thompson was enthused—but not enough to drop anonymity. 'As a contribution towards its solution, it is impossible to overrate Mr. Galton's laborious and scientific record relating to the subject, nor to overlook its importance. Had I done nothing more than elicit the production of this last work of his, I should have been amply content'.[50] Privately, Darwin offered encouragement. 'What a tremendous stir-up

[43] R. Jackson 2018; Tyndall 1865a,b.

[44] Thompson 1872b.

[45] In a rare slip, K. Pearson 1924, 131 claims that the paper was rejected by James Knowles of the *Contemporary*, but Knowles was rejecting 'Hereditary Improvement', later published as F. Galton 1873b—see Knowles to Galton, Nov. 1872, GALTON/3/3/11/8. Pearson also claims that George Grove at *Macmillan's Magazine* rejected 'Statistical Inquiries' but this could not be verified.

[46] A rare case in which untainted observational data may be better than experimental.

[47] Naturally Galton would send Guy an *Antecedents* questionnaire, to which he responded.

[48] *Spectator*, 1872-08-03.

[49] F. Galton 1872d. See Tredoux 2023a.

[50] Thompson 1872a.

your excellent article on prayer has made in England & America'.[51] Tyndall and Galton found other defenders but many detractors—notably Astley Cooper, the Duke of Argyll and James McCosh of Princeton—generating enough public interest on both sides of the Atlantic to justify a compilation *The Prayer Gauge Debate*, published a few years later.[52]

Given this connection between Galton and Tyndall, it is odd that the Irish physicist did not respond to the 'Antecedents' survey, though he was certainly sent it. Perhaps—we do not know—the Carlow man was that touchy individual recalled but not named in Galton's memoirs, 'a man of high scientific distinction, with whom I was well acquainted, who was of good social position, of whose family many details were already known to me, all of which were honourable, and whose biography has since disclosed no skeleton in the cupboard', but who 'was almost furious at being questioned'.[53]

Many of the other natural qualities of the respondents and their forebears analyzed by Galton are familiar today. These ranged over: physical height, which turned out to be nearly four inches taller than the population average, and half an inch more than their parents;[54] circumference of head, slightly larger than average but apparently negatively related to 'energy'; health, unusually good; mental perseverance, very good; memory, often but by no means always good; and independence of character, almost without exception. The head measurements of the scientists were verified by a whalebone hoop from their hats, the rest were self-reported.

In an unpublished notebook titled 'Special Peculiarities', Galton mused over the apparent relationship between height and head size in his scientists.[55] On closer examination this proves to be the world's earliest regression diagram. He has drawn the bivariate normal distribution, with the marginal distribution for each component. It has the characteristic ellipsoid shape that he would become so familiar with in later years. A regression line with a negative slope has been tentatively sketched in to follow the axis of the ellipsoid. See **??**, p. **??**.

Claims of excellent physical health, both in the scientists and their relatives, were overwhelming. Curiously, Galton did not publish James Clerk Maxwell's report that he was 'Often laid up before aet. 19, never since; never had a headache'. A wise omission in retrospect since, as Galton noticed, the infirmities that the respondents did report were often hereditary. Maxwell was fine until age 48, then died suddenly of abdominal cancer, as his mother had before him at the same age. A few months before he had advised Galton on the right choice of colours to achieve greatest separation to the eye of a reader.[56]

In his unpublished notes, Galton tabulated responses which expressed a 'marked' tendency by a subject to have particular qualities which we would associate with something like personality today.[57]

Table 6.2: Special Peculiarities.

Marked Quality	Count
Practical business habits	39
Steadiness	35
Curiosity about facts	26
Mechanics	26
Music	18
Religious bias	17

51 Charles Darwin to FG, 1872-11-08, DARWIN, DCP-LETT-8608.
52 Means 1876.
53 F. Galton 1908, 292-293.
54 F. Galton 1874c, 102.
55 'Tabulated Data for *English Men of Science*', GALTON/2/4/2/3.
56 James Clerk Maxwell to Francis Galton, 1879-09-04, GALTON/3/2/2/9. Maxwell died on November the 5th.
57 'Special Peculiarities', GALTON/2/4/2/2.

Disinterestedness	16
Social affections	14
Love of pursuit	14
Love of new-marvellous	14
Mathematics	14
Public spirit	12
Foresight	8
Constructiveness of imagination	7
Impulsiveness	5
Strong feelings	5

Special qualities like 'energy', that property which Galton thought might be negatively associated with head size, and 'practical business habits', have long been buried beneath notice but are worth disinterring.

There appears to be no modern equivalent for 'energy', which encompassed both mental activity and physical exertion. Galton described it, impenetrably, as 'secretion of nervous force'. Excessive amounts seem to produce something like hydraulic pressure, in which activity produces equilibrium and repose instead of fatigue. Nevertheless, judging from their submissions, the scientists themselves had no trouble understanding what he meant, and really were the 'vigorous animals' that Galton had supposed them to be in *Hereditary Genius.*

Secretions of nervous force fill up fully eighteen printed pages in *English Men of Science.* Strenuous travelling (Forbes and Hooker), all-day fishing and shooting (Gunther), adeptness at all field sports (Lawes), athletic excellence at Cambridge and endurance in the army (Alderson), pedestrianism, mountaineering and 'indefatigable deer-stalking' (Egerton), regular fifty-mile hikes without fatigue (Bentham), frequent twenty mile hikes before breakfast and mountaineering (Ramsay), hundred mile rowing exertions and rough travelling (Warrington Smyth), all-day rough shooting (Paget), unwearied running-walks in streets (Harcourt), prize-fencing and great adroitness at games and feats of agility in the navy (Archibald Smith), habitual day and night travel (Ansted), superior competitive running, leaping and walking (Hill), long pedestrian tours and strenuous mountaineering in the alps (Richard Owen), prodigious pedestrianism (Phillips), superior long-jumping at college (Cobbold), great endurance of fatigue when walking (J. H. Balfour), strong walking and rowing (Merrifield), extended daily dissection over multi-week stretches (Ferguson), excellence at cricket and rowing (Scott), superior fencing, swimming and high jumping (Haughton), prodigious mountaineering and pedestrianism (Brooke), uninterrupted work and pedestrianism (Prestwich).

Only a few confessed to what they thought was unexceptional bodily energy (Sanderson) or languor (Currey), and several of the rest who did so were controverted by Galton's own information (Herbert Spencer, J. Burdon-Sanderson and William Stanley Jevons).

Mental energy almost invariably accompanied physical energy and well-being. All this activity went far beyond purely published output. Many of the scientists had demanding business careers and, after long days at the office, kept going at night into the small hours, sometimes right through, at the activities they really enjoyed (Main, Farre, Andrews, Prestwich, De La Rue, J. Fred Bateman, J. H. Balfour, Cobbold, Lawes). Egerton said he 'never knew what it was like to be tired'. For Parkes, 'mental fatigue' was 'a sensation not known'. Cobbold claimed to be 'almost incapable of [mental] fatigue up to the age of thirty-eight'. Merrifield's freakish feats of longhand writing—thirty folios of seventy-two words or 2160 words an hour—must count under both mental and physical energy. De la Rue 'never thought of rest if I had anything in hand of interest'. Some confessed to constant restlessness (Forbes, Paget) but found 'wonderful relief in science' (Prestwich).

The surgeon Sir James Paget suspected that energy was the real characteristic of his co-respondents, not science per se. Shortly before the publication of *English Men of Science* he ran into Galton at one of the exclusive dining clubs of the Royal Society. 'I have had neither reading nor much writing, and only on Thursday any unusual amusement. That was at the Philosophical, where we had a pleasant dinner, and I sat between Hooker and Francis Galton. Galton tells me that some very striking facts come out of his enquiries about scientific men. One, especially, is their very high average of energy and capacity in mere business; the last thing the "vulgar" would expect of those whom they scoff-at as "philosophers." He hopes to publish his results in two or three months; but they will, I expect, chiefly show the need of similar enquiries among several other groups of successful or distinguished men: for I suspect the energy is usually the fact, the scientific pursuit the accident, of the successful career'.[58]

Allowance must be made for some degree of boasting and self-delusion in all these self-reports. Mental breakdowns of the kind that Galton himself, Herbert Spencer, and the non-respondent John Tyndall had suffered, were hardly unknown. Though the brewer-astronomer Richard Carrington declared his health 'excellent', the *Dictionary of National Biography* discloses that a 'severe attack of illness in 1865 left his health permanently impaired'. This may have been a stroke: he died in 1875 of a brain haemorrhage. But the self-reports are, on the whole, strongly corroborated by the prodigious careers of the subjects.[59]

We have already mentioned Galton's surmise that energy was negatively associated with the hat measurements he had solicited and verified. The smaller-headed among his somewhat large-headed scientists had an advantage in energy. His unpublished notes show the basis for the claim. Only upper and lower extremes are considered, to accentuate whatever difference in means might exist. He did not push the point too far given the limited (but suggestive) evidence. In Table 6.3 he has rated himself using a shaky E, a reference to his 'regular breakdowns', and perhaps Galton was a little envious of the uncomplaining sheer efficiency of his subjects, though his own output would eventually shade the bulk of those on the list.[60]

Table 6.3: Energy by Head Size.

>= 24 in.	Energy
Gassiot	
Grove	
Hirst	
Lord Hatherley[61]	
Osborn	E
Strachey	
Stokes	E
[7]	2

<= 22 in.	Energy
Haughton [S.]	E
Sabine	E
Richards	E
F. Galton	E (shaky)
Hooker	E

[58] Paget 1901, 268.

[59] 'Carrington, Richard Christopher' in Lee et al. 1901, 9, 181–3. Victor Hilts first noticed this, see Hilts 1975, 25. Carrington's biographer Stuart Clark does not shed any more light on this illness, Clark 2007, 119, and shows no awareness of the invaluable autobiographical material about his subject in Galton's survey, first recovered by Hilts.

[60] 'Special Peculiarities', GALTON/2/4/2/2. See chapter 3, p. 29, and Tredoux 2023c for Galton's history of breakdowns.

[61] W .P. Wood.

Lubbock	E
Parkes	E
Cobbold	E
Bastian	E
Bateman J F	E
Galton Douglas	E
Smyth War.	E
[13]	12

The sample size is so small that the data can be no more than suggestive and invite more investigation using representative data. The ratings of 'energy' are self-described but imprecisely defined. But for what it is worth, Fisher's exact permutation test (described in Appendix F, p. 431) rejects independence of energy and head size, p = 0.007224.

Today we might describe all this 'energy' as 'productivity' but that is a *consequence* of the quality rather than the quality itself, which may be a latent trait underlying multiple surface traits. These elite men of science appear to have been tightly wound, and the expressions of their 'nervous force' spilled over in multiple directions, some of which were not productive at all. It also overlapped considerably with that other quality which interested Galton, 'practical business habits'. These were far more prevalent than he had expected. At least 39 men claimed to have them, a plurality for any of the qualities he surveyed. This surprised Candolle, who suggested after seeing the preliminary results that no such association would be found on the continent.[62]

A continuum existed. There were many like Charles Darwin, disguised in the text as 'a most eminent biologist', who kept accounts and conducted his affairs methodically, corresponded efficiently and was a canny investor. Or they thrived in administrative settings, like Huxley and Captain Evans. When specially queried, the biologist Philip Lutley Sclater volunteered that his 'organ of order' was highly developed, driving him to constantly rearrange his collections. Then there were around twenty men who were outwardly heads of business firms by day and only scientists by night and on weekends. The anthropologist John Lubbock was, like the statistician William Newmarch, a banker. Sclater and another biologist, Frederick Currey, the zoologist John Gwyn Jeffreys and the statistician James Heywood had all practiced as barristers. Richard Christopher Carrington ran the family 'Royal Brewery' business in Brentford while making a reputation as an astronomer. His fellow astronomer William Lassell was the founder of the Lassell & Sharman brewery in Liverpool. Herbert Spencer had come up as a civil engineer before turning to philosophy and sociology. The engineers John Frederick Bateman, William Armstrong, William Fairbairn and Charles William Siemens turned their science into successful commercial enterprises.

Even the botanists John Miers, in South America, and James Bateman, in England, ran industrial concerns. Bateman also had a line in banking on the side. John Bennet Lawes turned chemistry to profit by starting the fertilizer business and experimental farm that would become the world-famous Rothamsted agricultural research station, home in future years to R.A. Fisher and J.B.S. Haldane. The statistician Robert Hyde Greg successfully ran the family business 'Quarry Bank Mill' in Manchester. The archaeologist John Evans was a partner in the stationers John Dickinson & Co. The astronomer Walter de la Rue was the head of the long-established family stationery and printing business in London, still in operation today as a paper manufacturer. The geologist David Forbes was a partner in the Birmingham nickel-smelting firm of Evans & Askin. John Stenhouse turned chemistry to profit through the London firm that he founded and operated for a while. The geologist Joseph Prestwich succeeded his father as a prosperous wine-merchant in Mark-lane, London. The X-Clubman William Spottiswoode

[62] Letter from Candolle to Galton, 1874-09-11, Tredoux 2023b.

(see Figure 6.6, p. 126) left Oxford to rescue the family printing business, as the Queen's Printer, after his father Andrew had nearly ruined it through 'unwise speculation'.[63] As Galton would write in his obituary, 'His administrative success became so great that it seemed to beget in him an insatiable desire for such labour', so that 'every scrap of his time was utilized'.[64] Then there were the physicians with very extensive private practices who might be counted as successful business men, including William Bowman, Thomas Watson, Arthur Farre, William Ferguson, Henry Holland, and Charles Brooke.

One striking fact that emerged from the responses was underplayed by Galton: the lack of affect in his subjects. 'The female mind has special excellencies of a high order, and the value of its influence in various ways is one that I can never consent to underrate; but that influence is towards enthusiasm and love (as distinguished from philanthropy), not towards calm judgment, nor, inclusively, towards science. In many respects the character of scientific men is strongly anti-feminine; their mind is directed to facts and abstract theories, and not to persons or human interests. The man of science is deficient in the purely emotional element, and in the desire to influence the beliefs of others. Thus I find that 2 out of every 10 do not care for politics at all; they are devoid of partisanship. They school a naturally equable and independent mind to a still more complete subordination to their judgment. In many respects they have little sympathy with female ways of thought'.[65]

A pair of questions in the survey directly asked the scientists to assess the origin of their own scientific bent, polling the relative influences of nature and nurture. 'Can you trace the origin of your interest in science in general, and in your particular branch of it? How far do your scientific tastes appear to have been innate?' Most respondents said yes to the second, and gave a variety of answers to the first.

The first question produced at least 7 kinds of influences: fortunate accidents (mentioned by 11), indirect opportunities and motives, such as possession of instruments or sporting habits of siblings (19), professional duties, such as service on expeditions or in charge of observatories (24), encouragement at home by scientifically-inclined parents or other relatives (34), influence of private friends acquired by chance (20), encouragement by tutors, a rare event (13), and travel in distant regions, even more rare (8). A few odd cases could not be classified (3). It is clear from the descriptions supplied that these influences mostly concern the specific subject eventually chosen by the scientists, e.g. physics rather than chemistry, and not the general orientation of the respondents toward science. Thus, several of the scientists (e.g. George Carey Foster, Charles Piazzi Smyth, Richard Strachey, Robert Main, William Stokes, Fleeming Jenkin) only became interested in their field after they were put in charge of a major institution or office dealing with it! Few general principles could be drawn from this diffuse information, except perhaps the predominance of Scottish examples under influential tutors and educators, due to their more progressive system.

The second question gathered many emphatic responses. 'Certainly innate' wrote Charles Darwin. 'Love of observation and natural history innate' according to Lubbock. Richard Owen was more specific. 'Homology innate, and derived from my mother'. The physicist Lassell made the link to early experience explicit. 'My tastes are entirely innate; they date from childhood'. Another physicist, Haughton, agreed. 'As far back as I can remember, I loved nature and desired to learn her secrets'. The surgeon Charles Brooke combined education and inheritance. 'Primarily derived from my father'. James Clerk Maxwell replied that he 'always regarded mathematics as the method of obtaining the *best* shapes and dimensions of things'. The mathematician and X-Clubber Thomas Hirst declared that 'My taste for mathematics appears to be innate. As a boy, I delighted in sums'. Another mathematician, Arthur Cayley, concurred.

63 F. Galton 1908, 184.
64 F. Galton 1883g.
65 F. Galton 1874c, 207.

'An early taste for arithmetic, and in particular for long division sums'. The chemist Harcourt stated that his interest was 'thoroughly innate'. 'Whenever I had the chance of turning from other studies to natural science, I always turned'. His colleague George Buckton was nearly as emphatic. 'Perhaps wholly innate. My first notions of chemistry were picked up from books'. Another chemist, John Hall Gladstone chose natural tendencies. 'They date from a very early period, and there was little to produce them in my early surroundings'. Galton marshalled a collection of these statements, adding up to a clear majority 56 cases out of 91.

Not everyone chose nature. At least 11 respondents thought their tastes were 'decidedly not innate', as Galton put it. The mathematician Robert Main said he was 'not aware of any innate taste for science'. The chemist John Bennet Lawes opted for chance. 'My pursuit of chemistry is entirely due to circumstances occurring after manhood'. The Zoologists William Crawford Williamson ('doubt very much their innate character') and John Gwyn Jeffreys ('do not consider them innate, but induced') stood for nurture. Several medical men thought their tastes 'Not at all innate' (George Humphrey), or 'cannot perceive that they were innate' (William Ferguson), or else thought them an accident (Wilson Fox). A remaining 24 replies could not be classified in either camp. Definite opinions, by discipline and respondent, are summarized below.
Innate 56 cases:

Physics & Math: Lassell, Haughton, Capt. Evans, Foster, De La Rue, Andrews, Stewart, Scott, Rosse, Maxwell, Hirst, Cayley. (12)

Chem & Min.: Harcourt, Buckton, Roscoe, J. H. Gladstone, Voelcker. (5)

Geology: J. Evans, Rupert Jones, Ansted, Warrington Smith, Egerton, Prestwich, Forbes. (7)

Zoology: Gray, Darwin, Lubbock, Owen, Cobbold, Mivart, Allman, Parkes, Newton, Huxley, Carpenter, Stokes, Thomson, Gunther, Sclater. (24)

Botany: Hooker, Ball, Bateman, Carter, Miers, J.H. Balfour, Henslow, Babington. (10)

Medicine: Paget. (1)

Statistics: Guy, Jevons. (2)

Mech. Sci.: Armstrong, Jenkin. (2)

Not innate 11 cases:

Physics & Math: Main. (1)

Chem. & Min.: Lawes. (1)

Geology: (0)

Zoology: W.C. Williamson, Jeffreys, Sanderson. (3)

Botany: Bentham. (1)

Medicine: Humphry, Ferguson, Watson, Fox. (4)

Geography: (0)

Statistics: Hatherley. (1)

Mech. Sci.: (0)

Galton concluded that a general taste for science was innate in most of his subjects, only becoming attached to a specific field because of the kind of 'accidental' influences dealt with above. His description of this resembles the 'imprinting' theory of animal behaviour popularized many years later by the ethologist Konrad Lorenz.[66] Men have a general tendency to fall in love, but once they fix on a particular woman to love, from many possibilities, they typically adhere to one after that selection. More general tendencies could underlie the scientific bent,

[66] Lorenz 1970.

such as a love of collecting—'common to children, idiots, and magpies' but 'of immense use to a man who wishes to study objects that must be collected in large numbers'—or fascination with mechanisms and precision instruments, which was certainly shared by Galton himself. Even a mania for hunting, shooting and fishing, which had at various times gripped both his cousin Charles and himself, could lead to zoology or geology.

The extent to which those innate tendencies were inherited was harder to estimate from the returns. At least a quarter had a parent with a similar taste, but skipping of generations probably took place. He gave discreet examples, always without mentioning names. He had a 'curious case'—really from Charles Darwin—about a great nephew and great uncle in [Darwin's] family sharing a 'passion for heraldry', where the uncle had died before the nephew was born. Another case—actually William A. Guy, whose statistics on the longevity of various classes Galton had used in his investigation of prayer—had a grandparent with a similar mania for statistics but no other family member, including his father, seems to have had it.

A 'lady now deceased' in Galton's own family—really his aunt Adèle (1784–1869) who had married the physician John Kay Booth—carried a family love for 'figures and tabulation' to 'so strange an extravagance' that he could have given pages of anecdote.[67]

But what was more commonly inherited was really *general ability*, enabling those with it 'to leave their mark on whatever subject it becomes their business to undertake'. An explicit restatement of his argument for general intelligence in *Hereditary Genius*.

The obvious objection to this sort of exercise is that the authority and opinions of individuals, however distinguished in any discipline, need not be trusted and amount to no evidence at all. This is exactly the argument offered by the eccentric philosopher Francis Lloyd, in a short tract which appeared a few years later. The self-described graduate of Halle and Athens summarized Galton's general work as pseudo-science.[68]

Anticipating this objection, Galton countered that his men of science were highly selected for acuteness of observation. Exaggeration and varieties of vain self-delusion were common enough among men in general, but could go either way. 'I am astonished at the unconscious vanity which I have elsewhere met with when making inquiries in heredity, shown by men who, owing enormously to natural gifts, wish to accredit their own free will with being the real causes of their success. One phase of this form of vanity is prominently illustrated by the late John Stuart Mill in his strange and sad autobiography, who declares (p. 30) that he was rather below par in quickness, memory, and energy, and that any boy or girl of average capacity and healthy physical constitution, who was properly taught, could make as rapid progress in learning as he did himself!'[69]

Though opinions might be biased in one direction or the other, Galton argued that scientists-—'cool and careful', 'especially manly, honest, and truthful'—were less likely to be so than men in general. Also, there is no reason to suppose that they would be *systematically* biased. The aggregate of their opinions would not be skewed. Biases would tend to cancel out when all responses were taken together. Where possible he had corroborated their evidence, and believed it was 'substantially correct'. He trusted the methods of statistics.

Many might disagree with Galton's explicitly-stated faith in his lions of science. But the arguments above are unproductive diversions and mislead about the nature of empirical explanation and discovery. *The evidence Galton accumulated about these traits describes what an alternative explanation would have to look like.* The reasons given by the scientists in favour of innateness rest on their own observations—that their proclivities were always present, before any experience that they could remember, and that their natural tendencies actively shaped

[67] See Tredoux 2023c, where Aunt Booth is dealt with at length.

[68] Lloyd 1876. The philosopher appears to have died in Smyrna a few years later, in distress, despite having persuaded Charles Darwin to send him a cheque for £10. For Lloyd, Francis, see page 456.

[69] F. Galton 1874c, 147-148. The identification of Mill was marked for deletion in the second edition which never materialized.

their experiences, rather than being the helpless playthings of their environment. Therefore credible alternative explanations will most likely involve subtle inchoate factors that precede intelligent understanding on the part of the individuals affected, and are not evident to their families either. They would have to apply to many different situations, at least as varied as the cases here, with powerful effects. They require positive production, not just positing—*habeas corpus.*

Of all the qualities above, beside high general ability, natural gifts like energy, health, steadiness, good business habits, independent thought and an innate taste for science seem to be critical. All these traits have to be present in balance to make a successful man of science at this rarefied level. Those who lack one or more of these qualities risk coming partly or completely unstuck in some way. They might become ineffectual tinkerers, cranks, monomaniacs, inventors of perpetual motion machines, absent-minded eccentric professors of little consequence, or perhaps what we would call 'autistic savants' with freakish specialized powers. There is an element of speculation here, but almost all of the assembled group had the required traits.

Galton turned next to examining nurture *irrespective of nature*, and the role of various kinds of education in promoting science. The specific nurture of the English Men of Science was most directly addressed by his survey questions about education, close in spirit to Candolle's concerns. These were pointed, since Galton was an active educational reformer, working over many years to introduce the teaching of geography into schools and universities, persuading the Royal Geographical Society to award prizes to pupils.[70] Reform was a novelty then. Natural science of any kind in a school setting was rare in the period that the scientists had been educated in, up to two or even three generations previously. Classical languages and literature and varying levels of applied mathematics dominated curricula. The universities prepared the bulk of the enrolled for the clergy, for the law or sometimes for both. But only about a third of these men had been to Cambridge or Oxford. Another third had been through other universities. The remainder had not been to universities at all.

Contemplate the geographer and Arctic explorer George Back, who had entered the navy before he was even 12 years old, then breezed confidently through the seamanship exams in record time, only to be captured during the Napoleonic Wars and gratified with further schooling as a prisoner in Paris. After that, Back was self-taught. But no more so than the botanist and logician George Bentham, who had never been to any school or university, but had gone on from tutors at home to self-education. Galton wanted to know if the scientists thought that their education, whatever it was, trained them in observation or not, and what they thought the merits, demerits and omissions of their individual experiences were.

The 87 responses with information about education showed that, like Bentham and Back, the scientists were largely self-educated in their fields. A preponderance were negative overall about their educational experiences. Darwin entered 'none whatever' under merit, and identified the classical education he had endured as a real hindrance to observation. Huxley declared his boyhood 'utterly wasted'. Lawes confessed that at Eton he did nothing and was 'always in scrapes'. Unwelcome dead languages dominated complaints: a 'blank waste of time' to Jevons, only 'nominally classic, practically nothing' to Salisbury, 'Latin! Latin! Latin!' to Ferguson, 'Latin through Latin—nonsense verses' to William Guy. Only odd cases like T. Rupert Jones and Richard Owen appreciated the classical training. Some wished for instruction in modern languages. e.g. Stenhouse, Cobbold, Heywood, Darwin, Egerton. Many longed for training in observation and more mathematics, e.g. Darwin, Egerton, Harcourt, Jeffreys, Williamson, Jones, Bastian, Marshall, Prestwich. They prized whatever freedom they had been given to pursue subjects on their own at school, sometimes with light guidance, but sometimes this led to a lack of discipline in their reading (Back, Stokes, Ball) and a 'want of system' (Miller,

[70] F. Galton 1870c, 1871b, 1878c.

Prestwich, Jenkyn, Ansted).

A minority reported that their schooling had some positive effect on their observational skills and that they had enjoyed their education overall. Of these some were sons of schoolmasters or educated extensively at home (Humphrey, J. Evans, Hill, Roscoe) while others just cheerfully absorbed all that was aimed at them and asked for more (Main, Jevons, Maxwell, Capt. Evans). But only the few mathematicians like Cayley and Maxwell among the respondents could fairly claim to have learned even the basics of their science formally.

Not satisfied with the abuse his respondents had aimed at the old methods of education, Galton made brimstone from their ingredients. 'It crushed the inquiring spirit, the love of observation, the pursuit of inductive studies, the habit of independent thought, and it protected classics and mathematics by giving them the monopoly of all prizes for intellectual work, such as scholarships, fellowships, church livings, canonries, bishoprics, and the rest'. But his respondents, at least, were *not* ultimately 'crushed' in that way. As he says, 'men of science are not made by much teaching, but rather by awakening their interests, encouraging their pursuits when at home, and leaving them to teach themselves continuously throughout life'.

The author is describing his own nature here. 'Being original, they are naturally less receptive; they prefer to fix of their own accord on certain subjects, and seem averse to learn what is put before them as a task'.[71] Which is to say that for the species of scientist he was studying—immensely gifted, energetic and sufficiently business-like to leverage commerce for support—no more than a minimal level of compulsory 'nurture' is necessary given the right natural material. More than that just frustrates and wastes the energy of everyone involved. Even more so for those not so superlatively talented but still capable. 'Aptitudes and tastes for occupations which enrich the thoughts and productive powers of man are as much articles of national wealth as coal and iron, and their waste is as reprehensible'.[72]

Saving wasted energy—or broadening access to the less gifted—does not fully explain the messianic zeal below the surface in all this. The lions wanted loose. 'This gigantic monopoly is yielding, but obstinately and slowly, and it is unlikely that the friends of science will be able, for many years to come, to relax their efforts in educational reform'. There is more than a hint here of a new religion, to be funded by the emerging professionalization of science, conducted increasingly by full-timers only.

The 'manly, honest and truthful' men of science—those 'massive, vigorous, capable-looking animals' on whom he loved to 'feast' his eyes—meant more to their innocent admirer than it is easy to credit today. Galton expected his readers, perhaps with his tongue just partly in his cheek, to warm to his hope for 'a sort of scientific priesthood throughout the kingdom, whose high duties would have reference to the health and well-being of the nation in the broadest sense, and whose emoluments and social position would be made commensurate with the importance and variety of their functions'.[73]

With the old priesthood faded into obsolescence, and prayer discarded, the hereditarian longed to redirect religious sentiments back toward mankind itself. The parishioners of that Church of Humanity, instructed by their new scientific priesthood, could then 'dwell on the undoubted fact that there exists a solidarity between themselves and what surrounds them, through the endless re-actions of physical laws, among which the hereditary influences are to be included. They know that they are descended from an endless past, that they have a brotherhood with all that is, and have each his own share of responsibility in the parentage of an endless future. The effort to familiarize the imagination with this great idea has much in common with the effort of communing with a God'.[74]

[71] F. Galton 1873c, 257.
[72] F. Galton 1873c, 223.
[73] F. Galton 1874c, 260.
[74] F. Galton 1872b. It is significant that this passage was completely reworded in F. Galton 1883d.

These were ideas that Galton and Alphonse de Candolle could agree on. They found that they could not quarrel for long. A warm but friendly correspondence continued to grow after the publication of *English Men of Science.* Papers and books were cordially exchanged. Criticism was muted. Often they made tentative plans to meet in person, since Galton regularly vacationed in Switzerland to soak in the waters and walk the hills, but it was 1886 before they succeeded. The friendship, one long uninterrupted agreeable-disagreement, continued up until the death of the Swiss in 1893 at age 86.

When *English Men of Science* finally emerged from the presses in early December of 1874, Galton had fallen ill with one of his regular complaints, and was attended by a Dr. A. Clarke. He spent most of the month recovering. His new work competed for attention with the likes of William Ewart Gladstone on Vaticanism; travel narratives by Clements Markham and Captain A. H. Markham, jostling with three years of adventures involving S. W. Cozzens in New Mexico and Arizona; hundred-page tracts on the rights of women; novels by Saxe Holm ('Draxy Miller's Dowry') and Lady Margaret Majendie ('Giannetto', the first of 9 exotic novels); 'Stray thoughts on London' by a 'Layman'; John Holms on the British Army in 1875; the recollections of the police detective 'Waters'; and the ruminations of George Henry Lewes on the 'Problems of Life and Mind: the Foundation of a Creed'.

The most perceptive review of the book was by one of its anonymized subjects, the logician and statistician William Stanley Jevons, who needed over three pages in *Nature* to do it justice.[75] Jevons found it difficult to overrate the 'exact and scientific' character of the work, but was on the whole slightly disappointed. He thought the issue of heredity already settled by Galton's earlier work and not worth going into again. 'I am not aware that anyone in the present day ever denies the hereditary character of personal peculiarities'. He bemoaned the lack of definite responses using Galton's 'very ingenious and highly scientific' method of inter-comparison using grades of the distribution, which would 'constitute a distinct advance in statistical inquiry' if the data could be collected.[76]

What Jevons wanted was a more precise statement of the relative powers of nature and nurture. He was not surprised that the scientists generally reported great energy since he supposed they were selected for that by their arduous careers. The reviewer did not disclose his own submission here: 'amount of brain work never remarkable', which Galton put down to a lack of proper perspective. The section on education resonated most with Jevons, and he quoted approvingly the disparaging remarks made by many about classical education—including his own anonymized comments, without disclosing that they were his! Overall the book was 'far more interesting in its details than in its general conclusions' which were hampered by the limitations of the sample, which was adequate for science as a whole but inadequate for representing specialized fields within it.

The Scotsman was parochially pleased with the fact that Galton had opened the book with a quotation from a Scotsman, Thomas Carlyle. They thought that the evidence for the inheritance of ability had been greatly strengthened.[77] It confirmed the saying, 'what can you get from a pig but a grunt?' They thought his methods for selecting his lions defensible. The idea that the scientists seemed to have slightly larger heads than the general population was intriguing.

Other notices were either bland summaries, as in the *Morning Post*, or a mixture of hostility and skepticism. An unsigned reviewer in the *Lancet* found the book 'fragmentary and discon-

[75] Jevons 1874.

[76] Chris Pritchard has suggested that Galton may have obtained the idea for intercomparison from Jevons' *The Theory of Political Economy* (London: Macmillan, 1871); see C. Pritchard 2018, 91. But it was already used heavily by Galton in *Hereditary Genius* (F. Galton 1869a) for cognitive ability, a short step from other mental characters. This precedes Jevons' work by two years and the remarks here suggest that the economist thought of Galton as the originator of the method.

[77] Anonymous 1875a.

nected'. The evidence was unsatisfactory because the respondents had been kept anonymous (much like reviewers in the *Lancet*). While the reviewer agreed that genius is in essence 'automatic activity of the mind', he dogmatically insisted that men of science certainly must have good memories—contrary to Galton's respondents, who reported that their memory varied in quality.[78]

Writing on the same day in the *Academy*, the barrister and future editor of the magazine, James Sutherland Cotton (1847–1917), scorned almost everything in the book.[79] He thought the method used was insufficient to establish anything definite. Like Jevons, he thought that far more scientists, in many more countries, would have to be surveyed, before drawing such general conclusions. Cotton was astonished by the frank confessions from the scientists that they valued abstract theories and facts more than people, and the deficient emotion identified by Galton. He was not convinced by failures to remember early influences that were not hereditary, arguing that those concerned may simply have forgotten or never noticed them due to infancy. The opinions on education given were incredible: 'the outside world may well be permitted to doubt whether the chief representatives of physical science in this country are good judges of the means by which their own subjects may best be advanced'.

An unsigned notice in the *Examiner* a month later questioned the value of statistical methods for this exercise in the sociology of science, given the restricted sample, but found the material on inheritance of scientific ability convincing.[80]

The reception of *English Men of Science* was short-sighted, and it remains the least appreciated of Galton's books. It was the early phase of what would become a long and very productive research programme. His data sharpened our understanding of what alternative explanations of topics like innateness and the traits of scientific men would have to look like. A century and a half of subsequent experience would confirm the more important of his conclusions. He added notable cases to the pedigree studies of the inheritance of ability first started in *Hereditary Genius*. He conducted the first statistical investigation of sexual selection using human case data. Some of the topics he covered, like 'energy', have still not been adequately explored today. Directly gathering better data about human heredity from circulars would reap large rewards in the future. The statistical technique of inter-comparison, frustrated in the end by this particular dataset, had been clearly defined and lodged in our consciousness. Regression and correlation, which he was still grasping at in his notebooks, would become central to applied statistics thanks to his later work. And now that we know the key to his anonymized sources we have been left a rich and perceptive biographical resource on some of the foremost minds of the Victorian era.

——oOo——

It is doubtful that many modern readers would share Galton's enthusiasm for rule by a scientific priesthood. Experience has suggested that professional scientists are no wiser than the general run of men. Among his lions were characters like the astronomer Charles Piazzi Smyth who wrote many learned treatises 'proving' that the great pyramids of Egypt were built under divine guidance, their proportions encoding cosmological constants, a personal obsession.[81] Scientists of his day were easy prey for the spiritualists, who were pleased to exploit, using simple magic tricks, overconfidence among their scientific participants. William Crookes, Alfred Russel Wallace and many others were victims. Granted, a fair number were not—including Galton, who attended numerous séances, and Huxley—but science per se provided no special protection.

78 Anonymous 1875b,c.
79 Cotton 1875.
80 Anonymous 1875d.
81 Smyth 1864.

Scientists played their role to the full in both world wars in the next century, on both sides, eager to serve any master. They were no less likely than other intellectuals to be seduced by the Soviet Union's fellow-traveller Potemkin tours. Julian Huxley, a distinguished grandson of the lion T. H. Huxley, was one of the many feted scientists who swallowed and regurgitated it all, only to be disillusioned later.[82]

But a more profound argument against the idea of rule by technical specialists was made by the philosopher Michael Oakeshott just after the Second World War. Oakeshott pointed out that civil associations are not problem-solving activities or corporations that participants join for a limited purpose and then go home from once the job is done. There is nowhere to go back to. People find themselves in civil association whether they like it or not. Emigration does not alter that. Technical knowledge is not called for in politics, however useful it may be in other activities like engineering. Rationalism applied to politics is a corrosive misrepresentation of the human condition.[83]

[82] Hollander 1981; J. Huxley 1932.

[83] Oakeshott 1991. The essay 'Rationalism in Politics' was originally published in the *Cambridge Journal* in 1947.

Figure 6.1: The Quincunx in a modern reconstruction, by Matemateca (IME USP), Creative Commons.

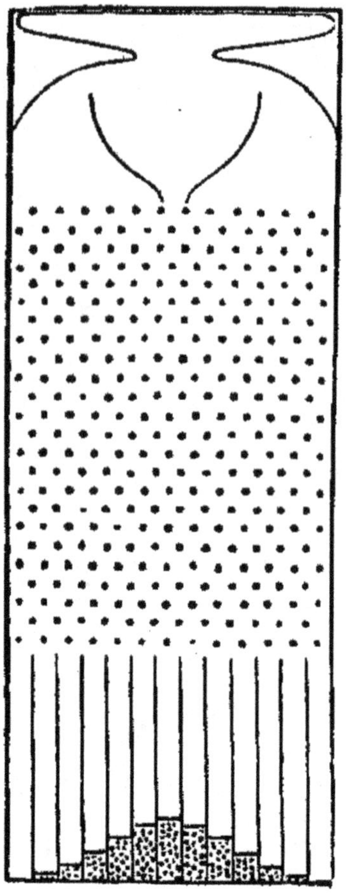

Figure 6.2: The Quincunx schematically.

Figure 6.3: George Bentham, botanist and logician. Nephew of Jeremy Bentham the utilitarian.

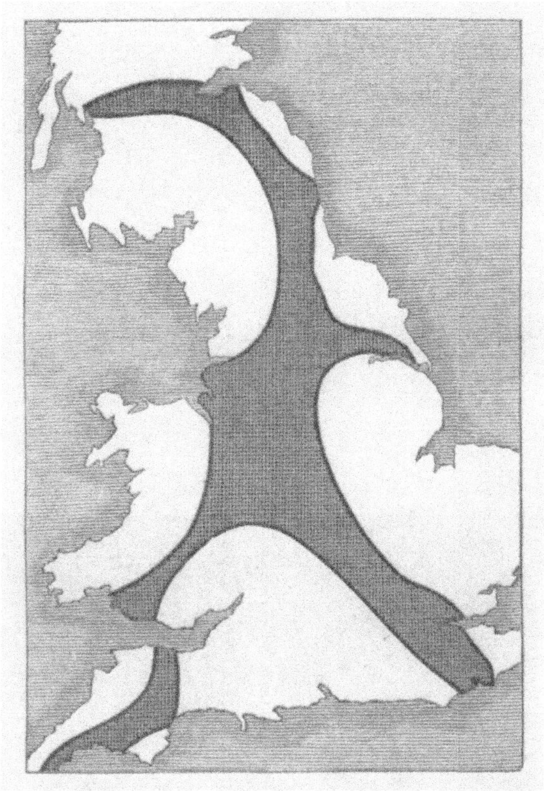

Figure 6.4: Birthplaces of the English Men of Science.

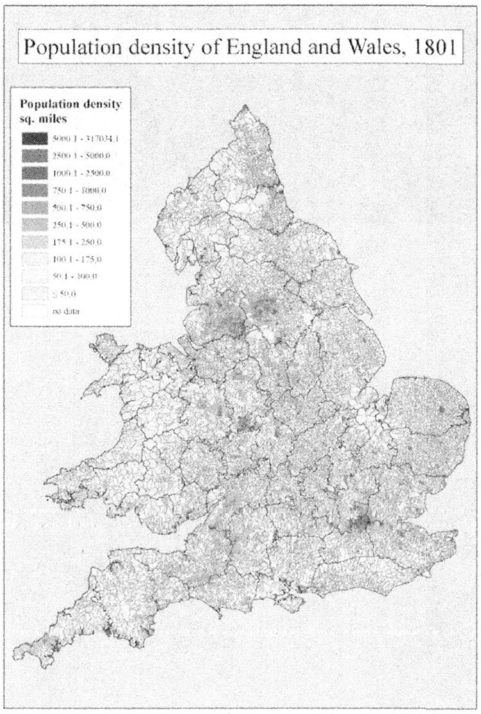

Figure 6.5: Population density of England and Wales in 1801.

Figure 6.6: William Spottiswoode (1825-1883), by G. F. Watts.

Chapter 7

Twins

In early November of 1875, a disappointing letter arrived at 42 Rutland Gate, Kensington. Galton had asked Meredith White Townsend (1831–1911)—the editor of the *Spectator*, a magazine practiced at pouring scorn on Galton's more original ideas—for more information about some identical twins.

9th Novr 1875

Sir,

I am sorry I cannot assist you in your very interesting enquiry. The twins referred to are the daughters of the Revd. H. Fox now I think about 13 years old. When I knew them or rather to be accurate when my wife knew them they were about six, and certainly were more nearly identical in appearance than any children I ever saw. Mr Fox has a living in Oxford but I am not on terms with him to enable me to send your questions to him, and the children's mother whom I knew is dead.

Yours [faithfully]
M. Townsend

A good many persons must know who the originals were of the brothers described in Whitty's "Friends in Bohemia". The sketch there given is most powerful & Whitty never invented anything.[1]

I believe that the most remarkable pair of twins now alive are two brothers named Robertson in Bradford. I don't know them, but I have been told that their resemblance, though they are full grown men baffles their most intimate friends.[2]

Just a day later, Townsend wrote with an update, including a casual reference to a pair of twins who were not just identical, *but had also been adopted and raised entirely apart from a young age.*

Spectator Office
10th Novr

Sir,

The clergyman, Mr Henry Fox, whose name I mentioned in my note of yesterday is now Vicar of Christchurch, Westminster. His twin daughters are 10, not thirteen.[3]

I am told that Mr Weldon headmaster under the Principal of Lou'bridge School, has twin sons of seventeen who are remarkable for a certain consciousness of identity. They have, or till lately had, a confirmed habit of speaking of themselves in the plural. "We are thirteen years old today". "We are not fond of animals" and so on.

[1] Edward Michael Whitty *Friends of Bohemia: Or, Phases of London Life* (1858).

[2] Annotation in pencil: 'see Mr Hertz'.

[3] Henry Elliott Fox (1841–1926). He took a BA at Cambridge in 1864.

It is quite probable that in answer to a direct application Mr Weldon might give you most interesting facts. The boys are I think at Rugby.

I have tried to learn for you something of a case in which two twin brothers separated through life refound one another at about 50 with their identity unimpaired, but I find real *evidence* unattainable. The brothers were named Wilson, were Liverpool men, merchants I think in later life, and were great-uncles of mine. I have heard their story all my life, but I can find since my Mother's death no-one who can give first-hand testimony.

Yours faithfully
M Townsend

Galton had become interested in twins while researching *English Men of Science*, published the previous year.[4] There he included an engaging anecdote about mistaken identity and 'passing' among a twin pair. 'I have been told of a case in which two twin brothers, both married, the one a medical man, the other a clergyman, were staying at the same house. One morning, for a joke, they changed their neck-ties, and each personated the other, sitting by his wife through the whole of the breakfast without discovery'. But as the book was going to press, Galton started to suspect that the story—actually about the twins Henry and George Creed—was not accurate. An unpublished notebook entry discloses that 'The vexation I felt at having published an incorrect statement was a strong motive to undertake the present inquiry [an empirical study of twins]'. In fact, as one of the sons of the twins concerned soon wrote to him, 'the necktie story was correct, except that it was at supper, not at breakfast'.[5]

Now his first article specifically about twins had just appeared in the *Fortnightly Review*.[6] The *Spectator* had complained of Galton's 'limited number of facts' about identical twins, yet mentioned in passing that 'we know of one such case, in which the most careful observation has detected nothing but an apparent "roundness" in the eyes of one twin, which on a change of attitude disappeared'. It seems that the scientist had written to the reviewer asking for some more information.[7]

By then Galton understood that the central aim of his emerging research programme was to distinguish the effects of nature and nurture, 'a convenient jingle of words, for it separates under two distinct heads the innumerable elements of which personality is composed'. This was the first deliberate application of that phrase to this problem, though the pairing occurs in Shakespeare's *Tempest*, where Prospero declares that Caliban is 'A devil, a born devil on whose nature Nurture can never stick'.[8]

In retrospect, Galton's understanding of the problem is remarkably sophisticated. Recall that by 'nature' he means 'all that a man brings with himself into the world', but he is careful to stipulate that 'natural gifts may or may not be hereditary'—that is passed on. For example, in modern terms, epistatic genetic effects from the specific *combination* of genes in a sexually reproducing individual cannot be passed on. Likewise, 'nurture does not especially consist of food, clothing, education or tradition, but it includes all these and similar influences whether known or unknown'. Here he means to gather together all those influences that are *not nature*.

This distinction between 'nature' and 'nurture' is close to our modern separation between genetic and non-genetic effects, but with the complication that, strictly speaking, Galton includes pre-natal effects under *nature*. Unfortunately, 'nurture' here suffers from the same defect

4 F. Galton 1874c, 13–16.
5 GALTON/2/4/3/10, Notebook containing cases of mistakes, f.30r; H. Creed to Galton 1874-11-09, GALTON/2/4/3/5.
6 F. Galton 1874c, 12–16.
7 *Spectator* 1875-11-06: 1387–1389. Galton's letter cannot be traced.
8 Circa 1610–11.

as 'environment', a modern term for *anything non-genetic*, including even random errors in development. Both labels invite a narrow and tendentious interpretation for a much broader class of effects.

At that stage Galton already suspected that twins could be used to investigate the relative powers of nature and nurture. He had no doubt that this could be done in principle, even though *both* nature and nurture are logically required for humans to develop. 'It is needless to insist that neither is self-sufficient'. It is easy to see that *sufficiently extreme* nurture can trump nature—'the highest natural endowments may be starved by defective nurture'—as noted previously.

For a more extended example, consider a collection of opaque water-mugs, of identical height, covered with lids. Each lid has a small hole for pouring its contents out, but not large enough to look through with the naked eye. A tiny hole lets air into the mug as its contents evacuate. Some are green, while others are red. By pouring out their contents, we are able to find out that some mugs have more water in them than others. One theory holds that all the mugs have the same capacity, but some were filled to the brim but others were hardly filled at all—nurture. Another theory holds that all the mugs were filled until they could hold no more, but that each mug has a *different capacity* due to an internal false bottom—nature. Either theory *could* be true, in the sense that the world could be that way. Even if the mugs really had different capacities, so that the entirely-nurture theory is strictly wrong, extreme water shortages could still validate its gist. If the only water available was a tiny pond, and the mugs were filled by dipping them in that pond, the water could run out before they are all filled. Some would be empty regardless of their capacity. Even if after inspection it becomes clear that the mugs were filled from an enormous lake with more than enough water for all, the matter is still not settled. Suppose an eccentric servant simply filled the red mugs more than the green mugs out of spite. By collecting only a scoop in the latter, even if the green mugs all had a higher potential to hold water, spite denied them the chance to realize it. Which answer is correct depends on the way that the world really is. There is no way to deduce the answer from first principles. It is a contingent fact which must be found out by experience.

One way to establish the effects of nature and nurture in *this* non-imaginary world is to see by how far differences in nurture can separate initially identical natures, represented by identical twins, over time. 'There are twins of the same sex so alike in body and mind that not even their own mothers can distinguish them. Their features, voice, and expressions are similar; they see things in the same light, and their ideas follow the same laws of association. This close resemblance necessarily gives way under the gradually accumulated influences of difference of nurture, but it often lasts till manhood'. Galton would considerably revise this last cautious assessment—that initial likeness would give way to experience—under the weight of evidence he set about collecting when *English Men of Science* was finally in press.

Twins were familiar objects of interest to all readers of novels and audiences at plays, but there were also twins in the Galton family itself. His father's sister Adèle (Mrs Booth, 1784–1869) had a twin brother Theodore (1784–1810), who had died of fever in Malta after an adventurous tour of the Mediterranean with his brother-in-law Francis Darwin (also an uncle of the naturalist).[9] Frank's own sister Lucy Harriot had identical twin sons, James Keir and John Lewis Moilliet (1836–1906, 1836–1909). Through his Cambridge friend Franklin Lushington (1823–1901), Frank had come to know of Franklin's distinguished cousins, the identical twins Vernon and Godfrey Lushington (1832–1912, 1832–1907).[10]

And, aside from his own medical training, Galton also knew that Darwin had dealt briefly with twins in his *Variation of Plants and Animals under Domestication* (1868). When considering 'internal' (hereditary) versus 'external' (environmental) effects on variation, his cousin

[9] See Tredoux 2023e.

[10] Lushington, Sir Franklin, see page 457.

cited a case showing obviously 'internal' peculiarities. 'Two girls, born as twins, and in all respects extremely alike, had their little fingers on both hands crooked; and in both children the second bicuspid tooth in the upper jaw, of the second dentition, was misplaced; for these teeth, instead of standing in a line with the others, grew from the roof of the mouth behind the first bicuspids. Neither the parents nor any other member of the family had exhibited any similar peculiarity'.[11] Darwin also cited a case from Armand Trousseau's *Lectures on Clinical Medicine*, which claimed personal knowledge of identical twins who suffered from rheumatic ophthalmia and asthmatic complaints at the same time, even when in different cities.[12]

For up-to-date knowledge about the physiology of twinning, Galton turned to a recent synopsis written by the Prague-educated obstetrician Ludwig Kleinwächter (1839–1906), published as *Die Lehre von den Zwillingen* ('The Study of Twins') in 1871. It had been known since at least 1850 (for example, in the work of H. Meckel) that there were two distinct types of twins: single-egg twins, and double-egg twins, a topic studied in some detail by Joseph Spaeth in Vienna.[13] Galton, who was also familiar with Spaeth's work, found Kleinwächter's book 'full of references' but marred by many misprints, and the author did not consider the questions the Englishman was specifically interested in. So in order to get more extensive information about real-life cases of twins, he resorted to a printed circular. It was a technique he had used with success as early as 1863, when he was investigating weather phenomena—his discovery of the anticyclone soon followed.[14] The twin circular he distributed is reproduced below.

> QUESTIONS ABOUT TWINS.
>
> By FRANCIS GALTON, F.R.S.
>
> In the course of some inquiries upon which I am engaged, connected with the resemblance between twins, it has come to my knowledge that you are very probably able to afford me valuable information. I therefore take the liberty of sending this Circular to your address, in which my purpose is briefly stated, and the questions to which I seek replies are contained.
>
> I propose to publish the result of my inquiry, with corroborative extracts from the communications I may receive; but the names of the individuals referred to, and of my correspondents generally, will be kept STRICTLY PRIVATE.
>
> *Object of the Inquiry.*— My object is to collect data for estimating the respective shares that "Nature" and "Nurture" ordinarily contribute to the Body and Mind of adults; meaning by "nature," everything that is inborn, and by "nurture," every influence subsequent to birth.
>
> The effects of Nature are clearly seen in persons, originally unlike and who *continue* to be unlike, although they are reared as nearly as possible in the same way; so the effects of Nurture might be traced in the gradual *extinction* of resemblances among those who closely resembled each other in childhood, but were afterwards reared in a more or less different manner.
>
> It is easy to obtain suitable instances for the first part of the inquiry, but it is only among twins that they exist for the second.
>
> I have explained myself more at length, and drawn provisional conclusions, in a work just published, "English Men of Science; their Nature and Nurture" (Macmillan & Co.), page 12. I take the opportunity of asking other questions, which relate to Heredity. The *strength* of the hereditary tendency to bear twins is not yet ascertained; neither is it known whether the tendency is transmitted *equally* by the male and female lines. If my correspondents will kindly reply to the questions about uncles, aunts, and cousins, they will give data to determine these matters.
>
> QUESTIONS.
>
> *.* Please address any communications with which you may favour me, to Francis Galton, 42, Rutland Gate, London, S.W.

[11] C. Darwin 1868, Vol. ii, 253.
[12] A. Trousseau (1801–1867) *Clinique Médicale* (1865) tom. i. 523.
[13] Spaeth 1860.
[14] F. Galton 1863a. See Tredoux 2023c.

(1.) Names of the twins, both Christian and Surnames? (It will, I trust, be clearly understood that *no names will be published.* I ask this question, to guard myself against entering the same case twice; to avail myself of corroborative evidence, if it should reach me; and to learn whether the twins are of the same sex or not, which is a point of some interest in respect to the frequency of strong resemblance.)

(2.) How far were they alike, in body and in mind;—in childhood, in boyhood or girlhood, in youth, and in adult life?

(If they were *never* very closely alike, the questions 3, 4, 5, and 6 must be disregarded.)

(3.) If they were closely alike at any age, give anecdotes to illustrate their resemblance; showing for instance, that near relations frequently mistook them. Test your estimate of their likeness, by considering details of it, thus, Were they of the same height and weight, and were their clothes of the same fit? Had they the same colour of hair and eyes? Had they similar powers of athletic performance? Were they alike in manual skill, as in drawing? Had they similar handwriting and intonation of voice? Were their special tastes the same, as for music and art? Were their dispositions similar, and their associations of ideas alike, as shown by their frequently making identical remarks, or by the same recollections occurring to both of them at the same moment? Was their state of health the same?

(4.) Up to what age were they educated together, and in what respects did their education and pursuits differ afterwards?

(5.) At what period did their close resemblance begin to diminish, and in what respects did they grow unlike in body and mind?

(6.) How far do you ascribe their increasing dissimilarity to the development, in due order of time, of the qualities they had at birth, but which had lain dormant, and how far to the effect of external influences?

HEREDITY.

Uncles and Aunts:—

(7.) How many Uncles and Aunts had they on the Father's side? Were there any cases of twins among them?

(8.) How many Uncles and Aunts had they on the Mother's side? Were there any cases of twins among them?

Cousins:—

(9.) How many Married Uncles had they on the Father's side ? How many children had those uncles (including all who may have died in infancy)? How many cases of twins among the children?

(10.) How many Married Aunts on the Father's side? How many children had those aunts (as above)? How many cases of twins among the children?

(11.) How many Married Uncles on the Mother's side? How many children had those uncles (as above)? How many cases of twins among the children?

(12.) How many Married Aunts on the Mother's side? How many children had those aunts (as above)? How many cases of twins among the children?

(13.) Can you give me the addresses of any persons known to you as being themselves twins or nearly related to twins, and who you think might be likely to respond to this Circular if a copy were sent to them?

The methods used were rough and ready. A modern survey would provide numerical scales for responses, instead of the free-form anecdotes that Galton accepted, which required extensive summarizing of prose for each response. But this may also have been a strength, since his more intelligent correspondents had greater scope as a result. The comprehensive list of attributes suggested here, incorporating both physical and mental or psychological features, goes far deeper than commonplace observations made about twins 'looking alike', reflecting Galton's general treatment of *behaviour* as a set of human features under evolutionary control in much the same way as bodily form. But the only questions that covered mental *ability* specifically were those about education and, less directly, 'pursuits'. He had no objective measures for mental abilities—those would come years later, after he helped to invent them—and would rely entirely on the judgment of his informants about how similar the twins were for all his attributes. He was aware that considerable variation in individual judgment would exist, but was only interested in aggregate results, where random variation would wash out if he got enough responses—but will be shown, there is some evidence that bias crept into the judgments of his respondents.

These questions only framed the discussion. Respondents would feel free to range over features they thought interesting, sometimes valuably. Galton had also revealed upfront that he would interpret initial differences which persisted despite experience as evidence of the potency of 'nature', which did not require the study of twins. He would take the erasure of similarity by experience to be evidence for the potency of 'nurture', for which twins had to be studied.

In the second section, with its seven questions about twins among forebears and other relations—uncles, aunts, cousins—Galton was trying to determine whether or not twinning itself was hereditary. The final question fished for names of other twins, 'who you think might be likely to respond to this Circular if a copy were sent to them', and would allow him to extend his reach.

Galton also constructed a second circular, which he distributed among doctors in charge of insane asylums, seeking information about links between cases of insanity among twins.

> QUESTIONS ABOUT TWINS.
>
> May I beg the favour of your informing me, whether in the course of your experience among the insane, you have met with any cases of twins suffering under the same form of insanity, so alike in minute details as to be comparable with that which has been described by a well-known authority, Dr. J. Moreau, in the strange account that I have translated, and given below?
>
> The motive of my inquiry is partly, but sufficiently, explained by the enclosed Circular, which I am sending to twins whose addresses I happen from time to time to learn, and which has already elicited numerous answers of great interest.
>
> FRANCIS GALTON,
>
> 42, Rutland Gate, London.
>
> Extract (translated) from p. 172 of *La Psychologie* par le Docteur F. Moreau (de Tours), Médecin de l'Hospice de Bicêtre, Paris. 1859.
>
> He speaks "of two twin brothers who had been confined, on account of monomania, at Bicêtre ... Physically the two young men are so nearly alike that the one is easily mistaken for the other. Morally, their resemblance is no less complete, and is most remarkable in its details. Thus, their dominant ideas are absolutely the same. They both consider themselves subject to imaginary persecutions; the same enemies have sworn their destruction, and employ the same means to effect it. Both have hallucinations of hearing. They are both of them melancholy and morose; they never address a word to anybody, and will hardly answer the questions that others address to them. They always keep apart and never communicate with one another."
>
> "An extremely curious fact which has been frequently noted by the superintendents of their section of the Hospital, and by myself, is this:—From time to time, at very irregular intervals of two, three, and many months, without appreciable cause, and by the purely spontaneous effect of their illness, a very marked change takes place in the condition of the two brothers. Both of them, at the same time, and often on the same day, rouse themselves from their habitual stupor and prostration; they make the same complaints, and they come of their own accord to the physician, with an urgent request to be liberated. I have seen this strange thing occur, even when they were some miles apart, the one being at Bicêtre and the other living at Saint-Anne."

Some twins could be located through suggestions based on Galton's own experience, his extended family and the recommendations of friends, including the wide circle he was in touch with through the professional societies he was active in. But the initial list was largely due to the work of the 'father of actuaries' Charles Ansell (1794–1881), whose survey *On the rate of mortality at early periods of life* had just been published by the National Life Insurance Society.[15] The society had commissioned the survey in 1871 to establish accurate premiums for endowments for children. Ansell listed facts about the longevity of 482 twins, listed under 'multiple births', gathered from a selection of over 50,000 children he had located. Galton's notebook shows that he obtained Ansell's address, perhaps from the publisher, and had written to him by early November of 1874 to request contact information for the twins. Ansell promptly replied on the 14th with a helpful list of the 190 twins for whom he had addresses—usually of parents or siblings—where both twins had reached at least the age of 3. The actuary requested discretion in mentioning the source of this personal data. With this start, Galton was able to distribute 600 of the 750 circulars he had printed. He asked his circle to distribute further copies for him, writing for example to George Howard Darwin, his cousin's mathematical son, 'I enclose a circular. If you happen to know any body (not probably a mutual friend) in whom you could launch it, with a word on my behalf, I should be very glad. I have lots of addresses but few are personally known to me, &, you know, that people won't answer a mere circular'.[16]

That was overly pessimistic. By January Galton could enthuse to George. 'My twin papers come in and some are very interesting. J. Wilson of Rugby is a twin and sends me lots of addresses. I got a most curious letter from Lady Eastlake, whose family abounds with twins, besides one treble and one quadruple birth. I feel saturated with midwifery and am haunted with imaginary odours of pap and caudle!'.[17] He recorded in his rough notes that he finally received 159 submissions from 125 correspondents, a response rate of just under 1 in 4. About 1 in 5 of his respondents had not been known to him at all, having been forwarded by other respondents.[18]

The twin study would continue from November of 1874 through to the following November, as replies and cases accumulated. Louisa recorded in her diary that Frank was 'occupied on enquiries about Twins' that December, despite falling ill. Her own illnesses were ever-present, changing only in amplitude. While the circulars were circulating, Galton was busy on other fronts. His growing preoccupation with all things hereditary is shown by his scathing review in that January's *Academy* of *Heredity; a Psychological Study of its Phenomena, Laws, Causes, and Consequences* (1874) by the French psychologist Théodule-Armand Ribot (1839–1916).[19] After calling out Ribot's 'prescientific' non-empirical approach and faulty logic with many examples from the text, he complained that the author had plagiarized material from several sources, including his own work. 'I find the tables and genealogies that I had compiled, after very considerable research and sifting, and which I published in *Hereditary Genius*, appropriated without a word of acknowledgment'. The book contained nothing of real value. 'I laid down M. Ribot's volume after honestly reading every line of it, with a weary sense of many wasted hours'.

Two weeks later, in-between processing the first replies to his twin circular, Galton raised the issue of plagiarism again. In the pages of *Notes and Queries* he objected that some verses which had been attributed to the poet Samuel Rogers were actually from *Phytologia* (1802), written by his own heretical grandfather Erasmus Darwin—once illustrious but by then, as today, largely neglected.[20] Gently but persistently reminding the world of their common grandfather was an

[15] Ansell 1874. Galton credits Ansell in F. Galton 1875g.

[16] Galton to G. H. Darwin, 1874-12-09. GALTON/3/3/7/8.

[17] Galton to G. H. Darwin, 1875-01-08. Elizabeth, Lady Eastlake (1809–1893) was an art historian and critic.

[18] GALTON 2/4/3. The study was therefore much more extensive than Karl Pearson supposed in K. Pearson 1924, where the figure of just 94 twins is inferred from F. Galton 1875g.

[19] F. Galton 1875f. See Tredoux 2023a.

[20] F. Galton 1875d.

interest shared with Charles Darwin.[21]

In March, back in London, it was time to circulate between the meetings of the scientific societies. At the Geographical, Galton attended the African trans-voyager Verney Lovett Cameron's 'Examination of the Southern Half of Lake Tanganyika', adding comments afterward.[22] Cameron's account suggested that the river Congo flowed out of the lake—a fact soon to be confirmed by Stanley's gruelling journey in the opposite direction down the river to the west coast. It was left unstated in the *Proceedings* that the explorer had fallen out with the Geographical, which he viewed as a cabal of interfering busybodies and armchair geographers. He had also taken up cause with Galton's former friend turned faux-enemy Captain Richard F. Burton. From the floor, Galton struck back, expressing pointed sympathy with Burton's 'disappointment' in being beaten to a second prize.[23]

A day later Galton was at the Anthropological, auditing a paper by Sir Duncan Gibb on 'Ultra Centenarian Longevity'. The emerging statistician pointed out that some criticisms of the remarkable ages—that the products of the probabilities of the unlikely events the cases combined was nearly zero—were flawed, since probabilities of two events cannot simply be multiplied unless the events are independent. Instead, the age distribution of the long-lived suggested that the data was worthless, since older ranges sometimes greatly outnumbered younger ones.[24]

Oscillating illnesses were interrupted by a trip to Fontainebleau in May of 1875. From there the Galtons went on for a long spell at Seeligsberg in Switzerland from June to August. Abroad it was Frank's habit to keep working on the project in hand, in-between hikes through mountains, with Louisa safely tucked into a hotel. Frank went back on his own to attend the meeting of the British Association at Bristol. He amused himself by inventing a speedometer for the bicycles only then coming into vogue, published in June's *Field* periodical.[25] It required the cyclist to count foot-strokes while keeping time by continually flipping a handlebar sand-glass calibrated to the size of the wheels, an idea originally suggested for walkers by his friend the mathematician Archibald Smith. Conveniently, a wheel of circumference n inches revolving m times in $^{10n}/_{176}$ seconds travels at m miles per hour. This sort of counting was an activity natural to Galton but would have bemused most of his readers.

After collecting all his twin data—collating from the long letters he received, sometimes from one of the twins but usually from their parents or siblings—there were 159 cases. In all of these he could at least judge overall likeness from the descriptions he received.

Table 7.1: Twin Likeness.

Girls alike	41
Girls partly alike	9
Girls unlike	18
	68
Boys alike	39
Boys partly alike	8
Boys unlike	16
	63
Different Sex alike	1
Different sex partly alike	10 (16)
Different sex unlike	17 (11)

[21] See Tredoux 2023c for Erasmus Darwin and his connections to the Galtons.
[22] F. Galton 1875c.
[23] See Tredoux 2023c for a full discussion of Burton, Speke and the Nile question.
[24] F. Galton 1875b.
[25] F. Galton 1875e.

28

The parentheses for different sexes show two different tabulations in his notebook, illustrating the arbitrary distinction between unlike and partly alike. It is also appears that fraternal twins were sharply under-represented in his replies, only amounting to 18%. This may have been because his networking method of locating twins was inherently more likely to turn up same-sex twins, who were more memorably twins to friends and relatives (twins of different sex are necessarily fraternal).

Of the 159 cases, Galton reckoned he had 'about 80' twin pairs showing 'close similarity' with substantial information about life-history and specific traits. Thirty-five of these cases had enough detail to allow close analysis, only one of which was a male-female pair (which he knew as well as we do must have been fraternal). There were twenty detailed cases of marked 'dissimilarity' from an early age. The rest were intermediate to a greater or lesser extent. Curiously he reported that his detailed cases of extremely unlike twins were almost all of the same sex. This is certainly not true of the total cases above, where unlike twins of different sexes are under-represented but still substantial. An additional reporting bias may have been involved, making it much less likely that twins of different sex were judged extremely unlike by his informants, perhaps because there was no need to exaggerate and press the point: mixed-sex twins are *obviously* not identical.

Galton supposed that the like twins were identical by nature, since that is what they started with initially. Likewise he supposed that the non-like twins were different by nature. Nowadays we know that these classes correspond to single-egg (monozygotic) versus two-egg (dizygotic) twins, an identity important to us because we know that the single-egg twins are genetically identical, whereas the two-egg twins are no more similar genetically than siblings. Galton did not know this, a point we will return to below. What mattered to him, and to his analysis, was the link between likeness and what he called nature or heredity—which we identify with zygosity but he did not.

A list of similar twin pairs survives in one of Galton's notebooks and is shown below, with their age and sex. Most pairs were adults, but there were cases as young as 6 and as old as 83.

— Arnall, Joseph and William Henry. 18. MM .

— Bailey, Isabel and Emma. Adult. FF.

— Bailey. MM.

— Baraclough, Mary and Sarah. 30. FF.

— Brandreth, William Harper and Frederic. 62. MM.

— Butler, Paul and Charles. 53. MM.

— C .... Adult. FF.

— Cane, Alicia Katherine and Edith Frances. 6. FF.

— Cobbold. 23. MM.

— —— (given by Cobbold). FF.

— Coles, Emily and Oscar. 13. MF.

— Cordes, Bertha and Adèle. 9. FF.

Cornish, Katherine Deacon and Helena Jane Kynnersley. 25. FF.

— Creed, Henry and George. Elderly. MM.

— Dale, Lewis Bruce and Wyndham Francis[26]. dead. MM.

— Davey, Eliza Conolly and Emily. 32. FF.

[26] Both died within 7 months of each other, from Bright's disease.

— Farquahar[27]. School age. MM.
— Fowler, John Wilfred and Joseph Gurney. 20. MM.
— Freeman, Mary Bell and Anne Eliza. 83. FF.
— Gröningen[28]. 28. FF.
— Guppy, William Good and Henry Brougham. 20. MM.
— Horner, George William and Maurice. 26. MM.
— Hull, Charlotte Frances and Charlotte Joanna. 18. FF.
— Iliff, Alfred and Harold. 12. MM.
— Jessop. 17. FF.
— [Euston] (by Jevons). MM.
— Kenney, Sophie[29] and Elizabeth. 58. FF.
— Kent, Emma Caroline and Charlotte Elizabeth. Middle Life. FF.
— Lambton, John and Frederick[30]. 20. MM.
— Lawford, George and William. 55. MM.
— Lehmann, Wilhelm and Adolph. 73. MM.
— Lucas, Mary and Martha. 20. FF.
— Lushington, Godfrey and Vernon. 43. MM.
— Macrae, Alexandrina Cornfute and Elizabeth Anne. 16. FF.
— Mansel, John Clavell and George Pleydell. 58. MM.
— Martin. 9. FF.
— Mildred, Mary and Martha. 79, 83 (both dead). FF.
— Moilliet, James Keir (1836–1906) and John Lewis (1836–1909). 38. MM.
— Nichols, Emily and Clara. 16. FF.
— Oldham, Ernest Haslope and Alfred Haslope. 26. MM.
— Patch[31]. Adult. FF.
— Philpotts, Sextus Howard and Septimus Buller. 34. MM.
— Neville-Grenville, Cicelie and Georgiana. 74. FF.
— Robertshaw, Calvin and Luther. 44. MM.
— Roundell, Richard Foulis and Charles Selborne[32]. 2. MM.
— Sansome, John and George. Adult. MM.
— Scott, Emily Augusta and Ellen Arabella. 15. FF.
— Shillitoe, Emily and Alice. Adult. FF.
— Sykes, Sophia and Lucy Isabella. 16, about 22 (both dead). FF.
— Dimsdale, Finetta-Esther and Caroline[33]. 51. FF.
— Thornley, Edward and Charles. 43. MM.
— Thornley, John and Alfred. 46. MM.
— Underwood[34]. 25. MM.

[27] Schoolboys at Harrow. Galton seems to have little more than general statements about them from Montagu Butler.
[28] Possibly Groningar.
[29] Married name Jeaffreson.
[30] 1855–1928, 1855–1929. Sons of George Lambton, 2nd Earl of Durham.
[31] Mentioned by Lady Strachey as living in India.
[32] 1872–1940, 1872–?
[33] Lady Tarleton and Mrs D.H. Mackinnon. Born 1824.
[34] From Mrs. Lane-Fox.

— Villiers, Edith and Elizabeth[35]. 34. FF.
— Wakefield, William and Lewis. 39. MM.
— Walters, David and Hugh. 20. MM.
— Watton, Frederick Leopold and Arthur Bernard. 6. MM.
— Welstead, Juliana Caroline and Mary Agnes. 42. FF.
— Wilkinson, Johnson and Osborn[36]. 53. MM.
— Wölfel, Johannes Carl Robert and Johannes Ludwig Gustav[37]. 20. MM.

The resemblance between the Lushington twins was especially striking. Vernon Lushington had responded directly to Galton's circular with a comprehensive account approved by his brother Godfrey.[38]

30 Dec 1874

Dear Galton

My brother Godfrey concurs in the inclosed replies to your Twins questions.

The Rev. George Self, (Saffron Walden) had twins twice running, but his family history has been a very sad one. His Wife died, and 3 of his children were carried off by a malignant scarlet fever. I mention this in case you sh[d] write to him.

With very kind regards to yourself & Mrs Galton

I am yours always

Vernon Lushington

P.S. I'll answer any further question you put to me, if I can.

(1). Vernon | Godfrey Lushington, born 8 March 1832 within 10 minutes of each other.

(2). Very closely alike, in boyhood, youth & adult life.

(3). Continually mistaken by acquaintances, and friends, occasionally even by near relations on a hurried or unexpected meeting, or under other special circumstances. The likeness has prevailed through life; i.e. thro' 42 years: but latterly Godfrey has worn a beard, while Vernon shaves, this, however does not prevent mistakes being still made.

Height. Godfrey. 5.9 1/4 Vernon. 5.8 1/2.

Weight. Both about 13 stone. Same clothes fit both, but Godfrey is rather the larger.

Hair, exactly the same colour (light brown).

Eyes — Blue Grey.

Powers of athletic performance practically the same, not naturally dexterous, but good power of endurance. Neither has any skill in drawing, or in any manual craft.

Handwriting not markedly similar, but both have the same character of legibility.

Voice. Both tenor, and very like in quality. Voices often mistaken.

Special tastes, not very marked but almost invariably similar. Both fond of music.

Dispositions very similar. Both share the same religious opinions, (which are not those commonly professed); both admire the same persons, books &c. Speaking broadly there is nearly identity of opinion.

Health of both good, and very similar.

35 Born 1841. Lady Lytton (painted at 21 by Watts, see Figure 7.1, p. 145) and Mrs Loch.
36 1822–1897, 1822–1906.
37 1855–?
38 Lushington to Galton, 1874-12-30, GALTON 2/4/3/5.

(4). Not parted until they were 14½ years of age, then Godfrey went to Rugby & afterwards to Balliol College Oxford.[39] Vernon was for 3 years in the Royal Navy, he then went to Haileybury (being designed for the Indian Civil Service) after a year there he was transferred to Trinity College, Cambridge.[40]

Both left the university about the same time, and adopted the profession of the Bar. They lived 9 years in the same lodgings together & married in the same year, 1865. Vernon has 3 children.

In 1869 Vernon became Secretary of the Admiralty, Godfrey Legal Adviser to the Home Office.

5. The closest likeness was in infancy; probably idiosyncrasies have been ever since developing (within certain limits). No marked dissimilarity can be specified.

6. The external circumstances of both have been very similar: but we attribute whatever likeness or unlikeness there may be, mainly to natural qualities.

7. [... no twins in the uncles, aunts etc. of the pair.]

Also included under 'close similarity' were the 16 year-old Macrae twin sisters, Bessy and Adie, who both appear to have inherited a slightly crooked little finger from their maternal grandmother. They were identical physically and, at least according to Galton's informant Dr. Finlay Murchison, mentally. They enjoyed swapping places undetected for their separate music lessons. On board ship, they became sea-sick at exactly the same time. 'External influences seemed powerless to produce any dissimilarity'.[41] For a moment it seemed possible that Galton had stumbled on exactly the same case mentioned by Darwin in his *Variation* (see above) but inquiries to the naturalist, and through him to his source the physiologist Dr William Ogle (1827–1912), showed that this was not the case—those were Ogle's own sisters.[42]

Very dissimilar cases were most likely fraternal twins, of which Galton had collected some twenty instances. For example, the remarkable pair Edward Pears Wilson and James Maurice Wilson (1836–1856, 1836–1931). They were sons of the Vicar of Nocton, Edward Wilson, who had been 17th Wrangler at Cambridge in 1825, and had educated them together. Edward had fallen victim to consumption in 1856, but his brother James, a Broad Church Darwinian, had been Senior Wrangler in 1859 and was teaching at Rugby school. He wrote to Galton that 'my brother Edward and I were complementary so to speak in point of ability. He was contemplative, poetical, literary to a remarkable degree, shewing great power in that line. I was practical, mathematical & linguistic. Between us we would have made a very decent sort of man'.[43]

To his regret, Galton was unable to use details from another case of dissimilarity, sent to him by a rather severe father of twins, the physician George Fowler Bodington (1829–1902). Despite being brought up exactly alike at all times, Walter and Arthur stayed dramatically dissimilar. 'Walter is an extremely stupid boy at lessons' with staring eyes and a vacant look, his father wrote, 'the despair of every teacher who has ever had him in hand'. On the positive side, he was 'an excellent mimic' and horse rider, though he cared rather more for horses and grooms than for other people. And Walter was always in trouble. 'If there is any dirty trick—any nastiness I mean—done amongst the children it is Walter & he alone who does it. (e.g. dipping a stick into a privy hole and daubing his brother with it)'. Galton's annotation reads 'Alas I can't quote this though it is much to the purpose'. Arthur on the other hand was clever and known in the

[39] Ed.: admitted in 1850.

[40] Ed.: admitted in 1852.

[41] Murchison to Galton, 1875, GALTON/2/4/3/5.

[42] Caroline Johnson and Amelia Mozley Ogle, see *The Correspondence of Charles Darwin*, 23; 1875. Galton had data for Alexandrina Cornfute Macrae (Nov. 1859–1942) and Elizabeth Anne Macrae (Nov. 1859–1954). He would have been intrigued by the fact that they were married in a double wedding in April 1887.

[43] Edward Wilson to FG, 1874-12-28. James Maurice Wilson to FG, 1874-12-31. GALTON/2/4/3/4.

family as 'pious Arthur' or 'the Bishop'. He was eager to please, even enjoying 'drawing-room life', knitting and needle-work, 'very much a girl indeed in his habits & ways'. Here some readers may feel a twinge of sympathy for Walter. Arthur Eaton Bodington (1864–1938) was educated at Caius College in Cambridge and became a physician. His father George went on to run an insane asylum in British Columbia, where Walter (1864–1943?) became a pioneer farmer on Sea Island, but only after wandering around the Wild West.[44]

William Allingham, the editor of the monthly *Fraser's Magazine*, was enthusiastic about Galton's project. The magazine offered to publish the results 'on the understanding that Fraser is not to be forestalled by the printed publication of the matter elsewhere', preferably 'without even being read to an audience' as 'a great many readers are satisfied if they can get an inkling of any topic that is talked about, so that even a brief summary w[d] take the edge off their curiosity'.[45] In due course 'The history of twins, as a criterion of the relative powers of nature and nurture' appeared in *Fraser's Magazine* for November 1875. It was sandwiched between 'A Ride through Bosnia' by 'V' and 'Ventilation in Hospitals' by Dr. William Gray. Victorian readers had a taste for solid material. With a circulation of about 6,000 the magazine was middling for a monthly—*Cassell's Magazine* sold perhaps 200,000, and even the *Cornhill* stretched to 18,000—but it commanded a readership that is thought to have been more educated and possibly more tolerant and open-minded.[46]

Galton argued that his twin data allowed two approaches. 'We might begin by enquiring about twins who were closely alike in boyhood and youth, and who were educated together for many years, and learn whether they subsequently grew unlike, and, if so, what the main causes were which, in the opinion of the family, produced the dissimilarity'. In short: do initially like twins become unlike through nurture? But he thought a converse approach more powerful. 'We can enquire into the history of twins who were exceedingly unlike in childhood, and learn how far they became assimilated under the influence of their identical nurtures; having the same home, the same teachers, the same associates, and in every other respect the same surroundings'. In short: do initially unlike twins become like via nurture?

The second approach addresses an obvious response to the first. Though twins may stay similar over time, they could be more or less fully formed within the home environment, where they are treated equivalently. Differentiating experiences after that may matter less. However, if that is the case then the same effects should be visible for initially unlike twins raised in the same environment, which they share far more closely than mere siblings would. If sharing that environment can make (or keep) twins similar, it should make initially unlike twins more like over time. Here Galton's reasoning was strongly influenced by Darwin's formulation in *Variation of Animals and Plants under Domestication* (1868/75): 'variations of all kinds and degrees are directly or indirectly caused by the conditions of life to which each being, and more especially its ancestors, have been exposed'.[47] Or, 'if it were possible to expose all the individuals of a species during many generations to absolutely uniform conditions of life, there would be no variability'.[48]

The traits eventually collated and tabulated for the twins were comprehensive, including height, weight, handwriting, hair and eye colour, expression, athletic powers, and voice. Also compared, though for fewer cases, were 'dispositions' (personality), mental powers and 'association of ideas', and even 'special ailments' or 'peculiarities'.

It is clear from his treatment, though he is not explicit about this, that Galton used physical resemblances to establish which pairs were identical, after which he tried to determine which

[44] Walter left reminiscences, including remarks about his relationship with his father, `https://search-bcarchives.royalbcmuseum.bc.ca/bodington-walter-eaton-1864-winnipeg-farmer-2`.

[45] GALTON/2/4/3/13.

[46] Ellegård 1971.

[47] C. Darwin 1875, 240-1.

[48] C. Darwin 1875, 242.

of those identical twins shared less easily established mental and behavioural characteristics. He was surprised to discover that handwriting hardly ever matched and so was not useful for equating twins.

For physical appearance, Galton had many examples of mistakes made by parents, siblings, close friends and tutors. Even the twins mistook each other at times. 'No less than nine anecdotes have reached me of a twin seeing his or her reflection in a looking glass, and addressing it, in the belief it was the other twin in person'. Galton's nephews James Keir and John Lewis Moilliet were not unusual in being punished at school for the sins of the other; at Cambridge, the porter refused to let either pass through the Trinity College gates when he was unable to distinguish the visitor, James Keir, from the college resident, John Lewis. One of the Wakefield twins was lost at the amusement gardens, but the inclusion of his brother in the search party repeatedly led to it being called off, on the supposition that the missing party had been found and was among them.

Galton counted nine cases of twins who developed exactly the same physical or mental diseases at the same stage, and found general references in the literature to more cases (though not all of these could be confirmed). 'The steady and pitiless march of the hidden weaknesses in our constitutions, through illness to death, is painfully revealed by these histories of twins'. Philosophizing followed. 'Necessitarians may derive new arguments from the life histories of twins'.

Even ideas seemed to connect in the same way in numerous cases, as in the German Lehmann twins. 'They make the same remarks on the same occasion, begin singing the same song at the same moment, and so on; or one would commence a sentence, and the other would finish it'. The Scottish twins Calvin and Luther Robertshaw—who had interchanged places while courting the same woman, ultimately claimed by Calvin, and unwittingly swapped carte-de-visite photographs—separately bought each other identical wine glasses during the same weekend, in different towns. Each Fowler twin would intermittently assume the other's cryptic subtleties of expression to become more like his brother than himself. Where there were differences in these cases, Galton thought they were in intensity or energy rather than in kind, 'in the key-note, not in the melody'.

Analyzing his cases, identical twins seemed to maintain a close resemblance throughout life, with some moderate divergence, which he attributed to the effects of non-shared diseases and accidents, or subtle differences in nature only expressed later in life. This was 'not at all what I had expected'—in *English Men of Science* he had expected divergences to steadily set in. Convergence of unlike twins under the influence of nurture did not happen at all. 'I have not a single case in which my correspondents speak of originally dissimilar characters having become assimilated through identity of nurture'. At this early stage he grasped the essentials of what we now call developmental genetics. 'The twins who closely resembled each other in childhood and early youth, and were reared under not very dissimilar conditions, either grow unlike through the development of natural characteristics which had lain dormant at first, or else they continue their lives, keeping time like two watches, hardly to be thrown out of accord except by some physical jar'.[49]

Taken together, Galton understood his results to show that life outcomes are by and large the results of nature eventually triumphing over experience, perhaps after some detours. 'Many a person has amused himself with throwing bits of stick into a tiny brook and watching their progress; how they are arrested, first by one chance obstacle, then by another; and again, how their onward course is facilitated by a combination of circumstances. He might ascribe much importance to each of these events, and think how largely the destiny of the stick had been governed by a series of trifling accidents. Nevertheless all the sticks succeed in passing down the current, and they travel, in the long run, at nearly the same rate'.

[49] F. Galton 1875j, 575, Bouchard and Propping 1993, 3.

But he added the crucial qualifier, alluded to above, that this is contingent on *how the world actually is*, and not *how the world might be*. 'There is no escape from the conclusion that nature prevails enormously over nurture when the differences of nurture do not exceed what is commonly to be found among persons of the same rank of society and in the same country'.[50]

The *Spectator* was broadly sympathetic but doubted whether Galton had enough data to have confidence in his far-reaching conclusions—after all he only had 35 detailed cases of close similarity of twins. The reviewer took exception to his ruminations about free will, suspecting that those thoughts encouraged doubt about Christian convictions. This provoked Charles Darwin, who had anticipated Galton's article with enthusiasm. 'Nothing seems to me more curious than the similarity and dissimilarity of twins'.[51] Now the naturalist wrote to his cousin on the 10th of November, 'I have this minute finished your article in Fraser & I do not think I ever read anything more curious in my life. It is enough to make one a Fatalist. I am in a passion with the Spectator who always muddles if it is possible to muddle'.[52] Although the *Aberdeen Press and Journal* seconded the *Spectator*—'Mr Galton's philosophising will be skipped by the judicious reader', since his 'facts are too few to warrant theorising'—the *Lancet* was intrigued by the results, especially the twin concordance for disease and mental derangement.[53] Later, Alfred Russel Wallace would write to say that the 'paper on twins is also wondrously suggestive'.[54] Similarly, that *coming man* of evolution, George Romanes, would write to say that 'I had no idea "nature" was of so much more importance than "nurture", & in no other way, I suppose, could this point be proved'.[55]

A version of the twins paper with minor corrections and some additions soon appeared in the more scholarly *Journal of the Anthropological Institute*.[56] Colonel Lane-Fox—whose wife was one of Galton's sources of information about twins—presided over the society, while Galton was a vice president. The most important addition was a final argument that anticipates the study of genotype-environment correlation by over a hundred years. 'Much stress is laid on the persistence of moral impressions made in childhood, and the conclusion is drawn, that the effects of early teaching generally, must be important in a corresponding degree. I acknowledge the fact, but doubt the deduction. The child is usually taught by its parents, and their teachings are of an exceptional character, for the following reason. There is commonly a strong resemblance, owing to inheritance, between the dispositions of the child and its parents. They are able to understand the ways of one another more intimately than is possible to persons not of the same blood, and the child instinctively assimilates the habits and ways of thought of its parents. Its disposition is "educated" by them, in the true sense of the word, that is to say, it is evoked earlier than it would otherwise have been'.[57] In modern terms, the nurture children receive is shaped by their own genes, passively or actively, and also by the genes of their parents and other relatives.[58]

The *Journal of the Anthropological Institute* also published a second more technical paper on the twin data, 'Short notes on Heredity &c. in Twins', which Galton read before the society on the 13th of November. It dealt with the inheritance of twinning itself.[59] From Ansell's actuarial data he estimated that the chance of a twin birth in Britain was about 1 in 50. But in Galton's circular data an aunt or uncle of a twin had a 1 in 20 chance of being a twin, 2½ times

[50] F. Galton 1875j, 576.
[51] Darwin to Galton, 1875-10-07, GALTON/1/1/9/5/7/19.
[52] Darwin to Galton, 1875-11-10, GALTON/1/1/9/5/7/20.
[53] *Aberdeen Press and Journal*, Wednesday 17 November 1875, 6; *Lancet*, 2 (109) 1875, 669-70.
[54] Wallace to Galton, 1876-03-03, K. Pearson 1924, 187.
[55] Romanes to Galton, 1876-01-28, GALTON/2/4/3/13.
[56] E.g. a reference to 'double-yoked eggs' was removed from the explanation for monozygotic twinning.
[57] F. Galton 1875i.
[58] Darlington 1953; Plomin 1994.
[59] F. Galton 1875g,i,j. The material was revised slightly for inclusion in F. Galton 1883d.

the expected rate. This suggests that the tendency to bear twins is itself hereditary. Modern evidence strongly supports that deduction.[60] Galton was the first person to reach it using hard data and statistical arguments. In current Western populations the twinning rate is about 1 in 80, close to Galton's estimate. About 25% of these are identical.

A third paper mentioning twins appeared in the same journal, immediately after 'Short Notes', titled 'A Theory of Heredity'. As we have already seen, this was Galton's alternative formulation of the biological mechanisms of heredity, which he proposed as a solution to the difficulties posed by Darwin's theory of 'pangenesis'.[61] But Galton was for once misled by his twin data.

As mentioned above, both extremely like and extremely unlike twins appeared to always be of the same sex in Galton's data. He deduced that *both* cases were probably single-egged twins, since if two-egged genesis produced extreme unlikeness, he ought to have plenty of cases of mixed-sex twins who were extremely unlike, yet he did not. This implied that twins from single eggs were probably of two extreme types, like and unlike, with the milder partly-unlike types coming from two-egged twins. However he conceded that this was a leap of reasoning without strong grounds. In this case, Galton was probably undone by multiple sources of bias due to the way in which his data was collected and judgments of similarity were made by his informants. With representative data, fairly judged, he would have had quite a few cases of extremely unlike twins of mixed sex, removing the impetus for his argument.

Following up on the later careers of the identical twins ironclads Galton's general conclusions. Johnson and Osborn Wilkinson, who had entered separate branches of the army, both rose to become generals, causing general confusion whenever they crossed paths, receiving each other's official documents. They had identical temperaments. In 1895 they jointly wrote *The Memoirs of the Gemini Generals: Personal Anecdotes, Sporting Adventures, and Sketches of Distinguished Officers*. A reviewer in the *Athenaeum* knew them in maturity. 'As like each other as the two Dromies'. 'There is no exaggeration in the extraordinary resemblance of the two brothers, as the writer of this review can testify from personal experience'.[62]

Similarly, the Guppy twin brothers, who were just 20 in 1875, followed identical trajectories. Henry Brougham Guppy qualified as a doctor at Edinburgh and served in the navy as a surgeon, going on to become a Fellow of the Royal Society, as well as a geologist and botanist in the Pacific. He wrote a biography of his brother William Good Guppy—who had also qualified as a surgeon at Edinburgh and served in the Royal Navy, but died of 'enteric fever' at Erzeroum in Anatolia during the Russo-Turkish War.[63] The Lambton twins, who were also 20 in 1875, show the vicissitudes of chance. John George Lambton, older than his twin by minutes, succeeded his father George, becoming 3rd Earl of Durham in 1879, but married a woman who quickly went insane and was institutionalized, producing no heirs. After his death in 1928, leaving only an illegitimate son by a dancer, his twin succeeded him as 4th Earl, but died shortly after in 1929, leaving six children. Both twins had served in the Coldstream Guards.

Curiously, little progress was made by other researchers on twins for the next quarter-century—despite Galton's popularization of his results in 1883, in his widely-read *Inquiries into Human Faculty*.[64] (See chapter 9, p. 159.) He was mostly absorbed in other directions himself. His notebooks on the twin data show him already groping for better measures of traits, which he eventually expended nearly endless energy and ingenuity on. However he did return to twins in his fingerprint researches, where he analyzed prints from 17 sets of twins.[65] (For

[60] Segal 1999.
[61] F. Galton 1875a. This is covered in detail in chapter 5, p. 85.
[62] *Athenaeum* (3597) 1896-10-03, 445-446. A dromy is a clone.
[63] Henry Brougham Guppy *William Good Guppy: His Life and Death at Erzeroum* (London: Virtue, 1879). Guppy also wrote several other books.
[64] F. Galton 1883d.
[65] F. Galton 1892a, 185. Contrary to the claim in Rende, Plomin, and Vandenberg 1990 that Galton did not

fingerprints, see chapter 11, p. 237.)

In 1905 Edward Thorndike would take advantage of the new mental tests and statistical techniques which resulted from Galton's later work to publish the next study on twins. In 1925 Herman J. Muller formulated an alternative version of the twin method, which compared identical twins adopted at an early age and raised entirely apart.[66]

Unfortunately the tantalizing letter to Galton from Townsend mentioning the Liverpool merchant twins, who had been separated and raised entirely apart, seems to have led nowhere. Muller had a few cases of these, but they are uncommon. This has led some to argue that Galton did not invent 'the twin method' at all, since he did not deal with twins raised apart. That conflates the general use of twins in behaviour genetics with one particular method for using them. It has even been argued that since Galton did not know about our modern sense of genes, which had not yet been formulated, he could not have used what we would call a twin method for weighing nature and nurture![67]

Similarly, it has been argued that since Galton did not know that identical twins were monozygotic and non-identical twins were dizygotic, even though he knew that twins were either single-egg or two-egg in origin, he was not consciously comparing genetically identical twins to genetically non-identical twins, and therefore was not performing twin analysis as we now understand it.[68]

However in that pre-genetic age Galton correctly understood, as pointed out above, that identical twins are *identical in nature* and that non-identical twins are *different in nature*—he was after all studying the relative influences of nature and nurture. Saying that twins are like or unalike because of their one-egg versus two-egg genesis tell us about the *causes and transmission* of their differences in nature, an interesting question but not to the point here, even if modern usage runs the two ideas together all the time. Despite knowledge in principle of the genetic causes since 1900, it was not until the present day that it was even feasible to do more than make an inference from an observation of similarity to the single-egg genesis: anything more requires sequencing the genes or some other genetic test. Almost all twin studies in the past have *not* performed genetic tests.

Setting nature constant and varying environment is an important piece of evidence, but hardly exhausts the use of relatives in behaviour genetics. One of Galton's methods set the shared (family) environment equal and natures different by using fraternal twins, the other equated both genes and environment. He had a key idea exactly correct here, which is to compare identical with fraternal twins. Modern designs have evolved far beyond even Muller's identical twins raised apart to include many kinds of genetic relationships, each one of which contributes information.[69] Since controlled experiments are widely thought to be unethical, every natural experiment has to be taken into account.

What Galton founded was not so much 'the twin method' as the *entire field of behaviour genetics* that the use of twins is incorporated into. After the pedigree studies in *Hereditary Genius*, the twin studies were the second major component of the comprehensive research programme that Galton was feeling his way toward. He was well aware of its deficiencies when it was just getting off the ground. He had to rely on the judgment of parents and relatives

return to twins.

66 Muller 1925. Muller used the case of Jessie and Bess Irwin first publicized by Paul Poppenoe a few years earlier, see Poppenoe 1922.

67 See Bulmer 2012. By arguments like these it may be proved *reductio ad absurdum* that Isaac Newton did not coinvent calculus with Leibniz because he did not have the concepts of real numbers and limits, which were only formulated in the 1800s and underpin the modern understanding. By the same token, Darwin could not have discovered evolution, which we now consider to be mostly selection at the level of the gene, also unknown to Darwin.

68 Rende, Plomin, and Vandenberg 1990.

69 Neale and Cardon 1992; Segal 2012.

about 'similarity' but he wanted less subjective measures of resemblance for traits, especially mental traits, in a form that would enable statistical analysis.

Over the coming years he would patiently work at devising measures, identifying unstudied physical traits like fingerprints in the process. The number of his cases was still very modest, but he would ultimately find ways to collect much larger data sets, so large that they would only be comprehensively analyzed over a hundred years later.[70] In processing the data, he would have to devise new methods of statistical analysis for comparing measurements of variables and drawing inferences from them. Often he would be surprised by the results.

Nevertheless the core conclusions that Galton reached even at these nascent stages have been borne out by subsequent research. A genius for close observation and discriminating judgment, grounded in clear-eyed use of empirical evidence, was enough to get Galton on the right track.

[70] R. Johnson et al. 1985.

Figure 7.1: Edith Villiers, twin of Elizabeth and among Galton's subjects. Painted in 1862 at age 21 by G. F. Watts.

Figure 7.2: T.H. Huxley lecturing at the Royal Institution in 1885, as imagined by Punch. *Tyndall appears in the bottom right.*

Chapter 8

Sweet Peas Regress

'Friday Night Discourses' at the Royal Institution were a weekly fixture during the first six months of the year. The lecture theatre in Albemarle Street, Piccadilly, was usually packed by high-minded 'members and friends' from the general public, attentive in their best dress. Ladies were not uncommon. Lady Palmerston would often bring her daughters. In 1869, Louisa and Frank attended John Tyndall's course of nine Saturday lectures on 'Light' at the Institution, which Galton joined that year. Frank gave his first Friday Evening Discourse there in 1874, introducing his work-in-progress on English men of science.

Three years later, at 9PM on Friday February the 9th, Galton was back to deliver another. His old colleague from *The Reader*, Sir Frederick Pollock, was in the chair that night. Lady Elizabeth Hamilton and her husband Lord Claud were in the audience. It was Lady Hamilton who had brought her daughter Louisa to a lecture in 1872, where she met Tyndall, lonely, dyspeptic and frequently sleepless. 'In short, he lacked a wife'.[1] They were engaged to be married in 1876.

The 1877 season was full. Two weeks prior, John Lubbock had described the 'Habits of Ants'. A few months later, T. H. Huxley would describe the 'History of Birds', James Bryce would canter through 'Armenia and Ararat', Frederick Pollock would scrutinize 'Spinoza', G. J. Romanes would flay apart the structure of 'Nerves', Mackenzie Wallace would reveal 'Intellectual Movements and Secret Societies in Russia', and John Tyndall would round off the season of 19 discourses with an improving disquisition on 'Putrefactive and Infective Organisms from a Physical Point of View'. Next to these, Galton's presentation on 'Typical Laws of Heredity' may sound dull, but it would be a triumphant display of innovation, illustrated by some highly original mechanical devices. Regression to the mean would be described for the first time. Natural Selection would be shown to be perfectly compatible with the 'law of errors'. His contemporaries would need years to absorb the new ideas fully. But Louisa was not in the audience. She had taken one of her frequent illnesses a few weeks earlier and was laid up in bed at Rutland Gate. By Easter they would despair for her life.

Three years earlier at the Institution, Galton had sung the praises of the ubiquitous bell-shaped 'Law of Errors' that he would rename the 'Normal Distribution'. He had relied on it heavily since 1865. But something had been bothering him for years. *Why* was the bell-curve so omni-present? The usual answer, that it was a necessary consequence of the 'Central Limit Theorem' wherever the causes of something were small, very numerous and independent, did not satisfy him. Many, perhaps all, natural phenomena were subject to the very *large* influence of a *single* factor, natural selection. How did the distribution shrug off such a large solitary influence and still reign serene?

He had the answer with him now, and could provide a convincing demonstration of it, using two advanced versions of the device he had first shown three years earlier, the Quincunx (see chapter 6, p. 103). These new devices were so ingenious that later writers would suppose that

[1] Caroe 1985.

they were not real at all, only *thought experiments* in lithographed diagrams. It was not so. The new Quincunxes were with him on the platform that night. Their separated stages of tumbling and accumulating lead shot would animate results from extensive experiments with the seed of sweet peas.

Galton had looked for the clues to the solution of his 'law of errors' conundrum in real experimental data. Plants are relatively straightforward to experiment on and control. Sweet Peas produce seed which can be measured by diameter, or more easily by weight. Since they are self-fertilizing, their generations can be charted precisely. In 1874 Galton tried growing some sweet peas at Kew Gardens, on the advice of its director, the botanist and X-Clubber Joseph Dalton Hooker. He could not easily do this in Kensington, at garden-less semi-detached Rutland Gate. He was usually away for a good part of each year anyway, rambling through the choicer spas of the continent, invariably followed by visits to family and friends in the country. That year Louisa recorded that they were 'prevented going away by domestic bothers, which have made the whole of this year sadly trying' (Frank's mother had died in February). They were not in London though, and it seems that Galton entrusted care of the sweet peas to the staff at Kew. The experimental strategy used is not known—it may have been similar to later trials—but he was disappointed. Somehow the entire crop was lost.

The second attempt at sweet peas engineered redundancy to guard against catastrophic losses. In April of 1875 nine growers were enlisted in different rural locations, from Nairn in the north of Scotland to Cornwall. Charles Darwin grew some at Down in Kent, and George Romanes in Cambridge. Darwin had been volunteered by his son, the mathematician George Howard, who was rapidly forming a close friendship with his numerate cousin.[2] The Galtons spent the summer in France at Fontainebleau, and in Switzerland at Seeligsberg, where Louisa recovered from another long illness. Before they left in early May, the peas had been subjected to rather drastic selection, artificial rather than natural, and distributed.

A large bag of seed, bought from a nursery, was sorted by weight into seven descending categories, K to Q. Each grower received seven packets of seed, one for each weight category, with the same planting instructions (seven 1.5×5 foot beds in parallel rows). There were ten seeds of equivalent weight in each packet. After several trials Galton satisfied himself that weight was a good proxy for seed diameter, but his results could be stated purely in terms of weight for any quibblers. He wanted to see the effect of this extreme selection on the distribution of seed produced by the offspring. He certainly expected the offspring to produce seed weights or diameters which followed the law of errors, as the original bag of seed did, but what was the relationship between the two distributions? He hoped that this would give some clue as to why the large single effect of selection, natural or artificial, did not invalidate the 'law of errors'.

The growers sent the seed of the offspring back to Galton for weighing, labelled with the category that the parents had come from, along with the plant foliage. Two of the nine crops failed, leaving seven complete sets, the offspring of 490 parent seeds. The sample size of the offspring seed is not known. Louisa noted in her record: 'Frank busy about the statistics of his sweet peas'. He later wrote that he took 'immense pains' about the analysis. This included measuring the foliage as a cross-check. One reason for these pains was a startling pattern that emerged from the data. The offspring of any packet of seeds had roughly the same 'dispersion' (which Galton called 'family variability'). Selection of parent weights had not altered this. 'I was certainly astonished to find the family variability of the produce of the little seeds to be equal to that of the big ones'.[3]

More interesting still, for each parent weight category, the offspring had rather different *mean* weights than the parent seeds. Seed derived from the heavier (larger) parents tended to

[2] FG to Charles Darwin, 1875-04-14, DARWIN, DCP-LETT-9928.

[3] F. Galton 1877d. The 'family' qualifier reflected Galton's (incorrect) idea that all siblings in a family receive the same 'stirp' of heritable information, see chapter 5, p. 85.

be lighter (smaller). Conversely, offspring from the lighter (smaller) parent seeds tended to be heavier (larger) on average than the parent category. The weights in each category had thus moved closer to the overall mean of the offspring. Galton called this 'reversion to mediocrity', after the biological idea of reversion to ancestral characters. Later this would become known as 'regression to the mean'. In the process he formulated and drew a regression line with a slope of roughly 1/3, relating the mean weights of offspring to their parents, expressed in terms of seed width. As we have seen above, he had already groped his way to the idea of regression four years earlier, when relating the heights of scientists to their hat sizes. Now he made the idea explicit, with a line he fitted using a simple smoothing process equivalent to the method of 'least squares'.

However Galton would not publish these detailed results for another eight years. 'The exact ratio of regression remained a little doubtful, owing to variable influences; therefore I did not attempt to define it'.[4] Indeed he never published a diagram of the regression line for pea weights, preferring in later years to diagram the much more complete results he got from human heights. The plot in Figure 8.2, p. 154, has been recreated from the summary of the data given a decade later by Galton, which is in terms of seed diameters in 1/100 of an inch.[5]

In the future, when Galton had considerably more data, he would realize that 'reversion to mediocrity' was a special case of a much more general law. For now this reversion was the key to the resolution of a specific mystery that had been puzzling him. If the means of each size category remained unaltered across generations, the variation among the combined offspring would steadily increase. The offspring of each size category were, as his data showed, equally dispersed, but necessarily more dispersed than the single-sized parents. Without balancing forces, the distribution would get inexorably wider over time, with extreme values becoming ever more extreme. Galton knew that accumulated distributions, where each component obeyed the Law of Errors, also obeyed the Law of Errors (see Appendix E, p. 427, for the details). This was a standard result from the *Law of Errors*, by his friend Sir George Airy—in the year before that book was published, Galton had accompanied the Astronomer Royal's expedition to Spain on the *Himalaya* to measure the solar eclipse under clearer skies.[6] Or, in language that Galton himself would invent, the sum of normals is normal. The sum of the parts would obey the Law of Errors but with increasing dispersion over time. But in practice, contrary to this theory, dispersion over the seed population stayed more or less constant, all other things being equal. One reason for this was now clear: the means of the offspring categories snapped back toward the overall mean, because of reversion to mediocrity. Another counterbalance came from reduced fecundity at the extremes: giants produce fewer offspring, as do midgets.

Presenting these ideas to a popular audience was a challenge. In May of 1876, Galton wrote to George Howard Darwin, who had been Second Wrangler in his year at Cambridge, asking for advice about his first attempt to write up the results.[7]

> May I take the great liberty of asking you kindly to look at a short paper I have just finished concerning my peas.
>
> "Experiments with plants, on the causes of statistical uniformity in successive generations".
>
> I should be truly glad of criticism before I send it in, while there is still opportunity to alter. I want to be (1) correct (2) intelligible.
>
> All the details will be in an appendix, with which I do not propose to trouble you in MS. I am very curious as to what you may say about the "selection".

[4] F. Galton 1885f.

[5] F. Galton 1889c, 226. Each child seed size shown is an average over multiple observations. Galton only gives percentages in each size bucket for the offspring of each parent size.

[6] Airy 1861; F. Galton 1861; Tredoux 2023c.

[7] FG to George Howard Darwin, 1876-05-10, GALTON/3/3/7/8.

Darwin's reaction was comprehensive and sharp enough to give Galton pause. A week later he responded with some chagrin.[8]

> How can I thank you sufficiently for the great trouble you have taken about my paper, and your criticism. I will wholly, or almost wholly rewrite and expand and not think of sending it in, in its present form.
>
> These confounded law of error ideas, which in themselves are so simple & clear but to express which no proper language exists, and which lie so completely out the every day lines of thought, are very baffling to deal with and to present. But I don't despair yet. However I won't bother you more until I have had another good drive at the thing, and then, as you kindly permit me, I will ask you again about it.

The paper was 'burked', as Galton later put it, in this early form and the manuscript has never been traced. He spent the rest of the year finding ways to get his ideas across more clearly. By February of 1877 he had two generations of sweet pea data to use.[9] This led to two modified versions of his Quincunx, to act as analogue mechanical models of the effects he wanted to describe. No advanced grasp of mathematics was needed to understand these. They gave visual evidence. Galton was baffled no more.

The original version of the Quincunx, devised in 1873, only showed the Law of Errors operating through a single stage of dispersion. Today this is still the most widely known design, copied many times by others. It was part of the Special Loan Collection of Scientific Apparatus at the South Kensington Museum and is described in the catalogue for 1876, as 'Apparatus affording Physical Illustration of the action of the Law of Error or of Dispersion'. By then it had acquired a button which could be used to jiggle a rake, preventing jamming in the funnel as the shot bounced down.[10] The first extension to the device added an intermediate catch and release stage, along with reversion to mediocrity. The second extension introduced additional selection of the shot as it passed down, to systematically punish the extremes of the distribution. Today this is called centripetal or 'stabilizing' selection.

A letter to George Darwin early in 1877, after Darwin had again criticized a new version of the write-up, explained how these thinking-aids worked.[11] For now Galton had adopted the word 'exponic' to describe the bell-curved shape of the Law of Errors, but would discard the word later.

> 42 Rutland Gate, S.W.
> January 12, 1877.
>
> My dear George,
>
> How can I thank you sufficiently. I am aghast at the trouble my unlucky memoir gives, and at the great pains you have taken to put clearness into it. I will certainly adopt your suggestions generally and rewrite the thing.
>
> Let me mention an illustration of one of the principles (Family Variation), which I think may interest you. You recollect that apparatus of mine with the shot;[12] well, suppose I want to show by a modification of it, how it comes to pass that when the ordinates of an exponic mountain subside, each of them, into an exponic hillock, as in the sketch, the sum of the hillocks is an exponic curve of larger modulus.
>
> In I,[13] I pour shot, and it makes an exponic heap at the bottom. In II,[14] I have cut the apparatus across at AB, and have interposed a row of vertical compartments with trap

8 FG to George Howard Darwin, 1876-05-17, GALTON/3/3/7/8.

9 Robert Olby 1966 misunderstands the scope of this data, which ultimately ran to three generations by the end of 1877.

10 F. Galton 1876a, 11.

11 Francis Galton to George Howard Darwin, 1877-01-12, GALTON/3/3/7/8.

12 The Quincunx.

13 The original Quincunx.

14 The two-stage Quincunx.

door bottoms that I can pull out and in to form a temporary landing for the shot, when I so desire. If these are open, the shot falls through and of course makes an exponic mountain at the bottom of II exactly as it did in I. But if they are closed, they intercept the shot and an exponic mountain (of less modulus) is formed on AB. [See Figure 8.3, p. 154]

Now I open the trap doors, successively; the shot in each vertical compartment rushes down and forms its own exponic hillock, and we have already seen what the sum of them will be. The ratio of the moduli of these heaps is self-evident (they vary as the square root of the indices which vary directly as the length of passage of the shot). For my Royal Institute lecture, I shall simply go into generalities to show what Reversion, etc., mean and how a law is possible, and shall hang up the formulae, but not speak a word about them.

Affectionately yours,
Francis Galton

The words 'suppose I want to show by a modification of it' in the above letter should not be taken to mean that the devices had not been constructed. They already had been, in some form, most likely by Tisley and Spiller, the precision instrument makers who had made the first, elementary, Quincunx. On the evening of the 9th of February, at the Royal Institution, Galton had his advanced Quincunxes with him on the podium. He knew that a parade of formulae alone would make many eyes glaze over. Visual aids were necessary. He had also included 'reversion to mediocrity' in the devices by means of an inwardly-sloping set of tubes, before the shot hit the pin orchard to be deflected this way and that. The *Morning Post* reporter was present and described the dimensions and operation of the extended device.[15]

> The formulae of the laws Mr. Galton has worked out were exhibited and briefly referred to as belonging to problems of the higher statistics. Their action, however, was illustrated by models. The simplest model may be described roughly in this way. A frame, 2ft. high, 18in. wide, and an inch and a half deep. The front is glass. The bottom portion of the case is divided in a single row into a number of compartments by little divisions running from the back to the glass front. All are the [same] height, about 4in. Above them are several rows of pins, projecting from the backs to the glass front. The rows are arranged so that the gaps of one row correspond in place with pins in another. Above these again is another series of divisions corresponding in number with the bottom series, and below each is a "shoot" or sloping tube, all of which converge to a central point. The top series of divisions is charged with pellets, shot, or small round seeds, each division having a sliding stop as its lower end. The stops are drawn in succession, and the contents of each division as liberated rattle down, striking the projecting pins, and thus being scattered, fall into the divisions beneath. When all the contents have been thus liberated it is found that a line drawn along the surfaces of the heap of pellets in the lower divisions forms a curve, highest in the centre and sloping down on both sides. Frequent repetition of the experiment produces the same curve. Whether the upper divisions are filled all to the same height, or whether they are so filled that a line drawn along the surfaces makes a curve, the result is the same. In this way the results of the processes of reversion and family variability were shown, the curve giving the degree of variation from the normal standard.

Surely no modern newspaper would engage readers with this degree of precision. Notice however that *how* the trapdoor stage ('top series') is filled with shot, by hand or by an elementary stage on top of it, is left undescribed. The depiction in Figure 8.7, p. 156, suggests that the model may have omitted this first stage, already shown by the elementary Quincunx, since it requires no thought at all and would make a physical device including it harder to carry.

Galton had an even more advanced Quincunx with him that night too. It was not a 'thought experiment' either. He had not described it to George Darwin, but a diagram of it is included in the written-up version of the lecture, see Figure 8.8, p. 157. It introduced a selection effect by means of a horizontally placed 'exponic' curve. The curve performed centripetal selection on the shot, draining away to waste a portion of the extremes at either end, with the appropriate

[15] *Morning Post*, Monday 1877-02-12, 3. Also syndicated in the *Welshman*, 1877-03-02, 6.

probability. The Law of Errors was preserved but with a narrower dispersion. The *Morning Post* reporter witnessed this device in operation too.

> The laws of productiveness and natural selection were explained with similar illustrations. The effect of these, together with that of dual parentage, help reversion in checking the step by step progress of dispersion in the race caused by family variabilities. It was shown that natural selection does not act by carving out each new generation to a definite pattern, irrespective of waste, but acts in strict conformity with the law of deviation. Individuals which deviate widely from a mean type either in excess or deficiency make but a small contribution to succeeding generations. The genealogical progress of a race was shown to consist in a constant outgrowth from the type centre and a constant dying away at the margins, and there is a tendency in the scanty remnants of all extraordinary exceptions to revert to the original type from which the race has sprung.

A report to the same effect appeared in the *Illustrated London News*.[16] 'The formula by which these laws are expressed was said to belong to the problems of higher statistics; but their actions were exhibited and made intelligible by working models, one of which illustrated "simple descent," in which the processes of reversion and family variability are alone concerned; while others explained the laws of productiveness and natural selection'. A notice in the *Academy* also described working models.[17] 'These conditions were expressed by a simple formula, and their combined action was fully illustrated by an apparatus, where reversion was represented by the passage of a heap of pellets through a row of converging shoots, and family dispersion by their subsequently falling through the teeth of a sort of harrow. It was explained that reversion acted precisely like an elastic spring, with greater tension the more it was stretched, and thus finally checked the step-by-step progress of dispersion due to family variability, which would otherwise cause the deviation of the race to continue indefinitely. Hence a condition of stable equilibrium was ensured, and perfect statistical resemblance between consecutive generations was maintained'.

Even putting the eye-witness reports above to one side, those who turn to the text of the Royal Institution lecture in its published form will also notice, in plain language there, consistent descriptions of working models.[18] It takes a series of mental contortions to believe, based on the text, that Galton was describing purely hypothetical devices. The original source of this mistake appears to be the statistician Stephen Stigler, in his influential *History of Statistics* (1986). Later authors have simply repeated the idea.[19] The fact that only the original quincunx has survived physically and not the more advanced ones—the patchy fossil record of scientific progress—has contributed to the confusion.

Stigler had some trouble building advanced quincunxes himself, and so concluded that they were most likely thought experiments, existing on paper only to elucidate concepts. The somewhat ambiguous letters to George Darwin above may also have played a role in this suggestion. He wrongly supposed that the manufacturer of the original quincunx was a 'tradesman'. Tisley and Spiller, of Brompton Road, were highly-skilled makers of precision optical equipment, and also produced specialized scientific instruments. Their workshop was not far from Rutland Gate. Galton himself was a compulsive inventor of instruments, a number of which went into production and received wide use in the field: a pocket heliostat, a precision whistle, an altazimuth, and many devices for automatically recording and transforming meteorological observations. Before those he had invented a printing electric telegraph when the field was still in its infancy.[20] In combination they succeeded. In 1892 the *Deutsche Mathematiker-*

[16] *Illustrated London News*, Saturday 1877-02-17, 23.

[17] *Academy*, 1877-02-17, 144.

[18] F. Galton 1877d.

[19] Stigler 1986. This is not to disparage Stigler, who has done more than any other modern writer to illuminate Galton's statistical context and legacy. Michael Bulmer also claims that there is no evidence that the advanced devices were constructed, c.f. Bulmer 2003, as does Chris Pritchard, c.f. C. B. Pritchard 2005.

[20] F. Galton 1849, 1850.

Vereinigung would list the Quincunx *with* advanced modifications in its *Katalog Mathematische und Mathematisch-Physikalischer Modelle, Apparate und Instrumente*, along with a detailed description by Galton.[21]

The published report of the lecture in the *Proceedings of the Royal Institution* treated more advanced cases in an appendix, but now with rigorous proofs.[22] The paper covered two parents (though without sexual selection) rather than just 'simple descent' from one, and went on to include reversion, reduced fecundity at the extremes of the population, 'family dispersion' among the offspring, and finally stabilizing natural selection at the extremes of the population. Galton introduced selection functions to describe the last process, as well as the effect of differential fecundity. By relating each stage formally to the prior stage, he was able to show the implications of supposing that the final population stays stable—that is, that it has the same mean and dispersion as the original population, and so does not evolve. That makes this 1877 paper the founding document of what we now call 'theoretical population genetics', preceding Fisher, Haldane and Sewall Wright by nearly half a century.[23]

—oOo—

Lou was seriously ill until June that year, and although she lived on, it was in a diminished state. The illness, she wrote in her yearly record, 'brought me nearer to Frank', 'my great support throughout'.[24] Her great support would continue one more generation of sweet pea experiments, for a total of three. However his attention was already transferring back to man. The peas were useful as far as they went, but better data would be required. New methods for understanding that data would be needed. Mankind itself might be improvable. He cast about for heritable human traits that could be measured. Precision whistles were already in hand. But his idea of anthropometry would not be limited to physical characters—hearing, height, strength, facial appearance—it would extend to intelligence, to 'good and bad temper', even to types of character, and the ability to see visions. Along the way Galton would explore synaesthesia and discover 'number forms'. And devise the first intelligence tests.

Diameter of Parent Seed.	Diameters of Filial Seeds.								Total.	Mean Diameter of Filial Seeds.	
	Under 15.	15–	16–	17–	18–	19–	20–	Above 21–		Observed	Smoothed
21	22	8	10	18	21	13	6	2	100	17·5	17·3
20	23	10	12	17	20	13	3	2	100	17·3	17·0
19	35	16	12	13	11	10	2	1	100	16·0	16·6
18	34	12	13	17	16	6	2	0	100	16·3	16·3
17	37	16	13	16	13	4	1	0	100	15·6	16·0
16	34	15	18	16	13	3	1	0	100	16·0	15·7
15	46	14	9	11	14	4	2	0	100	15·3	15·4

Figure 8.1: Galton's 1877 regression on sweet pea seed. Seed diameters, F. Galton 1889c.

[21] F. Galton 1892c.
[22] F. Galton 1877d.
[23] See chapter 13, p. 287.
[24] Entry for 1877, GALTON/1/2/1/8.

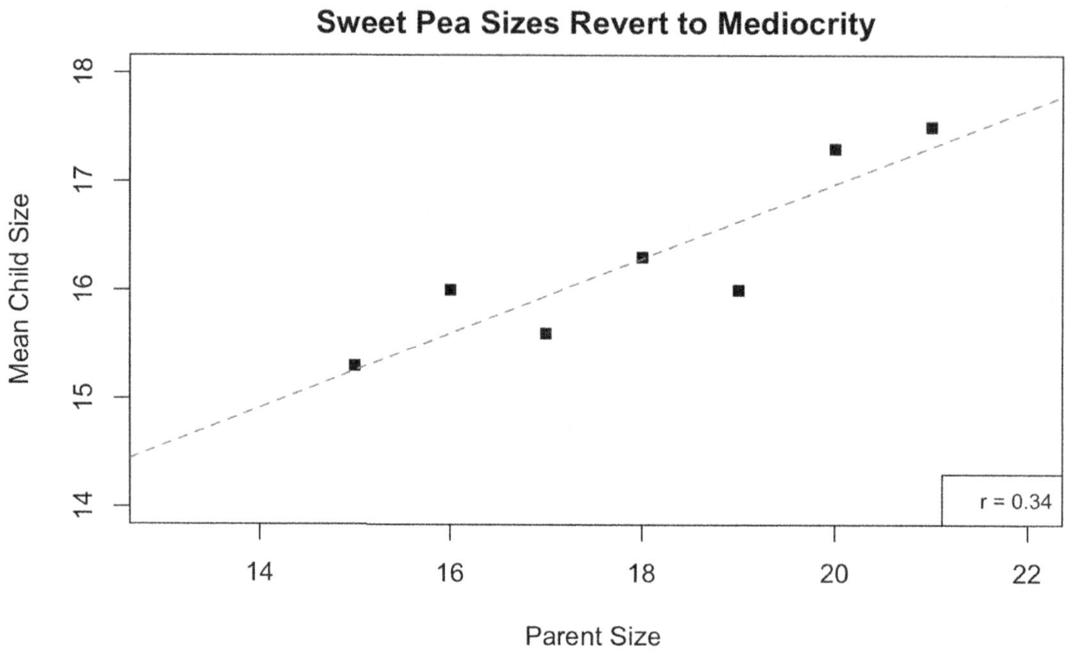

Figure 8.2: Galton's unpublished 1877 regression on sweet pea seed. Buckets are given as percentages, the sample sizes of offspring are unknown.

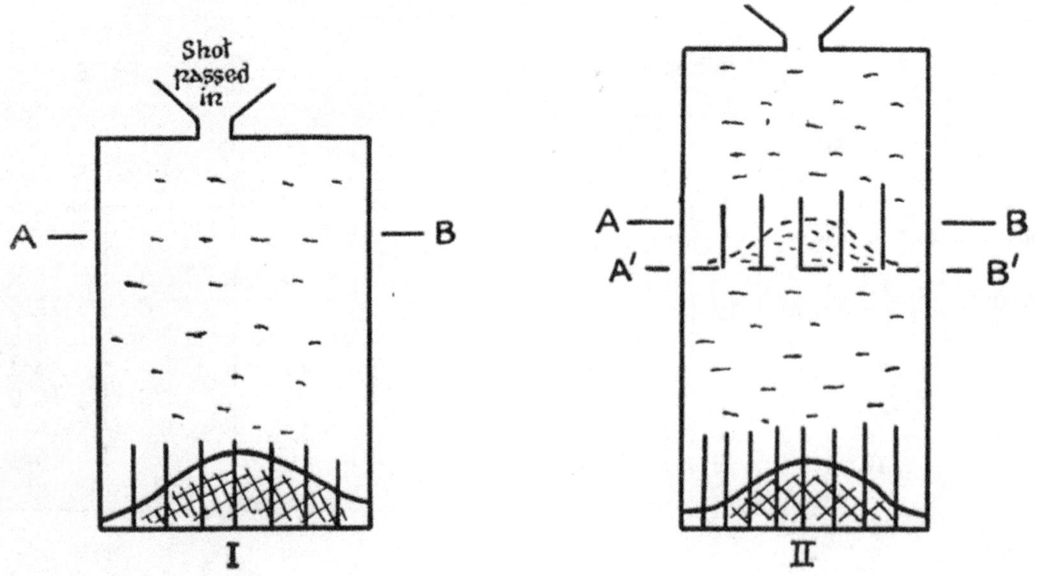

Figure 8.3: Redrawn for clarity: I. The original Quincunx. II. The combining Quincunx with trapdoors. FG to G. H. Darwin, 1877-01-12, K. Pearson 1930b.

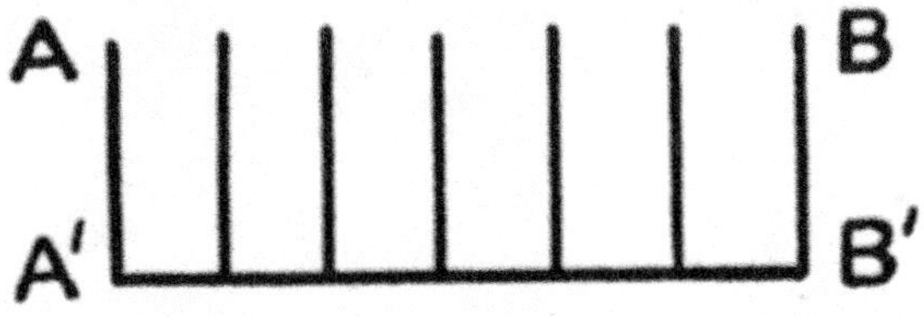

Figure 8.4: Quincunx compartments. Francis Galton to George Howard Darwin, 1877-01-12, K. Pearson 1930b.

Figure 8.5: Dispersal from trapdoor compartments with combination. Francis Galton to George Howard Darwin, 1877-01-12, K. Pearson 1930b.

Figure 8.6: Francis Galton at age 56 in 1878, at Vichy, K. Pearson 1924.

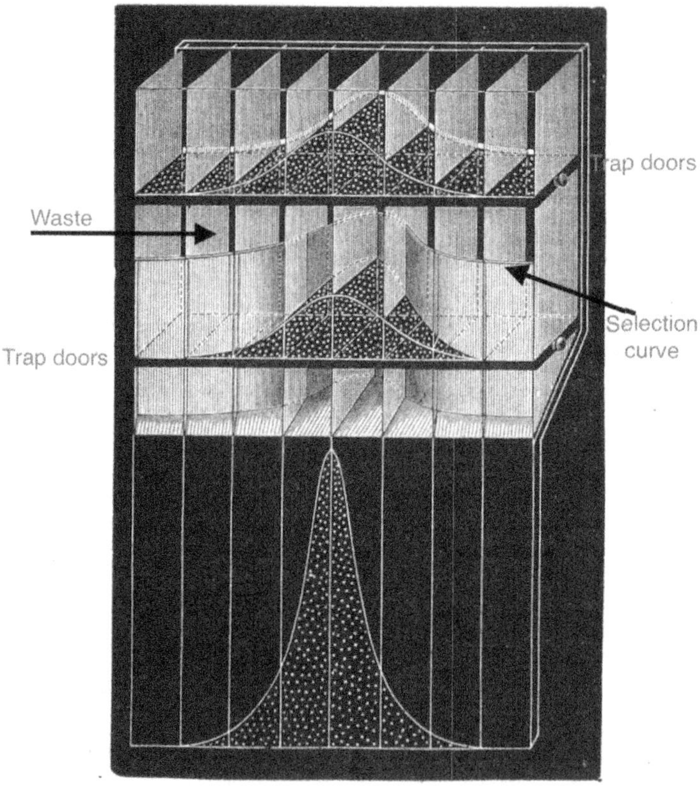

Figure 8.7: Quincunx with selection, annotated for clarity.

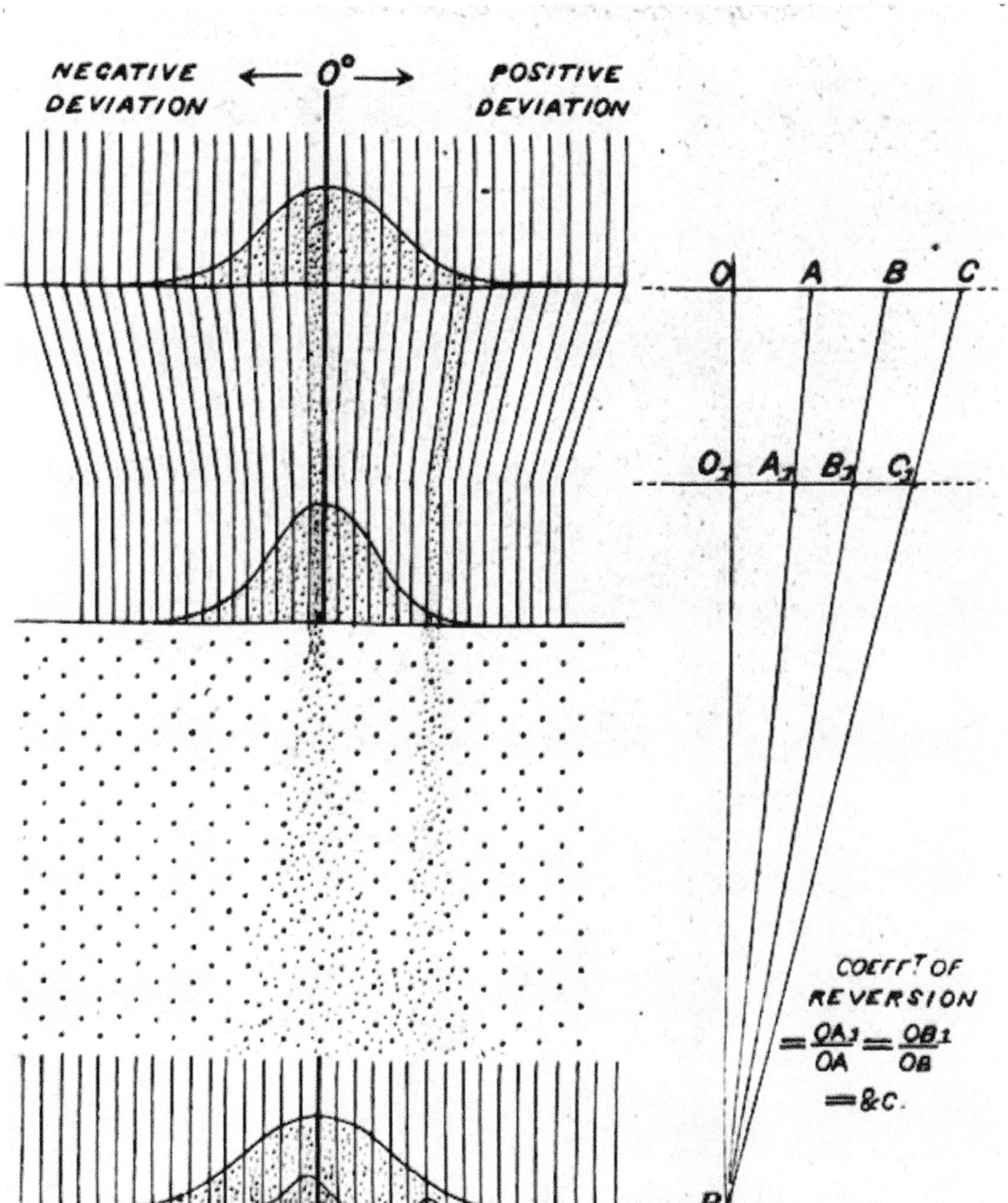

Figure 8.8: Quincunx with reversion and combination. The intermediate stage has trapdoor compartments.

Figure 8.9: Francis Galton at age 53 in 1875, K. Pearson 1924.

Chapter 9

Sounding Human Faculty

'It was anthropological evidence that I desired, caring only for the seeds as means of throwing light on heredity in man. I tried in vain for a long and weary time to obtain it in sufficient abundance'.[1] Galton had already made a start. Before turning to sweet pea seeds, he measured physical properties that varied in common experience, like hearing ability, height and weight.

Dr. William Hyde Wollaston (1766–1828) had calibrated the range of sound audible to humans in 1820, but Galton appears to have been the first person to systematically investigate hearing ability using a practical instrument that allowed precise measurements to be recorded easily. He wanted to see how races and individuals might vary in their ability to hear shrill notes, but soon extended his remit to cover animals and insects too. There were few things that he enjoyed more than inventing instruments and calibrating them. Progressive deafness in middle to old age was hereditary in his own family, and he would suffer from it himself as the years wore on, until he had to give up many public activities.

'Galton's Whistle' as it came to be known was developed sometime before 1876 (see Figure 12.1, p. 281). In that year it was exhibited at the South Kensington Museum (now the Victoria and Albert), just down the road from Rutland Gate, as part of a general demonstration of scientific instruments. By then he had several different designs, manufactured for him by Tisley and Spiller, and by Hawksley of Oxford Street. A pipe, no more than 0.156 inches in diameter, was fitted with a stopper on the bottom that could be screwed in or out to reduce or increase the depth of the instrument, which altered the vibrations per second emitted. Gradations marked on the outside were indexed to those vibrations at the given depth, which could range from 6461 to 84000. 'The whistles are of three forms: (1) a small cylindrical tube, which gives a pure note, but of small power; (2) a flat, wide and narrow whistle, of which the plug is a broad thin plate of metal; (3) an instrument which is externally a cylinder of 2 inches in diameter, but of which the effective part is merely an annulus; the plug of this is a cylindrical sheet of brass; it gives a powerful note, but not a pure one'.[2] Later, many variations on these designs would be produced by continental manufacturers like Edelmann (see Figure 9.1, p. 193).

Friends and family, young and old, were imposed on, with predictable results. It was proverbial that older people quickly lost the ability to hear the squeaks of bats, but many did not know their own limits. At parties in country houses, seniors could be flustered when those limits were exposed with exactness. At the South Kensington Conference, an exception was present in the form of the 62 year-old mathematician and musicologist Alexander John Ellis (1814-1890). The scholar would be the prototype for 'Professor Henry Higgins' in Shaw's play *Pygmalion* (1913) and thus the film *My Fair Lady* (1964), but his hearing far surpassed his short-term memory for names. 'With regard to the upper limits of audible tone, although I am rather an old boy, I may say that I heard all the high tones produced by Captain Douglas

1 F. Galton 1885f.

2 Museum 1877, 178.

Galton perfectly'.[3]

For more general testing—always willing to follow byways and explore side-alleys—more artifice was needed. Galton abhorred being noticed or conspicuous. So a long tube was disguised as a walking stick, with an India-rubber handle to squeeze air through it and sound the notes. With this in hand he worked his way through the Zoological Gardens, where he held it as closely as possible to the ears of the animals there until they were used to it, then squeezed. Pricked ears suggested they did hear, non-pricked ears were ambiguous. Cats, intensely selected to hunt squealing mice, appeared to have the best hearing of all. Small breeds of dogs were adept, but larger breeds less so. A visit to Switzerland in 1875 allowed for international comparisons. 'At Berne, where there appear to be more large dogs lying idly about the streets than in any other town in Europe, I have tried this method for hours together, on a great many large dogs, but could not find one that heard it'.[4] Ponies and cattle were more acute than horses. But insects and ants in particular would not respond. Perhaps he was helped here by his friend John Lubbock, being 'captivated by the ingenuity of his experiments on ants and bees' as he later recalled.[5] Still he kept tinkering.

Eight years later Galton was able to quadruple the range of his whistles by using hydrogen gas.[6] Variations on his more basic designs were in widespread use well into the 20th century, and can still be bought today as the common 'dog whistle'. He would use them himself in his long-running Anthropometric Laboratory.[7] The aural surgeon Sir William Bartlett Dalby recommended the instrument in 1885 as 'the best test that can be employed'.[8] The Cambridge Anthropological Expedition of 1898-9 to the Torres Strait islanders off New Guinea relied on the Hawksley variant, which was adjusted by jamming a solid rod, with a milled head, into the tube. The results suggested that the hearing of the islanders was no different from a control group in Aberdeenshire.[9] However that lay in the future, and in 1877 it was becoming clear to Galton that even with straightforward measurements like these, he lacked that crucial second dimension: family relationships. Only with an extensive data set including relatives could he find the insight he was after.

Physical height and weight were on the face of it a lot more straightforward to measure than hearing. But even here there was opportunity to invent improved instruments. Galton seized on the use of schools to gather anthropometric data as early as 1873, enrolling the Anthropological Institute in the process.[10] He had excellent personal contacts at multiple schools. The pupils were captive subjects there, and it was rumoured that intelligent and conscientious staff were already on hand to take the measurements. The stated intention was to collect representative data from all kinds of schools, pauper to upper-class.

The first school to respond was headed by Galton's friend the Rev. Frederic William Farrar, who had reviewed *Hereditary Genius* critically a few years earlier.[11] Marlborough College, a 'public' or rather independent school in Wiltshire, had 'peculiarly acceptable' facilities. The M.D. Walter Fergus was on staff, and the science master George Farrer Rodwell was a fellow of the Royal Anthropological Institute. They submitted measurements of the entire school of 540 boys, going beyond height and weight to include head circumference, chest, calf and biceps girth.[12]

---

3 F. Galton 1876b.
4 F. Galton 1876b.
5 F. Galton 1908, 178, Many insects, we now know, actually hear very well.
6 F. Galton 1883b,c.
7 See chapter 12, p. 261.
8 Dalby 1885, 67.
9 Pitt-Rivers 1901, 150-4.
10 F. Galton 1873e.
11 See chapter 4, p. 51. For Farrar, Frederic William, see page 452.
12 F. Galton 1874d.

It seems that Galton collaborated directly with Fergus and Rodwell, designing simple instruments for taking consistent measurements (see Figure 9.2, p. 194). He treated the measurements as a trial run for the analysis of a larger set from more schools, satisfying himself that Marlborough was homogeneous enough for the boys to be sensibly treated together, and that therefore the measurements followed the Law of Errors well enough. The schoolmasters had failed to find an obvious relationship between head circumference and 'intellectual vigour'. They supposed (correctly) that brain volume would be a better head measure, but they had no precise way to quantify the relationship. Galton did not address this finding in his remarks on their paper. A decade later he would provide the apparatus for sharpening the question.

By 1875 nine 'public' schools had supplied enough data to permit comparisons. The most obvious contrast to investigate was location. The supposed degeneration imposed by the industrial revolution on physique was a popular topic at the time. Five of the schools were based in the countryside—Marlborough, Clifton, Haileybury, Wellington and Eton—and four in cities—City of London School, Christ's Hospital, King Edwards in Birmingham, Liverpool College. Galton had attended King Edwards himself in the 1830s, under Dr. Jeune.[13] As soon as he started comparing the data he realized that the ages of the pupils were not comparable. They were stated in years, but should have been recorded in months or days. Thirteen year-olds in one school might be a few months older on average than thirteen year-olds in another because of delayed enrollment. He had to adjust for this as best he could.

Galton's analysis makes heavy use of his newly published method of 'statistics by intercomparison', summarizing the data sets by what he refers to as their 'ogival curve'.[14] We recognize this now as the *inverse* cumulative distribution but its use was brand-new at the time (see Figure 9.3, p. 194). To construct the curve, no absolute standard of measurement is needed, just the ability to say which of two elements is the larger for the quality at hand.

The key information, Galton points out, is given by the quartiles (a word he would introduce later), which lie on the horizontal x-axis of his curve. The median, or second quartile mark, divides the distribution into equal halves (or near enough). Likewise the first quartile has 1/4 of the values below it, and the third has 3/4 below it. Each quartile q can easily be mapped to the corresponding height or weight value, located on the vertical y-axis, by means of the function $y = Q(q)$ that is represented by the ogive curve.

By reducing the data sets to three quartile divisions, Galton could discount the extreme values at the tail, which otherwise give a misleading picture of the core of the data. To compare the country and city schools by height and weight, he averaged those three points within each group, then looked at the difference between the groups. This simple method gives a much better idea of the contrast between distributions than just comparing the means, since the variation is also taken into account, to some extent, by using the other quartiles. For 14 year-olds there was a consistent 8 pound weight advantage for the country schools, and a 1.4 inch advantage in height.

Because of the slight age advantage of the country schools, noted above, Galton adjusted these differences down *pro rata* to 7 pounds and 1.25 inches. He attributed the difference partly to retardation by their environment, since the boys tended to catch up later, and partly to total suppression by the same, since age did not seem to completely erase the gap. The conclusion was one that he and many others could sympathize with, since belief in the virtues of country living and the evils of polluted city life were widespread. The irony is that he did not consider the role of *nature* here, given that the schools drew from different classes. Both of those classes had more than enough wherewithal to get the minimal nutrition required, though they were perhaps subject to different environmental modifiers like pollution, but they were not

[13] See Tredoux 2023c.

[14] F. Galton 1875h.

necessarily equivalent by *race* (or genes, as we would say).[15]

At around this time, 1875, Galton succeeded in interesting the British Association for the Advancement of Science in the question, and an Anthropometric Committee was appointed. He would be a long-serving member on it, and contribute to their yearly reports, but he would not directly use any of the data they produced. Their work was fine as far as it went but it addressed more mundane aspects of human measurements that were of little use to his own more advanced interests. He needed information about heredity most of all, but could not get it, at least not yet. Characteristically he had already moved on to consider far broader aspects of human measurement.

In August of 1877, Galton was asked at the last minute to preside over the Anthropological Department at the British Association, which met at Plymouth that year. Lack of time meant that instead of a summary of other work in the field as a whole his address took the form of a substantial memoir describing his own emerging ideas. It can be read as a manifesto of what was to come. His subject was the classification of men into groups on the basis of similarity, either in mental characteristics, or physical aspect, or both.

'I especially desire to shew that many methods exist of pursuing the inquiry in a strictly scientific manner, although it has hitherto been too often conducted with extreme laxity'.[16] Recent developments on the continent—such as the work by Fechner on 'psycho-physics', e.g. the time taken for information to be conducted by the nerves to the brain—had made an exact study of mental phenomena feasible for the first time.

Unfortunately collection of such data on a large scale seemed beyond reach for the time being. Instead Galton turned to types of physical appearance, which he had been working on using composite photography. 'I wish to speak of the great convenience of photographs in conveying those subtle but clearly visible peculiarities of outline which almost elude measurement'.

If many photographs were sorted into a smaller number of similar groups, then the images could be blended using partial exposures, provided that they were standardized first so that the superposition was sensible. The blended composite photograph could be interpreted as an abstract or latent type underlying the individual images. Through painstaking experimentation Galton had worked out a practical procedure for constructing such images. He had already been successful in producing convincing composites, each of which could be taken for a real person but was actually an artificial creation from multiple elements. Yet the composite bore a *family resemblance* to the elements which went into it. Ordinary observers recognized the likeness straight away (see Figure 9.6, p. 196, for examples).

The remainder of 1877 and most of 1878 were taken up by energetic work on photography. The idea for the composites came originally from Galton's work with his cousin R. Cameron Galton on representing geographical information in maps twelve years earlier. Using a stereoscopic viewer, two images could be combined to give three dimensional representations of landscape topography.[17] However the plan was not adopted in practice by others. At the South Kensington Conference in 1876 he had returned to this theme of combining data in diagrams. He showed stereo 3-d images of models built for him by the Royal Engineers. He also exhibited a tramway-and-lens mechanism for making successive projected images fade smoothly into each other, which was a major challenge in the early development of motion pictures.[18] Maps could then display marked-up information in succession in a flowing way.

---

[15] The report of the post-reading discussion published in F. Galton 1875h gives a confused account of Galton's remarks, suggesting a Lamarckian idea of acquired characters that he certainly did not subscribe to. 'It was not maintained by Mr. Galton that the deficiency of size in the town group was altogether due to bad effects of nurture on the individual boys; on the contrary, he would be disposed to ascribe much of it to the town life of their parents, and probably of other ancestors'. This was not written by Galton himself.

[16] F. Galton 1877a.

[17] F. Galton 1865e.

[18] F. Galton 1877c.

In conversation, Herbert Spencer suggested the use of superimposed tracings, viewed as one against the light, to give the appearance of composite images. Some years before, Spencer had experimented with these for composing head outlines. That worked after a fashion but Galton soon progressed to stereoscopes again. Then at the prompting of Sir Edmund Frederick Du Cane (1830–1903, see Figure 9.4, p. 195), he decided to try using photographs.

It is not clear when Galton first met Du Cane, at that time the Supervisor of Prisons. It may have been during the scientist's twin survey of 1875. Du Cane was the father of fraternal twins, Eddie and Sabina, and supplied details of their similarities and differences.[19] The scientist recalled him as 'an extremely accomplished man, with high and humane views' who 'sympathised with not a few of the subjects on which I have been engaged'.[20] The Duquesne family were Huguenot refugees, an interesting fact in its own right to Galton, who had in *Hereditary Genius* praised the 'natural refinement', 'solid worth' and the 'culture and technical knowledge' that the heretics brought with them.[21]

Sir Edmund qualified as an engineer at Woolwich in 1848 and rose quickly through the ranks, showing great administrative talent—he certainly knew Galton's cousin Douglas, serving on many committees with him. The versatile engineer first came to prominence in the Great Exhibition of 1851 at Crystal Palace. After that he left to supervise the new Western Australian penal settlement, founded at the request of the colony itself. Du Cane would not have known at the time, though surely they discussed it later, that as he sailed down the West Coast of Southern Africa on the teak-built *Anna Robertson*, he passed by Galton's own scene of operations in the dessicated interior, drawing to a close that December. (Du Cane's watercolours recording four years of scenes among the convicts in Western Australia are sought after today.)

Returning from Australia in 1856, Du Cane massively fortified the coast around Plymouth in anticipation of a French invasion. Construction and administration of domestic prisons became a pressing problem when transportation abroad was abolished in 1867, so Du Cane addressed himself to it enthusiastically and economically. Wormwood Scrubs was his own design, and he supervised the building of it himself using convict labour.

Du Cane suggested that Galton consider photographs of criminal types, to determine if they had some distinctive appearance by type of crime. The prison service already used photography widely and faced significant technical challenges in indexing and searching their collection. Ever since his appointment in 1869 as Chairman of the Board of Directors of Convict Prisons, Du Cane had been developing an interest in systematic criminology, specifically in the causes and possible prevention of crime, hoping that the system could be improved through reform.

In 1875, at the Brighton meeting of the National Association for the Promotion of Social Science, Du Cane speculated that 'among a large number of criminals there are sure to be found a considerable proportion who have certain physical characteristics in common, which may or may not be attributed to the kind of life they lead, but which denote with certainty the habitual criminal ... and if that appearance is not a consequence of their being criminals or due to the same cause as the criminal disposition, then it might be inferred that either they were in some sense a separate caste, or else ... they are examples of the race reverting to some inferior type from which ages of civilization and culture have raised it'.[22]

In truth Du Cane was not at this point committed to any hereditarian explanation, considering it only an interesting possibility worth investigating, among many others. For an example of the latter, one of his early ideas from 1873 was to start a National Training School of Cookery, in order to impart skills to lower-class women, hoping that this would keep both them and their

[19] The catalogue of the Galton Papers at UCL dates this document, GALTON/2/4/3/13/7, as 1875, but the manuscript contains no date on it and may instead belong to a letter from 1880, GALTON/2/7/2/6/5. The only biography of DuCane is Hasluck 1973 but this is chiefly about the Australian period.

[20] F. Galton 1908, 261.

[21] F. Galton 1869a, xxiii.

[22] Du Cane 1876.

husbands from the temptations of theft (rather than simply providing them with more energy to steal with). Instead the school mostly attracted more genteel English women, who did not know how to cook either.

The idea of a distinctive criminal physiognomy had long been a folk belief. When he was exploring South-West Africa in the early 1850s, Galton remarked that the Namaqua tribe there had many features that were popularly thought at home to mark people as criminals. 'The greater part of the Hottentots about me had that peculiar set of features which is so characteristic of bad characters in England, and so general among prisoners that it is usually, I believe, known by the name of the "felon face;" I mean that they have prominent cheek bones, bullet shaped head, cowering but restless eyes, and heavy sensual lips, and added to this a shackling dress and manner'.[23] Now he could put this idea to the test.

So Galton turned his own face toward criminals. 'In previous publications I have treated of men who have been the glory of mankind, I would now call your attention to those who are its disgrace'. Through Du Cane he obtained identifying photographs of various classes of 'heinous criminals', anonymized at his own request but sorted by crime specialty. By May of 1877 he had thousands of these photographs in hand. He was also able to visit prisons in person.

Considering the example of 'magic lanthorns', which can blend projected images onto the same spot when used in concert, Galton fixed on the idea of partially exposing sets of photographs onto a single plate, to combine or average features from all. His first method for doing this required the reduction of component photographs, all of comparable shade and aspect, to a standard size, fixed on a cross drawn horizontally through the pupils and vertically through the nose and mouth. Assembling these together into a little booklet and fixing that a set distance from the camera he could briefly expose each photo in turn, flipping, uncapping for a set interval, and then capping the lens, until the full exposure time for the plate was reached. For a small collection, substantial similarity between the photos was needed to get good results, which were intermediate between the components. This worked especially well for families, where just two components could yield a composite easily recognized as an imaginary member of the family.

Through experience it emerged that just about *any* large collection of photographs could usually be sorted into about five groups, the last of them a miscellaneous collection. When hundreds were used there was no longer any need for careful pre-sorting, since each portrait contributed an almost vanishing amount to the whole (however as we shall see there were some failures, such as lunatics).[24]

Over time Galton would greatly refine and simplify the process he used for making composites, devising special instruments to accelerate the process. He also invented several instruments for viewing blended images without the need for photography, using systems of mirrors, angled and arranged to project the images onto the eye. Tisley & Spiller, as per usual, constructed them for him. His publisher Macmillan later suggested calling his new science 'photophysiognomy'.

At the time of the address to the British Association that August, many composites had already been created, but Galton was not yet ready to draw any definite conclusions. In the meantime he was, like Du Cane, prepared in principle to consider the idea that criminals were constitutionally different. He cautioned the audience against supposing that criminality was necessarily a debilitating illness or malfunction.

Anticipating biosocial theories of criminality and psychopathy by many decades, Galton suggested that criminality may be *adaptive* for that fraction of the population. 'The subject of vicious instincts is a very large one: we must guard ourselves against looking upon them as perversions, inasmuch as they may be strictly in accordance with the healthy nature of

[23] F. Galton 1853.
[24] F. Galton 1883d, 10.

the man, and, being transmissible by inheritance, may become the normal characteristics of a healthy race, just as the sheep-dog, the retriever, the pointer, and the bull-dog have their several instincts'. 'A man who is counted as an atrocious criminal by society, and is punished as such by the law, may nevertheless have acted in strict accordance with his instincts. The ideal criminal is deficient in qualities that oppose his vicious instincts; he has neither the natural regard for others which lies at the base of conscience, nor has he sufficient self-control to enable him to consider his own selfish interests in the long run. He cannot be preserved from criminal misadventure, either by altruistic or by intelligently egoistic sentiments'.[25]

The promise was that the physical nature underlying criminality might be exposed through the means of the composites, if those appeared to be distinctive—more ambitiously, perhaps even by type of crime. Both ambitions were to be disappointed. In April of 1878, after extensive experimentation, Galton was forced to conclude that when the images of criminals were blended, they seemed to dissolve back into average faces reminding viewers of ordinary people rather than distinctive rogues. And murderers seemed no different in composite than rapists and thieves.

Though the components were distinctive individually, averaging them eliminated that distinctiveness, as if each peculiarity was an idiosyncratic deviation from the same common centre point. 'It will be observed that the features of the composites are much better looking than those of the components. The special villainous irregularities in the latter have disappeared, and the common humanity that underlies them has prevailed. They represent, not the criminal, but the man who is liable to fall into crime. All composites are better looking than their components, because the averaged portrait of many persons is free from the irregularities that variously blemish the looks of each of them'.[26]

It was a special case of a more general phenomenon. It seemed that *any* set of blended photographs produced a face more attractive than the elements which went into it, as if beauty is an absence of peculiarity. However this does not imply that the face of an *average person* is more attractive. Blending the photographs yields a face that has all defects more or less eliminated, which would require an almost perfect natural development to show up in a real person, who would then be at the extreme right tail of the distribution of beauty. The *averaged* face is not the face of the average person.

If the criminals were disappointing, opportunities arose elsewhere. Lunatics at Hanwell Asylum in West London were tried, but proved very difficult to blend. This led to a memorable passage in Galton's autobiography. 'I could not make good composites of lunatics; their features are apt to be so irregular in different ways that it was impossible to blend them. I took a photographer with me to Hanwell, where it was arranged that the patients should sit two at a time on a bench. One of them was to be led forward and posted in front of the camera, while his place on the bench was filled by the second patient moving up into it, whose previous place was to be occupied by, a third patient. It happened that the second of the pair who were the first to occupy the bench considered himself to be a very mighty man, I forget whom, but let us say Alexander the Great. He boiled with internal fury at not being given precedence, and when the photographer had his head well under the velvet cloth, with his body bent, in the familiar attitude of photographers while focusing, Alexander the Great slid swiftly to his rear and administered a really good bite to the unprotected hinder end of the poor photographer, whose scared face emerging from under the velvet cloth rises vividly in my memory as I write this. The photographer guarded his rear afterwards by posting himself in a corner of the room'.[27]

In coming years Galton would use composites to investigate illness in particular. There was a

[25] F. Galton 1877a, 11-12.
[26] F. Galton 1878a.
[27] F. Galton 1908, 262-263.

popular supposition that consumptives in particular had a peculiar look, a 'diathesis' above and beyond simple malnourishment. In the early 1880s, in collaboration with a remarkable young physician, the quarter-Indian Dr. Frederic Akbar Mahomed (1849–1884), Galton exhaustively investigated this possibility for 'phthisis', as consumption or tuberculosis was then called—it was widely believed at the time that tubercular victims had especially narrow faces.

Dr. F. A. Mahomed remains an interesting figure today. In 1810 his accomplished grandfather Sake Dean Mahomed (1749?–1851)—originally from the Bengal region, of Muslim descent—opened the first Indian eatery in London.[28] The 'Hindoostane Coffee House' was located in George Street off Portman Square. Sake Dean's son Frederick, father of F. A., was a successful fencing master. The family rapidly assimilated into broader Irish and British society—in parallel, since Sake Dean Mahomed was a bigamist. His grandson graduated from Guy's Hospital in the early 1870s, practising there from the late 1870s onward.

Mahomed made his name through an investigation of 'Bright's Disease' of the kidneys, which he traced to long-term insidious effects of hypertension, i.e. pathologically high blood pressure. It was an explanation that he became overfond of, applying it more broadly with less success. His fascination with medical instruments like the wrist-attached sphygmograph for measuring blood pressure, which produced real-time pen-tracings on rolled paper of the recorded levels, surely gelled with Galton.[29]

Thus Mahomed was a rising star when he started his ambitious collaboration with the hereditarian. It is unclear exactly how they first met, but Galton had contemplated composites of the sick from the very start. At least 442 photographs of consumptive patients at Guy's, Brompton and Victoria Park Hospitals were taken, starting in January of 1881. A year later the result was a long and rather dull memoir, though with good illustrations of the somewhat sad and wan-looking composites. It was written up by Mahomed and published in the *Guy's Hospital Reports* for February of 1882. The conclusion was much the same as it had been for criminals. The collaborators graciously conceded defeat: 'our results appear to lend no countenance to the belief that any special type of face predominates among phthisical patients, nor to the generally entertained opinion that the narrow, ovoid, or "tubercular" face is more common in phthisis than among other diseases. Whether it is more common than among the rest of the healthy population we cannot at present say' (see Figure 9.7, p. 197).

Galton and Mahomed nevertheless saw some benefit from proving their own theories wrong, in a spirit that advocates of open reproducible science and enemies of the 'file drawer effect' would find pleasing. 'Our results are therefore negative, but it may be they are no less valuable; although we commenced our investigations with the expectation of establishing a "type" on a firm foundation, we shall be little less satisfied with them if they have succeeded in refuting an error'.[30] They hoped to take the general question of 'diathesis' up in the future, suspecting that despite many popular misconceptions 'it contains a nucleus of valuable truth'. The collaboration continued for two more years through the 'Life History Committee' that Galton hosted at 42 Rutland Gate, and will be treated more fully below. It was ended suddenly in November of 1884 by the death of the doctor from complications of typhoid, possibly caught from one of his patients at the London Fever Hospital. Mahomed was just 35 and at the height of his success.

Other applications of composite photography to anthropology appeared obvious. Families had already proved to be far more productive subjects, with a lot of common resemblance that showed clearly through in the composites. Where the degree of relationship was known, the contribution of the image to the composite could be proportioned accordingly, say by following the scheme of weights already sketched out in *Hereditary Genius*. It was a natural step from

[28] 1749 was the date volunteered by Sake Dean himself, but others have ventured, less precisely, c1759, though the concrete evidence for that is unclear. He is said to have been born in Bihar, administratively part of Bengal at the time.

[29] Goodhart and Jacobson 1886.

[30] F. Galton and Mahomed 1882, 492.

there to extended kin-groups, ethnic groups and races, each of which could be represented by composites, to capture their distinctive physical appearance, or 'type', where one was visible. The technique was tried on skulls with some success, and promptly followed up by John Shaw Billings, then at the War Department in Washington, DC, using various Indian tribes.[31]

The problem was that Galton had difficulty getting enough photos to analyze, despite advertising in various journals for donations of examples of roughly the right size along with pedigree details. The Jewish anthropologist Joseph Jacobs was able to organize a substantial collection of photographs from the Jews' Free School, at Bell Lane in Whitechapel. Jacobs enlisted Galton's aid in having composites made, which the scientist considered the best examples he had ever done. The subjects were 'dirty little fellows individually, but wonderfully beautiful' in composite form, though Galton was struck by the 'cold scanning gaze' he was subjected to by the locals, who appeared to be 'coolly appraising me at market value, without the slightest interest of any other kind'. (From an aficionado of objectivity, this was high praise indeed; in 1845, a slave dealer in the Sudan had valued him, on request, at £20.) A few years earlier, in *Notes and Queries*, he had suggested that the best place to look for characteristic Jewish physiognomy is the bas-reliefs on the Arch of Titus in Rome, slyly adding that it 'very decidedly' corresponds to an ancient medallion depicting Jesus.[32] Now Galton left interpretation of that physiognomy to Jacobs (who would later estimate Jewish intelligence using Galton's methods from *Hereditary Genius*, finding that it was roughly 4% higher on average than English).[33] Using old-fashioned silhouettes, a relic of the pre-photographic era, proved a surprisingly useful further simplification of the process, concentrating attention on family patterns in profiles only.[34]

Composite photographs promised then to reveal an underlying or latent reality, which was obscured by chance and circumstance but which could be approximated by precise means. Galton was struck as early as 1877 by a comparison with our own *mental* imagery and visual memory. It seemed likely that people would *differ* in just how accurate or inaccurate their 'pictorial imagination' was, and that those differences might be heritable. Some quick enquiries among friends and family suggested that this was indeed the case. In particular it seemed that there were people who were not 'visualisers' at all, while, more bizarre by far, a few others made mental pictures of *abstract concepts* like numbers and their relationships. More radical crossing of boundaries soon emerged.

At first Galton thought that human pictorial imagination might closely resemble composition of images through repeated sensory exposures. These could form the basis of 'generic images' in the mind's eye, capturing an underlying genus. Some reflection disabused him of this. On April the 25th of 1879—another Friday evening lecture—Galton argued before the Royal Institution that the undue weight given to isolated or initial impressions in 'blended memories' led to the sort of gross misconceptions which went under the name of 'general impressions'.[35] Quite unlike the precise composite portraits of families, and the coins of Alexander the Great and 'hideous' Cleopatra, that he projected to them from 'magic lanthorns' mounted on adjustable stands. He invoked German experimental psychology.

Ernst Weber's Law, adopted by Gustav Fechner, suggested that exponentially more repeated stimulus was needed to get the same response, or equivalently that the effects of a stimulus went up logarithmically with its intensification. Initial impressions were much more potent in forming mental images than subsequent ones. Combining them led to an exponentially decaying average of images, which were not aggregated in the systematic way that Galton had composed

[31] F. Galton 1885c.
[32] F. Galton 1880a.
[33] F. Galton 1885c,e; Jacobs 1885, 1891.
[34] F. Galton 1887c.
[35] F. Galton 1879a.

his photographs to find the 'pictorial statistics' they went into. Even the most facile visualizers, artists and writers were limited.

Galton ran up his flag at the close, 'those who are not accustomed to original inquiry entertain a horror of statistics. They cannot endure the idea of submitting their sacred impressions to cold-blooded verification. But it is the triumph of scientific men to rise superior to such superstitions, to devise tests by which the value of beliefs may be ascertained, and to feel sufficiently masters of themselves to discard contemptuously whatever may be found untrue'.

By the second half of 1879 our cold-blooded verifier of impressions was closely focused on the question of differences in visualizing power and the ways in which the mind's eye used images—the 'faculty of visualizing'. An exchange in the *Spectator* at the close of the previous year had caught his eye. A dismissive obituary of George Parker Bidder (1806–1878), a famous prodigy of mental-arithmetic, had drawn protests from friends and family, containing details that read like an omitted case study from *Hereditary Genius.* No idiot savant, Bidder was not just a demonically-efficient mental calculator but also a highly successful businessman and engineer. This despite being no more than the son of modest stone-masons in both lines, who had exhibited their fascinating boy to the public at the neglect of his education. This general intelligence was supplemented by a powerful special talent for visualizing and retaining information. 'If he saw or heard a number it seemed to remain permanently photographed on his brain'.

Bidder's eldest brother—a nonconformist, possibly a member of the Plymouth Brethren—was said to have memorized almost the entire bible. Another was an actuary. His barrister son George P. Bidder Q.C. had been a 7th Wrangler at Cambridge and had inherited a portion of his father's talent for enormous mental sums, adding the ability to play simultaneous blindfold chess.[36] It was the son who drew attention in the journal to his father's 'mental pictures', in which 'figures seemed to stereotype themselves without an effort', combined with 'almost inconceivable rapidity of motion', which the barrister partly shared. Even grand-daughters and grandsons seemed to have inherited the visualizing tendency, 'when I hear anything remarkable said or read to me, I think I see it in print'.

Galton soon initiated a correspondence with Bidder's son in April 1879 to ask for more details. It was evident that visualizing was a faculty that differed greatly between people, and seemed to be heritable. The question was how to measure it, and gather the corresponding data.

Some rough notes were kept on the 'visualising powers of women' drawn from relatives and the wives of friends, such as Mrs. George Romanes and Mrs. Archibald Smith etc. Here we discover that Galton's wife Louisa 'cannot call up the features at all, but [only] the general form, attitude, and shape of heads' and 'fails in illumination & completeness, also in person's features'.[37]

To obtain more extensive data on the distribution of visualizing powers, Galton composed another circular questionnaire. He was up in Birmingham in November of that year distributing it, accompanied by a lecture, to the Philosophical Society.[38] Almost a dozen replies would result from this visit alone. Respondents were asked to describe how they visualized a definite object, such as their breakfast table earlier that day. How clear or dim was the image? Was it sharply defined or blurred? Were some elements more complete than others? Were the colours distinct and natural? How broad was their mental (photographic) 'field of view'? Further questions addressed their ability to recall print from pages, their ability to simulate movement of furniture in a room, recall wall-paper and carpet patterns, the features of persons well-known to them,

36 Robertson 1878. For reference, the eminent logician Bertrand Russell was a 7th Wrangler. F. D. Mitchell 1907 claims that the elder brother was a Unitarian, but Rouse Ball has him a Plymouth Brother, Ball 1922, 275.

37 GALTON/2/7/2/7/8.

38 F. Galton 1880b.

even at different aspects, and so on. Other senses were touched on: smell, tone, taste. Later Galton would use a revised version of this questionnaire, complicating interpretation.

The questionnaire was distributed to people of high ability, starting with the circle of scientists that Galton knew personally. Almost all of the initial respondents claimed to have little to no visualizing ability: 'they looked on me as fanciful and fantastic in supposing that the words "mental imagery" really expressed what I believed everybody supposed them to mean. They had no more notion of its true nature than a colour-blind man who has not discerned his defect has of the nature of colour. They had a mental deficiency of which they were unaware, and naturally enough supposed that those who were normally endowed, were romancing'. The astronomer John Herschel even responded that the analogy was nonsensical, like examining the relationship between knitting and 'a closely knit argument'.[39] This prompted Galton to quickly check non-scientists for comparison. The results reassured him that visualizing really did exist, and pretty broadly. More careful interrogation of scientists eventually revealed a number who confessed to visualizing.

Huxley owned up in conversation to having 'Strong visualizing powers. Phantasmagoria like Johan Muller, a perception of people & faces never seen before down a street (but vision clearest at one side of field of natural view). Can see blood corpuscles in retina clearly whenever looking up into sky. Audibly a "musical prismatic spectrum"—can't give a better name—used to have it much stronger when younger'.

Flinders Petrie, soon to become a highly distinguished archaeologist and Egyptologist, but then no more than a frequent contributor of letters to *Nature*, worked out sums using a mentally visualized slide rule. He could visualize just the section he needed to perform the computations, 'one of the most striking cases of accurate visualising power it is possible to imagine'.[40]

William Spottiswoode said that he 'can see algebraic formulae with much ease & to the extent of half a page at a time. It is not easy to describe the size [or] any peculiarities of the letters. Has not any special facility in seeing geometric figures. Has acquired a great aptitude for taking in contents of curiosity shops with apparently a careless glance, it is apparently a useful faculty to him. Can carry in his mind's eye with great clearness the look of a room & the sort of furniture it will want so that he buys furniture with much confidence'.[41]

Richard Francis Burton wrote that 'In my case whenever I have a mental sum to do I close my eyes and see the figures. They are Arabic (1, 2, 3) never Roman (I, II, III) and in dark upon dull yellow ground. As a boy I could play chess blindfolded, and saw the board as clearly as if it were before my eyes. All this I simply connect with local memory which when strong is capable of being strengthened. On the other hand I can remember dates only by referring them to some familiar event. A pure date is an impossibility to me'.[42]

Soon there were about 100 replies from adults to analyze, 19 of which came from exceptional Fellows of the Royal Society and the rest from those who had shown 'distinction in various kinds of intellectual work' (Frank's farming brother Erasmus was included for good measure). Included in the final selection were Charles Darwin and his sons George and Horace, George Busk from the X-Club, the philosopher George Croom Robertson, Sir Edmund du Cane, the astronomers Major John Herschel and William Huggins, the economist W. Stanley Jevons, the philologist F. Max Müller, the biologist G. J. Romanes, the physicist Arthur Schuster, the astronomer Charles Piazzi Smyth, the anthropologist Edward Burnett Tylor, the logician John Venn, the surgeon Sir James Paget, cousin Douglas Galton the engineer, and John Ball the naturalist. Some tantalizing names were omitted for obscure reasons, such as the eminent painter Sir Fredrick Leighton. There was a promise from one respondent to forward the questionnaire

[39] Herschel to Galton, GALTON/2/7/2/2/5.

[40] F. Galton 1883d, 66. See also GALTON/2/7/2/4/31. For Petrie, William Matthew Flinders, see page 457.

[41] Loose MS Notes, GALTON/2/7/2/6/4.

[42] Richard F. Burton to Francis Galton, 1879-10-29? Tredoux 2016, Vol. 1.

to the Rev. Charles Dodgson (Lewis Carroll) but sadly nothing seems to have come of that.[43]

1. *Airy*, Osmund. 1845–1928. Historian, son of the astronomer G.B Airy.
2. *Alfitt*, S.
3. *Balfour*, Francis Maitland. FRS. 1851–1882. Biologist.
4. *Ball*, Henry. Probably the Birmingham gun-maker.
5. *Ball*, John. FRS 1818–1889. Member of the Alpine Club and naturalist.
6. *Bevan*, Rev. James Oliver. 1843–1930. Historian, prolific author.
7. *Browne*, (Sir) James Crichton. FRS. 1840–1938. Psychiatrist.
8. *Bruce-Clarke*, William. 1850–1914. Surgeon.
9. *Brunton*, (Sir) Thomas Lauder. FRS 1844–1916. Physician.
10. *Burton*, (Sir) Frederick William. 1816–1900. Painter.
11. *Busk*, George. FRS. 1807–86. Zoologist and X-Club member.
12. *Butler*, Rev. Arthur Gray. 1831–1909. Headmaster of Haileybury, Galton's brother-in-law.
13. *Carter*, James S. Alpine bootmaker and tourist outfitter.
14. *Causer*, J. Woodford. 1856–1944. Methodist minister.
15. *Cayley*, Charles Bagot. 1823–1883. Poet, younger brother of Arthur Cayley the mathematician and Senior Wrangler.
16. *Coates*, George. 1851–1935. Physician.
17. *Collens*, Edward. Possibly the chemist of the Birmingham Vinegar Brewery Company.
18. *Darwin*, Charles Robert. FRS. 1809–1882. Naturalist.
19. *Darwin*, (Sir) George. FRS. 1845–1912. Physicist.
20. *Darwin*, (Sir) Horace. FRS. 1851–1928. Engineer.
21. *Dismon*, J. S.
22. *Distant*, William Lucas. 1845–1922. Entomologist.
23. *Du Cane*, Sir Edmund. 1830–1903. Engineer and prison reformer.
24. *Duckworth*, (Sir) Dyce. 1840–1928. Surgeon.
25. *Ellis*, Rev. John Clough Williams. 1833–1913. 3rd Wrangler, rector of Gayton.
26. *Erle*, Twynihoe William. 1828–1908. Theatrical scene painter.
27. *Farrer*, Frederick Willis. 1829–1909. Painter.
28. *Fayrer*, Sir Joseph. FRS 1824–1907. Physician.
29. *Galton*, (Sir) Douglas. FRS 1822–99. Engineer and sanitary reformer.
30. *Galton*, Erasmus. 1814–1909. Farmer.
31. *Garnett*, Richard. 1835–1906. Poet and scholar.
32. *Gore*, George. FRS. 1826–1908. Chemist and electro-metallurgist.
33. Gore-Booth, Capt. James. 1840–1913. Engineer.
34. *Haselee*, Richard. Member of the Birmingham Philosophical Society, Spring Hill Congregational College.
35. *Hawtrey*, George Procter. 1847–1910. Schoolmaster, actor and playwright.
36. *Henslow*, Rev. George. 1835–1925. Botanist and curate, son of Darwin's mentor.
37. *Herschel*, Major John. FRS. 1837–1921. Engineer and astronomer.
38. *Hills*, Edmond Grove. FRS (later). 1864–1922. Grandson of Judge William Grove.
39. *Hooker*, Brian Harvey Hodgson. 1860–1932. Mining engineer, emigrated to Australia. Son of the X-Clubber Joseph Dalton Hooker.
40. *Howlett*, Rev. Frederick. 1821–1908. Astronomer.
41. *Hudson*, F.
42. *Huggins*, (Sir) William. FRS. 1824–1910. Astronomer.
43. *Jevons*, W. Stanley. FRS. 1835–82. Economist and statistician.
44. *Kelly*, Wesley. Spring Hill College.
45. *Lance*, F. W.
46. *Larden*, Walter. 1855–1919. Alpinist. Schoolmaster at Cheltenham College. Prolific author.
47. *Lee*, C. H.

[43] Burbridge 1994, from whom the list of respondents has been adapted.

48. *Lefroy*, Sir J. Henry. FRS. 1817–1890. Military engineer and surveyor of terrestrial magnetism.
49. *MacAlister*, (Sir) Donald. 1854–1934. Physician and mathematician.
50. *Macdonell*, George Paul. 1855–95. Barrister and magazine writer.
51. *Manton*, John Shakespeare. 1820–1911. Button manufacturer and member of the Birmingham Philosophical Society.
52. *Marshall*, John. FRS 1818–1891. Surgeon.
53. *Mathews*, William. 1828–1901. Alpine Club member and botanist.
54. *Mitford*, Major H. B. Birmingham Philosophical Society.
55. *Moore*, (Sir) Norman. 1847–1923. Physician.
56. *Muller*, Friedrich Max (Rt Hon.). FRS. 1823–1900. Philologist.
57. *Murphy*, John Joseph. 1827–1894. Philosopher of science.
58. *Von Osten Sacken*, Carl Robert. 1828–1906. Entomologist.
59. *O'Connor*, E.
60. *Paget*, Sir James. FRS. 1814–1899. Surgeon.
61. *Paget*, Stephen. 1855–1926. Surgeon.
62. *Plarr*, Gustave. 1819–1892. Mathematician.
63. *Pollock*, (Sir) Frederick. 1845–1937. Barrister.
64. *Preece*, (Sir) William Henry. FRS 1834–1913. Electrical Engineer.
65. *Pryce*, Rev. Robert Vaughan. 1834–1917. Congregationalist minister, later lecturer in logic and mental and moral science in Cheshunt College, then Principal of New College, University of London.
66. *Rawson*, Sir Rawson William. 1812–99. Statistician.
67. *Richmond*, George. 1809–1896. Painter.
68. *Robb*, John Montagu Caesar. 1857–1924. Son of Captain John B. Robb R. N., 46 Rutland Gate. Studied engineering at Heidelberg, became a game farmer in Liphook. His mother was Mary-Ann Boulton, sister of M. P. W. Boulton.
69. *Robertson*, George Croom. 1842–1892. Philosopher, Editor of Mind.
70. *Roget*, John Lewis. 1828–1908. Painter.
71. *Romanes*, George John. FRS. 1848–94. Biologist.
72. *St Clair*, Rev. George. 1836–1908. Minister and historian, active in the Ethnological and Geographical Societies.
73. *Saunders*, A.
74. *Schuster*, (Sir) Arthur. FRS. 1851–1934. Physicist.
75. *Shute*, Richard. 1849–1886. Philosopher.
76. *Siemens*, (Sir) C. William. FRS. 1823–1883. Electrical Engineer.
77. *Simon*, Rev. David Worthington. 1830–1909. Dissenting minister from Spring Hill College.
78. *Smith*, William Anderson. 1842–? Writer on Highland life, cod-liver oil manufacturer, naturalist.
79. *Smith*, Basil Woodd. 1831–1902. Magistrate, involved in the Royal Institution.
80. *Smyth*, Charles Piazzi. FRS. 1819–1900. Astronomer.
81. *Southall*, William. 1825–1886. Pharmacologist and drug manufacturer, member of the Birmingham Natural History Society.
82. *Stacey*, Jeannette?
83. *Stewart*, John Alexander. 1846–1933. Philosopher.
84. *Sully*, James. 1842–1923. Psychologist.
85. *Tedder*, Henry Richard. 1850–1924. Librarian, Athenaeum Club.
86. *Todd*, Edgar. Possibly a Congregationalist minister, Spring Hill Academy.
87. *Tonks*, Edmund. 1824–1898. Barrister and foundry owner.
88. *Tylor*, (Sir) Edward Burnett. FRS. 1832–1917. Anthropologist.
89. *Venn*, Rev. John. FRS. 1834–1923. Logician.
90. *Waller*, Rev. Thomas Henry. 1832–1920? C. of E. Rector of Waldringfield.
91. *Ward*, James. 1843–1925. Psychologist.
92. *Watson*, E.W. Spring Hill College.
93. *Wedderburn*, Sir David. 1835–1882. Barrister and statesman.
94. *Whitford*, Rev. B.? Clergyman.

95. *Willink*, Henry George. 1851–1938. Barrister at Chancery and Alpine Club member.
96. *Willoughby*, Rev. William Charles. 1857–1938. Missionary.
97. *Winson?*, Rev. H. W.
98. *Yardill*, F. F.?[44]
99. *Yeo*, Gerald F. FRS. 1845–1909. Surgeon.
100. *Young*, Charles J.
101. *Young*, Julian. Son of Charles J. Young.

Galton was able to rank his respondents for each question. The scientists had a noticeably lower median ranking than other professions. He ventured that 'an over-readiness to perceive clear mental pictures is antagonistic to the acquirement of habits of highly generalised and abstract thought, and that if the faculty of producing them was ever possessed by men who think hard, it is very apt to be lost by disuse'.[45] Even though a number of the scientists were superb visualizers, they dominated the ranks of the very worst. Charles Darwin was not among those, for though he could not call up an image of a single sentence of a printed page, had no power for geometrical figures, and could not remember numbers at all, let alone produce images of them, he took great pleasure from mental scenes of landscapes. He also had a moderately complete view of the morning's breakfast table in perfect colour, though with a small field of view. But unlike Spottiswoode, he never paid any attention to furniture, and he confessed to never having played chess.[46]

Galton applied his 'method of intercomparison' to show how one could compare the *distribution* of visualizing ability between different groups, going far beyond mere contrast of averages. For a comparison group he used 60 schoolboys from the Charterhouse science class (later that collection grew to 168). Both distributions could be described using octile ranks (deliberately downplaying the tails) then collated at those points. This is the first application of a now universal technique to a psychological subject.

The data-reduction philosophy is explicit. 'There are many who deny to statistics the title of a science and say that it is a mere collection of facts. For my part I think that there is such a thing as a science of statistics, though its field is narrowed almost to a point. Its object is to discover methods of epitomising a great, even an infinite, amount of variation in a compact form'.

It was an argument that Galton had made before when urging the abolition of 'Section F' at the British Association, whose nominally 'statistical' proceedings had always been about the presentation of masses of facts rather than any rigorous theory of statistics.[47] Here a succinct comparison of the distributions suggested that schoolboys had on the whole somewhat better visualizing ability than the scientists. Galton supposed that age tended to suppress the visualizing faculty, independently of specialization.

Not only did the faculty for visualizing vary between individuals, Galton concluded that it was highly heritable (see below) and hazarded that it differed between races too. Some races like the neotenized Bushmen of South-Western Africa, among whom he had personal experience, were greatly gifted in his estimation—going by their deft rock drawings from memory, produced without the aid of formal training. Likewise the Eskimo appeared to be capable of mentally mapping very large territories. He was magnanimous enough to rate the French superior among Europeans. Though these claims are hard to evaluate today since little work after Galton's time seems to have been done on the question one way or the other, they cannot be dismissed out of hand.

44 Not identified. Apparently from Spring Hill College, Birmingham.
45 F. Galton 1880b.
46 K. Pearson 1924, 194.
47 F. Galton 1877b.

It was also George Bidder Q.C. who introduced Galton to a phenomenon that he would be the first to analyze and describe systematically in print: 'number forms'. Mentally, Bidder arranged series of numbers in geometrical forms of circles and ascending curves, and pictured these involuntarily when performing mental arithmetic. He appended a helpful drawing, which Galton reproduced several times in coming years, shown in Figure 9.8, p. 197. Inquiries revealed that several other people saw such 'number forms' in their heads. Each was unique to the individual, sometimes grotesquely detailed. Galton himself did not see any forms, and found the peculiarity very diverting. Victorians were just too embarrassed to admit to secret vices of that sort without encouragement.

After gathering numerous examples, which rapidly accumulated after a letter to *Nature*,[48] Galton delivered a lecture on number-forms to the Anthropological Society on the 9th of March 1880. Dignified people were invited who were prepared to admit in public to seeing their personal forms, exhibited prominently as posters: George P. Bidder Q.C. himself; the botanist Rev. G. Henslow; the physicist Arthur Schuster; John Lewis Roget, son of the thesaurus creator; the astronomer Basil Woodd Smith; and the geographer Colonel Henry Yule (see Figure 9.11, p. 199).

The seers of number forms spoke from the floor to explain their figures. Bidder explained that 'not only numbers, but almost all subjects of thought and memory present themselves to his mind in a visualised form'. He did not at all associate his mental pictures with remembered sense impressions. The Rev. Henslow even came prepared with an exhibit of his own watercolours. Colonel Yule remarked that, like Molière's prosaic Monsieur Jourdain, he had been entertaining number forms for 40 years without knowing that they were special. None could say where they got their forms *from*, only that they had been there since childhood.

Galton deduced from shared features of the forms, especially where they were discontinuous, that they represented a form of translation between clashing names for numbers and their decimal representation. Names often had duodecimal base-12 origins, rather than base-10. And as he assured his audience, 'I have the strongest evidence of its hereditary character after allowing, and over allowing, for all conceivable influences of education and family tradition'. Later he estimated that the prevalence in the general population of number forms was between 1 in 25 and 1 in 30. He had 22 families in which forms were observed in multiple members, which was very unlikely by chance given the base rate above. In eight cases this went beyond the immediate family. In at least four cases similar shapes were observed.

Many of Galton's respondents did not just see number forms, they also associated *colours and patterns of colours* with numbers, in elaborate and mutually incomprehensible ways. There were exotic cases of coloured hearing: some saw specific colours when they heard certain sounds. This syndrome is now called *synaesthesia*, in which senses cross over. Cases had also been published on the continent of people who associated colours with sounds—in 1873 by Bruhl and in 1881 by Bleuler and Lehmann—so Galton did not dwell on these except to note that the tendency seemed to be highly variable, invoked by vowel sounds, and 'very hereditary'.[49]

There were also extreme cases where visualizers became visionaries, seeing apparitions and phantasmagoria. Many heard voices as well. Galton supposed that there was a *continuous spectrum* from mild cases of number forms to harmless and fleeting visions, then ultimately to extremes in which the sufferers became completely delusional. His respondents all appeared to be sane, but some of them secretly lived in a world in which apparitions were always imminent if not constant companions. He neglected the possibility that they might later develop full-blown schizophrenia and were just at an early stage. One who certainly did not succumb to that was the surgeon Sir James Ridson Bennett (1809–1891) who according to his obituary was calm and not known for imagination, yet confided in Galton that more than once he had heard 'the

48 F. Galton 1880c,d,e,f.
49 F. Galton 1883d, 147–148.

well-known postman's knock, followed by the entrance into his study of a man dressed in a fantastic medieval costume, perfectly distinct in every particular, buttons and all, who, after a brief time, faded and disappeared'.[50]

Historically, figures with guiding stars like Napoleon are plentiful. At that very time, the Mahdi 'seer', nemesis-to-be of 'Chinese' Gordon, was taking over the Sudan. Galton was willing to entertain the idea that the visionary tendency could also, like visualizing in general, vary between races, as in the Gypsies.[51] However that may be, the use of a spectrum rather than dichotomous yes/no classes has been vindicated today by multiple Genome-Wide Analysis Studies, which show an additive genetic basis for a wide variety of schizophrenia-like conditions.[52]

That confirmation raises the broader question of Galton's methods. He was well-aware of the obvious objections: we have to rely on self-reports based on introspection, and we do not have a direct way to know that a 'strong', 'clear', or 'distinct' impression for Arthur Schuster follows the same standard as the Rev. Henslow. Galton tried to control for that at first by limiting his formal analysis to the distinguished 100 on his list, even though he had many more cases—including over 180 from women, who seemed to enjoy introspection and sharing the results as much as the Charterhouse schoolboys did. As with the subjects of *English Men of Science* he was hoping that superior ability among his subjects would improve the accuracy of their self-observations. Within a few years he had so many total cases, elicited from all over the world by his journal and magazine articles, that he could use evidence from all his sources without hesitation.[53]

Galton was blazing trails, not surveying with a theodolite. Moving quickly, he followed where the evidence seemed to lead. In capable hands, his methods worked well. They had quickly led him away from his original idea that mental images were analogous to imperfectly blended photographs. In fact they were *not*, since his respondents would often revel in imagery while having poor recall, and imagine things, as Galton thought he could do himself with spheres, in 360° aspect. Alfred Binet, who would later follow Galton's lead on intelligence tests, was soon to offer further confirmation through experiments on calculating prodigies, showing that their recall was order-dependent, therefore not actually read from images.[54]

In parallel with his enquiries into mental imagery, Galton had started experimentally investigating the ideas furnishing his own mind. The surviving notebooks suggest these exercises started around September 1st of 1878, when the Galtons were abroad in Tours. They were repeated when the couple returned to Leamington on September 25th, then back in London at Rutland Gate on October 17th, and again on January 13th of 1879. He called them 'psychometric experiments'. Even ideas could be subject to measure and number. We have already observed him toying with an early version of this exercise while walking down Pall Mall—in fact he was walking home from an afternoon at the Athenaeum Club—encountering a random object and then noting the first two ideas that came into his head as a result. It was important to arrest ideas immediately and examine them. Later he estimated that he encountered about 300 objects to associate with on the way. Repeating the exercise suggested that the bulk of these associations recurred and were not new. There were fewer ideas knocking around in his head than he liked to think.

When seated more comfortably at home or abroad, Galton set about these experiments more systematically. It took some trouble to find the right method. Eventually he settled on working from a list of 75 words, from 'Abbey' to 'Almanack', recording in a notebook the first two idea-

[50] F. Galton 1908, 274. See Obituary, Sir James Risdon Bennett, *British Medical Journal*, 1891-12-19, 1336.

[51] It is now well-established that African Americans have markedly higher rates of schizophrenia: Frisby and O'Donohue 2018, 603ff. There are also persistent claims that Ireland has higher rates: Scheper-Hughes 1979; Torrey 1980; Torrey and Miller 2001.

[52] Plomin 2018.

[53] F. Galton 1883d.

[54] Galton reviewed Binet's book in F. Galton 1894c.

associations he could 'seize' and examine. A stopwatch, to which he had attached a convenient control he had invented, helped to time the interval between seeing a briefly revealed word and the first association. He repeated the exercise several times and tabulated the results from 300 (4×75) trials. Conveniently, this matched the number of objects seen on his quantitative stroll down Pall Mall. The experiments confirmed that very similar ideas tended to associate. Out of 505 ideas only 289 were unique, and a number repeated in 2, 3 or even 4 of the trials. On average he took $2\frac{1}{4}$ seconds—of course he worked this out—to come up with the first idea after seeing the word.

Some of the notebooks containing these experiments—Galton found the experience repellent and had to force himself through them—are still extant today. They are at least as interesting for their clues to his world of ideas as they are for their initiation of what later became the *word-association test* of personality. 'They lay bare the foundations of a man's thoughts with curious distinctness, and exhibit his mental anatomy with more vividness and truth than he would probably care to publish to the world'.[55] Though he preferred not to publish the details, modern readers surely cannot resist the temptation to rifle through them. The records of all four word-association tests follow. They were conducted as numbered, each on the same list of 75 words given below, with the recorded timings in seconds.

1. 1878-09-01. Tours.
2. 1878-09-25. Leamington.
3. 1878-10-17. Rutland Gate.
4. 1879-01-13 and 14. Rutland Gate.

1. Abbey: 4. Tintern. Bolton.
2. Abbey: 3½. Kirstal [sic]. Bolton (I had forgotten the name).
3. Abbey: 1 ½. The face of an abbess. Name of Kirkstall.
4. Abbey: 2½. Kirkstall as a name.[56] Ilkly road towards Bolton abby[57] as a place.

1. Aborigines: 5. Dr. Hodgkin. Australians at Corroboree.[58]
2. Aborigines: 2. Dr. Hodgkin. Portfolio of pictures of abor.
3. Aborigines: 1¾. Dr. Hodgkin. a lot of black fellows at Corroboree.
4. Aborigines: 2¼. Dr. Hodgkin[59] — Australian aborigines & Sir Ch: Nicholson.[60]

1. Abscess: 5. Squeers. Outpatient room Birm. Hosp. Sense of abscess in one of my own arms.
2. Abscess: 2. Squeers cutting the abscess.
3. Abscess: 1¼. View of one (on myself) ripe to the cut. Name of Squeers.
4. Abscess: 2¾. Bishop of [Labuan][61] & abscess in neck. Squeers & the boy.[62]

1. Abyss: 4. Looking down a chasm. Quintus Curtius.
2. Abyss: 2½. A chasm. Quintus Curtius.
3. Abyss: 1½. A ragged chasm. Quintus Curtius.

55 F. Galton 1879b.
56 Kirkstall Abbey in Leeds.
57 Sic. Bolton Priory or Abbey, near Wharfedale in North Yorkshire.
58 A convocation of Australian aborigines.
59 Thomas Hodgkin M.D. (1798–1866) known to Galton from the Anthropological Society.
60 Sir Charles Nicholson (1808-1903).
61 Perhaps Francis Thomas McDougall (1817–1886).
62 Charles Dickens, *Nicholas Nickleby* (1839).

4. Abyss: 3. Quintus Curtius[63] — a view of a gap — some Latin syntax line ending with '[Abyssus]'.

1. Academy: 3. des Beaux Arts of Florence. Fitzwilliam Museum at Cambridge.
2. Academy: 2¼. *Academy* newspaper. A faint image of modern Athens.
3. Academy: 1¼. of Arts. Leighton's name (just now spoken of as probable president).[64]
4. Academy: 2. Newspaper (saw it in imagination) words ['della arte']

1. Accoutrement: 4. Cavalry horse blue & gold saddle gear. Ballad of Leonora.
2. Accoutrement: 2. Saddle gear of a cavalry horse.
3. Accoutrement: 1¾. The black blue & gold sabretache of a cavalry officer.
4. Accoutrement: 2. Sabretache of a cavalry officer as hanging in place but without wearer or horse.

1. Ace: 3½. Besique. Inside a room sitting down to cards.
2. Ace: 2. A card. The words 'ace of St. Paul's' kept recurring (they seem nonsense).
3. Ace: 2. The words ace of diamonds. deuce ace.
4. Ace: 2. Cards in a spread out pack, the black above the aces of some of the cards being visible not the ace itself.

1. Acheron: 4. Styx. Cavern by Naples (indistinct).
2. Acheron: 2¼. A zig zag gorge in [C...]. The word Styx
3. Acheron: 1½. Styx (merely the word).
4. Acheron: 2. Naples & Monti (volcanic thrown up suddenly) a cavern out of which a small river came—the word styx.

1. Acid: 4. Labor. at Birm. where I learnt chemistry. Slight acid taste in mouth.
2. Acid: 2. Some laboratory or other. A sour taste. A fiz. Story of negro [kind] & Seidlitz blow up.
3. Acid 3: 1¼. Slight acid taste. A fiz with Seidlitz.
4. Acid 4: 2. A sense of making a wry face & a taste of acid.

1. Acrobat: 3. Twisting round a pole in some high building. Zazel.
2. Acrobat: 2. A tight rope and some one turning round it. Zazel.
3. Acrobat: 1½. A man turning head over heels in space. Zazel.
4. Acrobat: 1¾. The aquarium prepared for Zazel[65] — the word "Dare" (it being the theatrical name of some acrobat now advertised).[66]

1. Actor 1: 4. Irvine [sic]. Name of Charles Keane.
2. Actor 2: 2½. Hamlet. An actor growing into Irvine [sic].
3. Actor 3: 1½. Charles Keane (only the name).
4. Actor 4: 2. Irving's name[67] — 'as when some well graced actor leaves the stage'.

[63] Roman historian. Perhaps the Lacus Curtius is the abyss.
[64] The painter Sir Frederick Leighton (1830–1896).
[65] The acrobat Rossa Matilda Richter (1863–1937), stage name Zazel, who was fired out of a cannon into the Royal Aquarium in 1877.
[66] Leona Dare aka Susan Adeline Stuart (1854–1922), the trapeze artist.
[67] The dominant actor of his day, Sir Henry Irving (1838–1905), a friend of Richard Francis Burton and Bram Stoker.

1. Actuary: 3. Street where Ansell lives. Confused sense of entering into some offices ? in Birmingham.
2. Actuary: 2¼. Colenso's cousin. Confused imagery.
3. Actuary: 2. Mr. ..... (who has gone mad). Ansell.
4. Actuary: 1½. ? name Colenso's relative.[68] The square where I went to call on Ansell.[69]

1. Adder: 3. Boulton's garden. Zoolog: gardens. Drachenfels (I thought of them & cld. not at first recall my adventure with the snake).
2. Adder: 2. Boulton's garden.
3. Adder: 2. Puff adder (the word). Boultons' garden.
4. Adder: 2½. The words puff adder. Boulton's garden walk near which was the mound frequented by the adder.[70]

1. Adjutant: 3. Saw bird stalking to fuse. Thought of Cape Town.
2. Adjutant: 2. The bird stalking. Idea occurred, but c^{d} not recall words of gagging in waggishness.
3. Adjutant: 1½. The bird stalking up to cartridge, 'gagging in his waggishness &c'.
4. Adjutant: 1½. The bird stalking about 'gagging in his waggishness' a long billed fowl.[71]

1. Admiral 1: 3. Sherard Osborn. Portsmouth dockyard.
2. Admiral 2: —. Admiral Hay. A [...] admiral's uniform. Sherard Osborn.
3. Admiral 3: 1¾. Name Sherard Osborn. Cooper Key.[72]
4. Admiral 4: 2. [Inclefield] Thackeray's ballad about little Billy.

1. Advocate: 4. Scottishness generally (in look and manner). Sir David Dundas.
2. Advocate: 3. A quick thought of many things, none definite. Portia (the name).
3. Advocate: 2. Image of the back of wig and shoulders of a barrister pleading.
4. Advocate: 2. 'Devil's' then some haze & I stopped before it cleared.

1. Afternoon: 3. Lotus eaters 'always afternoon'. Felt it was an afternoon at the time of experiment (in Tours).
2. Afternoon: 3. 'Seemed always afternoon'. Then a pause, & nothing clear succeeded.
3. Afternoon: 1½. 'Seemed always afternoon'.
4. Afternoon: 1. Land when 'always afternoon'.[73] 'Cool of the evening' Ld Houghton.[74]

1. Abasement: 3½. The open foundations of a house being built. A vague sense of David.
2. Abasement: 3. A sense of bending down. King David. A ground (basement) floor.
3. Abasement: 1. The word 'ground floor'. Image of foundations of a newly building house.
4. Abasement: 2. Down to basement floor. David.[75]

68 Possibly Francis Colenso (1862–1910).
69 Charles Ansell, the actuary who had supplied data for Galton's twin study of 1874-5.
70 Possibly Montagu Boulton, who lived for a while in Palestine when Galton was there circa 1846. However there are no puff-adders in Palestine.
71 Francs Bacon, a quote used in Samuel Johnson's Dictionary.
72 Sir Astley Cooper Key (1821–1888).
73 Tennyson
74 Richard Monckton Milnes (1809–1885) who knew Galton from the RGS and other societies and was once a close friend.
75 King David.

1. Abbreviation: 3½. Pitman's name. Much confused idea in which Spencer's name occurred.
2. Abbreviation: 1½. Book about writing by abbreviation it looked like Odell's shorthand.
3. Abbreviation: 2. The new way of brief writing. Image of Odell's book in shorthand.
4. Abbreviation: 1½. Pitman's system (saw a little square book with short-hand characters in a label on its side).

1. Abduction: 3. Gibbon Wakefield. The recent Irish case.
2. Abduction: 3. A running away with. Gibbon Wakefield.
3. Abduction: 1. Gibbon Wakefield.
4. Abduction: 1¼. Gibbon Wakefield.[76]

1. Abhorrence: 4. Ill defined picture of Macbeth's witches (arranged like the 3 fates)
2. Abhorrence: 2½. Mental attitude of uplifted hands & averted face like Martha in raising of Lazarus.
3. Abhorrence: 1½. Turning away head & lifting hands as in picture of Martha in raising of Lazarus.
4. Abhorrence: 1¼. Attitude like Mary in raising of Lazarus (half felt and half seen).[77]

1. Ability 1: 2½. A lawyer-like man working deftly.
2. Ability 2: 3. A sense of doing things [easily]. Nothing more definite.
3. Ability 3: 2. Very vague sense of facility in action (nothing more definite in time).
4. Ability 4: 3. A sense of slickness – nothing came definitely.

1. Ablution: 4. Very general sense of brooks & water. Especially Gehazi in Jordan.
2. Ablution: 1¾. Washing hands in Turkish ritual.
3. Ablution: 2½. Washing hands Turkish fashion. Then washing face and ears.
4. Ablution: 1. Jordan—saw & felt hand washing above elbows, like Arabs.[78]

1. Abnegation: 3. Shaking the head in refusal. Something about women's abnegation.
2. Abnegation: 3. Nodding head in refusal.
3. Abnegation: 1½. Nodding off.
4. Abnegation: 1. Nodding off a proposal (saw & felt)

1. Abnormal: 4½. Deviating from straight line (half imagery, half muscular sense). Coll: of Surgery teratology.
2. Abnormal: 4. Words 'out of the common way' (nothing more definite in all this time).
3. Abnormal: 3½. Puzzled all this time & nothing definite came.
4. Abnormal: 5. Nothing definite.

1. Abominable: 2½. Sense of withdrawal & disgust (the taste was faintly affected).
2. Abominable: 2. Sense of dislike. The psalm 'they are all abominable'.
3. Abominable: 1¼. Mental attitude of disgust & some disagreeable smell 'they are all abominable'.

[76] Edward Gibbon Wakefield (1796–1862), the Australian colonizer who was imprisoned for abducting a 15 year old girl.

[77] See above.

[78] Galton lived in Jordan for 6 months in 1846.

4. Abominable: 1½. 'No not one'[79] felt & acted with mouth & nose.

1. Abrasion: 4. Skin off hand & upper arm, with a faint pain.
2. Abrasion: 1¾. A rubbed sore.
3. Abrasion: 2. Sense & fright of a painfully rubbed & raw surface.
4. Abrasion: 1½. Skin scrubbed (felt) on arms.

1. Absence: 2. Void through a departure. 'absence makes the heart grow fonder'.
2. Absence: 1. Going away, 'absence maketh the heart'.
3. Absence: 1¼. 'Absence makes the heart'.
4. Absence: 2. 'makes the heart grow fonder' nothing definite.

1. Absolution: 3. A priest in chancel waving hands. The sense of being absolved.
2. Absolution: 2½. A priest waving hands.
3. Absolution: 1¾. Faint image of a priest waving hands.
4. Absolution: 1½. Attitude with hands in front enclosing the clasped hands of penitent kneeling.

1. Abstemious: 2½. Putting away things. Sancho Panza's dinner at Doctor.
2. Abstemious: 2. A sense of hunger.
3. Abstemious: 2. A sense of leanness & empty stomach.
4. Abstemious: 2. A sinking in stomach & sense of leanness—also a grim visage.

1. Abstruse: 3. Sense of difficult consideration of some undefined subject.
2. Abstruse: 2. General feeling of it. F. fond of 'abstruse subjects'.
3. Abstruse: 1. Master F. fond of abstruse subjects.
4. Abstruse: 1½. Mrs. French about me.[80] Sylvester's name.

1. Abundance: 3. A pouring out [...]. Something about 'abundance of corn & oil'.
2. Abundance: 1¾. A pouring out. Surplus of, ab'ce of, corn & wine.
3. Abundance: 2. A cornucopia pouring out. Abundance of corn & wine.
4. Abundance: 1½. A cornucopia — ab:ce of corn & wine (Biblical).

1. Accelerating: 3. Words 'accelerated velocity'. Sense of falling through then symbol v^2 presented itself.
2. Accelerating: 1½. Accelerated velocity. An open page of Carr's synopsis.
3. Accelerating: 1¼. Accelerated velocity.
4. Accelerating:1¾. A wheel rolling swiftly down hill—words 'accelerated velocity'.

1. Acceptable: 2¼. A girl accepting an offer. Vague idea of a man in kid gloves.
2. Acceptable: 2. A bland smilingness. Something about engagement in marriage.
3. Acceptable: 1¼. Mere words 'acceptable gift of the Lord'.
4. Acceptable: 2. An accepted lover—'acceptable gift of the Lord'.

1. Accession: 1½. The word 'King George'. Image of a coronation began to form.
2. Accession: 1¼. Words 'Queen Victoria'. Altar at Westminster Abbey.

---

79 Psalm 14.

80 Galton was schooled by Mrs French in the little schoolhouse next to 'the Larches' in Birmingham.

3. Accession: 1¼. Of sovereign to throne. Queen Victoria.
4. Accession: 3. Of Queen Victoria—then a puzzle & nothing definite.

1. Accidental: 1. Railway accident insurance.
2. Accidental: 2½. The metaphysical view. Made a blunder with the stop watch & attention became distracted.
3. Accidental: 1¾. 'Accidental cause'. The metaphysical meaning was beginning to form itself.
4. Accidental: 2½. Insurance — poisoning.

1. Acclamation: 1½. Vox populi. Vote by acclamation at the Geogr. Society.
2. Acclamation: 1½. Voting by acclamation. A people's acclaim in Tennyson Ode to Duke of Wellington.
3. Acclamation: 1. Vote by acclamation. 'High acclaim' words in Tennyson Ode on death of Wellington.
4. Acclamation: 2. Vox populi — 'people's high acclaim'.

1. Accommodation: 2. Money lending matters. Erasmus about 'accommodation land'.
2. Accommodation: 1½. Accommodation land. A W.C.[81]
3. Accommodation: 1½. Thought of a W.C. & the absurdity of the idea kept others out of my head.
4. Accommodation: 1¾. A commode. — accommodation land & Erasmus.[82]

1. Acerbity: 3. Vinegar. Saying sharp things.
2. Acerbity: 2¼. Vinegary expression & taste. Old Cooley (I cd not at first remember his name).
3. Acerbity: 1½. Vinegary look in Uncle Hubert's face.
4. Acerbity: 2¼. Vinegar. Old Cooley.[83]

1. Achievement: 3. A sense of action (nothing [...] I made a mess of it)
2. Achievement: 3¾. A chivalrous fighting. The mere word 'glorious' occurred.
3. Achievement: 1½. Coat of arms — tried at real meaning but failed in time.
4. Achievement: 1½. Armorial bearings.

1. Absolute 1: 2. Capt. Absolute. Sense of command. M. Boulton's metaphysics.[84]
2. Absolute 2: 3. Captain Absolute. Boulton's metaphysics (I had not formed any general idea).
3. Absolute 3: 1. Captain Absolute.
4. Absolute 4: 1½. Captain Absolute.[85] The absolute in logic.

1. Abstract: 1½. 3 things I know, can't recall them.
2. Abstract: 3. A taking away. Lawyers deeds.
3. Abstract: 1¼. Of dictionary. Index.

[81] Water closet.
[82] Land originally used for grazing livestock, but then bought for building houses on.
[83] William Desborough Cooley (1795?–1883) the geographer.
[84] Matthew Piers Watt Boulton, grandson of the inventor and childhood friend of Galton, who wrote on metaphysics as 'M.P.W. Bolton'.
[85] Sheridan *The Rivals*.

4. Abstract: 2. Thought — appearance of a deed abstract.

1. Actuality: 2. Things existing (a good general idea).
2. Actuality: 4. Could not get a clear idea all this time.
3. Actuality: 2½. 'Caught in the act' (I puzzled a good deal).
4. Actuality: 3. Not definite enough for [description]

1. Adequate: 3. (Too abstract, no clear idea).
2. Adequate: 5. Vague, the word 'enough' kept recurring.
3. Adequate: 2½. 'enough' (word equivalents were being hunted in vain).
4. Adequate: 2. (Again made a mess)

1. Advantage: 2. 'Forebore his own advantage' Tennyson.
2. Advantage: 3½. Scene in a hilly country of a goat in a hillock butting down other goats. Forbore his own advantage.
3. Advantage: 1¾. 'Forbore his own advantage'.
4. Advantage: 1¾. Forbore his own advantage—sense of being on the top of a hillock defending oneself against assailants.

1. Aesthetics: 2. Sully.[86] The walk from Seeligsberg when I talked over aesthetics with Wilson.
2. Aesthetics: 3½. Grant Allen's name, but did not get further than Grant. That made me think of Grant Duff & his house at Twickenham.
3. Aesthetics: 3½. Grant Allen.
4. Aesthetics: 1¾. Grant Allen.[87] Walk with Wilson at Seeligsberg.

1. Acquaintance: 1½. A bowing acquaintance.
2. Acquaintance: 3½. 'Sh^{d} acquaintance be forgot'. A general sense of handshaking and bowing.
3. Acquaintance: 1¼. Sh^{d} auld acquaintance be forgot.
4. Acquaintance: 1. Should auld acquaintance be forgot.

1. Acquirement: 3. Gain. [Gadaigne] & rubbing the fingers like Evard the Courier.[88]
2. Acquirement: 7. Latin & Greek (the idea [wouldn't] fix).
3. Acquirement: 4½. A puzzle ends in the 'great acquisition in a country house' (I mistook the word).[89]
4. Acquirement: 1½. Greek & Latin view of a middle class school in some suburban row of houses.

1. Acquisition: 7½. 'A great acquisition at a country house'. Mrs Hertz name occurred.
2. Acquisition: 7. 'A great acquisition'.
3. Acquisition: 1½. (see above) gain. A miser like Quentin Matsys.[90]

86 James Sully (1842–1943).
87 Charles Grant Allen (1848–1899) the author and Darwinian.
88 Montagu Boulton's courier on the Nile adventure of 1845. See Tredoux 2023c.
89 This is the word 'acquisition' below.
90 The Flemish painter, 1466–1530.

4. Acquisition: 1. 'Great acquisition in a country house' in the 'Casual's' history.[91]

1. Activity 1: 2. Sense of running. Activity in *English Men of Science.*
2. Activity 2: 7½. (can't fix my ideas they run in words)
3. Activity 3: 1¼. General sense of nimble changes of position.
4. Activity 4: 2. A running about hither & thither—something about quicksilver.

1. Adhesive: 2½. Plaister. An acquaintance sticking to one.
2. Adhesive: 4. Adhesive plaister. Friends who can't be shaken off.
3. Adhesive: 1¼. Plaister.
4. Adhesive: 2. Sticking plaister—postage stamps.

1. Adjudication: 2. Sense of judgment. Hills as judge in Alexandria.
2. Adjudication: 2. Sense of judgment. The late Russell Gurney.
3. Adjudication: 1½. Mental attitude of a judge. Russell Gurney.
4. Adjudication: 1½. Seat of delivering judgment—image of Sir J. Stephen.[92]

1. Administration 1: 1½. General idea. The words 'chemin de fer'.
2. Administration: 4. Of Gold Coast. Bartle Frere.
3. Administration: 2. Mental attitude. Sir Bartle Frere.
4. Administration: 2. Feeling of solid official establishment—name of Grant Duff.[93]

1. Admission: 1½. The foolish virgins, 'too late, too late, you cannot be'.[94]
2. Admission: 3. To theatres.
3. Admission: 1½. Into a theatre.
4. Admission: 2. To a theatre—by.

1. Adoption: 2½. Something including 'Foundling Hosp.'.
2. Adoption: 2. Of a baby. Jinx's baby.
3. Adoption: 1¼. Of a son. Adoption by grace.
4. Adoption: 1½. Son by adoption. Mania (I had just been hearing the story of the 2nd mania having acquired that name by adoption).[95]

1. Adoration: 2. Heart praying. Magi.
2. Adoration: 2½. The Magi. Hands lifted up.
3. Adoration: 1. Of the Magi. Faint sense of uplifted hands.
4. Adoration: 2. Magi—attitude with mouth in an O and breathing out an Ohh!

1. Adulation: 2. Sense of active & passive soft soldering. Thought of Adèle.
2. Adulation: 4. Praise. Soft solder. A sense of licking.
3. Adulation: 1¾. A sense of licking and slavering.
4. Adulation: 2. A licking with the tongue—Mrs. H.[96]

[91] Perhaps the 'Amateur Casual' James Greenwood, *A night in the workhouse* (1866): 'what in the higher grades of society is called "quite an acquisition" ' (of a thief).
[92] Sir James Fitzjames Stephen.
[93] Sir M.E. Grant Duff (1829–1906).
[94] Tennyson.
[95] Likely melancholy.
[96] Possibly Mrs. Hills, daughter of Justice Grove.

1. Adulteration: 2½. General sense. Reference to wine.
2. Adulteration: 4. Adultery. Wine & beer.
3. Adulteration: 2½. Adultery. Débus and the claret.
4. Adulteration: 1½. Of wine—adultery.

1. Advance: 1½. Marching with soldiers. 'The guards & at them'.[97]
2. Advance: 2½. Ray Lankester's name (? as suggested by analysis).[98] Slight sense of forward motion. 'Charge Chester charge'.[99]
3. Advance: 1. As spoken to troops in battle.
4. Advance: 2. Afghan campaign—order given to advance.

1. Adventitious: 3. General idea. Word 'particle' in grammatical sense.
2. Adventitious: 3½. Adventitious particle. Claimant to an estate.
3. Adventitious: 1¼. A stranger. 'Welcome little stranger'.[100]
4. Adventitious: 2½. Thought vaguely of adventurers—particles.

1. Acoustics: 2. Science of. My whistles. Tisley & Spiller.[101]
2. Acoustics: 3. Tyndall. A stretched string vibrating.
3. Acoustics: 1¼. Tyndall.
4. Acoustics: 2¼. My whistle. Tisley & Spiller.

1. Adage 1: 2. 'Like the poor cat i' the adage'. Bell-the-cat.
2. Adage 2: 3½. A scroll worn by some people like beadsmen—very indistinct image.
3. Adage 3: 1. 'Like the poor cat i' the adage'.
4. Adage 4: 1¼. Like the poor cat i' the adage.[102]

1. Addition: 3. A sum. '... (addition) is as bad ...'.
2. Addition: 3½. Multiplication is vexation &c.[103] Something else? what.
3. Addition: 1¼. Multiplication is vexation &c. I thought wrongly addition came with the verse.
4. Addition: 2. '... multiplication is as bad ...'

1. Adjective: 2. Throwing stones.
2. Adjective: 4½. Something thrown at. I cd not get any other notion in time.
3. Adjective: 1¼. Something thrown at. The mental attitude of throwing.
4. Adjective: 2. Some sort of idea of its real meaning—then something known at.

1. Adjunctive: 4. Piecing in.
2. Adjunctive: 4. Something joined to. the German word 'füzen' occurred or rather the unformed idea of it.
3. Adjunctive: 2. Piecing to. The German word 'füzen'.

[97] Attributed to the Duke of Wellington, at Waterloo.
[98] Sir Edwin Ray Lankester (1847–1929).
[99] Sir Walter Scott *Marmion*.
[100] Said of newborns.
[101] The instrument makers.
[102] Shakespeare, *Macbeth*.
[103] Nursery rhyme. 'Multiplication is vexation; / Division is as bad; / The Rule of Three doth puzzle me, / And Practice drives me mad.'

4. Adjunctive: 2¾. Could not realise—got the phrase 'joined to'.

1. Animal: 3. Something hairy. 'animal, vegetable, mineral'.
2. Animal: 2½. As a term of reproval.
3. Animal: 1. 'Animal vegetable & mineral'.
4. Animal: 2. Animal, vegetable or mineral? — Smell of wild animals.

1. Advantageous: 3. Blackberry jam (shd have been shrimp paste—the idea was Dick Gaisford Haileybury). Benigna Gaisford.[104]
2. Advantageous: 3. Blackberry Haileybury jam, as connected with Dick Gaisford. Advantageous marriage.
3. Advantageous: 1. Shrimp paste & Dick Gaisford.
4. Advantageous: 2. marriage—thought of Romanes.[105]

1. Adversity: 3. 'Sweet are the uses of adversity'. Something else ?what.
2. Adversity: 3½. Sweet are the uses of adv. then a pause & nothing else occurred.
3. Adversity: 1¼. sweet are the uses of adv.
4. Adversity: 1½. 'Sweet are the uses of adversity'.[106]

1. Advice: 2½. 'Oh gentle dreams it gars me greet'. thought of Louisa.
2. Advice: 3½. Good advice. 'Oh gentle dreams &c.'
3. Advice: 1. 'Oh gentle dreams &c'.
4. Advice 4: 1¼. Good advice—'oh gentle dreams it gars me greet.'[107]

1. Aeriform: 2½. Vapour. Captive balloon at Paris.
2. Aeriform: 3. Gas. Captive balloon at Paris.
3. Aeriform: 2. Gas. A balloon.
4. Aeriform: 2. Gas, vapour; Paris balloon.

1. Affability 1: 3½. Aunt Booth's establishment at Birm.[108] Sweetness. Sir R. Murchison humbugging.[109]
2. Affability: 4. Pecksniff & Pecksniffery. Ch. Evans blandly speaking (I had recently seen him).
3. Affability: 1. A face (Bishop of Bond St.) with a humbugging friendly expression.
4. Affability: 2. The face and face action of an affable man—Bishop of Bond St.

1. Affection: 2. 'The sense of affection'. Louisa's name occurred.
2. Affection: 2. Louisa's friends viz. Mrs R. G.[110]
3. Affection: 1½. Feminine gracefulness. Ly. P.[111]

[104] Children of Stephen Henry Gaisford (1818–1895). Dick was educated at Haileybury. Arthur Gray Butler, Galton's brother-in-law, was headmaster there.
[105] George Romanes (1848–1894) the evolutionist. He married Ethel Duncan, the daughter of a Scottish merchant, in February 1879.
[106] Shakespeare, *As You Like It*.
[107] Burns.
[108] Aunt Adèle Booth née Galton.
[109] Sir Roderick Murchison, frequently President of the RGS, known for courting the well-to-do.
[110] Emelia Russell Gurney (1823–1896). [Ly. P.]
[111] Possibly Lady Amy Pelly?

4. Affection: 1¾. The sentiment.

1. Alliance: 2½. Royal marriage. carrying on thought as above. Ld P.
2. Alliance: 3. Giving of hands. Altar and Sir L. Py.[112]
3. Alliance: 1. Ly. P.[113]
4. Alliance: 1¼. Laying a hand in that of another person — a heart sensation of trust.

1. Affinity: 2½. G. Darwin & cousin marriage. Confused general idea including rabbits.
2. Affinity: 2½. Chemical. G. Darwin's next of kin.
3. Affinity: 1½. In love matters. Selective affinity. The lady roped to the gentleman at Zermatt.
4. Affinity: 1¼. Affective affinity, sense of. Zermatt tying together story.

1. Affliction: 2. 'Affliction sore, long'. General idea.
2. Affliction: 2½. 'Affliction sore, long ...'.
3. Affliction: 0¾. 'Affliction sore long'. Eyes lifted up Carlo Dolce fashion.[114]
4. Affliction: 1¼. Affliction sore long time I bore.[115]

1. Affluence: 2. Sense of easy distribution of wealth. The words £3,000 a year occurred.
2. Affluence: 1. A pouring in from a cornucopia. 'heart affluence' (Tennyson Memoriam).
3. Affluence: 1½. A pouring out of wealth. Albert Grant.
4. Affluence: 1¼. 'Heart affluence in discursive talk'.[116]

1. Affront 1: 2. Frown & sword drawn. Bonn duel.[117]
2. Affront 2: 3. A frowning affronted person. Sense of savagery.
3. Affront 3: 1¼. A frown & hand in sword.
4. Affront 4: 2. A frown & a [drowsiness] of a male figure.

1. Agent: 2½. Edward Wilmot.
2. Agent: 2½. Edward Wilmot. In neither case c^{d} I think of the name in time.
3. Agent: 1½. Ed. Wheler. Edward.
4. Agent: 1. Edward Wilmot.[118]

1. Ague: 2. Louisa at Chambord (we were on the point of driving there Sept 3/78). My ague.
2. Ague: 2½. My own ague. Tertius & Grace going to Palestine.
3. Ague: 1¼. 'as dim & meagre as an ague fit'. My own ague. Mrs. Hills.
4. Ague: ¼. My own (& feeling of it) –'is dim and meagre as an ague's fit'.[119]

1. Aide de Camp 1: 2. Galloping officers in cocked hats.

[112] Sir Lewis Pelly.
[113] Possibly Lady Amy Pelly?
[114] The Italian painter Carlo Dolci (1616–1686).
[115] An inscription commonly found on gravestones.
[116] Tennyson, *In Memoriam.*
[117] Possibly a reference to Charles Astor Bristed's duel with the notorious Marquis de Gallifet near Strasbourg. Bristed was at Cambridge with Galton and lived near Baden Baden at the time of the duel.
[118] 1808–1864. The land agent for the Duke of Newcastle. He married Emma Elizabeth Darwin, daughter of Galton's uncle, Sir Francis Sacheverel Darwin.
[119] Shakespeare, *King John.*

2. Aide de Camp 2: 3. Knowles at Aldershot. Officers in lifted hat riding rapidly about.
3. Aide de Camp 3: 1½. Galloping about of mounted officers. Capt. Knowles at Aldershot.
4. Aide de Camp 4: 2. Knowles at Aldershot.[120]

1. Aigle: 2. Napoleonism. Aigle in Rhone valley.
2. Aigle: 2½. Aigle near Bex. The French eagle.
3. Aigle: 1¼. by Bex.
4. Aigle: 1. Aigle by Bex[121] — Napoleon 1st eagle in arms.

1. Albino: 1½. Albinos in Africa. Cretinism.
2. Albino: 2½. Robert Lowe.
3. Albino: 1¾. An abstract albino of. White ferrets a smelling in a cage.
4. Albino: 1¼. Lowe[122]—a view but indistinct, of one.

1. Album: 2½. Mrs. M. (a book of album appears outside)
2. Album: 2½. Mrs. M. Lucy Wheler's books.
3. Album: 1¼. Mrs. M. (I recollected I had thought of this before). Scrap books bought [...].
4. Album: 1½. Mrs. Moilliet—a thin quarto vol.[123]

1. Alderman: 2. Alderman White. Aldermen generally. R. Gurney.
2. Alderman: 4. Alderman White. A city dinner.
3. Alderman:1¼. Alderman White. A turtle shaped man.
4. Alderman: 2. White[124]—view of inside of Mansion House at a dinner.

1. Alembic: 2½. Laboratory at Birm. Retort (not alembic) on stand.
2. Alembic: 1½. Laboratory surrounding the laboratory I worked in in Birm.
3. Alembic: 1½. The thing itself in Birm. Laboratory.
4. Alembic: 1½. View of the thing—the laboratory at [Birm.][125] gradually shaped itself around it.

1. Alluvium: 2. View of a shining river. My plan for lecture & illustrations. Delta of Rhine (Lake Geneva).
2. Alluvium: 2. Buckland's name.[126] Valley of Rhone at L. Geneva. Generalized idea.
3. Alluvium: 1½. Delta of Rhone.
4. Alluvium: 2. Of Rhone valley—much haze nothing clear after.

1. Almanack: 3. A generalized almanack. Almanack de Gotha.
2. Almanack: 1¾. Almanack de Gotha. Almanack like that of De La Rue but generalized.
3. Almanack: 1½. De Gotha—thought of Petermann.[127]

---

120 General Knowles (1797–1883).
121 A place in Switzerland.
122 Robert Lowe (1811–1892) the statesman, who was an albino.
123 Galton's sister Lucy married into the Moilliet family.
124 Alderman Sir Thomas White who became Lord Mayor of London.
125 Birmingham.
126 Possibly William Buckland the geologist. Another candidate is Frank Buckland the naturalist.
127 The geographer August Heinrich Petermann (1822–1878).

Influences here ran from high to low, filtered through his reading, family milieu, schooling and the broader cultural memory of his class and kind. Shakespeare (he had been a member of the Shakespeare Society of Stratford after returning home from the Middle East), Francis Bacon, Burns, Sheridan, Tennyson (whom he had met on the Isle of Wight), Dickens, Biblical jingles, places, pictures, the versatile people of Victorian Science, family, acrobats fired from cannons, random doggerel and stock phrases. All jostled in his head but tumbled out consistently when interrogated. Surprisingly, there were no remnants from his early obsession with Byron.[128]

The transcript shows that for a small number of words Galton drew a blank. He could not think of anything within a reasonable space of time. When he inspected these failures they turned out to be mostly abstract words. He put the failures down to an inherent difficulty on his part in dealing with abstract ideas. The highest minds, he was sure, were best at grasping abstract ideas, whereas he could hardly cling to their skirts. Clements Markham, who knew Galton well from the RGS, wrote in his private reminiscences that his colleague was without any trace of vanity.[129]

Categorizing the recorded idea-associations, he found that they split roughly into thirds: ideas prompted by the *sound* of words; mostly visual *sense-imagery*; and *histrionic* associations in which he saw himself doing something. He then realized that one could also group the words by the type of idea they were most likely to provoke. Their origins were a bit surprising. Far from being dominated by early-childhood experiences, only 39% could be traced back to his boyhood or youth, the rest coming from manhood (46%) and fairly recent events (19%). However associations with earlier origins did tend to recur more often.

The subconscious mind seemed to be directing operations behind the scenes. 'Perhaps the strongest of the impressions left by these experiments regards the multifariousness of the work done by the mind in a state of half-unconsciousness, and the valid reason they afford for believing in the existence of still deeper strata of mental operations, sunk wholly below the level of consciousness, which may account for such mental phenomena as cannot otherwise be explained'. Ironically, though this theme would become an obsession for certain schools of 20th century psychology, it would be left alone by the school that his own work led directly to, the 'London School' of Differential Psychology.

Galton started down this line of enquiry not just because he was intrigued by the idea of measuring something popularly thought to be evanescent, but because he hoped to find another class of heritable traits. But that proved to be a dead end. He could experiment on himself in this way, but it was far harder to experiment on others and trace family relationships, nor could many conduct self-experiments as plausibly. And there was another problem: large swathes of people from similar backgrounds seemed to share an overwhelmingly common set of ideas. When asked to tag random scrawls with stories in parlour games, they tended to draw from this limited set (the 'Babes in the Tower' etc.). Other lines of investigation were more promising.

By publishing his imagery researches, Galton had already provoked an outpouring of material from the general public, who seem to have been overjoyed by this validation of their idiosyncrasies. Material accumulated quickly. Dr. James Key from the resort town of Montagu in the Western Cape of South Africa initiated a lasting correspondence containing elaborate watercolours of his personal colour associations, and alarming phantasmagoria seen by his wife (see Figure 9.14, p. 201). The painter Mrs. Mary Elizabeth Haweis sent him her many curious drawings in which words were represented by very specific faces. She told him about her recurring hallucinations involving masses of pink roses floating in space invariably effaced by a shower of gold sparks. Many others followed in the same vein, though with less skill. Soon there was more material than he could use in articles.

[128] Tredoux 2023c.
[129] RGS Archives, CRM 47.

In 1883, gathering together multiple strands of his miscellaneous papers since *English Men of Science* and before, Galton brought out a collection titled *Inquiries into Human Faculty and its Development.*[130] To obtuse reviewers the book looked like a bulky but loose and ill-assorted bundle of quirky ideas, united only by a personal obsession with counting and measurement, added to a fondness for irreverence. Read more carefully, an underlying research programme can be seen coming together, with strongly connected themes at one remove. The papers on Domestication of Animals, Gregarious and Slavish Instincts, Efficacy of Prayer, Hereditary Improvement, Statistics by Intercomparison, Twins, Galton's Whistles, Composite Portraiture, Generic Images and Number Forms, Psychometric Experiments, and Visions of the Sane were all included. Some of this material had been edited, with parts reorganized, but there was a lot of new material on imagery in particular. There were also entirely new treatments of topics, only referred to in passing until then, linking these together: Energy, Sensitivity, Character, Criminals and the Insane, Features in Portraits, Early Sentiments, Enthusiasm, and an expanded treatment of the Nature-Nurture relationship.

There were also sections on a theme that Galton had taken up only recently and would expend enormous effort on in subsequent years: Anthropometric Laboratories with special equipment for testing humans on a wide variety of dimensions, mental as well as physical, and associated instruments for doing this. He proposed Anthropometric Registers, which could be maintained less formally by families themselves using the information they could easily record, including not just births and deaths but also heights, weights, illnesses and series of photographic portraits by age.

The final third of *Human Faculty* was taken up with a far-reaching expansion of 'Hereditary Improvement', introducing the word *eugenics* for the first time. But it was really an idea that Galton had introduced as early as 1865 when he first started applying himself to ethnology (see chapter 2, p. 7, and Appendix C, p. 399). Taking Darwinism to its logical conclusion, if natural selection produces animals, including humans, with certain traits, the tables can be turned and humans can be specially selected (by themselves, in a calculated act of evolutionary self-consciousness) for desirable traits, altering the species deliberately instead of letting things fall as they may. This is closely analogous to sexual selection, by which a species already sets its own internal standards for propagation.[131]

To Galton, improvement of the species was not just a good idea but a positive duty. 'Man has already furthered evolution very considerably, half unconsciously, and for his own personal advantages, but he has not yet risen to the conviction that it is his *religious duty* to do so deliberately and systematically'.[132] Large sections of *Human Faculty* made the argument that instilling religious feeling was eminently feasible, while other sections promoted the argument that traditional (received) religion was open to question and therefore ripe for replacement. This new *scientific* religion could appeal to the same underlying sentiments—the need for moral purpose, the millenarian state of grace—without carrying the burden of superstition. But there were some complications in working out the details.

To understand Galton's conceptualization of and solution to these complications one must first take into account his idea of *race*, which is now obscured by a radical change in usage. Galton's meaning followed his cousin. Darwin's notorious book was actually titled *On the Origin of Species by Means of Selection or the Preservation of Favoured Races in the Struggle for Life.* The second part of the title, with its 'favoured races', is usually glossed over today. Here a 'race' is any breeding group, of greater or smaller extent, depending on the context. That group carries hereditary information, which we call genes and understand fairly well, but

[130] F. Galton 1883d.

[131] Provided that sexual selection is not just a disguised form of regular fitness selection, and really favours arbitrary characteristics with no other signalling value, of the kind that a runaway process might launch by accident, as R.A. Fisher argued, Fisher 1930.

[132] F. Galton 1883d, 198. Emphasis added.

Galton, as described above, thought of as 'gemmules' and was actively involved in attempting to understand. For Galton, races were distinct if and only if the hereditary information they carried and propagated as a breeding group was different. From a modern point of view, a novel genetic mutation would introduce that sort of difference. That difference need not be physically visible in body morphology, skin colour and so on, unlike most modern usage of the word 'race', and is not limited to broad categories like 'black', 'white' or 'East Asian'.

One way of improving the species could be by selective breeding of only the most desirable members. Here desirability, inherently subjective, includes but is not limited to intellectual ability, extending to bodily qualities, so as to produce a complete package that is healthier, more vigorous and able to thrive under increasingly demanding modern conditions. But Galton distinguished between two kinds of selection: one which produces no lasting change in the population (race), and one which does. 'So long as the race remains radically the same, the stringent selection of the best specimens to rear and breed from, can never lead to any permanent result. The attempt to raise the standard of such a race is like the labour of Sisyphus in rolling his stone uphill; let the effort be relaxed for a moment, and the stone will roll back'.[133]

So long as the race remains radically the same. Galton's views here have been widely and persistently misunderstood as the idea that *regression to the mean*—his discovery after all—necessarily entails that selection alone cannot produce lasting changes in a population. That is certainly not his argument, and is back-to-front. We will defer a full discussion of this recurring theme for later.[134]

Human Faculty was published by Macmillan on April 24th, 1883. The book would sell briskly, with 247 copies gone by early June. A further 31 copies were distributed to reviewers and agents, along with 48 presentation copies to Galton's broader circle: George Darwin, John Lubbock, General Strachey, Herbert Spencer, Croom Robertson, Moncure D. Conway, and three of his principal informants on imagery, Dr. James Key of Montagu, the Rev. George Henslow and Mrs. Haweis.

Sister Emma was evidently upset by the chapter on the efficacy of prayer. And, it is reasonable to expect, the rest of the family too. Her brother holds his ground calmly, and soothes over the dispute by recalling the joys of Epping Forest.

42 Rutland Gate SW
May 5 / 83

Dearest Emma

At length, I have simultaneously in hand the 3' worth of stamps that I owe you and a note to send them in, so here they are. We hear that Cyril uses his cup daily, so the present has been a success.

I am very sorry that the chapter in my book to which you allude gives you any pain or might give any others, but I am sure you will recognise that it deals with a mere matter of fact & observation, also that it is one's duty to be truthful. If my facts and observations are wrong, they will be shown to be so, and no one w^{d} be more ready to acknowledge it than I should be but I must say, that when it was first published the large amount of controversy that it raised did not succeed, to my mind, in shaking my conclusions in any way.[135] We must be reasonable, even though sentiment or education may be at first contradicted & more or less shocked. Just think of the horror that the practical beliefs even of yourself & of those who think in the same way, would have excited a few centuries ago. Why you would have been actually burned as heretics! I

[133] F. Galton 1883d, 306.

[134] See page 215 and following.

[135] This shows that Emma was referring to the chapter on prayer, which was reprinted from *Fortnightly Review*—see F. Galton 1872b. In the second edition, F. Galton 1907a, the chapter was one of three left out at the insistence of the publisher, Everyman.

take the responsibility of writing my views because I am conscious of having given as much serious & honest attention to the subject as anyone, & I have friends who have urged me to write. It is one of the few services that a man situated like myself can do, to take up an unpopular side when he knows it to be the true one.

No reviews are yet out of the book. I have not heard much of it yet from friends, but what I have heard has been very appreciative. However I feel in my own mind sure of getting here & there, & for a while, into hot water.

Three or four days ago Louisa went for a night to Harrow. I had a dinner engagement in London, and spent the next day in a most enjoyable excursion to Epping Forest. You have no idea how interesting it is to any one who leaves the roads & scrambles about it. I got between 5 & 6 hours steady walking in as wild a bit of country as one could find anywhere with beautiful clumps of pollard beeches, here & there, & hardly saw a soul, nor a single pedestrian like myself. Sir Fowell Buxton is one of the 'verderers' & takes great interest in it.[136] It is a long strip between 2 lines of railway. that are only 2 miles or so asunder, like the arms of a tuning fork & there is a big road right down its middle, nevertheless space remains on each side for complete solitude.

[Doodle of wooded wedge between railway tracks omitted.]
Hyde Park corner looks so open now, with broad roads, & the Duke's statue temporarily on the ground is very disconsolate and sad in appearance.

I dined with Bowman last night (the Sec^{y} of the Royal Institution) & went to Scott's lecture there in 'Forecasts' afterwards.[137] We were at the private view of the R. Academy yesterday, of which you will have seen plenty in the papers. We shall be so pleased to have you a month hence. Louisa has been getting curtains up in the drawing room. Ever very affectionately

Francis Galton

The *St James Gazette* is early to print on the 12th of May, grasping at once that Galton is redirecting inquiries from the evolutionary past to the evolutionary future.[138] But if records are to be kept of family tastes and aptitudes, they suggest lodging them in advance for official safe-keeping at the General Register Office in Somerset House. Otherwise there might be ascertainment bias. 'At present when a gentleman is sentenced to be hanged he exhibits a surprising alacrity in making known that his sisters, his cousins, and his aunts have all been afflicted with homicidal insanity'. They decline to comment on Galton's statistical inquiries into prayer, suggesting instead that his 'very remarkable book' shows energy that is not always wisely directed, as in his studies of associations of ideas, the point of which escapes them.

A few weeks later the *Saturday Review* pants at the sheer breadth of the material and the ingenuity on display, but doubts whether Galton's attempts to engineer the human race will succeed, given the strong dislike shown by so many of 'being done good to', and their obvious willingness to neglect their own lives, let alone those of future generations.[139] They especially scoff at the idea that encouraging academics to breed will improve the race, given that any man of ability will long since have left college.

The resolutely sober *Scotsman* detects the same 'laborious accumulation' combined with 'acute speculation' shown previously by the author.[140] While they accept that nature has been

[136] Verderers administered Royal Forests used as common land.
[137] Sir William Bowman (1816–1892), who had accompanied an adolescent Frank Galton on a medical tour of Europe in July 1838. See Robert H. Scott, 'Barometrical Disturbances', *Proceedings of the Roy. Inst.*, 36, 139–42.
[138] Anonymous 1883q.
[139] Anonymous 1883j.
[140] Anonymous 1883l.

shown to dominate nurture, they doubt that humanity would tolerate instruction in its 'matrimonial arrangements'. The new religion Galton proposes combines a 'melancholy vagueness' in its 'splendid aim' with a 'bland indefiniteness' in its means and a 'rich remoteness' in its results. They suggest that most would instead counter 'what has posterity done for me?'

A sharp reviewer in the *Pall Mall Gazette* notes that Darwinism is now spreading into every department of thought and action.[141] Two major themes in the book are noticed: commonality of human types and variability of human individuals, both of which are neatly captured by composite photography. The results from twins are found convincing and suggest 'a certain awful scientific fatalism'. But the material on association and visualizations is too exotic to accept uncritically.

In the *Academy*, the radical evolutionist Grant Allen is (unsurprisingly) in sympathy with Galton's aims and methods, but agrees with previous reviewers in doubting whether notoriously selfish individuals would be so altruistic as to make their breeding arrangements with the long-term good of the race in mind.[142] The well-informed *Westminster Review* makes no comment on eugenics, but warmly welcomes the new psychology of individual differences as a branch of the not much older discipline of experimental psychology as pursued by Fechner and Wundt.[143] The reviewer in *Mind* goes further, inferring from the demonstrated importance of individual differences that general statements about human psychology now need to be qualified accordingly, a surprising anticipation of the 'London School' philosophy that would be developed in the next century.[144]

The *Quarterly Review* prefers to ignore the author's 'desultory' inquiries into psychology. Instead it brings the focus back to eugenics as a religion, its major concern. It is so pleased that Galton is willing to entertain the rudiments of some religion at least—perhaps even, it hopes, ultimately to reconcile religion and science—that it is prepared to pass over his attacks on prayer, and forgo quibbling about specific doctrines.[145]

Curiously the most pointed criticism of Galton's attacks on religion comes instead from the biologist and ardent Darwinian George Romanes in *Nature*.[146] After graduating BA in 1871, Romanes had won the Burney Prize at Cambridge two years later, when the statistical prayer controversy was fresh, for an essay on 'Christian Prayer and General Laws'. He had argued that the mere two-year lifespan advantage for clerics found by Galton might multiply into sixteen years if only half actually prayed, with one half of those prayers for their own health, and say one half again of those are answered by other physical benefits than a longer life—a multiplier of eight altogether. The logic of this argument, which Romanes recapitulates now after a decade, is obviously fallacious to any statistician—what hypothetical multiplier need one stop at?—but fortunately Galton takes no notice of it. Romanes is strongly enthusiastic about both the eugenic aims of the author ('the prospects of the race would be greatly brightened') and the eugenic policies proposed by him ('well considered').

The *Spectator* is likewise in sympathy with Galton's methods and most of his concerns—except those to do with prayer—but following so many others finds improvement of the breed a feeble basis for faith, and reproduction of previously-celibate fellows of the universities faint hope for the outcome.[147]

It is striking that almost all of the contemporary reviewers of *Human Faculty* do not raise any objection to Galton's eugenic ambitions. They doubt only the effectiveness of his proposed

[141] Anonymous 1883n.
[142] Allen 1883.
[143] Anonymous 1883e.
[144] Anonymous 1883o.
[145] Anonymous 1883f.
[146] Romanes 1883.
[147] Anonymous 1883a.

methods.[148] The lone exception appears in the *Guardian*, where the reviewer anticipates many later arguments. 'Who is to decide whether a man's issue is not likely to be well fitted "to play their part as citizens"?' The celibacy jury this implies is opposed to 'common sense and good feeling'. They think the doctrine indistinguishable from an anti-religion.[149]

Eventually a cheaper reissue of *Human Faculty* was brought out by Everyman in 1907. The text was unaltered but three problematic chapters were omitted at the insistence of the publisher: 'Enthusiasm', 'Possibilities of Theocratic Intervention' and 'Statistical Inquiries into the Efficacy of Prayer'.[150] In the interim, Galton had gone in search of a much larger trove of data, an exercise which would generate many new ideas and expand his reach even further.

[148] Anonymous 1883b,c,g,h,i,k,m,p; Jacobs 1883.

[149] Anonymous 1883d.

[150] A copy of the first edition with extensive annotations survives in the Galton Papers at UCL, GALTON/2/11/4/2. This revised edition was never published.

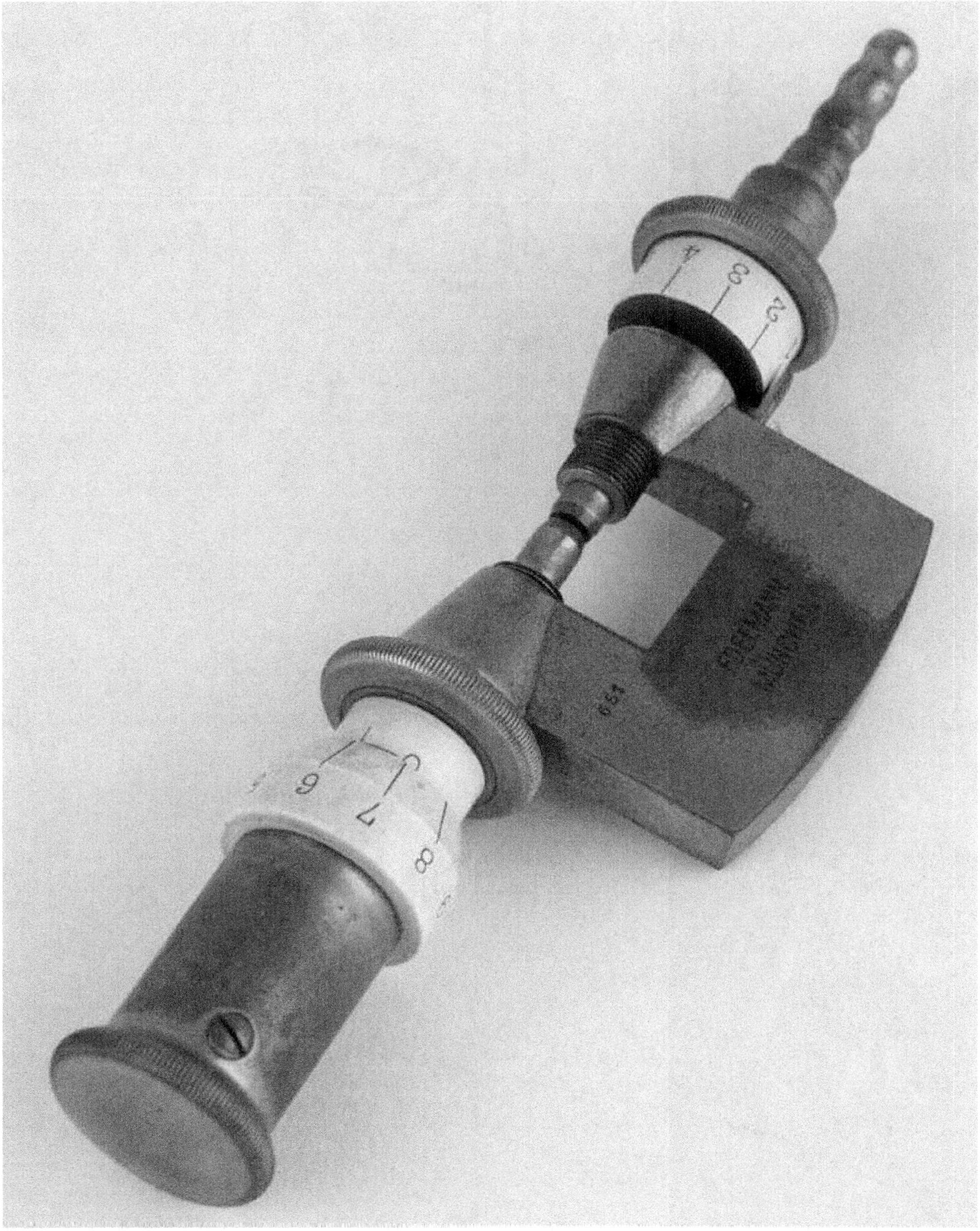

Figure 9.1: Galton's Whistle as manufactured by Edelmann, Science Museum Group, Creative Commons.

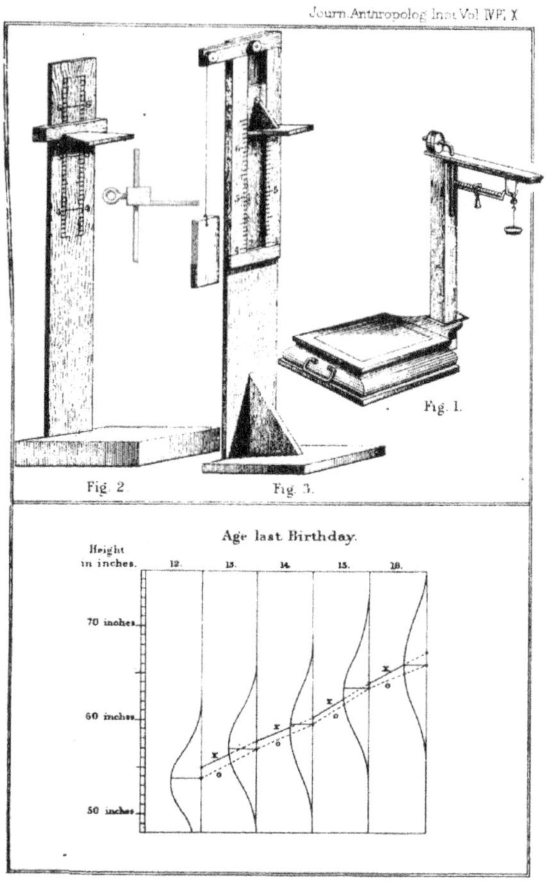

Figure 9.2: Instruments designed by Galton to standardize measurements of school pupils.

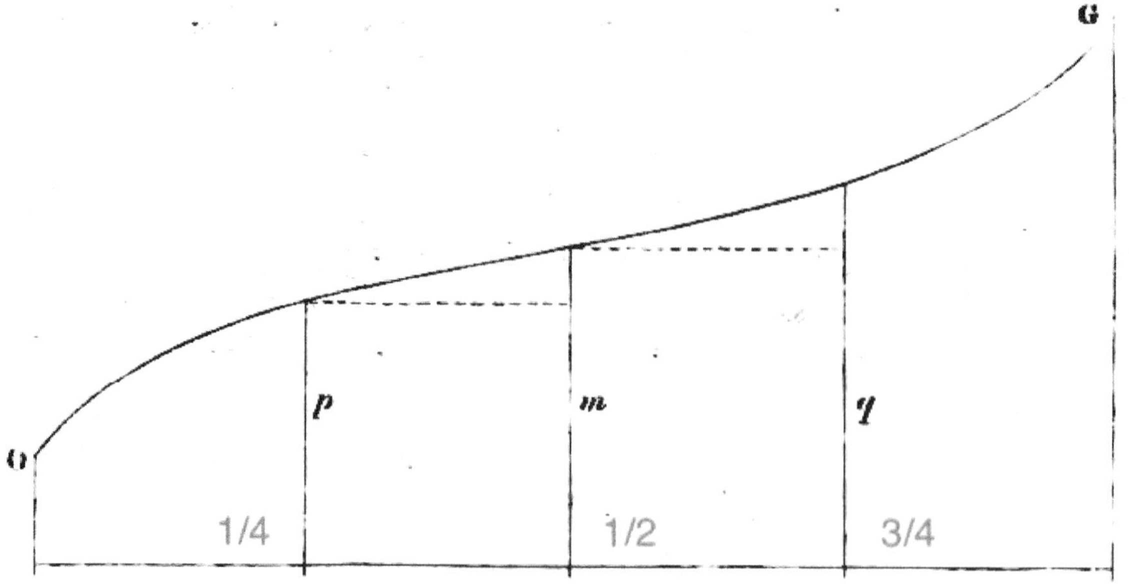

Figure 9.3: Galton's ogive, or inverse cumulative normal distribution. The first quartile is p, the median is m, the third quartile is q.

Figure 9.4: Sir Edmund du Cane in middle age.

Figure 9.5: Dr. Frederick Akbar Mahomed (1849–1884).

Figure 9.6: Composites of families, K. Pearson 1924.

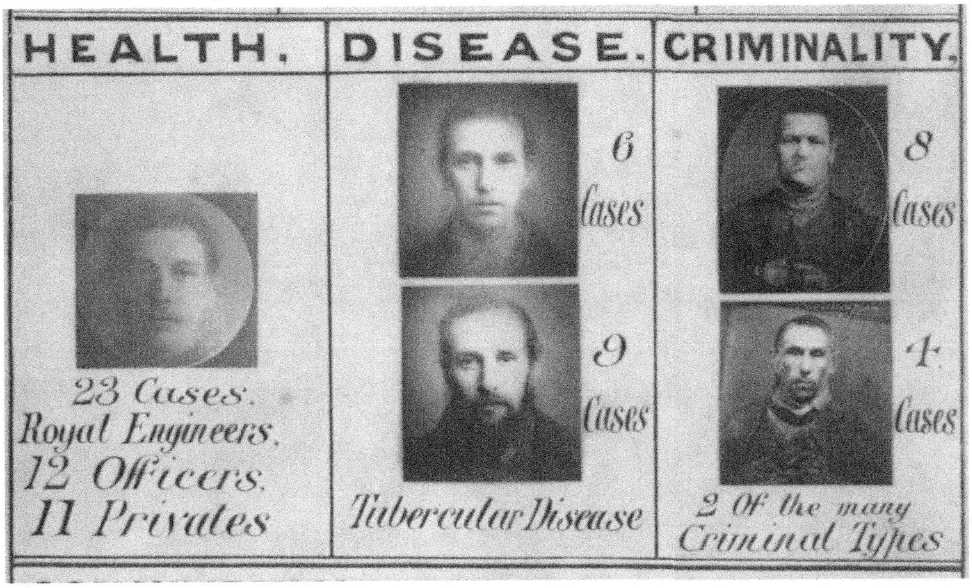

Figure 9.7: Composite photographs, health, disease and criminality, from F. Galton 1883d.

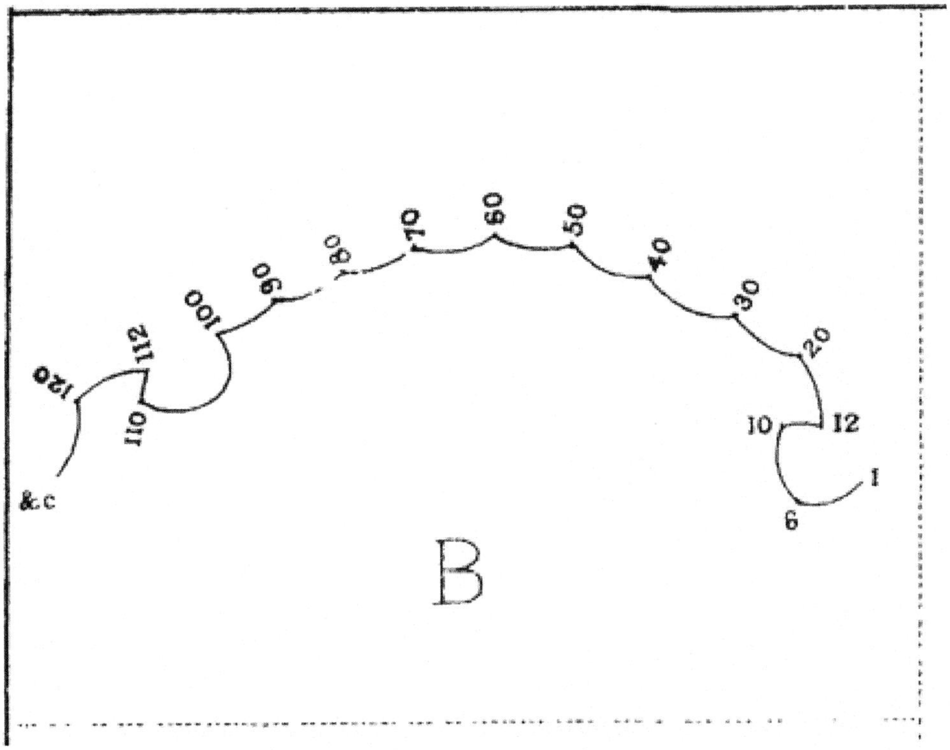

Figure 9.8: George Bidder's number form, from F. Galton 1883d.

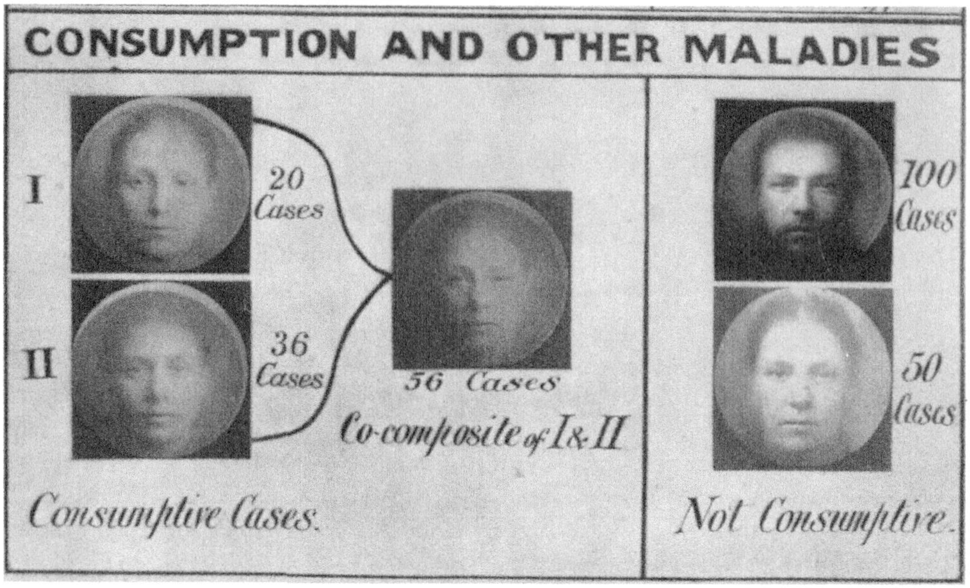

Figure 9.9: Composite photographs, consumption and other maladies, from F. Galton 1883d.

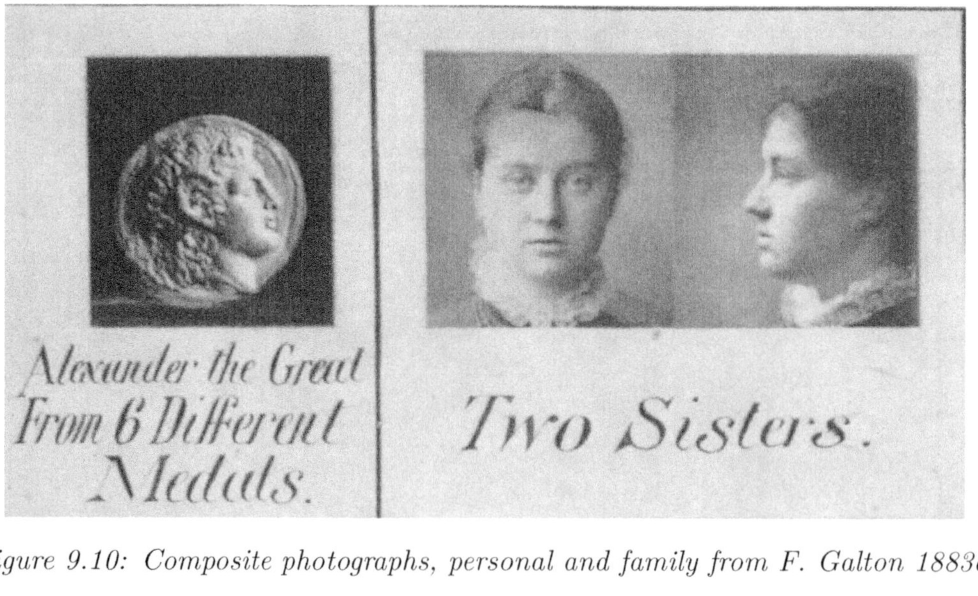

Figure 9.10: Composite photographs, personal and family from F. Galton 1883d.

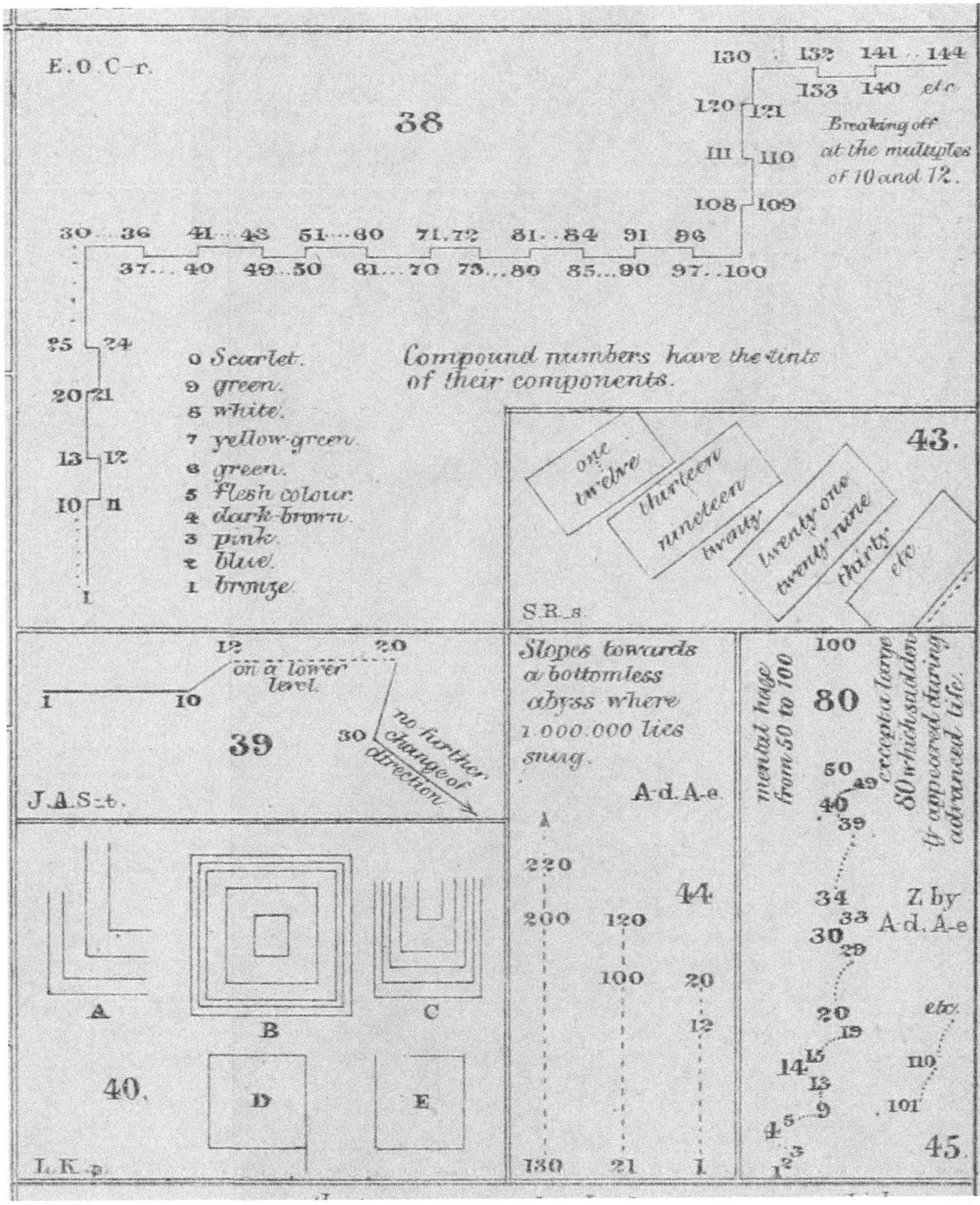

Figure 9.11: Various number forms, selected from many examples, F. Galton 1883d.

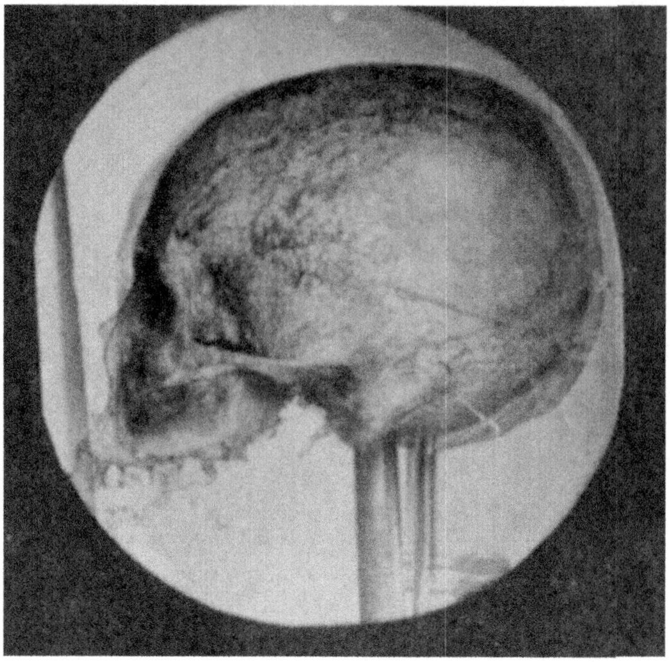

Figure 9.12: Composite skulls, from K. Pearson 1924.

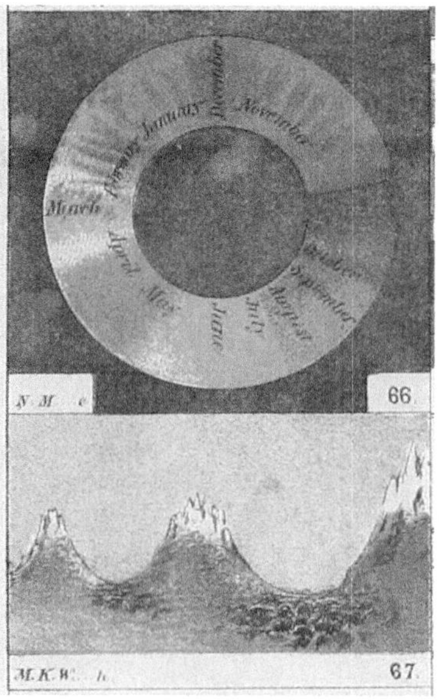

Figure 9.13: Mrs. Elizabeth Kempe Welch mentally pictures the months of the year in full colour, F. Galton 1883d. Her daughter Lucy (1869–1958) would become a noted artist.

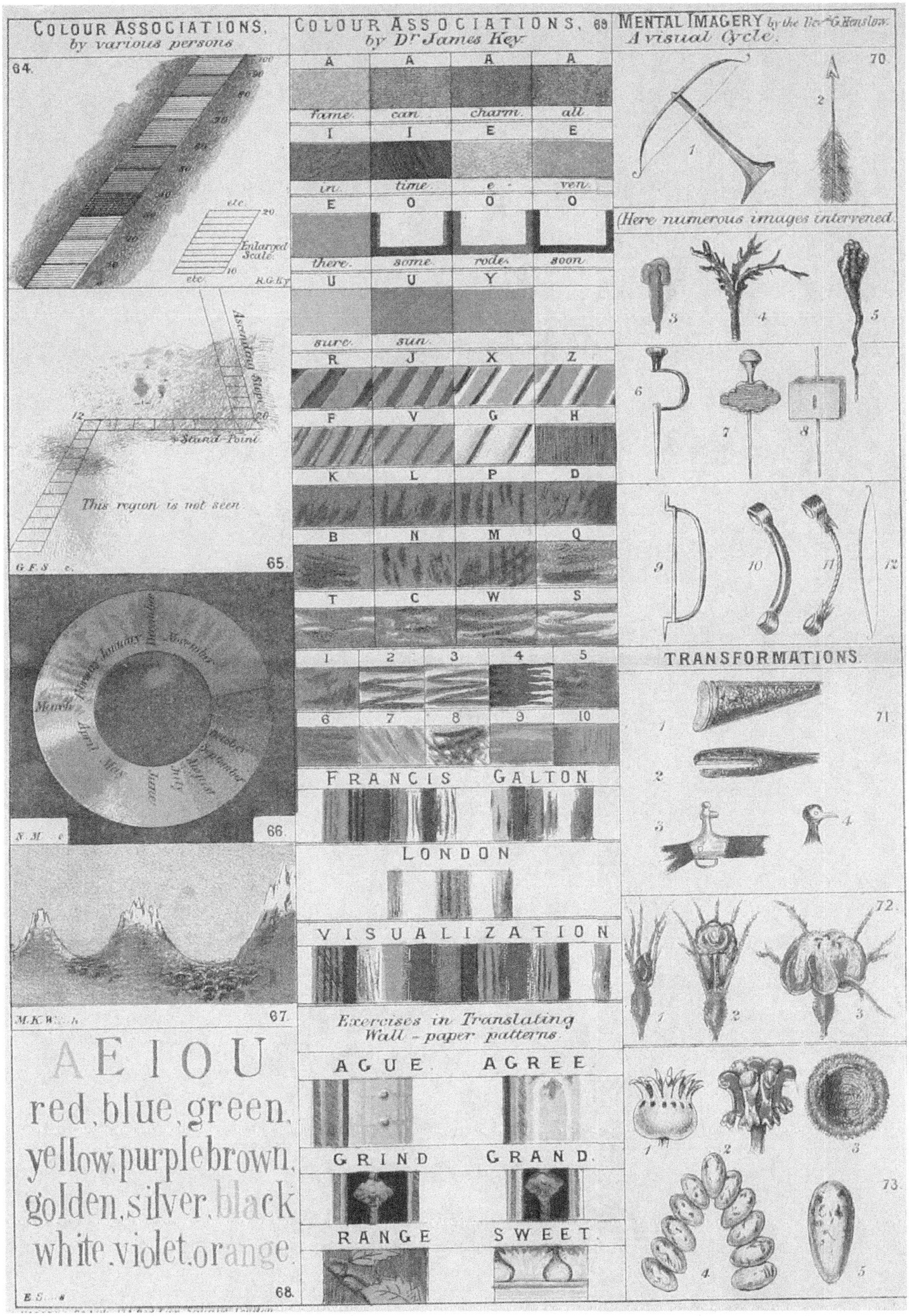

Figure 9.14: Various colour associations, F. Galton 1883d.

Figure 9.15: Francis Galton at around age 60 in the early 1880s, K. Pearson 1924.

Figure 9.16: Francis Galton at around age 60 in the early 1880s, K. Pearson 1924.

Figure 9.17: Francis Galton at age 60 in 1882, by Gustav Graeff. About 20,000 brushstrokes, all counted.

Chapter 10

Natural Inheritance

The most obvious place to get data about humans and their characteristics is directly from the life history of people themselves. The natural means is the circular technique that Galton used since the early 1860s when he started collecting weather data. But it was not obvious that people were even keeping records of the kind of data that he needed, or could scrape it together from half-remembered incidents and correspondence, let alone be willing to share it with strangers. As early as January of 1882, Galton was appealing in *The Fortnightly Review* for the public to keep systematic photographic chronicles throughout their own lives, and the lives of their children.[1]

The lustre of a purely photographic life history greatly faded by the following year when it became clear that only limited information could be gleaned even from large collections of composite photographs. Two months later Galton was back in the *Fortnightly* arguing that other 'anthropometric and medical' facts should be recorded alongside the photographs. The public could be assisted by Anthropometric Laboratories, where they could go to get rigorously tested in standard ways, and have the data recorded. At a minimum this included basic measures like height, weight, chest girth, capacity of lungs, hair and eye colour, along with birthplace and race, including those of parents. This would help to fill in the extensive blank spaces in the developing map of human abilities. In time Galton would construct exactly such a laboratory and measure very large numbers of people. In the meanwhile, until that was established and data accumulated, other ways had to be found.

The intent was to go well beyond the sort of facts comprehended by the Anthropometric Committee of the British Association.[2] Galton had been a member of the committee since its inception in 1875, where he was joined by the likes of Dr. F. A. Mahomed (whom we have met already) and the great archaeologist, ethnographer and notable eccentric General Augustus Henry Lane Fox Pitt Rivers (1827–1900), who had contributed to Galton's twin studies. Towards the end, Dr. Charles Roberts (1851–1901), a wealthy surgeon 'of spare figure and intelligent face',[3] joined the committee and contributed a lot of his own data. But Galton was frustrated by its limited objectives—the basic measures above. By the time its 'final report' was delivered in 1883, by which time he was its chairman, £280 had been spent collecting data from some 53,000 people, 'of both sexes and of all ages', though many returns were only partially complete. The reluctance of people to complete even the basic information requested discouraged the committee from expanding to more data.

What data the committee did get was enough for Galton to puzzle his colleagues with his novel statistical methods, especially medians and quartiles. Of particular interest were 'stature, weight, and complexion (colour of eyes and hair) of men in different counties as indicating their racial origin, and the influence of soil, climate, occupation, and other sanitary surroundings'.

[1] F. Galton 1882b.

[2] F. Galton 1882c.

[3] See *Plarr's Lives of the Fellows*, RCS: E003107. `https://livesonline.rcseng.ac.uk/client/en_GB/lives.`

The returns showed that the 'Scotch' were 68.71 inches tall, whereas the Irish (really, Irish immigrants in England, Scotland and Wales) were 67.9 inches, the English 67.36 inches and the Welsh only 66.66 inches tall. Again, the Scotch were heaviest, at 165.3 pounds, with the Welsh second at 158.3 pounds, whereas the English and Irish at 155 and 154.1 pounds lagged well behind. The committee speculated that mixture between the shorter neolithic inhabitants of the isles and successive waves of taller settlers contributed to these racial differences in height.

Heights also differed sharply by occupation. The professional classes averaged fully 69.14 inches in height, as opposed to 67.08 for the labouring classes. Likewise, Fellows of the Royal Society stretched to 69.76 inches. If this disposed the professions and scientists to look down on their social inferiors they were humbled by the rural population of Galloway in Scotland, where the average was fully 70.5 inches. The general run of criminals averaged only 65.6 inches, just above Hertfordshire labourers at 65.35. Town and country were sharply separated, so that Edinburgh and Glasgow only reached 66.35 inches, puny compared to the Galloway giants. However the committee was aware that environmental effects like climate and industrial poisoning were confounded by different geographical settlement patterns and subsequent sorting of populations.[4]

Suggestive though racial and regional differences might be, the glaring limitation of the Anthropometric Committee's approach was the lack of information about family relationships between its subjects. Without this, nothing definite could be learned about the laws regulating the transmission of even straightforward characteristics like height, eye and hair colour. To get this crucial data, for multiple generations, the first plan Galton adopted was to conduct his own survey, but to limit it to medical men and their extended families. In the *Fortnightly* of August 1883, shortly before the final report of the Anthropometric Committee was presented at the British Association, he proposed a competition. He would offer £500 in prizes—in those days nearly five years of wages for a skilled tradesman, but today roughly equivalent to a tidy £30,000—for the best detailed records medical men could submit.[5]

Galton explained that his interest concerned data related to 'all that can throw light on the physiological causes of the rise and decay of families, and, consequently, on that of races. Some diseases are persistently hereditary, and others are not. They are variously found in different varieties or sub-races of men, and these have various other attributes including various degrees of fertility. We cannot as yet foretell, but we may hope hereafter to do so in a general way, which are the families naturally fated to decay and which to thrive, which are those who will die out and which will be prolific and fill the vacant space'. Since the future evolution of humanity depended on the results, he had little patience for those who would prefer to conceal their family maladies, 'as though timidity of utterance could hush thoughts'. 'It seems to me ignoble that a man should be such a coward as to hesitate to inform himself fully of his hereditary liabilities, and unfair that a parent should deliberately refuse to register such family hereditary facts as may serve to direct the future of his children, and which they may hereafter be very desirous of knowing. Parents may refrain from doing so through kind motives; but there is no real kindness in the end'. However that reluctance was real and had to be overcome.

This was all really a solicitation for comments, but it quickly gathered momentum. Galton had hoped that the medical men would be driven by scientific zeal, have more detailed family knowledge than the general run of people, fewer scruples about sharing that information with others, and greater skill in describing facts exactly and dispassionately. No doubt his own medical studies and occasional amateur practice contributed to this enthusiasm. He had at first proposed collecting data about race, conditions of life (environment or nurture, as in town versus country residence), form and feature (basic anthropometric facts), health (lifespan,

4 F. Galton 1883a.

5 F. Galton 1883f. National Archives monetary converter, see https://www.nationalarchives.gov.uk/currency-converter/.

diseases and accidents, insurance), vigour (as in energy, or capacity to get work done), keenness of physical sensation, artistic capacities (left undefined), intellect (as determined by competitive victories in youth and absolute success in life), character (in modern terms, personality), and whatever else respondents might care to note about pursuits and interests. But by the end of 1883, Galton had been persuaded that the competition should be opened to the much more numerous general public, who would have to be helped though by a more definite structure for their submissions.

To assist respondents, Galton compiled a book-length collection of tabular forms, *Record of Family Faculties* (for a sample form see Figure 10.2, p. 226). Released in December of 1883, this contained a preface explaining the intent and the terms of the competition.[6] The donor retained flexibility in determining the proportions of the £500 in total prizes he would award. The competition was limited to Great Britain, where facts could be more easily verified, though the book would be available to American readers too. Each of 14 'full-blooded' relatives had a separate section in the roughly 70 pages provided—see Figure 10.2, p. 226, for an example. Respondents had to situate themselves within this fixed hierarchy if they had their own children or grandchildren. A binary numbering scheme invented by Galton encoded degree of kinship precisely (males were always even numbered, females odd), and a redundant alphabetic code (e.g. 'FF' for father's father) was handy for error checking. An appendix allowed for hereditary peculiarities passed down to children to be noted as needed. Additional documentary evidence was encouraged. Female relatives were recommended as great sources of family information, and certainly in the case of the author's own family this was true: Frank's sisters had amassed tomes of material in 'Galton Books' that survive down to the present day.

The judge of prizes made it clear up-front that he was interested in the total contribution of relatives, and not in the occasional notable but remote ancestors that genealogists tended to fixate on but actually contributed little. By filling in the somewhat demanding schedule, respondents would be furthering the cause of science (no empty promise as it turned out) and possibly all of mankind. 'Our present ignorance of the conditions by which the level of humanity may be raised is so gross, that I believe if we had some dictator of the Spartan type, who exercised absolute power over marriages, assigning A to be the wife of B, and C to be the wife of D, and who acted with the best intentions, he might possibly do even more harm than good to the race'.[7]

The competition was announced in January of 1884 in the *Times* and *Nature*. The closing date for submissions was set at May 15.[8] In parallel, the scientist had been chairing the 'Life History Sub-Committee' of the British Medical Association since early in 1882, collaborating once again with Dr. F.A. Mahomed. The result of a series of meetings at Rutland Gate was an individual *Life History Album*, also issued in December of 1883. This was for recording events as they happened, rather than cobbling them together retrospectively from relatives as in the *Record of Family Faculties*.

Many of these albums were sold and completed by their owners—they occasionally resurface filled-in as curiosities in antiquarian bookstores. Mahomed, who died less than a year after it was completed, but not before he started filling in his own copy of the book, supplied much of the momentum behind the project. However the book was issued at his insistence under the name of a reluctant Galton, who was not convinced of its coherence—a second edition was called for in 1902, which Galton reorganized more to his taste.[9] In truth it anticipated much later developments in centralized medical records, forming a portable individual record that could be carried by its subject through life and submitted to the British Medical Association in

[6] F. Galton 1883i.
[7] F. Galton 1883i.
[8] F. Galton 1884b,c.
[9] F. Galton 1902.

an anonymized form if the subject wanted to do so. Enclosed charts allowed subjects to track their developmental progress against established norms to detect serious deviations.[10] It was a longer-term project for posterity—the initiators could not hope to get useful data from it in their own lifetimes.

While Galton waited for the fruits of his *Family Faculties* competition, he was already busy planning the Anthropometric Laboratory that he had been promoting for years. This would allow him to firm up some of the physical measurements at least, using standardized procedures and instruments that he could test and design himself. There was also the possibility of using physical apparatus to assess intellectual ability on a numerical scale. This would allow him to go beyond counting prizes and using reputation as a measure of intellectual distinction. In November of 1882 at the Anthropological Society he had presented an instrument that he believed could do something like that.[11] It measured the delicacy of perception of weight through a relative comparison, just-perceptible *differences*.

Apparently identical gun cartridges held concealed amounts of lead shot, which increased in a *geometric* scale, following the discoveries of the German psycho-physicists Weber and Fechner. For each subject it was determined what the smallest difference in weight was that could be detected. Because it was a relative measure, it removed the element of strength, testing only delicacy. 'The trials I have as yet made on the sensitivity of different persons confirm the reasonable expectation that it would, on the whole, be highest among the intellectually ablest'. Women appeared to be less delicate in this sense than men, and the supposedly superior sense of touch of the blind did not hold up to his experiments at an asylum for them. Croom Robertson, the editor of *Mind*, was intrigued by the instrument and collaborated with Galton for some time after on the design of instruments intended for better-isolated senses.[12]

A perfect opportunity presented itself. The International Health Exhibition, sponsored by the Prince of Wales, would open in May of 1884 in South Kensington on the (then) grounds of the Royal Horticultural Society. It would run for several months. Millions of visitors were expected, all within strolling distance of Rutland Gate. The railway companies would offer discounted tickets. Galton was promised generous space to array his instruments and persuade the high-minded among the general public to submit themselves to measurement. He worked out that a reasonable fee of three pence per examinee (roughly £1 in today's money) would more than cover the operational costs. A small booklet describing the tests would also be on sale for a penny. The instruments would be donated by him at his own expense, while the Exhibition would supply the woodwork enclosure and the security. No names were supposed to be recorded in the laboratory copy. Each subject would receive a carbon copy of the results, which the laboratory would accumulate for analysis.

On the 28th of May, after the Family Faculties competition had closed, but shortly before the Health Exhibition opened, Galton was able to outline his philosophy of measurement in the Senate House at Cambridge. He had been invited to deliver that year's prestigious Rede Lecture, an occasion that went back hundreds of years. T. H. Huxley had delivered it the previous year. Before that, Matthew Arnold had spoken, and almost all the Victorian notables—Charles Darwin was a rare exception. Depending on which newspaper one read, there was either a sizeable crowd that Wednesday, or a somewhat smaller one than usual due to the exams.[13] Galton maintained from the outset that all human faculties were measurable, with enough effort. This even included character or temperament, which we would call 'personality' today. His twin researches had revealed that 'free will' was really an exercise of a given predisposition underlying the choices made. All that was needed was to find incidents (which we might call

[10] Goodhart and Jacobson 1886; Hughes 2013.
[11] F. Galton 1882a.
[12] F. Galton 1883h.
[13] *Cambridge Chronicle and Journal*, Friday 30 May 1884, 4. *Evening Mail* 1884-05-30, 3. The lecture was never published in its entirety, but appeared in part in F. Galton 1884e.

items) on which people made different choices, thereby revealing their underlying character, and aggregate a substantial number of those different decisions. For example, temper was a quality well-understood to vary by individual and by family and those differences could be quantified in this way.

However, much simpler and more well-known physical measurements were also necessary, and the soon-to-open Anthropometric Laboratory at the already-underway Health Exhibition would go about collecting those. Johns Hopkins University in Baltimore had made great progress in measuring students there voluntarily. 'I am audacious enough to suggest the establishment of at least a plain Anthropometric Laboratory here at Cambridge. I should consider it one of the best works of my life if the remarks I have the privilege of addressing publicly to this distinguished audience, were to lead to the introduction of a new requirement in the University, that of a systematic but simple measurement of every undergraduate, in his matriculation and again before taking his degree'. The audience reacted with laughter and applause, leading to prolonged applause when Galton explained that he believed this would identify defects in development early on, and perhaps allow them to be corrected.[14]

To demonstrate that many people did not in the least mind being measured, Galton gave the example of a unique series of records covering many years that he had recently uncovered at a venerable wine and coffee merchant, Berry's, on St. James Street in London. Customers, many of them from the more privileged classes, were weighed for their amusement when they came in, while sitting in a chair on ancient beam scales. Ledgers stretched back over three generations and gave multiple entries for the same individuals. Since ages could be determined from the peerage, Galton had been able to satisfy himself, using 139 individuals in the ledgers, that in the Regency era the upper classes had put on flesh much faster than in present Victorian times. Previously they had peaked a decade earlier at age 45 rather than 55, which he put down to their gross and dissolute way of life.[15]

As the scientist later wrote in the *Fortnightly Review*, he did not reveal to the audience that he was wearing an instrument himself while delivering his lecture. Concealed under his coat, strapped by a band to his chest, was 'Maret's pretty little pneumocardiograph', which with each heart beat or other movement blew puffs of air through an India rubber tube. He had not hooked it up to a concealed continuous paper-roll-scratching recorder, but was sure that with some assistance that would have been no trouble and would not have bothered him.[16]

The Health Exhibition proved to be enormously popular, realizing well over 4 million visitors from its opening on May 8 to its closing at the end of October (see the view from *Punch*, Figure 10.7, p. 230). The Anthropometric Laboratory was housed in its East Corridor annexe, in a latticed enclosure 8 foot wide and 36 foot long. There were many competing attractions. Most popular of all were the giant artificial fountains of the Water Pavilion, which shot jets of water 120 feet in the air, illuminated in multiple colours by electric arc lights, controlled by a buried room full of engineers, choreographed from a tower. The city's conquest of cholera through cleaner water was on display.

Life-size reconstructions of crowded and ill-lit buildings from the plague-ridden era of De Foe and Pepys fascinated the crowds. A vegetarian eatery, run by a sect said to be headquartered in Manchester, was overshadowed by the first full-menu Chinese restaurant in the city. Flush toilets, water filters, electric lighting, room ventilation tubes, and Japanese manufactures—judged to be marvellously improved—were on show. Not outdone, the Anthropometric Laboratory attracted many more visitors than it could handle, eventually measuring 9337 paying customers, just covering its costs: 4726 adult males and 1657 adult females, the remainder being children.[17]

14 *Cambridge Chronicle and Journal*, Friday 30 May 1884, 4.

15 Galton had earlier that year published a description of this data set and analysis in *Nature*, F. Galton 1884h.

16 F. Galton 1884e.

17 F. Galton 1884g.

All were measured on 17 different tests. Daily operations were run by a doorman and a Serjeant Williams, with Galton himself occasionally assisting.

Each person took between 20 and 30 minutes to pass through, obscurely seen by the waiting crowd through the latticework as they took their tests. After their easier details were taken, they went through successive stations.

1. Eye color, for which standard specimens of artificial eyes were provided. Hair color standards were displayed but not recorded because of dyes.
2. Eyesight for each eye. For keenness, i.e. the greatest distance at which diamond type could be read, and for color-sense, by matching shades of wool.
3. Judgment of length by finding the middle of a piece of wood, with the pre-calibrated error revealed. Similarly for squareness, by aligning a movable arm to a pre-set line to form a square, and revealing the pre-calibrated error.
4. Hearing in vibrations per second. This was the highest audible note of a Galton whistle, as the hall was too noisy to do anything more elaborate.
5. An exhibition of instruments for measuring sensitivity of touch, since it was not practical to test this delicately in the sausage-press of the exhibition.
6. Capacity to breathe in inch-displacement. Measured by a commercial spirometer, into which one emptied the lungs after a deep breath to raise a vessel suspended in water.
7. Swiftness of a blow with a fist. Initially this was a rod to strike at, but deal and oak rods were repeatedly broken by clumsy subjects striking obliquely. 'It was a matter of surprise to myself, who was born in the days of pugilism, to find that the art of delivering a clean hit, straight from the shoulder, as required by this instrument, is nearly lost to the rising generation'. The instrument was replaced by one of Galton's own invention, with a stirrup which the subject used to strike into space, and trailed a piano wire behind it, detaching when in full motion, by which time the speed had already been measured using a spring mechanism against a known rate.
8. Strength of pull, in the motion of an archer against resistance, and squeeze with both hands. Both of these were instruments made by the Salter firm.
9. Sitting height and standing height in shoes, with a heel thickness measure, in inches. A measuring rod was used.
10. Arm-span in inches, using a pair of attached sliding rods.
11. Weight in normal indoor clothes, using a commercial balance.

Almost from the start, Galton regretted not measuring head circumference (which he expected to be linked to intellect) because of the practical difficulties of head-dress and hair-dos. The only measure he could associate with intellectual ability was judgment of length and squareness, as he was not able to test delicacy of touch perception either, again for practical reasons. He was well aware that he did not have 'random specimens', as he put it, but had to make do with inferences about differences between those he did have. He surely regretted not knowing the family relationships among his subjects. Analyzing the full data set would take many years. In later years the data set would be expanded even further, so that it was not fully analyzed until 1985![18]

Preliminary results, obtained using Galton's new percentile methods, confirmed well-established stereotypes about the sexes. There was for example hardly any overlap in the distributions for strength. A woman who ranked at the 90th percentile for strength of squeeze relative to other women, ranked only at the 5th percentile relative to the men. That would explain why men typically did the squeezing. For all measures, the male percentiles corresponding to the

[18] R. Johnson et al. 1985.

very highest female percentiles were: Pull, 4; Squeeze, 7; Breathing capacity, 10; Height, 14; Weight, 26; Swiftness of blow, 26; Keenness of sight, 37. An evolutionary dimorphism, and a neat illustration of the power and simplicity of the new percentile method for comparing distributions. 'The population of England hardly contains enough material to form even a few regiments of efficient Amazons'.[19]

The lab proved the concept. It showed that the public would willingly submit to fairly extensive measurements in large numbers. Attention was aroused across the country and there were immediate plans to set up similar laboratories, for which purpose Galton catalogued his collection of instruments and arranged for a manufacturer to produce them.[20] The archaeologist Flinders Petrie wrote offering a long series of helpful practical suggestions for taking measurements more accurately.[21] Galton would later set up another lab of his own close to the site of the original, with which it is often confused. The data collected in this initial exercise was interesting, but required the invention of new statistical methods—to reveal, say, the links between social class or occupation and the measures taken—before it could be fully analyzed in coming years. Some of those methods would be invented by Galton himself, but only once other suggestive data had been harvested.

While the Health Exhibition was underway, Galton was able to judge the returns for his *Family Faculties* competition. He had received 'very little trash' and over 150 worthwhile responses from all ranks of society, comprehending over 5,000 individuals. Over a dozen of the entrants were Quakers, whose submissions were judged by Galton to be among the best. The few weak submissions fixated on isolated notable ancestors through the male line only, which most but not all of the genealogists who entered were guilty of. Fully 40 people were awarded £7 prizes (about £460 today), while 44 received £5 each (roughly £330 today). The prize winners included the notable historian Stanley Lane-Poole, and as Galton had anticipated, many women—such as Lady Elizabeth Cust, who would later publish her history of the Cust family; the widow of Galton's old friend the late mathematician Archibald Smith; the wife of the Master of Clare College in Cambridge, Hester Atkinson; and the wife of a prison governor. 'I see before me work that will require some years to complete'. But this was unduly pessimistic. The prize money had been well-spent. Within a year this data set, with its family relationships, would allow Galton to formulate what is now the most widely-used statistical technique in all the sciences.

Galton already had some idea of what he would find in his 'R.F.F.' data set, as he would come to call it. He had given no indication of this hypothesis when running his competition, so that he did not bias the responses in favour of it. He expected to find the same phenomenon he had observed in his peas, which he had then called reversion but would now start calling regression. By the time that the British Association met at Aberdeen in September of 1885, he was ready. He had a convenient platform as that year's President of the Anthropological Section, which was inaugurated in the Song School on Thursday the 10th of September. It was a lecture, on 'Types and their Inheritance', rather than an address, 'to place beyond doubt the existence of a simple and far-reaching law that governs hereditary transmission, and to which I once before ventured to draw attention, on far more slender evidence than I now possess'.[22] Large diagrams were suspended to puzzle the audience. He concentrated on the height data he had obtained from his R.F.F. respondents.

Because Galton had family relationships baked into the design of his survey, he could compare the heights of parents and children, the easiest case to work out. Height conformed well to the 'Law of Errors', as shown conclusively by the Health Exhibition data, and was in any case

[19] F. Galton 1884f.
[20] F. Galton 1887b.
[21] GALTON/3/3/16/20.
[22] F. Galton 1885a.

what one would expect, being the result of an accumulation of many small influences. There were 930 usable adult children in the set, from 205 families. Unlike the case for sweet peas, where he had first selected size ranges and then deliberately bred from them, he had a complete sample of parent-child heights. The difference in height between (usually shorter) mothers and (usually taller) fathers had been solved pragmatically by just multiplying the height of the mother by a factor of 1.08, and averaging to give a 'mid-parent' height. Not only did these heights 'revert to mediocrity' in the children, he could give a precise numerical estimate of the fraction. In deviations from the median height, children could expect to have 2/3 the deviation of their mid-parent. Tall parents were likely to have somewhat shorter children. Likewise short parents were likely to have somewhat taller children. These estimates were obtained by a graphical method in which some smoothing was required. He had also obtained a separate data set about brothers, through personal contacts, on which he had confirmed his results, but he did not have time to enter into details.

With the aid of a Cambridge mathematician Galton had obtained a formula allowing one to work out the 'coefficient of reversion' exactly, using two estimates obtained from the data set at hand: the amount of 'race variability', and the amount of 'co-family variability'. The calculations showed a close match to Galton's estimates, based on smoothed data. He was satisfied that there was no assortative mating for height in his set, which simplified working out the needed variance of the mid-parents. 'I may be permitted to say that I never felt such a glow of loyalty and respect towards the sovereignty and magnificent sway of mathematical analysis as when his answer reached me, confirming, by purely mathematical reasoning, my various and laborious statistical conclusions with far more minuteness than I had dared to hope, for the original data ran somewhat roughly, and I had to smooth them with tender caution'.

There is no sign that the Aberdeen audience—a large attendance according to the newspapers—realised just how far-reaching Galton's innovation would be in coming years. Sir Lyon Playfair regretted that he was in a poor position to thank the President, given that he was below mediocrity in height himself, but hoped that he might be redeemed by slightly taller descendants (laughter). To repay the speaker, who had counted 7% of the audience fidgeting during Playfair's lecture the previous day, Playfair had counted fidgets during the exposition of regression, but graciously conceded that he could only detect 2% of the audience dozing off (more laughter). Finally Sir John Lubbock asked the speaker if his law applied to other peculiarities of the parents. Galton assured him that it did.[23]

The method Galton used to make his statistical breakthrough would be called 'Exploratory Data Analysis' today.[24] In his hands it was more powerful than in most. Determining that there was regression in median heights was easy enough, as it had been for the sweet-pea seeds. He placed the mid-parent heights into nine buckets, obtained their medians by ranking them, and calculated the median heights of the children, graphing the nine pairs of points in a scatterplot. As always he used the median as a convenient estimate of the mean, since it required no calculation and he was confident that the data obeyed the normal law of errors, in which the median *is* the mean. It was easy to fit a line to the plot using a straight-edge to get the best fit. The more rigorous method of 'least squares' to fit points with minimum error was already known but not yet understood in this context (see Figure 10.9, p. 232).

However Galton realized that the distributions of data supplied much more insight than their means or medians alone, a position he had been advocating in print and in scientific forums for years. So he cross-tabulated the data, using inch-sized buckets, to see both of the height distributions and the way in which they were related to each other *jointly* (see Figure 10.10, p. 232). Next he used a technique that he had been honing ever since his days as an African explorer, as a surveyor and map-maker. He found contours of approximately equal values.

[23] *Aberdeen Free Press*, Friday 11 September 1885, 6. See also F. Galton 1885g.

[24] Tukey 1977.

Applied to weather data in the early 1860s, this had revealed anti-cyclones through isobars of equivalent pressure. More recently he had used it to show how the world could be partitioned into *isochronic* postal or passage times from London—see the striking result using coloured 'isochrones' in Figure 10.11, p. 233.[25]

The cross-tabulated heights in Figure 10.10, p. 232, can be read as a map of counts. Similar counts can be connected together into contours. In any cell, a count corresponds to a probability, since it can be divided by the total number of observations to get a frequency. The joint tabulation maps to a *joint probability distribution.* This is much clearer if they are smoothed, because random perturbations make them seem more different than they really are. Aggregating adjacent cells is a simple way to smooth them, because it allows spurious excesses and deficits to cancel each other out. Galton settled on aggregates of four adjacent cells, enough to smooth locally without dissolving the data entirely. He carried this data around with him in a notebook, puzzling over it. While waiting for a train at a station near Ramsgate he saw it.[26] With some judicious fitting by eye, distinct regularity pops out of the number map. Ellipses—the squashed circles or 'conic sections' that were a prominent part of his long-forgotten mathematical training at Cambridge—connect similar clusters of points.

The fit is by no means perfect, but Galton had a freakish talent for accurately teasing out patterns of this kind, which he would have attributed to his own Darwin-family hereditary particles. As a result he could draw the shapes in Figure 10.12, p. 233. As the diagram shows, an ellipse joining points of equal probability has two axes corresponding to the way in which the circle was squashed: a major axis from left to right, and a minor axis from top to bottom. The next step is to notice that for a child to gain 2 inches in height (moving 2 points to the right of the origin O) the parents must gain 3 inches in height, a ratio of 2/3. Thus the line OP has a slope of 2/3. Crucially, OP intersects the ellipse at a tangent NY. This is more than a curiosity because it allows us to see that regardless of the ellipse constant, which increases or decreases its area, the tangent point would have the same slope. So it does not matter which particular ellipse is considered, the slope relating one height to the other can be determined.

The more crucial breakthrough was to notice that one could just as well deduce the heights of parents from the heights of their children, and that this shows regression to the mean as well, but with a different slope. In Figure 10.12, p. 233, this is the line OM. Galton could calculate the slope of this line from the median heights, except now he was first putting the heights of the children in buckets, then calculating the corresponding median of the mid-parents for each bucket. Now the slope of the regression was roughly 1/3 rather than 2/3. Previously in the case of the sweet-peas Galton had supposed that the regression he had observed was due to the mechanism of descent, but this suggested that it was more general. To test this out, Galton gathered a second 'Special Data' set, but now for heights only.[27] Through trusted personal contacts he was able to circulate a 'card of enquiry' on which the heights of *brothers*, between 24 and 60 years old, were to be recorded in order of birth. Only cases where accurate measurement was possible were to be recorded. He stopped when he had 783 brothers from 295 families. Because he had limited the inquiry, he had more confidence that these measurements were accurate and consistently made. The subjects were related to one another, but they were not of course descended from one another, they were both male and there were no transmutations to mid-parents. This served both as a replication, with data he had more trust in, and a generalization, since the relationship had changed. It turned out that the regression between older and younger brothers was also about 2/3 for his data set.

Now that the patterns he had noticed were replicated, Galton wanted a more formal derivation of them. It was over forty years since he had studied mathematics at Cambridge, and he

[25] F. Galton 1881a,b.
[26] F. Galton 1908, 302.
[27] This was Galton's name for it, by which he meant it was specifically for height, not extraordinary.

was now well into his sixties. Refreshing his knowledge of ellipses required some trips to one of the well-stocked libraries within easy reach of Rutland Gate. He was sitting in the library of the Royal Institution one evening in 1885 when the Scottish chemist James Dewar (1842–1923) noticed him labouring. Dewar had invented the vacuum flask, but was no mathematician. However his son-in-law J. D. Hamilton Dickson (1849–1931) had worked on electricity and transatlantic cables under Lord Kelvin in Glasgow. After returning to Cambridge to study mathematics, Dickson had made Fifth Wrangler in 1874. Now the Scot was installed in Peterhouse as a mathematics tutor, when he was not discoursing in fluent Japanese, and relished problems. Galton agreed with Dewar that having Dickson work on it in an abstract form would save himself a great deal of revision. It would also be an independent confirmation that the surprising results he had obtained were sound.

Dickson's solution confirmed that the ellipses Galton had observed were a necessary consequence of the way that the heights were jointly distributed. He gave a formula for that distribution in which the ellipses define lines of equal probability (see Appendix G, p. 433, for a rehydration of the mathematics involved in Dickson and Galton's methods). He derived a simple relation between the two regression slopes, which depended on the variability of the data. Galton was able to use a formulation of this relation to check his regression estimates. They closely agreed. Respect glowed.

It is curious that Dickson did *not* explicitly provide a method for calculating regression coefficients from data, which is the activity most modern victims of 'Statistics 101' courses will associate with regression analysis. Galton did not use or need one in practice, since he could plot medians against medians and fit a line, a far easier procedure before the advent of calculating machinery. It was obvious that a well-known method like least squares could be used to do this more precisely, and within a year Cornell University was offering courses on how it could be done. An alternative formula for calculating the slope from the variances can easily be extracted from Dickson's proof, provided one knows whether the regression is positive or negative. It was explicitly given by Galton in an address to the Anthropological Society in early 1886, though it contains an obvious misprint.[28]

The second conceptual breakthrough that Galton made was to recognize that regression is not tied to descent, which is just a special case. This is what he had meant when he had told the Anthropological Society that, back in 1877 when he first described regression, 'I was then blind to what I now perceive to be the simple explanation of the phenomenon'.[29] The 'Special' data set that he had collected on the heights of brothers—from personal contacts of which we know little—allowed him to regress the height of one brother on another. He had far greater confidence in the accuracy of these heights than he did for the R.F.F. data set, since he had greater control over the measurement procedure, which he had specified and was not subject to the vagaries of female headdress and hairstyles. The heights were clearly connected, but only imperfectly, even though one was not descended from the other. If one knew the height of an exceptionally tall brother, then another brother was most likely shorter. The difference was that now, unlike the parent-child case, the variabilities of the brothers (say an elder brother versus a younger brother) were the same. From the worked out formulae, this implies that the regression direction does not matter: the coefficients are the same. Checking the coefficients calculated from the data, this was certainly true. Respect glowed white hot.

Once he had a means for predicting the heights of offspring from parents, parents from offspring, or brother from brother, Galton immediately became more ambitious. What if one knows the heights of multiple antecedents? Grandparents, uncles, nephews, cousins .... How could one utilize all of these to come up with a composite estimate? Some preliminary investigation showed that this was not a straightforward sum of series. 'The possible problems are

[28] F. Galton 1886d; Stigler 1986.

[29] F. Galton 1885f.

evidently very various and complicated, I do not propose to speak further about them now'. Those problems would have to wait a few months more until he considered eye color.

In truth Galton could easily have found the mathematical solutions returned to him by Dickson, given no more than a few months. In formulating the problem he had stated most of the solution already. The required techniques involved some differentiation, routine algebra and knowledge of the formulae for the normal distribution and ellipses.[30] He preferred to downplay his own role wherever possible, but letting Dickson tackle the problem really was a better use of both men's time. Concepts mattered more than algebraic manipulation. There is a striking example in the formulae that Dickson helped to derive. A simple manipulation of one of these shows that the regression coefficients can be equated by dividing them by their variances. A straightforward trick that was not apparent in 1886 to either Galton or Dickson. It would take another three years for Galton to make that conceptual breakthrough, which is not about mathematics per se.[31]

The technical details of the regression material, including Dickson's proofs, were written up and read to the Royal Society on January 26th of 1886. Far more than the Anthropological Society, this was an audience with the skills to appreciate the details. George Gabriel Stokes (1819–1903), in his day a Senior Wrangler at Cambridge, was in the chair. Though Stokes was a pioneer of multivariate calculus, he did not pick up the threads of Galton's unfinished multivariate problem. Over the next three years there would be scant understanding, even in mathematically adept circles, of the far-reaching possibilities of the analysis.

The concrete results shown by Galton for height were noticed far more widely at first than the theory itself. For instance, his data set had shown very little assortative mating for height, which led some to suppose that this was generally the case. It was even taken up by the *Shetland News*, read perhaps by the grit-toothed islanders among whom Galton had hunted seals forty years earlier.[32] In the future, larger data sets would show moderate assortative mating and somewhat more pronounced regression. However the concept of regression to the mean, where it has been comprehended at all, has led to persistent popular misunderstandings about its nature.

Failure to spot regression to the mean recurs so often it seems to be a universal law of human behaviour. Examples could be multiplied endlessly. They involve selection of a bad condition to which some intervention is applied, after which the condition is observed again. Usually it has improved. Therefore the policy is supposed to work.

Take for example traffic accident hotspots, which are well-known. To try and improve these, subliminal messages may be incorporated into billboards, encouraging motorists to drive carefully. Sure enough, the next time accidents at the hotspots are measured, they have improved. Large sums of money are appropriated for more billboards and gratefully swallowed by advertising companies. Nobody notices that accidents have (inexplicably) worsened at 'cool-spots'.

The trouble of course is that an extreme measure will be less extreme the second time it is observed, whether higher or lower than the mean, simply because the observations are not perfectly reliable. An improvement may mean nothing at all. This is why experimental controls are needed, for those will show improvement (or worsening) with no intervention at all.

The flip-side of failing to spot regression to the mean is seeing it everywhere, forever. This is less common since it requires at least some familiarity with the technical terms involved, but many notorious examples exist—consider supposedly inexorable regression to business

30 Though Stigler 1986, 289, describes Dickson's proof as 'rather severely geometrical', I am unable to discover much geometry in it. The reader may evaluate Appendix G, p. 433.

31

$$\frac{\beta_{Y|X}}{\beta_{X|Y}} = \frac{\sigma_Y^2}{\sigma_X^2} \implies \frac{\beta_{Y|X}}{\sigma_Y^2} = \frac{\beta_{X|Y}}{\sigma_X^2}$$

32 Saturday, 1886-01-23, 3.

mediocrity—and they may well recur forever.[33] The root of the misunderstanding is the idea that regression keeps on *cumulating* in a series of observations until one is left with the (same) mean. This is not in general the case and requires some specific assumptions.[34]

If one measures the 100m sprint time of an athlete, an extremely fast value will probably be slower the next sprint, assuming we know nothing else about the athlete, but that does not imply that hundreds of successive observations will reveal the athlete to be no more than a mediocre member of the general population, let alone show him proceeding geometrically toward the population mean. In fact the athlete's time will spread around whatever the notionally true time for that athlete is, which may well be considerably higher than the mean of the general population, or lower. The result will be a normal distribution around that true time, with times closer to the mean more frequent than those further away. Then again, if we know nothing specific about any given athlete, beyond which population he comes from, then our best estimate of his true time really is the mean of that population. This underscores that reasoning about 'the mean' requires clear definition of the corresponding population. In the particular case of evolutionary selection, it is the resulting breeding population.

It is one of the ironies of reputation that Galton himself has been associated by some with the idea of successive, cumulative, or 'universal' regression to the mean, *which he certainly did not suppose to be the case in general.* If 'universal regression' really is some sort of inexorable statistical law it necessarily applies to *all kinds* of variation and if Galton had believed in it then he could not have attached any value to eugenic measures. *Reductio ad absurdum.* To some extent, Galton's at-times vague descriptions invited misinterpretation here, most prominently by his energetic and influential biographer Karl Pearson, who even went so far as to doubt whether Galton really understood regression![35] If there is misunderstanding here, it is biographical and egregious. Few people grasp the concept of regression as clearly and intuitively as Galton does, precisely because he does not reach it by algebraic methods only.

The 'type' is a key idea for Galton. It is an ideal form. By this he means no more or less than the (normal) distribution of a trait, knowledge of which requires both the mean and the variance. He had already argued for the (relative) stability of types since 1869, when closing *Hereditary Genius.*[36] What he had shown more recently was that regression to the mean is intrinsic to the stability of the type distribution, *since it is exactly equivalent to the assumption that the distribution does not change.* For example, if the variability of the mid-parents is *not* required to be equal to the variability of their children, then there will be *no necessary regression to the previous mean*, a new mean may appear. To say one is to say the other, as Galton and Dickson had demonstrated mathematically.

So regression to 'the' mean is really about the stability of 'the' type, and how difficult it is to change that type by natural or artificial selection. There is no *mathematical* reason why the type must be stable, a fact that Galton was acutely aware of. It is an *empirical* question. After additional context has been established, we will return in more detail to the impact of type-changing difficulties on Galton's views about evolution, and his ideas about 'organic stability'.[37]

Once the fundamental results for height were worked out, Galton moved on to consider other characteristics he had asked for from the 'R.F.F.' hopefuls. Eye colour is perhaps less subject to measurement error but trickier to analyze than height, because it is a discrete rather than a

[33] Secrist 1933.

[34] For a history of the concept of regression to the mean and several alternative derivations of it from simple assumptions, see Stigler 1997.

[35] K. Pearson 1924, 80ff. Others, e.g. Bulmer 2003, have followed Pearson's mistaken interpretation, but I must emphatically disagree with them.

[36] F. Galton 1869a, 368; F. Galton 1885a,f, 1886d.

[37] See page 293 and following.

'blending' trait.[38] Galton thought that hazel-coloured eyes are a minor exception to this, but he preferred to simplify things by distinguishing light from dark coloured eyes, with a residuum including hazel.

To reason as he had before about continuous rather than discrete quantities, Galton decided to model the *proportion* of eye colours in a given family. In retrospect this is a startling and under-appreciated leap. It is equivalent to modelling the *probability* that an individual in the family has a particular eye colour, light, dark or other: 'a given fractional chance that each member of a family will inherit either a light or a dark eye-colour, must be taken to mean that that same fraction of the total number of children in the family will probably possess it'. (Today we call this logistic regression, and use a transform to map the probability back to a discrete classification, but Galton did not need that step.)

Next Galton broadened his scope to multiple factors. What if one knows the eye colours of parents, grandparents, nephews, cousins and more? How would those be related? In the R.F.F. data he could extend to grand-parents and still have just enough data points to work with.

For this analysis Galton formulated what later came to be know as the 'ancestral law', which systematically relates ancestral traits to those in offspring. Its principal purpose is really explanatory. Later he would confide that he first 'stated it briefly and with hesitation', 'because it was then unsupported by sufficient evidence'.[39] It had originally been 'suggested by general considerations', by which he meant that it followed analytically from his theory of hereditary transmission, and was not derived from data.

Prediction was *not* a primary aim of the ancestral law, but estimates were obtained from it to cross-check if it held up in practice when compared to observed data. They seemed close, a fact that Galton was somewhat naively pleased with. There was no prior standard for closeness to meet, only post-hoc assessment. (Since the 'law' has been poorly understood in the past, a full technical explication of it is given in Appendix H, p. 443.)

A confounding factor had to be dealt with first: what if the proportions of eye colours in the population were not stable and changed over time, say under the influence of sexual selection? Popular belief was that the English population was darkening over time. Accounting for such a secular change would greatly complicate any analysis of inheritance. Fortunately, graphing the distributions by generation suggested that the proportions were stable enough, and popular belief could be set aside. This allowed Galton to predict the proportion of eye colours, given data about the eye colours of relatives. Comparing predictions derived in this way with the observed colours allowed the resulting 'law' to be tested. There appeared to be close agreement between observation and prediction for eye colour.

Nothing gratified Galton more than neat results showing up in (what he took to be) the real world. Using this kind of reasoning he could derive estimates of the proportions of eye colours in descendants, if something is known about one or more of the ancestors, for many different cases. Using 78 families from the R.F.F. data set, he found good agreement between his estimates and the empirical data he had collected about eye colour. He gave errors (predicted % of eye colour against actual % of eye color) for the three cases of known eye colour proportions: two parents only, 4 grandparents only, 2 parents *and* 4 grandparents. He was pleased that errors decreased noticeably in the last case where more information was known (see Figure 10.13, p. 234).

As Appendix H, p. 443, points out, but bears repeating, Galton's method here is *not* the much more general multiple-regression technique that Karl Pearson would later invent, where *parameters of a model are directly estimated from whatever data set is at hand*, and any relationship at all is amenable to the analysis. It is a mistake to conflate Galton's approach with later formulations which have a different intent. It clouds rather than clarifies understanding

[38] F. Galton 1886c.

[39] F. Galton 1885f, 1886c, 1889c, 1897b.

of his reasoning.[40]

After *Natural Inheritance*, Galton went in search of other data sets to confirm what he had found. Moths seemed suitable for breeding, but after some years he ran into difficulties due to low fertility (most likely from inbreeding or perhaps disease) and the 'disturbing effects of differences in food and environment on different broods'. So that path was abandoned. Small mammals with short generations were contemplated but that proved unnecessary when a large data source already to-hand popped up. Sir Everett Millais (1856–1897), a hunting son of the pre-Raphaelite painter Sir John Everett Millais (1829–1896), had kept a comprehensive record of the pedigrees of basset sporting hounds. Sir Everett had been an influential early breeder and promoter of the dogs in England. His *Basset Hound Club Rules and Stud-Book* covered the years 1874-1896, with records for some 1,000 hounds overall, and coat-colour details for 817 dogs. Those coat-colours could be conveniently contrasted as tricolour vs. non-tricolour.

As with eye colour in the R.F.F. data, Galton tested whether or not his ancestral technique, given known facts about a few generations of ancestors, produced estimates of tricolour versus non-tricolour proportions matching the observed proportions in the basset hound studbook.

The results seemed like a decent match (see Figure 10.14, p. 234) a fact which impressed Karl Pearson all the more because the statistician thought Galton's approach mistaken in more than one way. Even after Pearson applied his own corrections, he found it didn't make much difference at this level of precision—which actually suggests in and of itself that the concordance should not be made too much of.[41] It is now well-known, though it was not then, that simple formulae using sensible fixed weights (not estimated from the data) can yield surprisingly good results for data sets which inherently contain a great deal of spurious random noise anyway.[42] There are now many techniques for overcoming such noise, without which results may degrade on fresh data sets—the central problem of machine-learning.

The *Family Faculties* data was to yield one final investigation of heredity, this time of personality, specifically 'good' and 'bad' temper.[43] Galton had asked his respondents to rate the temper of their relatives, and found that he had 1,981 responses, which he thought plenty for his purposes. But first he had to create equivalence classes for the various terms his respondents used. Here it seems Tolstoy was approximately right in the first line of *Anna Karenina* (1878). Each of Galton's good tempered families is good tempered in only a few ways, whereas each bad tempered family is bad tempered in myriad ways. Or perhaps his correspondents were connoisseurs of bad temper. Only 15 terms could be considered descriptions of 'good' temper (e.g. 'gentle', 'placid', 'sunny'), whereas over three times as many terms (46) referred to 'bad' temper (e.g. 'aggressive', 'arbitrary', 'bickering', 'capricious', 'captious', 'choleric', 'contentious', 'crotchety' etc.).

Galton was inclined to trust these ratings. 'The accurate discernment and designation of character is almost beyond the reach of anyone, but, on the other hand, a rough knowledge and description of its prominent features is easily practicable; and it seems to me that the testimony of a member of a family who has seen and observed a person in his unguarded moments and under very varied circumstances for many years, is a verdict deserving of much confidence'. When he examined the distribution of the people whose temper is labelled as good, bad or neither, taken together he found roughly the same number of people at the extremes, and twice as many in the neutral category. This 1:2:1 ratio gratified him, as it conformed to the expectation one would have if two factors determined temper, and a mixture of them cancelled

[40] K. Pearson 1896; Stigler 1986. Here I respectfully disagree with Bulmer 2003, who offers reformulations of Galton's Law, 'as it should have been', to match more modern understanding. Bulmer's suggestion that Galton should have been doing multiple regression in the first place requires him to have invented it in the first place, a rather stern requirement for 1885.

[41] K. Pearson 1930a, 43.

[42] Dawes 1979.

[43] F. Galton 1887a.

out. A nice regularity.

'Rude tools' were needed to suit this 'rough material'. Regression equations would not be possible without numerical measures. To determine whether or not temper was inherited Galton simply compared the offspring of the 43 families where both parents were good tempered against the offspring of the 25 families where both were bad-tempered, ignoring neutral cases. The good tempered parents yielded offspring where 30% were good-tempered and 10% bad. From the bad-tempered families 52% of the offspring were bad-tempered and 5% good. Working in the other direction, from children to parents, Galton found 46 families where at least two of the children were good-tempered. Of the 333 parents, uncles and aunts related to these, 23% were good-tempered and 18% bad-tempered. Likewise there were 71 bad-tempered families yielding 633 parents, uncles and aunts, of whom 29% were bad-tempered and 18% good-tempered. He took this as strong evidence in both directions that temper was indeed inherited.

The obvious objection to Galton's method here is that it is possible that temperament is shaped by upbringing, and that good or bad tempered parents can somehow transmit that quality to their children through rearing. This is rather harder to argue when proceeding from children to uncles and aunts, unless one speculates that even more remote connections can transmit lifelong characteristics through only occasional contact, if any. But *possibilities* should not be confused with *probabilities* here. Given any collection of evidence we must search for the most likely and parsimonious explanation.

As we have argued above for *Hereditary Genius*, any explanation of the data in terms of rearing must also explain both the persistence of temper throughout life, the large variance of that temper within families, and its uncannily regular diminution, tracking degrees of hereditary relationship. The randomization effects of heredity were already well-comprehended even in Galton's day, and something similar would be required here, along with some plausible mechanism to explain the regularities tracking relatedness. Absent that, Galton's explanation is more natural and likely by far.

Over a century of subsequent work has emphatically confirmed Galton's straightforward conclusions, using twin and adoption studies in ways that he had himself suggested. Those methods are able to rule out the sort of far-fetched objections mentioned—somewhat like fishing with dynamite—though it must be said that a sufficiently motivated true-believer can cling to any theory almost indefinitely.[44]

All of the above statistical investigations into inheritance were gathered together by Galton and published in his book *Natural Inheritance*, which was submitted to Macmillan sometime after the middle of 1888. The author conceded that his readers might not appreciate the necessary intrusion of statistics. 'Some people hate the very name of statistics, but I find them full of beauty and interest. Whenever they are not brutalised, but delicately handled by the higher methods, and are warily interpreted, their power of dealing with complicated phenomena is extraordinary. They are the only tools by which an opening can be cut through the formidable thicket of difficulties that bars the path of those who pursue the science of man'.[45]

In New York, *The Critic* agreed with most of the popular press, on both sides of the Atlantic, that the mathematical material was 'anything but attractive reading; and Mr. Galton, with all his talent, has not been able to free his book from this somewhat repellent character'.[46]

There is some additional material in the book, not published elsewhere and among the least read of all Galton's output, covering inheritance of the 'artistic faculty' and disease, again based on the R.F.F. data. Galton believed that these traits could be treated in the same way as (almost) any other trait, no matter how complex.

Artistic ability, though it looks difficult to measure, can be made tractable by using inter-

[44] Polderman et al. 2015.

[45] F. Galton 1889c, 62-3.

[46] Anonymous 1889a.

comparison, so that it is scored relatively, in terms of deviations from the mean of the normal distribution. This avoids the need for direct measurement using an absolute (hard to agree on) scale. Galton no longer feels the need to show that the faculty is inherited. 'A man must be very crotchety or very ignorant, who nowadays seriously doubts the inheritance either of this or of any other faculty'.[47] Instead he tries to show what may be less obvious, that regression to the mean applies here too. Assuming that there is a continuous measure of the artistic faculty that is normally distributed, it will have some unknown mean P, which Galton's argument does not need to estimate. Some brief checks suggest there is no assortative mating for artistic faculty—a fortunate thing, he thinks, otherwise one might be married to a 'grand piano in constant action'—so one can reason about the mid-parent as in the case of stature. Then, using some reasonable assumptions, Galton calculates that 60% of the children should be above an expected threshold of artistic ability, and 40% below. This is, he feels, acceptably close to the observed data, which has 64 above and 36 below (for the technical details, see Appendix I, p. 447).

Disease was somewhat harder to handle satisfactorily because of the severe limitations of his sample size and survivorship bias. But by limiting himself to consumption (tuberculosis, which is less ambiguous and more likely to kill its victims in time to be recorded) he saw some evidence for transmission of the disease through tainted mothers. For once the normal distribution did not emerge from his tabulations. Instead he got a 'double-humped curve' of 'consumptive taint', suggesting at least two distinct causes, which he interpreted as an underlying susceptibility plus infection. Reasoning about regression to the mean led him to suppose hypothetically that a continuous trait, 'consumptivity', was present in varying degrees in the population and this trait was normally distributed. Next he used an argument like that for artistic faculties to check if the observed percentages of 'consumptivity' were consistent with expected percentages under regression to the mean. However he had to make some ad-hoc assumptions to get agreement and cautioned against placing 'too much stress' upon the result.[48]

While he was completing the above investigations, the ever-curious and energetic scientist had added a new class of problems to his burgeoning interests. Alphonse Bertillon's anthropometric laboratory in Paris had announced at the 1885 Prison Congress held in Rome that it could identify criminals accurately by combining multiple physical measurements: head length and breadth, left foot length, middle-finger length of the left hand—classified simply as large, medium or small in each case—in addition to the 'mugshots' of faces and profiles now commonly associated with criminals. The system was already adopted by the police departments of France, Germany, Spain, Italy and Denmark.[49] By late 1887 some 60,000 criminals had been measured and recorded in Bertillon's lab alone.[50]

Louisa's diary recorded that Frank was 'Much interested in Bertillon's method of identifying prisoners by measurements'. He may have been only the second Englishman to see it in detail. Edmund Robert Spearman had been there the previous year and described it in a letter to the *Globe* of October 1887, which seems to have piqued Galton's interest.[51] The inquisitive scientist had been delving in Bertillon's techniques ever since, scouring publications, corresponding with Spearman, and querying the Frenchman himself on points of detail, as Galton's Anniversary Address to the Anthropological Society of January 1888 explains.[52] During an invited 'Friday

47 F. Galton 1889c, 155.

48 Galton had considerable difficulty calculating what the probable error for 'consumptivity' among the children would be. The $\frac{1.5}{1.7}$ ratio used for stature and artistic faculties seemed far too high. Experimenting, he found a probable error of 0.8 brought reasonable agreement in the expected percentages, but could not independently justify this estimate.

49 *Aberdeen Evening Express* 1887-09-29: 2.

50 F. Galton 1892a, 155.

51 *Globe* 1887-10-01, 3.

52 F. Galton 1888b.

Evening Lecture' at the Royal Institution on May 25, ranging over the general topic of 'Personal Identification', Galton described Bertillon's system in detail, but he had not yet seen it in person. In September, on the way back from a cure at Vichy, he visited the lab for the first time.[53] He briefly noted, as he had already confided to the Anthropological Society in January, that the measurements taken were probably not independent, making some combinations far more likely than others. 'No attempt has yet been made to estimate the degree of their interdependence. I am therefore having the above measurements (with slight necessary variation) recorded at my anthropometric laboratory for the purpose of doing so'.[54] If they were interdependent, that meant adding measurements did not contribute as much new information as might be supposed.

Interdependence of bodily measurements also troubled the Anthropologists. Given some bones from the skeleton of a man, how does one estimate his stature? This was a question that had long puzzled anatomist-palaeontologists like Richard Owen, who often had only one bone to reconstruct a skeleton from. Obviously taller people tend to have bigger feet, and those with big feet tend to be tall. However there were many competing methods in use for doing this, using multiple bones.

Galton was in the audience at the Anthropological Institute in June that year when the archaeologist Lieut-General Augustus Henry Lane Fox Pitt-Rivers (sighted several times here already) presented results from excavating burial pits on his own estate, Cranborne Chase. Pitt-Rivers wanted to use stature inferred from several bones to distinguish different races. However the General was moved to complain that 'a uniform system of estimating stature from the long bones is a matter of the most urgent necessity' since the 4 inches of variation caused by the four English methods he tried exceeded the average differences in inferred height between his races. He resolved this prickly dilemma by descending to a French standard.[55]

At this Galton pricked up his ears. 'It seemed to me that the anthropologists had not discussed their facts in the best statistical manner, and that they ought to have adopted a different form of treatment to any they had hitherto tried'.[56] The multiple bones used for estimation of stature are obviously not independent. More generally, there is a relationship between *any* two physical measurements of a person. But how to quantify it?

After his R.F.F. data was exhausted, Galton revived his Anthropometric Laboratory in January of 1888.[57] It was rebuilt and run at his own expense on land provided rent-free by the 1851 Exhibition Committee, with egress through the adjacent South Kensington Museum. That year the laboratory would measure over 1,000 subjects after quickly discarding the 3d fee initially asked. Galton was prepared to cover all costs just to get the data. About 350 men yielded the Bertillon head and limb measurements needed. He started plotting these measurements against each other on graph paper and smoothing. Soon the elliptical shape that he had found when investigating the inheritance of height appeared. Suddenly it was apparent that these were just special cases of a much more general problem.

It is likely that Galton had already reached this stage of his investigations when he presided over a session of the Anthropological Society on November 13, 1888. His friend Edward Burnett Tylor (1832-1917, see Figure 10.15, p. 235) was presenting a paper 'On a method of investigating the development of institutions applied to laws of marriage and descent'. Tylor had collected data about many primitive societies and argued that it showed a strong 'adherence' between the complexity of societies and patrilineal organization. It appeared that increasing sophistication

[53] In an interview with the *Pall Mall Gazette*, published in November of 1888, Galton said that he had seen the Bertillon system detect a criminal three weeks earlier, Anonymous 1888. F. Galton 1908, 251, gives the impression from its order that Galton had visited Bertillon before the Royal Institution lecture but this is not so as F. Galton 1888f mentions visiting the lab some time after the lecture.

[54] F. Galton 1888e.

[55] F. Galton 1888a.

[56] F. Galton 1890d.

[57] Not March, contrary to Louisa's record.

led to an abandonment of matrilineal modes. Galton praised Tylor's collection of data as a definite improvement on a botched attempt by Herbert Spencer, and offered a remarkable insight.

> Dr. Tylor's memoir dealt both with a method and with conclusions; it was of the method only that he (Mr. Galton) would now speak. It consisted in ascertaining the degree in which the concurrence of certain customs was exceptionally frequent. He thought that the degree of interdependence, to which the various degrees of exceptional frequency testified, might with advantage be expressed in terms of a scale, in which 0 represented perfect independence, and 1 complete concurrence. By doing so, the values of the various concurrences would become more clear. As an example of what he meant, he would refer to a scale used in certain psycho-physical inquiries and discussed in Fechner's book, where the true significance of the various percentages of success and failure was tabulated.[58]
>
> It was extremely desirable for the sake of those who may wish to study the evidence for Dr. Tylor's conclusions, that full information should be given as to the degree in which the customs of the tribes and races which are compared together are independent. It might be, that some of the tribes had derived them from a common source, so that they were duplicate copies of the same original. Certainly, in such an investigation as this, each of the observations ought, in the language of statisticians, to be carefully "weighted." It would give a useful idea of the distribution of the several customs and of their relative prevalence in the world, if a map were so marked by shadings and colour as to present a picture of their geographical ranges.

The remark demonstrates that Galton is already thinking before mid-November of quantifying associations using a simple numerical scale, here from 0 to 1, or 'perfect independence' to 'complete concurrence'.[59] He also points out that finding an association is complicated by the possibility that it is an artefact, not caused by the factors supposed but rather by a confounding influence, in this case geographical adjacency between societies, which may lead to mere copying of customs.

Tylor understood at once. 'The difficulty raised by Mr. Galton that some of the concurrences might result from transmission from a common source, so that a single character might be counted several times from its mere duplicates, is a difficulty ever present in such investigations, as for instance in the Malay region, where groups of islands have enough differentiation in their marriage systems to justify their being classed separately, though traces of common origin are at the same time conspicuous. The only way of meeting this objection is to make separate classification depend on well marked differences, and to do this all over the world'. But it would not be that easy to solve.

The full breakthrough to correlation followed shortly after this, presumably by late November. Galton described it clearly in his Presidential Address to the Anthropological Society in January of 1889.[60]

> As the lengths of the various limbs in the same person are to some degree related together, it was of interest to ascertain the extent to which they also admit of being treated as independent. The first results of the inquiry, which is not yet completed, have been to myself a grateful surprise. Not only did it turn out that the expression and the measure of correlation between any two variables are exceedingly simple and definite, but it became evident almost from the first that I had unconsciously explored the very same ground before. No sooner had I begun to tabulate the data than I saw that they ran in just the same form as those that referred to family likeness in stature.

The special cases could be unified by dividing by the probable error (we would use the variance), giving the general form of a 'co-relation' coefficient, a name that was soon assimilated with a non-mathematical one already in use, 'correlation'.

[58] Gustav Fechner (1801–1887), see Fechner 1860.

[59] This would soon be called a 'correlation'. The astute reader will notice that independent variables must have a correlation of zero, but a correlation of zero is insufficient for independence.

[60] F. Galton 1889h. His account was fleshed out a year later in F. Galton 1890d, incorrectly supposed by Stigler 1989 to be the first description of the circumstances surrounding the discovery of correlation.

The simplification permitted by the correlation coefficient has proved universally irresistible, so that it is now along with regression itself possibly the most widely used statistical technique in all the sciences. It is a straightforward extension of Galton's regression coefficient, which is *standardized* to become the correlation coefficient, simply by dividing by the variance (in modern terms) of the dependent variable. In a stroke this resolves the asymmetry of the regression coefficient, and allows a value like 0.6 to be understood equivalently regardless of the units involved in the measurements that are correlated. Mathematically it is an elementary idea: it is just a simple division.[61] This was obviously no mathematical breakthrough but rather a conceptual shift. In principle anyone who had read Galton's 1885 papers on regression and understood division and variance could easily have formulated it, but doing so required a mental leap that nobody else had made.

In the case of E. B. Tylor's sophisticated societies, we now call the confounding factor *spatial autocorrelation*. Here Galton had clearly perceived a logical problem before the corresponding formal concepts had even been fully worked out, though he was already more than half-way there by thinking in terms of a single scale of relationship and a completely general treatment of 'association'. The methodological conundrum first noticed by Galton in his informal remarks took on a life of its own in anthropological and sociological circles, where it is well-known as 'Galton's Problem'. It is said to have tormented the anthropologist Franz Boas.[62]

There is some confusion about when exactly and where Galton made his correlation breakthrough. Karl Pearson thought that an anecdote in Galton's *Memories* described it: 'the circumstances under which I first clearly grasped the important generalisation that the laws of Heredity were solely concerned with deviations expressed in statistical units, are vividly recalled to my memory. It was in the grounds of Naworth Castle, where an invitation had been given to ramble freely. A temporary shower drove me to seek refuge in a reddish recess in the rock by the side of the pathway. There the idea flashed across me, and I forgot everything else for a moment in my great delight'.[63]

Against this, Louisa's comprehensive record shows that the Galtons only visited Naworth (near Carlisle in Cumbria) in 1889, a year later. Another account by Galton from 1890 does not mention Naworth and explicitly ties correlation to the personal identification studies without drama.[64] But that is not necessarily incompatible with this anecdote, since he was not writing in 1890 for the same sort of audience as he was in 1908. Here I respectfully differ from Stephen Stigler, who follows Victor Hilts in supposing that Galton was really describing his discovery of *inter-comparison* at Naworth, not correlation, and that the incident probably happened in the 1870s, when the idea of inter-comparison had first been published.[65] If Louisa's record is correct then the Galtons did not visit Naworth at any time before 1889.

The flash of revelation about deviations of statistical units is more plausibly about the use of standardization to obtain correlation, which was far more closely associated with the laws of heredity in Galton's thinking at the time. He had been clear from the start that inter-comparison required nothing more than ranks, and he had the essence of that idea in the 1860s already.[66] The incident may simply have happened elsewhere in the country, perhaps in the

61 Assuming that we regress y on x to get the regression coefficient β, that σ_{xy} is the covariance between x and y, and that $\sigma_x = \sqrt{\sigma_{xx}}$:

$$\rho_{xy} = \frac{\beta}{\sigma_y} = \frac{\sigma_{xy}}{\sigma_x \sigma_y}.$$

62 Naroll 1961; Raoul Naroll and Roy G. D'Andrade 1963.

63 F. Galton 1908, 300; K. Pearson 1924, 393.

64 See Louisa's Record in Tredoux 2023a; F. Galton 1890d.

65 See F. Galton 1889h; Hilts 1975; Stigler 1989.

66 As explicitly stated in F. Galton 1875h.

grounds of Stanmore Hall, where the Galtons visited the Hollonds in 1888. The (mistaken) appearance of Naworth in the anecdote anchors the incident close in time to 1889, when the Galtons actually did visit the castle.[67]

It was too late to include correlation in *Natural Inheritance.* Galton worried that someone else might point it out instead after reading the book, so a description—where the correlation coefficient was called a 'co-relation'—was rushed into the *Proceedings of the Royal Society.* It appeared typo-ridden in public before the book was released by Macmillan in December of 1888.[68]

The occasional use of mathematics in *Natural Inheritance* severely limits its popular audience. The now-obsolete notation and statistical idiom is an added barrier to modern readers. But it would be highly influential. Many were slow to catch on to Galton's new ideas at first, including the man who would go on to become his chief disciple, Karl Pearson (to whom correlation came to mean welcome liberation from concepts like causation). But that was over three years later. On his first reading, as he recalled at the close of his life, it had been rather different, 'I find my copy of *Natural Inheritance*, read and annotated forty years ago, defaced by many marginal notes expressing anger at Galton's analogies'. 'But these notes were written before I had read and grasped the value of much of the later work in the book'.[69]

A more immediately receptive reader was the Cambridge mathematician John Venn (1834–1923), who was able to inform Galton privately of some of the numerous typos and other mistakes which crept into the text (errata for subsequent printings). Venn nevertheless wrote a fulsome review for *Mind.* He thought the book was 'a model of the way in which statistics should be employed. These have been acquired, tested and put to use with the most scrupulous care and skill, and the trust yielded to them is just as great as it should be, and no greater'. The mathematician's review concentrated on a lucid description of Galton's work on the inheritance of stature, managing to accurately describe the gist of the work without involving any of the symbolism used in the text.[70]

Another enthusiastic reader was the mathematical economist F.Y. Edgeworth, whose grandfather Richard Lovell Edgeworth had been a member of the 'Lunar Society' with Galton's grandfather Samuel Jr. The economist reviewed the book for *Nature.*[71] He found it an 'imposing scientific edifice' constructed 'on a foundation of geometrical reasoning, and culminating in the clouds of biological hypothesis'. He commended the 'masterly use of the calculus of probabilities' and thoroughly approved of the attention given to the importance of variability in its own right, not just as a nuisance. Though some of the methods deployed in the text (see the detailed exposition of Galton's four solutions to Problem 2 in section G.2, p. 438) were not necessarily the best possible, they gave consilient results. Though he did not mention it, Edgeworth had himself already used variance partitioning techniques similar to two of Galton's methods.[72]

The stature data obtained by Galton closely agreed with other sources familiar to Edgeworth, including Italian data. He was less certain of the eye colour and temperament data, which seemed less suitable to exact treatment than stature and subject to differences between observers as to the meaning of their terms. But the eye colours at least were tractable and he found the close agreement between Galton's predictions and the known facts 'indicative of a real law'. He left evaluation of Galton's theories about the processes of heredity and evolution to other specialists, who could clarify whether the concepts introduced were appropriate, but

[67] Otherwise Galton has to make two gross mistakes and not just one. But *caveat emptor*: on the same page of *Memories* Galton has his regression studies on peas starting in 1885 rather than 1875, just one of many slips.

[68] F. Galton 1888c.

[69] K. Pearson 1930a, 61, Stigler 1986.

[70] Venn 1889.

[71] Edgeworth 1889.

[72] Stigler 1986, 294.

the reviewer cautioned that those opinions ought to be backed by evidence and analysis at least as statistically sophisticated as Galton employed.

On the other side of the Atlantic, John Dewey was also enthusiastic, inviting statisticians in all fields to pay attention to the new methods, especially for their focus on distributions rather than mere averages.[73]

Critics or no critics, *Natural Inheritance* and its elaborate machinery would have to fend more or less for itself from now on. In the meanwhile Galton had already moved on to the problems posed by personal identification and description. Although these had prompted the formulation of correlation, they would take him much further afield in coming years—apparently far removed from his main interests, but the basis for yet another element of his lasting reputation.

£500 IN PRIZES

Are offered for EXTRACTS from the FAMILY RECORDS of Competitors, to be sent before MAY 15. *Full particulars of the conditions are published in the* " Record of Family Faculties " (*Macmillan and Co.,* 2*s.* 6*d.*), *which consists of Tabular Forms and Directions for entering Data, with an Explanatory Preface, by FRANCIS GALTON, F.R.S., Author of* " *Hereditary Genius,*" *&c.*

EXTRACT FROM THE " *TIMES* " OF JANUARY 9

" A new and instructive recreation is offered by an eminent man of science to the families of England. It may be called the game of 'Family Records.' It is at least as attractive as most of the games, puzzles, and strings of questions with which many families now beguile their leisure and weary their acquaintances, and it is likely to be far more useful in its results. . . . . The inquiry he proposes to institute is in many respects a novel one. It is, perhaps, as well not to take it too seriously at first, lest those who are invited to cooperate should be frightened rather than attracted by it. If the plan is to succeed, as we sincerely trust it may, it must recommend itself by various motives to various minds, and we are convinced that there are many who would consent at once to play an attractive and interesting game, though they would be very likely to decline if they were asked to take part in a serious scientific inquiry. Which of the two classes of contributors to Mr. Galton's 'Records' is most likely to win the prizes is a different question."

Figure 10.1: Record of Family Faculties flyer, 1884.

[73] Dewey 1889.

2 FATHER ....

1. Date of birth. Birthplace.

2. Occupation. Residences.

3. Age at marriage. } The place for this entry is at 4 in next printed page.

4. do. of wife } The place for this entry is at 3 in next printed page.

5. Mode of life so far as affecting growth or health.

6. Was early life laborious? why and how?

7. Adult height. Colour of hair when adult. Colour of eyes.

8. General appearance.

9. Bodily strength and energy, if much above or below the average.

10. Keenness or imperfection of sight or other senses.

11. Mental powers and energy, if much above or below the average.

12. Character and temperament.

13. Favorite pursuits and interests. Artistic aptitudes.

14. Minor ailments to which there was special liability { in youth. in middle age.

15. Graver illnesses { in youth. in middle age.

16. Cause and date of death, and age at death.

17. General remarks.

Figure 10.2: Record of Family Faculties *form for the father. Similar forms were provided for other direct relatives.*

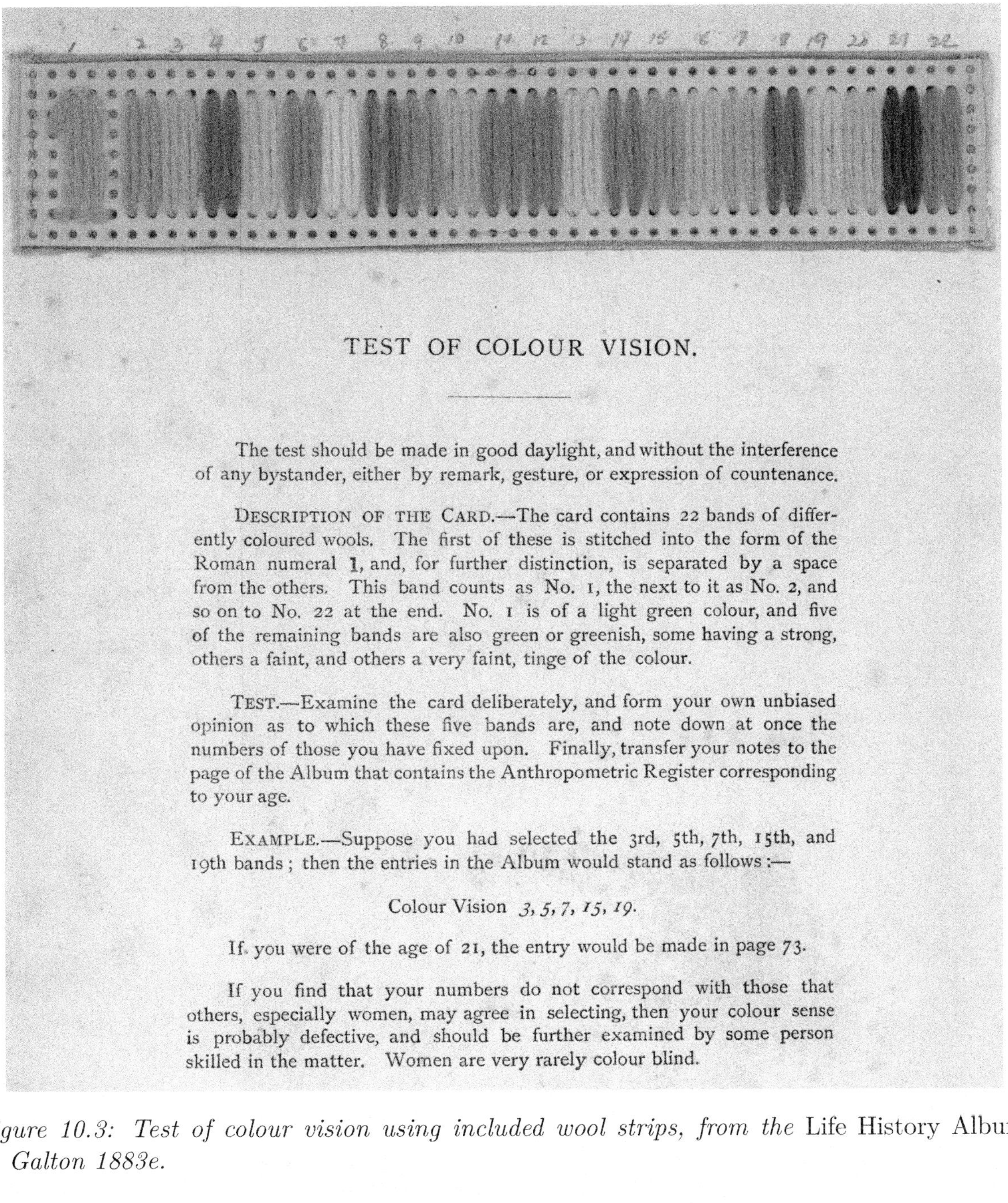

TEST OF COLOUR VISION.

The test should be made in good daylight, and without the interference of any bystander, either by remark, gesture, or expression of countenance.

DESCRIPTION OF THE CARD.—The card contains 22 bands of differently coloured wools. The first of these is stitched into the form of the Roman numeral I, and, for further distinction, is separated by a space from the others. This band counts as No. 1, the next to it as No. 2, and so on to No. 22 at the end. No. 1 is of a light green colour, and five of the remaining bands are also green or greenish, some having a strong, others a faint, and others a very faint, tinge of the colour.

TEST.—Examine the card deliberately, and form your own unbiased opinion as to which these five bands are, and note down at once the numbers of those you have fixed upon. Finally, transfer your notes to the page of the Album that contains the Anthropometric Register corresponding to your age.

EXAMPLE.—Suppose you had selected the 3rd, 5th, 7th, 15th, and 19th bands; then the entries in the Album would stand as follows:—

Colour Vision *3, 5, 7, 15, 19.*

If you were of the age of 21, the entry would be made in page 73.

If you find that your numbers do not correspond with those that others, especially women, may agree in selecting, then your colour sense is probably defective, and should be further examined by some person skilled in the matter. Women are very rarely colour blind.

Figure 10.3: Test of colour vision using included wool strips, from the Life History Album, *F. Galton 1883e.*

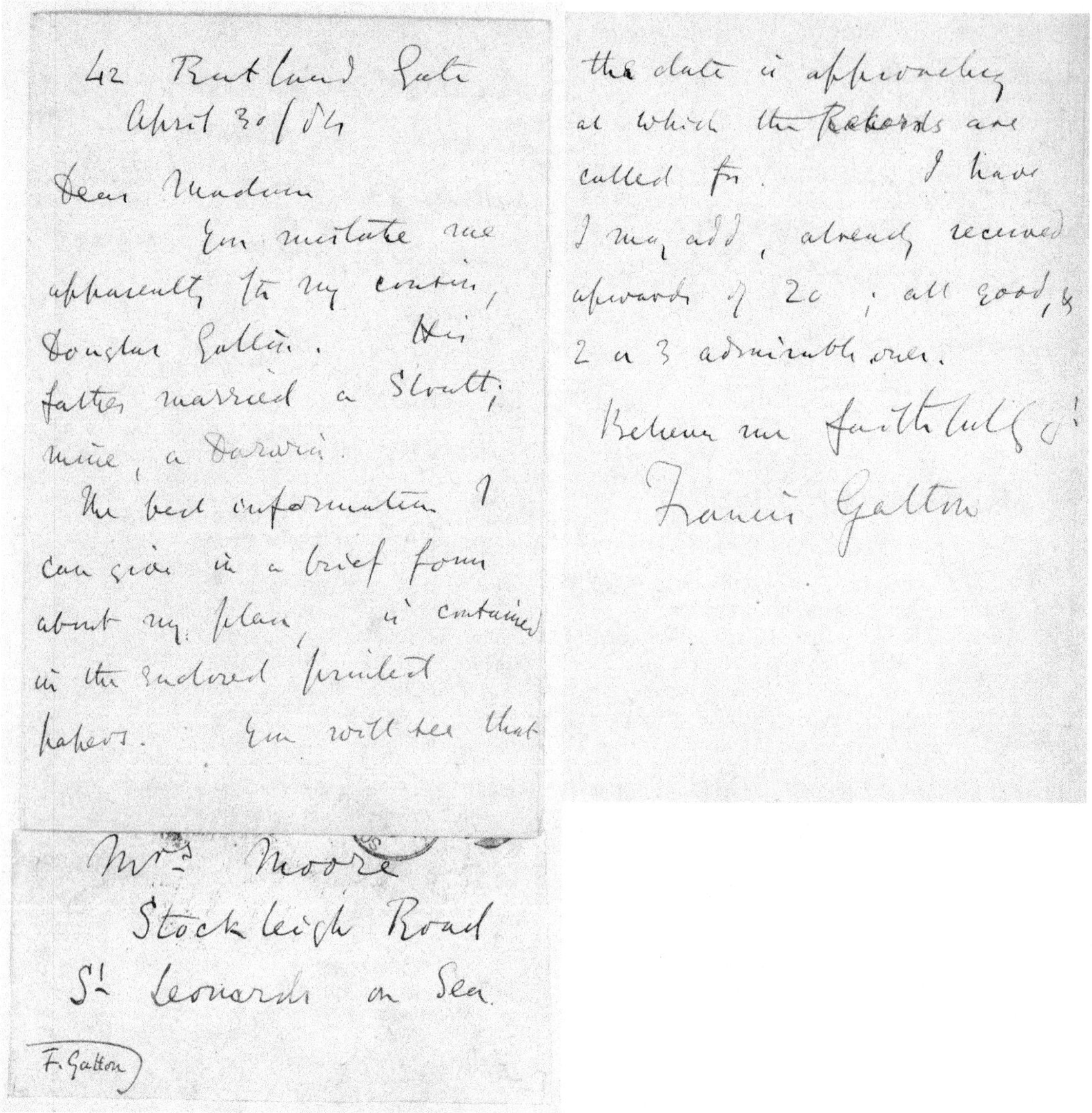

42 Rutland Gate
April 30/84

Dear Madam

You mistake me apparently for my cousin, Douglas Galton. His father married a Strutt; mine, a Darwin.

The best information I can give in a brief form about my plan, is contained in the enclosed printed papers. You will see that the date is approaching at which the Records are called for. I have I may add, already received upwards of 20; all good, & 2 or 3 admirable ones.

Believe me faithfully yours

Francis Galton

Mrs Moore
Stockleigh Road
St Leonards on Sea

F. Galton

Figure 10.4: Letter from Galton to Mrs. Moore, 1884-04-30, regarding the Record of Family Faculties *competition.*

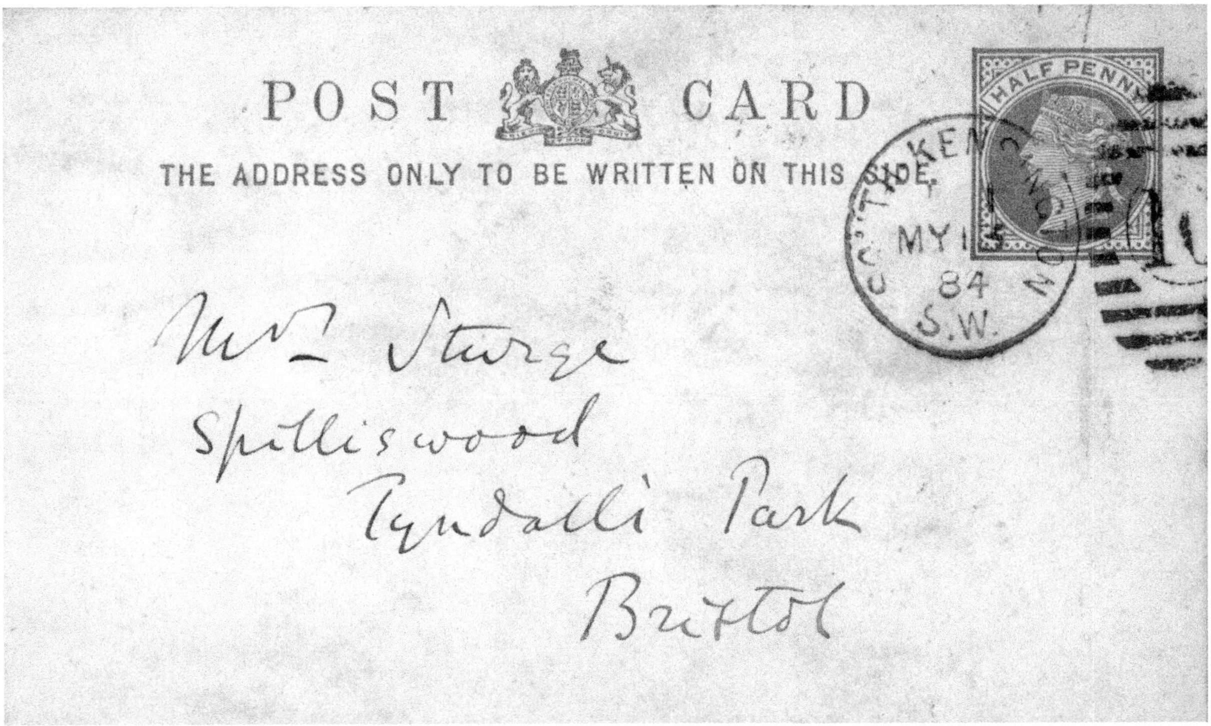

Figure 10.5: Postcard from Galton to Mrs. Charlotte Sturge née Allen (1817–1891), 1884-05-13, verso. The address should read 'Chilliswood'. The Sturges and Allens were Quakers. From the author's collection.

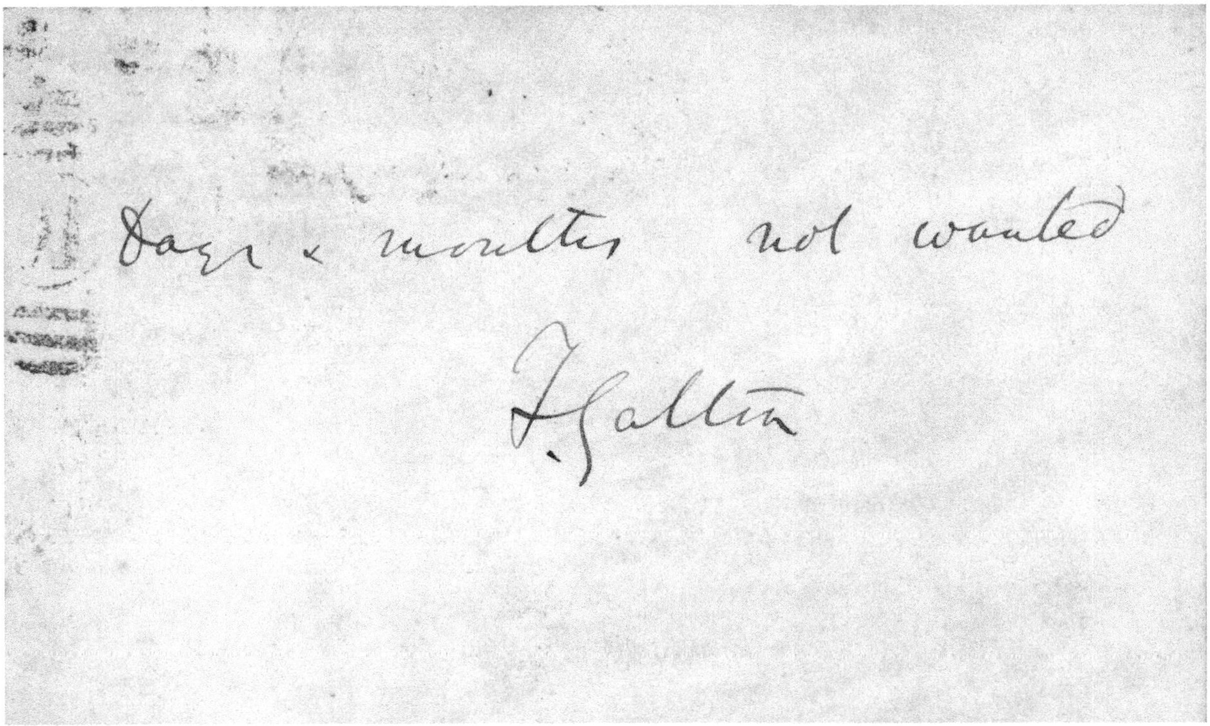

Figure 10.6: Postcard from Galton to Mrs. Sturge, 1884 0 13, recto. 'Days & months not wanted' for dates.

Figure 10.7: The International Health Exhibition of 1884 as seen by Punch.

Figure 10.8: The (First) Anthropometric Laboratory at the International Health Exhibition of 1884, K. Pearson 1924. Not to be confused with the second (1888-91) and third (1891-5) laboratories located adjacent to, then inside, the South Kensington Museum.

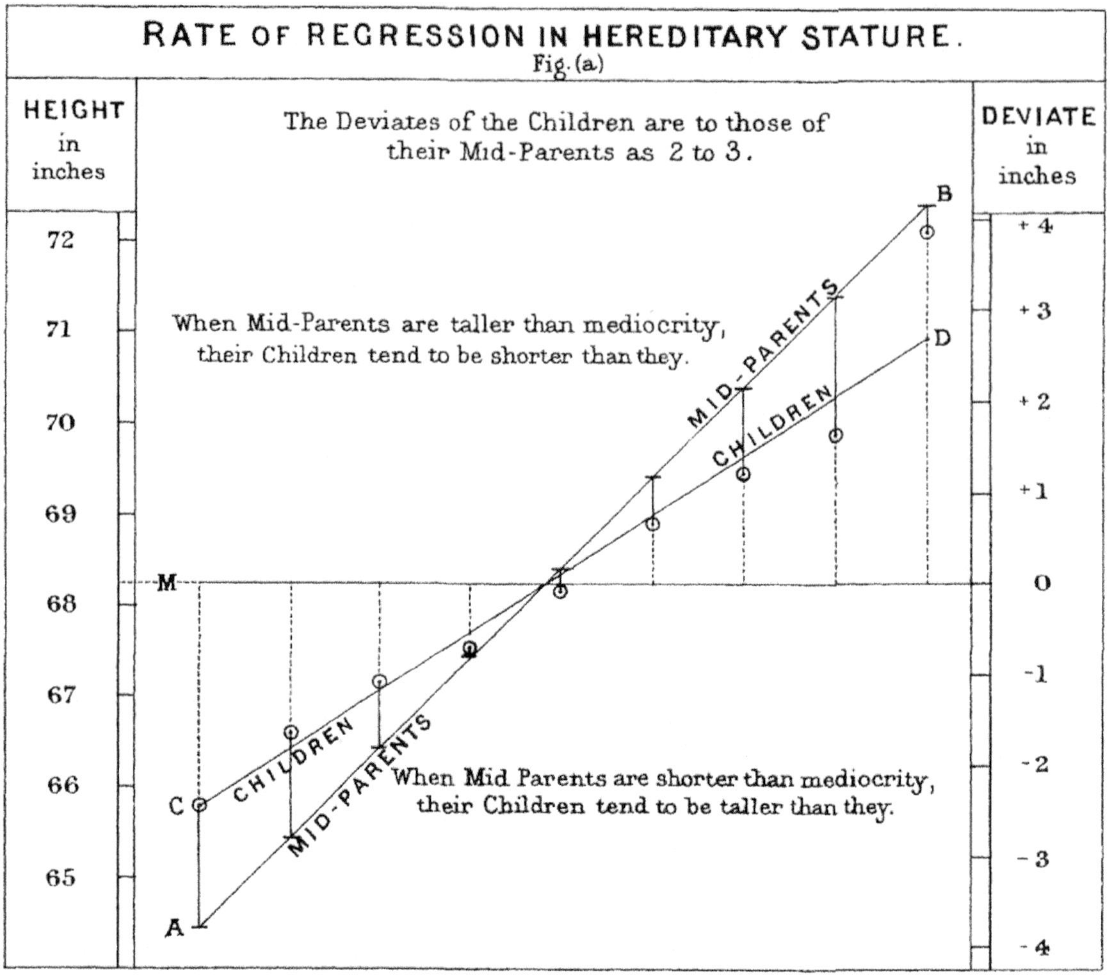

Figure 10.9: The heights of children regress from their mid-parents toward the mean, F. Galton 1885f. Galton used this simple illustration repeatedly in his lectures and papers.

NUMBER OF ADULT CHILDREN OF VARIOUS STATURES BORN OF 205 MID-PARENTS OF VARIOUS STATURES.
(All Female heights have been multiplied by 1·08).

Heights of the Mid-parents in inches.	Heights of the Adult Children.														Total Number of		Median
	Below	62·2	63·2	64·2	65·2	66·2	67·2	68·2	69·2	70·2	71·2	72·2	73·2	Above	Adult Children.	Mid-parents.	
Above ..	..	..	..	..	..	..	..	..	..	..	..	1	3	..	4	5	..
72·5	..	..	..	..	..	..	..	1	2	1	2	7	2	4	19	6	72·2
71·5	..	..	..	..	1	3	4	3	5	10	4	9	2	2	43	11	69·9
70·5	1	..	1	..	1	1	3	12	18	14	7	4	3	3	68	22	69·5
69·5	..	..	1	16	4	17	27	20	33	25	20	11	4	5	183	41	68·9
68·5	1	..	7	11	16	25	31	34	48	21	18	4	3	..	219	49	68·2
67·5	..	3	5	14	15	36	38	28	38	19	11	4	..	..	211	33	67·6
66·5	..	3	3	5	2	17	17	14	13	4	..	..	..	..	78	20	67·2
65·5	1	..	9	5	7	11	11	7	7	5	2	1	..	..	66	12	66·7
64·5	1	1	4	4	1	5	5	..	2	..	..	..	..	..	23	5	65·8
Below ..	1	..	2	4	1	2	2	1	1	..	..	..	..	..	14	1	..
Totals ..	5	7	32	59	48	117	138	120	167	99	64	41	17	14	928	205	..
Medians ..	..	..	66·3	67·8	67·9	67·7	67·9	68·3	68·5	69·0	69·0	70·0	..	..	..	..	..

Figure 10.10: Heights Cross-tabulated.

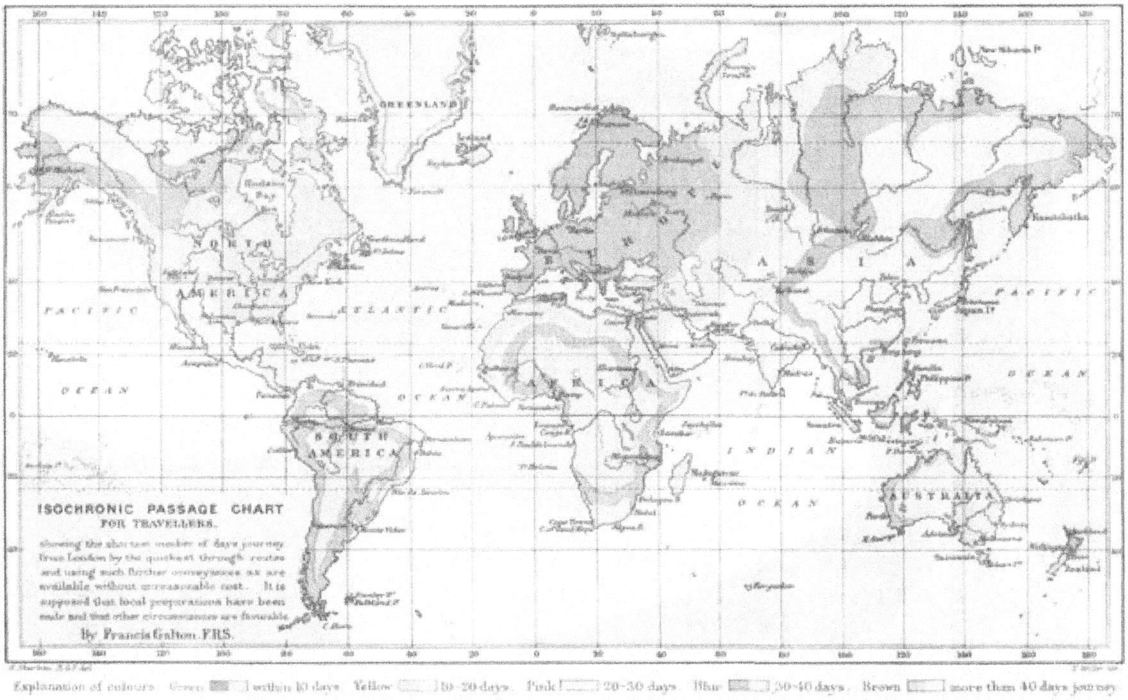

Figure 10.11: Colour-coded Isochronic Postal or Passage Chart, 1881.

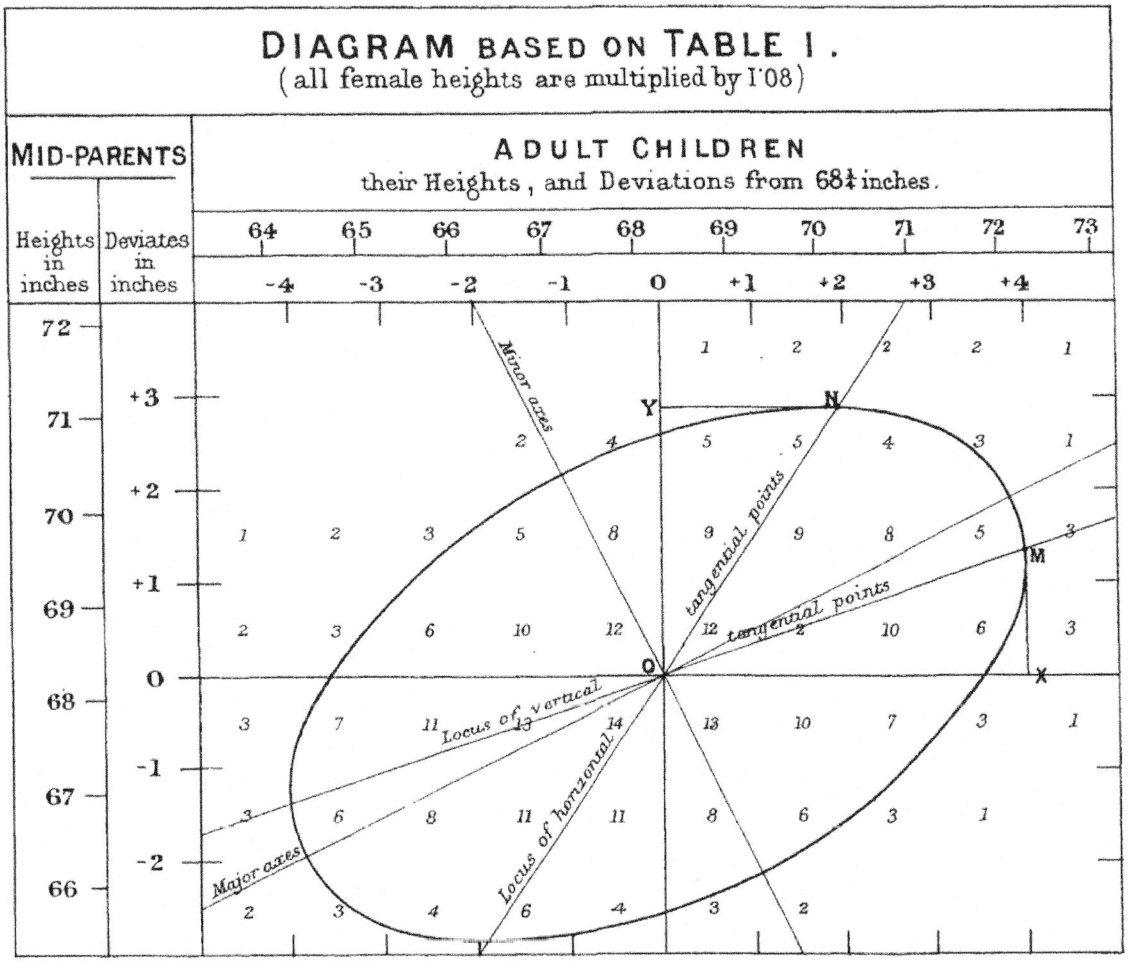

Figure 10.12: Ellipses connect smoothed heights.

Table VII.

Number of Errors of various Amounts in the 3 Calculations of the Numbers of Light Eye-coloured Children in the 78 Families.

Data employed referring to	Amount of Errors.					Total Cases.
	0·0 to 0·5.	0·6 to 1·1.	1·2 to 1·7.	1·8 to 2·3.	2·4 and above.	
I. The 2 parents only	19	30	18	5	6	78
II. The 4 grandparents only	16	28	10	10	14	78
III. The two parents and 4 grand-parents	41	17	8	4	8	78

Figure 10.13: Percentage eye colour prediction errors.

No. of tricolours in parents.		Number of tricolours in grand-parents.				Total tricolour offspring.	
		4	3	2	1	Calculated.	Observed.
		a	*bcei*	*dfgjkm*	*hlno*		
2	All cases	119	119	28	11		
	Coefficient......	0·91	0·83	0·76	0·68		
	Tricolour calc'd.	*108*	*99*	*21*	*8*	*236*	
	„ observed	106	101	24	8		239
1	All cases	37	158	60	6		
	Coefficient......	0·66	0·58	0·51	0·43		
	Tricolour calc'd.	*24*	*92*	*30*	*3*	*149*	
	„ observed	20	79	36	4		139
0	All cases	..	..	18	6		
	Coefficient......	..	..	0·26	0·18		
	Tricolour calc'd.	..	..	*5*	*1*	*6*	
	„ observed	..	..	7	2		9
				Grand totals		*391*	387

Figure 10.14: Basset hound coat colour predictions, from F. Galton 1897b.

Figure 10.15: The Anthropologist Edward Burnett Tylor (1832-1917).

Figure 10.16: Francis Galton at age 65 in 1887, K. Pearson 1930b.

Figure 10.17: Francis Galton at age 66 in 1888, K. Pearson 1930a.

Chapter 11

Fingerprints

We read of the dead body of Jezebel being devoured by the dogs of Jezreel, so that no man might say, "This is Jezebel," and that the dogs left only her skull, the palms of her hands, and the soles of her feet; but the palms of the hands and the soles of the feet are the very remains by which a corpse might be most surely identified, if impressions of them, made during life, were available.[1]

Since the measurements that Alphonse Bertillon relied on for identification of criminals were correlated and therefore weaker than the Frenchman supposed, Galton was moved to look for other means of identification. The general problem of recognizing differences had long been on his mind. How *much* difference for example is there between two portraits, or two photographs, or any two shapes at all? If one stands far enough away from two pictures displayed side by side they eventually become indistinguishable to the eye, and that distance might serve as a measure of similarity. From a different angle, his work on composite photographs had continually shown the family commonality underlying faces, greatly complicating definite identification of individuals by appearance.

Determination of identity had been on the public mind ever since the sensational conviction of Arthur Orton in 1874 for perjury while impersonating Sir Roger Tichborne, who had drowned at sea in 1854 off the coast of Brazil. Encouraged by a clairvoyant, Tichborne's own mother thought her son had returned, corpulent and coarse with a funny accent. There was a personal connection of sorts, since Orton, a rough-living Australian butcher, had for several months fraternized in South America with Galton's friend Richard F. Burton in the late 1860s.[2]

One method which now caught his attention had been mentioned several times in different places over the years. Fingers appeared to have patterns in their skin 'furrows' which could be recorded in printer's ink or dust, and clearly read with minor magnification once photographed. The patterns looked distinctive (see Figure 11.2, p. 255). Some had speculated that they could be used to definitely identify individuals. Galton had taken some prints himself sometime in the 1870s, and he had also been sent examples of hand prints taken from twins in 1880 and likely before that. But at that time the prints did not impress him and he had other priorities.[3]

Galton now made enquiries to find out what had been definitely done so far. Numerous references at second-hand suggested that a substantial study of the physiology of fingerprints had been conducted in 1823 by the Czech Jan Evangelista Purkyně (1787–1869). After taking up a professorship in physiology at Breslau (due it is said to the enthusiasm of Goethe and Humboldt) the Czech described a classification system based on the recurring patterns found in the prints. With the help of the Royal College of Surgeons librarian, Galton was eventually able to locate and have a translation made of this exceptionally rare Latin disquisition, *Commentatio de examine physiologico organi visus et systematis cutanei.*[4] However Purkynje made no mention of using prints for identification, nor did he deal with their persistence over time.

---

[1] F. Galton 1892a, 113.

[2] See Tredoux 2016.

[3] From Ellen Stevenson, 1880-08-14, GALTON/2/7/2/8/13.

[4] 'A Commentary on the Physiological Examination of the Visual Organs and the Cutaneous System'. This

Another source emerged from over the Atlantic in the pages of the *American Anthropologist*. Prof. Otis T. Mason of the Smithsonian had referred to 'dermal topography' in some comments on a paper, 'Anthropological Notes on the Human Hand' by Frank Baker, which had been read before the Anthropological Society of Washington on May 4, 1886. Galton wrote to Mason in early February of 1888 to ask for more information. His letter suggests that he had been actively looking for information about fingerprints for several months at least.

> 42 RUTLAND GATE, LONDON, S. W.,
> Feb. 9, 1888.
>
> Dear Sir
>
> Pray pardon my asking a question of you. It refers to the following passage in your remarks, p. 74 of the AMERICAN ANTHROPOLOGIST. "He then referred to dermal topography and its use by the police and others as a means of identifying persons."
>
> I should be particularly obliged for references on this matter. It is said to be largely used in China, but I have failed to get this verified as yet. I know of cases of its use in India, but by no means extensively. Has it ever been in general use by the police of any country? What evidence is there of its having been efficacious? What parts of the skin have been used for the purpose? And how is the examination or impression made? I am the more anxious to learn as I find a flat disagreement on one important point bearing on this subject in published accounts by two different persons.
>
> I expect to have an opportunity for making some rather extensive experiments in this matter, and hardly know whether it is worth while to begin them. On this account I should be the more obliged for such information as you can give me, as a brother anthropologist (I am president of our Anthropological Institute). Believe me, truly yours,
>
> FRANCIS GALTON.

On enquiry, it turned out that Mason's information was from a man who would later that year help to found the National Geographic Society, but it fell well short of use by a police force. 'Mr. Gilbert Thompson, of the U.S. Geological Survey, states that in 1882 he was in charge of topographical parties of the U. S. Geological Survey, in the vicinity of Fort Wingate, New Mexico, and had occasion to make disbursements by means of written orders upon the sutler at that post, with whom he had deposited personal funds. "In order to protect myself against any alteration in the amount as written, and being without any other resource, I adopted the additional precaution of making an ink impression of my thumb at the top and bottom of the orders, at the left hand, and I wrote in ink on the upper impression the number of the order and on the lower the amount in figures corresponding to the written value of the order." Mr. Thompson found the plan he had adopted thoroughly efficacious'.

Thompson had used his own print for tamper-detection and not identification as such, but there was no shortage of talk over the years about finger prints and identification. Mason could report to Galton that a scheme to control admission of Chinese immigrants in USA using thumb prints had been scorned by the authorities, who thought it impractical, doubted that the patterns were stable, and preferred photographs. The unlucky suggestion can be traced back to the San Francisco Customs House where Special Agent Oliver Lyman Spaulding and Detective Harry Morse had been seconded by the Treasury Department. Morse had already suggested using fingerprints in the early 1880s. Moreover Franklin Lawton from the San Francisco Mint told Spaulding that he had heard from a Detective John Maloy of prints being used in a court

appears to have been his second or 'habilitation' PhD thesis, the first being awarded at Prague 1818. Breslau is now Wrocław in Poland. Various transliterations of 'Purkyně' are used. Galton writes it as both 'Purkinje' and 'Purkenje'.

case in Albany in the 1850s. Supposedly Maloy (later the Chief of Police there from 1870 until his death in 1884) had won the case on the basis of a bloody print found on paper, subsequently matched to the suspect. There has never been more than hearsay about this supposed use, which had no recognized basis in law at that stage, but this has not stopped its propagation in popular histories.[5] Lawton had persuaded the successful photographer Isaiah West Taber (1830–1912), then based in San Francisco, to photograph, enlarge and exhibit prints. The photographer worked from ink impressions.[6]

Other mentions of prints showed up in sources like *The World of Wonders* (1883), whose entry for 'Thumb Portraits' was widely serialized in many newspapers throughout the country.[7]

> If the "ball" or cushion-like surface of the top joint of the thumb be examined it can be seen that in the center—as, indeed, in the fingers also—is a kind of spiral formed of fine grooves in the skin. The spiral is, however, rarely if ever quite perfect. There are irregularities or places where lines run into each other here and there. Examining both thumbs, it will be seen that they do not exactly match, but the figure on each thumb is the same through life. If the thumbs of any two persons are compared, it will further be found that no two are alike. There may be, and generally is, a "family resemblance" between members of the same family, as in other features; there are also national characteristics, but the individuals differ. All this is better seen by taking "proof impressions" of the thumb. This is easily done by pressing it on a slab covered with a film of printers' ink, and then pressing it on a piece of white paper; or a little aniline dye; India ink—almost anything may be used. The Chinese take advantage of this to identify their important criminals, at least in some parts of the empire. We photograph their faces; they take impressions from their thumbs. These are stored away, and if the delinquent should ever again fall into the hands of the police, another impression at once affords the means of comparison. The Chinese say that, considering the alteration made in the countenance by hair and beard, and the power many men have of distorting and altering the actual features, etc., their method affords even more certain and easy means of identification than our plan of taking the criminal's portrait. Perhaps we might with advantage take a leaf out of their book.

Two 'thumb impressions' were clearly shown (see Figure 11.5, p. 256). The Chinese theme would show up repeatedly but, as later research established, mistook a symbolic mark for a real means of identification by comparison of prints, which Chinese finger-dabs were typically too indistinct to permit.[8] 'Thumb Portraits' may have been the source for Mark Twain's 'recollection' in *Life on The Mississippi,* published that same year, of a fortune-teller who took fingerprints, supposedly having learned the art from a French jailer who kept a register of prisoners.

Neither Galton nor his Smithsonian informant were aware of Thomas Taylor (1820–1910), who in 1877 made the earliest *verifiable* suggestion of forensic fingerprinting. Taylor was born in Perthshire, Scotland, and trained in physics and chemistry at Glasgow University, followed by art and drawing at the British School of Design. He had joined the Department of Agriculture in Washington DC as a microscopist, where he 'exhibited on a screen a view of the markings on the palms of the hands and the tips of the fingers, and called attention to the possibility of identifying criminals, especially murderers, by comparing the marks of the hands left upon any object with impressions in wax taken from the hands of suspected persons. In the case of murderers, the marks of bloody hands would present a very favorable opportunity'. This remarkable anticipation slipped out of sight until 1971, perhaps because the *American Journal of Microscopy and Popular Science* called the idea 'a new system of palmistry'.[9] Taylor would go on to become an influential mycologist and phytopathologist

[5] There is no mention of this in *History of the Police Service of Albany: From 1609 to 1902* (1902) by William Henry Paddock, which is not otherwise overflowing with newsworthy information.

[6] F. Galton 1888e misspelled his name 'Tabor' when mentioning him in passing, which many copied forever after. See also Cole 2001, 122-7.

[7] Galton found references to it in *Science* 8, 1886, 166, 212.

[8] F. Galton 1892a.

[9] Taylor, Thomas 1877.

However there was more information to be found closer to home. In 1880 *Nature* had printed an exchange of letters on fingerprints between Henry Faulds, a missionary surgeon in Japan, and William Herschel, formerly a colonial administrator in India.[10] Curiously Galton had encountered Faulds before, by way of Charles Darwin, then promptly forgotten all about him.

Dr. Henry Faulds (1843–1930, see Figure 11.12, p. 259) was one of those delightfully eccentric characters who enlivened the Victorian scene, brimming with original ideas, a talent for scientific observation, diverse interests and very peculiar personality traits. Born in 1843 in the small town of Beith, Ayrshire, into a Presbyterian family then successful in trade, he was well-educated until the business folded. Work in menial jobs saw him through medical training in Glasgow, briefly under the renowned Joseph Lister. Deeply religious, somewhat at odds with his Darwinian interests, he worked abroad as a Presbyterian missionary doctor. In India at Darjeeling he quarrelled with the church authorities after only a year. Finding a place in Japan from late 1873 to the mid 1880s, he was a surgeon at the hospital within the Tsukiji segregated area for foreigners, near Tokyo.

It was at excavations by the naturalist Edward Sylvester Morse (1838–1925) of the Omori 'shell mound', a midden near Yedo Bay, that the severely myopic Faulds first stumbled on fingerprints. The naturalist's agnosticism provoked Faulds, and the missionary 'felt called upon by God to appear wherever Morse lectured and to attempt to controvert his views'.[11] The controversialist noticed (as he tells it) at close-up the patterns in signature impressions left by Japanese potters in their work. Potsherds of their pieces lay interspersed within the shell mound.

It is hard to pin down the precise date that this happened. Faulds' publications prior to 1879 make no mention of fingerprints, and on November 11 of that year he commented at a meeting of the Asiatic Society of Japan that it 'had occurred to him that the finger markings in primitive pottery might be made to contribute some faint ray of light. The furrows on the tips of one's fingers form a very distinct pattern. In all the fingers of one man's hand they might be found to run downwards obliquely from left to right. In another the thumb only might show another pattern. In another still, all the fingers might be different from this, and so on, so that it was not impossible that a new means of reaching some legible race marks might be added to science by a careful comparative study of these familiar finger-point patterns. At present the facts known to him in this connection were simply puzzling, but law must underlie them'.[12]

After gathering some samples from his students, Faulds hit upon the idea of using impressions of fingerprints to find their owners. As this was not mentioned to the Asiatic Society, it probably happened after Nov. 11, 1879. Exactly how thoroughly he investigated it is hard to determine, as we have only his own description of this to go on, and he supplied little detail until 1905, after which he had already become embroiled in public controversies about his role.[13] Attempts to interest others in fingerprints through a printed circular produced no results, so he wrote to Charles Darwin early in 1880 seeking help.[14]

Tsukiji Hospital Tokio
Japan Feby 16th 1880

Sir,

As an ardent student of your writings I trust I may venture to address you on a subject of interest. I allude to the rugae or furrows on the palmar surface of the human hand. These form

[10] F. Galton 1908, 252, Faulds 1880; William J. Herschel 1880.

[11] Schwantes 1953, 126. Faulds strongly believed that religion and evolution were compatible, and that the latter did not refute the former.

[12] Faulds 1879.

[13] There is generally a lack of independent corroboration for many of Faulds' claims, and the secondary literature that has built up around this topic repeats his own stories, e.g. Beavan 2001; Wilton 1938.

[14] Faulds to Darwin, 1880-02-16, GALTON/3/2/1/24.

singular and intricate patterns which vary in detail with each individual but may be classed according to their leading lines without much difficulty. Now, I have been led to study these in prehistoric pottery & I am engaged in what proves a somewhat difficult task—collecting data from all quarters of such ancient impressions & comparing them with similar impressions of living men of all races. Already I see some glimpses of light but facts are hard to get in sufficient numbers. The few monkeys &c. which I have got show similar but somewhat different markings & if man's origin has been from organised 'dust' perhaps a comparative study of lemuroids &c. may yield results of real value. I hope for this & have bethought myself of your powerful aid. A word or two would set observers working everywhere.

I enclose a filled up form & can send a number to any one who wishes them. The practical value—and Englishmen will look for that—is in the work of identification in medico-legal studies, thus

1. Copies of palmar impressions of convicted criminals—as photographs now are used—the latter become unlike the original, the rugae, never.

2. In cases like that of 'Tichborne?' Are his rugae of the Tichborne or Orton type? for heredity rules here marvellously.

3. In cases where mutilated remains are found & various people are missing. The rugae again may be compared with that of the parents &c.

4. Where impressions exist of bloody fingers by a murderer, or prints on fresh paint or drinking glasses, windows &c. by a robber, &c.

The Chinese I find since entering on this enquiry, take impressions of this kind from criminals, as our servant girls used to seal their letters with their thumbs. (The Japanese as the ancient Egyptians, use their fingernail.)

Hoping you may find this to be of interest.

I remain

Your obliged reader

Henry Faulds

Surgeon Supt. of Hospital

Included with the letter was a sample printed circular showing hand imprints in ink collected for an individual, as well as some other details like hair colour and race. Faulds' idea here that one could fix identity through the fingerprints of *relatives* (the Ortons vs. the Tichbornes), a generally hopeless task, shows that he had little actual experience of taking prints at this stage beyond a few individuals.

Darwin courteously replied that he was 'most unfortunately situated for offering you any assistance' since he lived in the country 'from weak health' and seldom saw anyone. However he promised to forward the letter to Galton, 'the most likely man that I can think of to take up the subject to make further enquiries'.[15] Darwin kept his word and forwarded the enquiry from Faulds to his cousin that day though without presuming that Galton himself would be interested. 'The enclosed letter & circular may perhaps interest you, as it relates to a queer subject. You will perhaps say hang his impudence. But seriously the letter might possibly be worth taking some day to the Anthropolog. Inst. for the chance of someone caring about it. I have written to Mr. Faulds telling him I could give no help, but had forwarded the letter to you on the chance of its interesting you'.

Galton's response suggests that he was at least aware of fingerprints and had some cursory experience with them, though some mystery remains. No trace can be found now of this early foray, which was never mentioned again.[16]

42 Rutland Gate, S. W.
April 8 / 80

My dear Darwin

[15] Darwin to Faulds 1880-04-07, reproduced in Faulds 1912, 22-23.

[16] K. Pearson 1924, 195.

I will take Faulds' letter to the Anthro. and see what can be done; indeed, I myself got several thumb impressions a couple of years ago, having heard of the Chinese plan with criminals, but failed, perhaps from want of sufficiently minute observation, to make out any large number of differences. It would I think be feasible in one or two public schools where the system is established of annually taking heights, weights etc., also to take thumb marks, by which one would in time learn if the markings were as persistent as is said. Anyhow I will do what I can to help Mr Faulds in getting these sort of facts and in having an extract from his letter printed. I am so glad that my 'visualising' inquiries seem interesting to you. I get letters from all directions and the metaphysicians and mad-doctors have been very helpful.

Very sincerely yours,

Francis Galton

Our united kindest remembrances to you all

Once he read it in detail, the Faulds letter seems to have made little impression on Galton. Perhaps this was because the surgeon gave no considered evidence for his claims, besides which Galton was then at his busiest scientifically. It was passed on to the Royal Anthropological Institute as Darwin suggested, but they do not appear to have published any part of it, or to have followed up on it. The letter would later be returned by the Institute to a surprised Galton, who had evidently forgotten all about it, in 1894.[17]

The physician-missionary was disappointed but pressed on regardless, publishing a letter later that year in *Nature* titled 'On the Skin-furrows of the Hand'. In it he described, as he had to Darwin, how he had noticed human and later animal fingerprints, and supposed racial differences in the patterns. He disclosed how to take impressions using printer's ink, and mentioned some conjectures about the use of prints in ethnological classification, in forensic identification of criminals, and in determining identity through prints of relatives. Most importantly, he mentioned in closing the 'for-ever-unchangeable finger-furrows of important criminals'.[18]

Faulds gave two concrete instances where he had used prints forensically to establish the identity of people at crime scenes. Here we have only his descriptions of the events to go on. The first involved theft of surgical alcohol from a bottle in his hospital, which he was able to trace through a set of ten greasy prints on the bottle to one of his employees. The second involved a sooty palm print left on a hospital wall by a burglar, which he was able to show did not match someone accused of the burglary. At least, that is the way the second incident has been commonly told, following Faulds' own description in *Nature*. However, many years later Faulds corrected this second case, stating that the sooty palm print contained no discernible fingerprints! In fact, he based his analysis on the general outline of the hand, and this second case has nothing to do with fingerprint analysis. Nobody appears to have noticed this crucial correction. Faulds' own reports about his use of fingerprints should be treated with caution.[19]

Some of Faulds' ideas about fingerprints were on the right track, but others were not. For example it is true that fingerprints do not change *in certain particulars though not others*, but Faulds gave no reason exactly why he believed that to be true. The careful reader would have noted that early in his letter Faulds stated that 'I was led, about a year ago, to give some attention to the character of certain finger-marks'. Therefore since 1879, but in later publications the date would become more vague, even reaching back at times to 1878 or 1877.[20]

How could Faulds have reached such far-reaching conclusions about the permanence and uniqueness of fingerprints on the basis of just one year of experience? This was picked up at once by William James Herschel (1833–1917, see Figure 11.6, p. 256), who followed-up in *Nature* that he had personally used fingerprints in India for identification 'for more than twenty years'. He doubted that the method could be used to reliably identify the race of an

[17] Darwin to Galton, 1880-04-07, GALTON/3/2/1/22.

[18] Faulds 1880.

[19] Faulds 1923, 46.

[20] Faulds 1905b, 1923.

individual. As Herschel diplomatically put it, 'the conclusions of your correspondent seem, however, to indicate greater possibilities of certainty'.[21] Once again this suggests very little practical experience on the part of Faulds, who would have seen the difficulties at once with a moderately large collection of prints.

After the above exchange, actual use and study of fingerprints for identification seems to have sunk without a trace, aside from the American speculations. Now eight years later it was rapidly revived. Galton hastened to get in touch with both Faulds and Herschel to find out more. It appears from a subsequent exchange of letters that he was unable to get hold of Faulds at the time.[22] The surgeon had finally left Japan under a cloud in 1886 and was back in England by then, struggling to make a living. Herschel, who had retired from the Indian service in 1877 and was also back in England, proved easier to get in touch with. It turned out that he had a small but precious collection of fingerprints taken in India, stretching back to 1859.

Herschel was a man of some ability from a remarkable family. His musical German-born grandfather Frederick William Herschel (1738–1822) had been 'King's Astronomer' to George III, landing a knighthood. His father Sir John Frederick William Herschel (1792–1871) was at least as distinguished as a scientist, also a fellow of Royal Society and already a subject of Galton's investigations into pedigrees of genius. Unlike Sir John's younger sons, William did not take up astronomy or science, opting for India instead as a rising administrator. Early on he was responsible for managing subcontractors for a 'road-metalling' project in Bengal. Personation, fraud and perjury were rife. Stand-ins would often be hired to write examinations. Signatures on deeds were routinely denied and contested by the litigious population. Verisimilitude was not, as Galton later put it, universally easy to obtain.[23]

> The difficulty of identification is increased by the roaming habits of the natives, many of whom travel great distances for pilgrimages, petty commerce, or change of employment, so that witnesses may not easily be found to identify them. Again, while the natives of India and of Egypt have beautiful traits of character and some virtues in an exceptional degree, their warmest admirers would not rank veracity among them. It is not insinuated that false testimony is unknown in English courts of justice, or in England generally; indeed I find, on a rough attempt at a vocabulary (made for quite another purpose), that more than fifty English words exist, which express different shades and varieties of fraud; but if a map of the world were tinted with gradations of colour to show the percentage of false testimony in courts of law, whether in different nations or communities, England would be tinted rather lightly and both Bengal and Egypt very darkly.

One element of the problem was distinguishing people by appearance. Although some post-modernist commentators have made much, to use their abstract-noun argot, of the 'supposed homogeneity' of 'others', whether Chinese immigrants in America or Indians in India, Galton, Herschel and 'others'—when they were not busy 'subjecting the body' to 'identification'—were just accurately reporting shared experience.[24] It was explicitly a difficulty that *Europeans* had in distinguishing unfamiliar people, who really did appear less variable than the populations they were used to. 'Whatever difficulty may be felt in the identification of Hindoos, is experienced in at least an equal degree in that of the Chinese residents in our Colonies and Settlements, who to European eyes are still more alike than the Hindoos, and in whose names there is still less variety'.[25]

It occurred to Herschel that a hand impression in ink might serve as a more impressive form of signature from sub-contractors. One Rajyadhar Konai was the first to give an impression,

---

21 William J. Herschel 1880.
22 Inferred from Faulds to FG, 1884-12-03, GALTON/2/9/6/13/17.
23 F. Galton 1900c.
24 Cole 2001.
25 F. Galton 1892a, 152.

now immortalized in the history of fingerprinting.[26] At first Herschel only thought to awe the natives, but after noticing the distinctive patterns in the prints, he became more ambitious. It was more convenient to take finger impressions only, rather than the whole hand. This became a personal hobby. As it dawned on him that the prints could cut through much of the mendacity that the administration encountered, he tried to interest the colonial government, but officials said they were wary of making far-reaching changes so soon after the Mutiny of 1857. He had trouble interesting anyone beyond his own family in the technique, so for the time being he could only accumulate a substantial collection. However by 1877 he had acquired enough power as Magistrate and Tax Collector at Hooghly (near Calcutta) to put fingerprinting into practice for his own use. The administration of government pensions was plagued by 'remarkably vital' recipients, so they were registered. Prisoners were recorded too, since it was not unknown for stand-ins to be paid to serve sentences, or for corpses to be conveniently produced instead. These fingerprint records impressed the locals, who did not object to being recorded. One man even chopped the third joints of his fingers off to escape conviction, unsuccessfully it turned out.

Encouraged by his success, Herschel sent an official letter to the acting Registrar for Bengal, James Austin Bourdillon (1848–1913), to once again recommend adoption of the system. The letter detailed the persistence and distinctiveness of the patterns, now established over some two decades for multiple individuals, and explained the benefits of establishing identity. Unfortunately Bourdillon was distracted at the time, as he confided to Herschel many years later, so the idea did not go further than some preliminary inquiries. The letter also found its way to the Inspector of Jails for Bengal, Henry Beverley, until recently the Registrar himself. Reliable identification of prisoners was obviously a boon, but Beverley did not understand the system. 'He said he did not see what I wanted him to do'.[27]

All this meant that Herschel retired the next year with a respectable collection of prints taken over a long period, which he took back to England with him. When Faulds' letter of 1880 appeared in *Nature* he had been drawn out, only to retreat again into the background. Now in 1888 matters developed quickly. Herschel was corresponding with Galton by March 2nd, immediately sharing prints from his Indian album. 'I should wish indeed that in any use made of the method the fact of my having had it in full official practice in India in 1877 after nearly twenty years experimenting, sh^{d} ... be recognized'.[28]

This was invaluable to the scientist, who was keenly aware of the need for solid evidence of the permanence and individuality of fingerprints, collected over a substantial period of time. By May 25th Galton had thoroughly canvassed the prior work. His Friday Evening lecture to the Royal Institution on 'Personal Identification and description' closed with 'those fine lines of which the buttered fingers of children are apt to stamp impressions on the margins of the books they handle'. He mentioned Faulds, Herschel and the American instances. He showed his audience projections of Herschel's own prints, taken 18 years apart but essentially unchanged. He had also mastered the technique of taking impressions well enough to demonstrate it live. 'Professor Dewar and Mr. Galton pressed each a finger tip on smoked glass, and showed the imprint of same on a screen by means of a projection lantern, to demonstrate the difference in the spiral lines of the two fingers'.[29]

Identification was not the only interest here. Galton also hoped 'to discover independent features suitable for hereditary investigation. It has long been my hope, though utterly without direct experimental corroboration thus far, that if a considerable number of variable and inde-

[26] Shortly before his death, Herschel wrote a concise summary of his fingerprinting history, see William James Herschel 1916.

[27] Herschel to FG, 1888-03-02, GALTON/2/9/6/13/32. This unflattering recollection is omitted in William James Herschel 1916.

[28] Herschel to FG, 1888-03-02, GALTON/2/9/6/13/32.

[29] Dublin Review (20), 184.

pendent features could be catalogued, it might be possible to trace kinship with considerable certainty. It does not at all follow because a man inherits his main features from some one ancestor, that he may not also inherit a large number of minor and commonly overlooked features from many ancestors. Therefore it is not improbable, and worth taking pains to inquire whether each person may not carry visibly about his body undeniable evidence of his parentage and near kinships'.[30] This hope would drive the enormous investment of time and effort that Galton devoted to the subject for the next five years.

The scientist aimed to establish some basic propositions that he considered prerequisites if 'finger furrow' patterns were to be reliable. It had to be *proved* rather than just asserted that fingerprints persisted over a long period of time (decades, at least), were unique to an individual, and could easily be classified, stored and matched in large numbers. For the first point, Herschel's collection was critical, but more detailed and careful analysis had to be done, with more follow-up prints collected from the subjects for whom earlier prints existed. For the second and third points, he assembled a very large collection of prints, principally thumbprints from the Anthropometric Laboratory. Wherever he went he also carried a small fingerprinting apparatus with him, imposing himself on those he met for their prints.

By November of 1890, some 2,500 thumb prints were available. They provided enough material to justify reading before the Royal Society the very first scientific paper devoted to the use of 'The patterns in thumb and finger marks' for *identification*.[31] (Though the Laboratory was forced to close temporarily in February 1891, when it lost the use of the land loaned to it, new premises were found within the galleries of the Science Collections of the adjacent South Kensington Museum, where it reopened on August 3, larger and better-lit.) The paper concentrated on classification of the patterns for indexing, and a substantiated case for persistence over time, using the newly expanded longitudinal collection started by Herschel.

Purkyně's classification of patterns into 9 types proved to be inadequate when applied to a selection of 1,000 prints. Instead Galton constructed 10 classes based on 'essential points of difference', of which 25 subdivisions based on 'secondary points' seemed to cover almost all the patterns. For the first time Galton raised the use of 'minutiae', those irregular localized deformations within patterns, showing some examples (see Figure 11.7, p. 256). These minutiae rather than the higher-level patterns are the basis for a detailed comparison of prints for identity over time, since distinctness and persistence are closely related. Using 25 to 30 points of reference, prints separated by up to 20 years could be shown to be essentially identical, even if altered in other respects by ageing and accidents. The paper closed with some observations about the evolutionary significance of the patterns that we will return to in a subsequent chapter.[32]

A few months later a follow-up paper for the Royal Society dealt specifically with 'Methods of indexing finger-marks'. This positioned fingerprinting as way of greatly increasing the practical efficiency of *Bertillonage*, since it added what that system lacked, an almost completely decorrelated measurement. Galton had greatly simplified his primary classification by now into just three classes: primaries (arches), loops and whorls. These were easily numbered, with a dual number to capture the orientation (in later iterations he would change these to letters). However a practical problem remained to wrestle with: the population had a great number of loops, leaving a big card tray to search through, which begged for more indexing. For now he recommended scanning to match minutiae with a magnifying glass while quickly flipping through cards.[33]

Policemen may not have been regular readers of the publications of the Royal Society or of

[30] F. Galton 1888e.
[31] F. Galton 1890h,i.
[32] See chapter 13, p. 294.
[33] F. Galton 1891b.

the summaries printed in *Nature*. But they might have read Galton's popular articles, printed in *Lippincott's Monthly Magazine* and the *Nineteenth Century*. These were short enough to maintain interest but long enough to cover essentials. In November of 1892 his first book on *Finger Prints*, as he now called them, was brought out by Macmillan.[34] It includes a comprehensive history of previous mentions and uses of fingerprints, considering Faulds (unfortunately misspelled 'Fauld' now) in passing and Herschel in detail. A translation of part of Purkyně's scarce *Commentatio* is given.

An entire chapter is devoted to practical advice for actually taking and reproducing prints, since Galton had found this tricky to do well. For ink-based methods, the key is to get the ink on the ridges but not in the furrows, and to roll the thumb. A summary of the physiology is given for completeness, not for new information. Previous schemes for classifying the patterns and indexing them are updated and elaborated at length. Letters are now used instead of numbers. A tabulation of 100 prints is demonstrated. Evidence for persistence of prints is extended to 389 comparisons covering twenty year periods for every age group, a further iteration of Herschel's initial data set.

What is completely new in the book and a crucial step in the history of the subject is a strong statistical argument for the uniqueness of prints. As Galton had already argued, identity of prints is based on the comparison of minutiae, with the broad patterns playing a minor role as a disqualifier. He shows that a single print provides strong evidence by calculating the chances against two prints coinciding in all details. The key to his argument is a simple model for the effects of minutiae.

If a square of paper of some size is dropped onto a picture of a print, it will obscure the area under it. Lines will enter and exit it. The minutiae present within the obscured square act to alter the number of those lines, through junctions, forks and islands. If the square is really small, say just enough to cover one or two ridges, then it is easy to interpolate the missing part, with probability close to 1. The larger the square is, the harder this will become, as more opportunity for alterations due to obscured minutiae occur, and the number of ridges to complete grows. By trial he found that a square with a side of 12 ridges or more was impractical to complete, i.e. a probability closer to 0 than 0.5.[35] It follows that for some intermediate size between these two extremes the chances of interpolating the square will be 0.5. After more trials he found that a square of 6 ridges seemed to offer that 0.5 guessing chance. This he verified, again empirically, using three different methods. The results strongly suggested that the squares were independent, since knowledge of surrounding squares could not improve the interpolation to more than guesswork. This is entirely contingent on the nature of fingerprints as they happen to occur among humans, and dependent on the accuracy of Galton's observations.

A typical print can be divided into at least 24 six-ridge intervals, giving a chance of 0.5^{24} of guessing all of them, assuming that they are independent. However, when guessing a particular square, some information is present from the general direction and number of the incoming and outgoing ridges which would not be known in a completely blank chequerboard. Galton conservatively estimated the chances of guessing these to be 1 in 20 and 1 in 250 respectively. For simplicity say 0.5^4 and 0.5^8. This gives a probability of $0.5^{24+4+8} = 0.5^{36}$ or call it 1 in 64 billion. If the world population is say 16 billion, and each person has ten fingers, then the chances of any two random fingers matching, drawn from the entire world, would be no better than say 1 in 4 (using the binomial distribution). When two fingers are used, this probability squares to become radically smaller, and so on.

As Galton pointed out, finer consideration would reduce the chances even further, and he was able to show that even if one compares two prints and finds that they match for a few

[34] F. Galton 1890j, 1891a, 1892a.

[35] Galton actually argues that the probability of guessing a 12-ridge square is practically 0 but this is dangerously close to circular (it is to be proved that a print cannot be guessed) and not necessary for his argument.

minutiae the remaining minutiae are still highly unlikely to match. An in-depth analysis of Bertillonage demonstrates that it can classify a collection of hundreds of thousands of subjects, though without unimpeachable identification. This becomes massively extended by adding fingerprints to the system as an additional element for indexing, gaining also a definitive means of identification.

Two more wholly original chapters bring evidence to bear for the first time on questions that had kept Galton interested in finger prints for so long: the extent to which they are hereditary or not, and their racial peculiarity. His greatly expanded collection of prints confirmed that the patterns were indeed hereditary to at least some extent. The argument used anticipates later methods for comparing observed with expected events. Galton formed what would now be recognized as a contingency table for fraternal pairs he split into 'A' and 'B' at hazard.

Table 11.1: Fraternal Fingerprints.

	A children			
B children	Arches	Loops	Whorls	[Tot. B]
Arches	05 (01.8)	12 (12.3)	02 (04.9)	19
Loops	04 (05.8)	42 (39.5)	15 (15.7)	61
Whorls	01 (02.4)	14 (16.2)	10 (06.4)	25
[Tot. A]	10	68	27	105

The idea is to test whether or not the patterns for A and B are independent. The marginal totals divided by the grand total (105) give the probability of each row or column. For example the probability of an arch for B children is 19/105. If these are independent then the probability of any cell is the product of the marginal probabilities. Thus P(A=Arch and B=Arch) = P(A=Arch) × P(B=Arch). These *expected* probabilities under independence can be compared with the *observed* probabilities, which for each cell is the observed number divided by the grand total. Or simper still, multiplying each cell probability by the grand total gives the cell total we should compare, given in brackets in the table. Taking arches as an example, we calculate $105 \times (10 \times 19/105^2) \approx 1.8$. The table suggests that arch-arch and whorl-whorl pairs deviate a lot from the expected totals.[36] More precise statements of this argument would have to await Karl Pearson's development of the χ^2 probability distribution a few years later—Galton was anticipating the method—which describes the sum of the squared differences between the observed and expected cells, and does not involve multiple comparisons. By another test, prints from 105 fraternal 'couplets' agreed rather more closely to each other than 101 randomly drawn prints did.

Galton also had 17 sets of twins, whose patterns coincided largely but not invariably. But he did not have enough evidence to quantify these relationships more finely. By contrast he now had enough prints from English, Welsh, Hebrew, Negro and Basque sources to abandon his earlier speculations about the usefulness of fingerprints for anthropological purposes. In the aggregate there appeared to be slight differences, but they were too weak to be useful in practice: 'inquiries were continued in varied ways until hard fact had made hope no longer justifiable'.[37]

Before *Finger Prints* had appeared, the British Association of 1892, meeting at Edinburgh that year without Galton, had appointed a committee to investigate improvements of the systems of identification then used. They were prompted largely by a report on the possibilities of *Bertillonage*, by then in use, according to its founder, in modified forms in the USA, Belgium,

[36] Beware though of multiple comparisons!

[37] Here he had practical help from Frederick Howard Collins (1857–1910), a member of the Indexing Society which Galton had joined in recent years.

Switzerland, Russia, Tunis, Romania, as well as in many South American countries and in British India. Naturally Galton was asked to join the committee, which was tasked to report its findings to the Home Office, Army, Navy, India and the other colonies.

While the committee deliberated, Galton proceeded apace with his own investigations. It was one thing to deal with fingerprints taken under ideal conditions, but in the real world conditions often are far from ideal. Prints taken by mere mortals are often blurred and indistinct. Forensically obtained prints, from imprints on glass and the like, are particularly likely to be imperfect. This raised the question whether or not imperfect prints can be reliably classified, or will human judgement product different results? Likewise, will identification still be feasible, so that they stand up in court?

A supplementary chapter *Decipherment of Blurred Finger Prints*, issued early in 1893 in pamphlet form, summarized Galton's experiments on the matter. New examples of prints taken imperfectly and separated by long periods were available from follow-ups to Herschel's Indian collection (see Figure 11.8, p. 257). Galton found that while blurred prints presented difficulties, they were not serious. Some disagreement in classification occurred, but it was uncommon, though he conceded that a trained expert was required for this. Judgements of identity based on the minutiae, revealed clearly by photographic enlargements even in the 'wintry daylight of London', typically had enough points of comparison to hold water for anyone versed in the art. The system was demonstrably resilient.[38]

Shortly after the pamphlet was published, Galton wrote to *The Times* reporting *inter alia* more evidence to hand. He now had another 964 prints of all ten digits of Indian prisoners in Jeypore, blurred but accurately dealt with.[39] He was convinced the system was not only practicable but also would achieve considerable cost-saving if it replaced the antiquated methods then in use by the courts for identification of prisoners, where policemen wasted days of time on the uncertain process of recollecting faces previously known to them, and many unnecessary trials of the wrong people wasted financial and human, not to mention moral, capital.

Drawing up the report of the BA committee prompted a second visit to Bertillon's laboratory in Paris, where Galton was measured himself on April 19, 1893, on his way home from a holiday on the Riviera. He had already suggested to Bertillon in July of 1891 that the Paris laboratory ought to try using fingerprints, but the Frenchman was reluctant. 'Thank you for your recent letter regarding digit prints. I am very willing to add your procedure to anthropometric records, especially for children. But I fear some practical difficulties in cleaning the fingers after printing, etc. Then will my uneducated agents take the necessary zeal to learn your method? I believe that you often go through Paris, could you, on your next trip, devote a morning to me at the Depot, for a fitting on the hoi polloi?'[40] By the time of Galton's visit in 1893, Bertillon had still not added fingerprints to his repertoire.

Given Galton's presence, and the weight of reasoned evidence marshalled in his publications, extensively popularized in the press, it was no surprise that the report of the British Association Committee, completed by July of 1893, called out the possibilities of fingerprinting as an addition to *Bertillonage*. This led later that year to further consideration of fingerprinting by a committee chaired by Charles Edward Troup (1857–1941), an unusually able and energetic civil servant educated at Oxford and the Middle Temple. Troup had been asked on the 23rd of October, by the Home Secretary H. H. Asquith, to find the 'best means available for identifying habitual criminals'.

The new committee's brief included the 'Anthropometric system' of Bertillon and 'the suggested system of identification by means of a record of finger marks'. Galton was not on the committee himself but was called to give evidence, and the committee made several in-person

[38] F. Galton 1893a.

[39] F. Galton 1893b.

[40] Bertillon to FG, 1891-06-15, in K. Pearson 1924, 144.

visits to the Anthropometric Laboratory to see it taking prints, filing them and locating them in its index. On a further visit to Pentonville Prison more than a hundred prisoners were fingerprinted.

At the close of 1893 the scientist could take stock. 'Finger marks fascinated me on account of their unique persistence, & the belief that they might be (which they are not) of ethnological &c. value. Being anxious to get to the bottom of this & to use their peculiarities to the uttermost, I have given perhaps an undue amount of attention to them. However it seems, after all, likely to lead to useful practical results'.[41]

The report Troup tabled in mid February of 1894, which was published as a 'Blue Book', recommended use of fingerprinting in conjunction with Bertillonage for classification and indexing, provided that some concerns about the practicality of indexing very large numbers of people could be resolved. For identification they also recommended a hybrid system, noting that they were even prepared to recommend using fingerprints alone, save for the fact that Galton himself had not proposed that. Despite initial resistance the new system was put into use in the courts as early as 1895.[42] However the Anthropometric system would be speedily replaced all over the world, despite its wide adoption, once the advantages in practice of fingerprinting proved themselves. It was much harder to get consistency between measuring offices for anthropometric measurements.

Events had been moving faster in less deliberate (they might say less ponderous) jurisdictions. A Croatian émigré Juan Vucetich[43] had already adopted fingerprinting in Argentina, where he was in charge of the Police 'Office of Identification' at La Plata. In 1891 Vucetich had read in the *Revue Scientifique* a translation in précis of Galton's Royal Society papers by the French naturalist Henry Crosnier de Varigny. He started taking fingerprints of criminals and in 1892 was able to obtain a confession from a mother who had killed her two children, using a bloody imprint as an incentive—succeeding where the more ordinary means of torture applied to the woman's lover had failed. Like many of the people who would propagate fingerprinting, Vucetich devised a modified classification and indexing scheme to suit his needs. It had to be refined over time with experience, but would become an object of Argentine pride and dominate the Latin world.

Bertillon added fingerprints to his measurements in 1894. Eventually they would completely replace *Bertillonage* even in France, much to the annoyance of Bertillon himself. It is said that when Vucetich, who had remarked on this switch, visited him in Paris, the irate Frenchman either slammed the door in his face (according to Bertillon's own relatives) or physically assaulted the Croatian (according, one assumes, to Bertillon's enemies).[44]

In India, Edward Richard Henry (1850–1931), Chief of the Bengal Police, followed the example of the Army there and put Galton's system into practice, with minor modifications of the classification and indexing system to subdivide more common patterns found in India. Henry corresponded extensively with Galton and visited the Anthropometric Laboratory several times in October 1894 to see the system in action. Henry's official manual for the print system used in India fully acknowledges this debt to Galton: 'many of his terms have been adopted, his definitions accepted, and his suggestions followed whenever practicable. As will be explained, it has been found desirable to increase the number of type patterns, and to fix them by definition with a view to eliminating as much as possible transitional cases, and a system of classifying not dependent upon the employment of suffixes has been adopted'.

One of Henry's changes was to count ridges to disambiguate prints when indexing them, but this technique, which Herschel had discussed with Galton in 1888, had already been noted

41 GALTON/2/4/18/3/1.
42 Anonymous 1895.
43 Ivan Vučetić, 1858–1925.
44 Rhodes 1956.

in the Royal Society paper of 1890, and reiterated in a letter to Henry.[45] Subsequently Henry introduced the system in South Africa and then took over its use in England, where it became associated with his name.

It is unclear exactly what Henry Faulds had been doing about fingerprints between 1880 and 1894, when Troup submitted his report. After 1886 he left Japan for good, after quarrelling with the church about his involvement in journalism, to the detriment of medical practice, writing for outlets like the *Chrysanthemum*. He settled in England, where he wrote a book describing his experiences in Japan. It is curious that this book, *Nine Years in Nipon*, contains no mention of fingerprints, or even of the pottery fragments that formed an essential part of Faulds' retelling of the story. There are many other details, including memorable meals that Faulds had, interesting places he had visited, and descriptions of plants and animals; but nothing on fingerprints.[46] Whatever collection of prints Faulds had undertaken in Japan was not pursued, with no more than occasional observations (at the very most) once he returned to England. He could have enlarged his collection there indefinitely. Nor is there any reasonable explanation for his absence regarding fingerprints in the professional journals or even the popular magazines. He founded an unsuccessful journal of his own, *East Asia*, after his return to England. It seems to contain no mention of fingerprints.

Faulds later said that he was briefly in contact with Scotland Yard in 1888 to advocate the use of fingerprints for forensics, but evidently nothing came of this. The police may have considered Faulds a harmless crank, an impression that might have been reinforced by his aggressive personality. With hindsight it is clear that his difficulties included a lack of evidence. People would not believe that fingerprints were permanent or unique until the evidence was substantial—after all, bodily features can change remarkably with age.

Official bodies are conservative by nature and require a great deal of evidence and persuasion to budge. But Faulds was squarely outside the scientific mainstream, even beyond the pale of the general reading public, unaware of Galton's extensive work since 1888. He had seen none of the popular or scientific articles, nor had he seen Galton's two books on the subject. His later accounts showed some awareness that this was a problem for his story, and he claimed to have been troubled by doubts, worried that fingerprinting might be misused to wrongly convict people if the method was not sound. This is hard to reconcile with his first flush of enthusiasm, and contradicts his claims that he spent a great deal of time advocating its use by the police.

Now in 1894, after reading newspaper reports about the Troup report. Faulds was furious that his thunder had been stolen by others, so he wrote another letter to *Nature*.[47] A pattern was established here that would recur many times in the future. He insisted on his priority as founder of the science of fingerprinting. He referred to Herschel, calling the former administrator's statements about using fingerprints in India allegations, and challenging him to produce an official document he had sent in 1877: 'a copy of that semi-official report would go far to settle the question of priority, as its date is nearly two years previous to my having noticed the finger-furrows'. Faulds wanted to know details. 'What precisely did he do, and when?'

The letter backfired on Faulds. Herschel promptly published a pointed reply, including a copy of the 'semi-official report' in question, dated 'Hooghly, August 15, 1877'. Referring to Faulds' original research, Herschel was skeptical: 'To the best of my knowledge, Mr. Faulds letter of 1880 was, what he says it was, the first notice in the public papers, in your columns, of the value of finger-prints for the purpose of identification. His statement that he came upon it independently in 1879 (? 1878) commands acceptance as a matter of course. At the same time I scarcely think that such short experience as that justified his announcing that the finger-furrows

[45] F. Galton 1890i; FG to Henry, 1894-10-05, GALTON/2/9/6/13.
[46] Faulds 1885.
[47] Faulds 1894.

were "for-ever unchanging"'.[48]

On seeing this letter, Galton wrote to Faulds. Unfortunately only the reply from Faulds has survived, but it reveals that he was asked to share his data, such as it was, but refused. His explanation, that he 'may do something with my material', suggests that he had so far not done much.[49]

Fenton, "Potteries"
Dec 3nd 1894

Dear Sir

I have just got a copy of "Nature" containing Sir W. Herschel's letter & am much obliged to you for calling my attention to it. The matter of indexing to which you allude was simply that the incorrect spelling was repeated in the index for your book, which by the way I got a copy of & appreciate very greatly. I have lent it to all our sergeants and inspectors in this district & they have been much interested. My address has always been correctly given in the Medical Directory. As I may do something with my material by & by, please excuse me from lending it to you for your publications. You must always deserve the greatest credit for your lucid and masterly exposition of the subject.

I am Dear Sir

Yours Sincerely
H Faulds

Faulds did not reply then to Herschel—it is remarkable that Galton actually had to call his attention to Herschel's response in *Nature*—and published nothing further on fingerprints until 1904. Galton himself published one more book on the subject, *Fingerprint Directories* (1895).[50] This contained lengthy extracts from the 'exceptionally able' and 'lucid' Troup Report and further refinements of Galton's classification scheme to deal with intermediate and more common patterns. Then he largely withdrew from the field, devoting his attention once again to biometry and genetics, and to the surge of interest in eugenics after the turn of the century. He was content to watch the practical prosecution of fingerprinting by Henry, Vucetich and others from a distance, though he still collected prints when he had the chance. However if Galton thought that Faulds had been finally dealt with, he was very much mistaken.

In 1905, when fingerprinting was already well-established all over the world, Faulds re-emerged and proceeded to publish several books and pamphlets on the subject, along with a number of articles in popular journals. He hoped to write himself back into a position of prominence in the field. Even though Herschel had readily agreed that, to the best of his knowledge, Faulds was the first to publish in public on fingerprints, and even though Galton readily acknowledged the same, Faulds would always return to this in his after-the-event writings as if it was still in dispute and required some demonstration, which he could always supply triumphantly. He would not take 'yes' for an answer. Now he refused to accept Herschel's document of 1877, insisting in a letter to the *Birmingham Post* that the recipient was not identified by name (in fact, he was identified by his official position). His *Guide to Finger-Print Identification* went so far as to simply leave out any mention of Galton's influential books in its timeline of events.[51] Reactions swiftly followed.

[48] William J. Herschel 1894.

[49] GALTON/2/9/6/13/17. Lambourne 1984 also notices this admission by Faulds. By 'indexing' Faulds is complaining that his name was spelled 'Fauld' in the text and index of F. Galton 1892a, something he raised many times over subsequent years.

[50] F. Galton 1895. Notice that now 'finger print' is 'fingerprint'.

[51] Faulds 1905a,b.

In the *Times Literary Supplement*, Arthur Shadwell describes Faulds' book as 'desultory' and complains that it is larded with too many unnecessary quotations with no relevance. Faulds himself is identified as 'a man with a grievance' given to 'tedious, silly-clever sarcasms' and Shadwell warns us that 'if Mr. Faulds thinks his case will be assisted by writing of this kind he must have a pretty poor opinion of his readers'. Of Faulds' original work on fingerprints, Shadwell writes: 'It does not go very far. The thing was not taken up; Mr. Faulds does not appear to have applied it practically himself, and though he subsequently worked out a system of classification, he has not published it. From his own account, it is not the system worked out several years later by Mr. Galton and published in 1892, nor the system adopted in Bengal in 1894, extended to the whole of India in 1897, and subsequently to Great Britain'.[52]

Galton's own review, also published in the *Supplement*, credits Faulds as 'a zealous and original investigator of fingerprints' when stationed in Japan, and says that he appeared to be 'the first person who published anything, in print, on this subject', but that his proposals 'fell flat' because they were unproven. 'It was necessary to adduce better evidence than opinions based on mere inspection, of the vast variety in the minute details of those markings, and ... for purposes of criminal investigation, it was necessary to prove that a large collection could be classified with sufficient precision'. Galton then refers his readers to the controversy in *Nature* between Herschel and Faulds, and says of Faulds that 'he overstates the value of his own work, belittles that of others, and carps at evidence recently given in criminal cases'.[53]

Faulds had indeed launched a quixotic campaign against the use of single-print matches in criminal convictions, even going so far as to sign up as an 'expert witness' for the defence in a trial which included a single-print match. As it happened, Faulds did not testify in the trial. The defence did not have to resort to the prickly surgeon.[54] Faulds repeatedly refers to this one-print problem in his writings, reiterating the fact that he had advocated taking all ten prints of criminals in his letter of 1880. He gives the impression that others did not take all ten prints, but this is not true: Galton had collected all ten since 1891 but only needed three to index his own collection. Faulds seems to have confused the indexing scheme used by Galton with the number of prints actually taken, but none of this has anything to do with the use of a single print match for *identity*.

Given a single (latent) finger print taken from a crime scene, if all ten prints of a criminal are available then this increases the chance of a positive match, but taking fewer does not increase the probability of a false match: that depends entirely on the likelihood that one print will match another by pure chance. Galton had established that a perfect match between two single prints was highly unlikely to happen by pure chance, and if enough points of comparison, i.e. minutiae, were present, a single print would certainly be good enough.

Faulds was simply and flatly against using single prints at all, and further alienated himself from Scotland Yard as a result. In a letter to *Nature* in 1917, an aggressive response to a brief book Herschel had written on fingerprinting the previous year, he repeated this claim. 'A most curious confusion has arisen from an original police blunder that no two single finger patterns are ever alike, for which, I think, Sir William himself is mainly responsible. I am quite sure that there is no scientific basis for such an assertion. My syllabic system of classification, applied to a large collection, would enable such an assertion to be severely tested, but I know of no other method in existence which could do so'.[55]

Faulds insists then that *all ten prints* should match. The disappointed surgeon has an unshakeable faith in his own abilities and in the correctness of his 'system', even if he is plainly proven wrong by research in a field that he does not keep pace with. If forensic fingerprinting re-

[52] Shadwell 1905.
[53] F. Galton 1905a.
[54] Beavan 2001.
[55] Faulds 1917b; William James Herschel 1916.

ally had required a ten print match it would have been stillborn. But aside from identity, Faulds does not even understand the *indexing* problem. In his own words, 'Galton was supremely anxious to have his natural facts, his finger-print records, arranged precisely in similar parcels, so that one would not be excessively rich in records compared with its neighbour. Now, what does it matter to the keeper of records, or even to the tax-payer, whether one class of patterns is big or little?'[56] It matters a great deal, as the Troup Committee points out.

With age, the conviction that he had been cheated grew till it consumed the surgeon. Paranoia coloured monomania. He even went so far as to petition Winston Churchill (then the Liberal Home Secretary) for a knighthood. He was prompted, apparently, by Galton's own elevation. When this honour was refused, he had his M. P. put questions in the House of Commons to Churchill, asking for an explanation. He published a stream of books and pamphlets, with titles like *The Hidden Hand* and *How the English fingerprint method arose?* (apparently through the work of none other than Henry Faulds).

When Karl Pearson wrote to *Nature* shortly after Galton's death, asking the public for biographical information, Faulds wrote to him. 'I don't feel well-used in the matter, but have no desire to rake up old bones', helpfully enclosing a pamphlet doing exactly that.[57] He would rake for years to come. In 1921 he even published and edited a short-lived journal, *Dactylography*, financed from his scarce means, to continue this self-promotion.[58] Embittered and financially embarrassed, no longer able to practice medicine due to infirmity, he died still unrecognized in 1930.

Even in death it was not over. Faulds had acquired an untiring champion in the form of the enthusiastic Scottish barrister George Wilton Wilton (1862.xii.15–1964.iv.04), who tried manfully to convince the world that his fellow Scot really did invent fingerprinting, even as we know it. Wilton's *Fingerprints: History Law and Romance* (1938) was really almost entirely about Faulds, and mostly based on Faulds' own material.[59] Even though he was moved to describe one of Faulds' self-vindicatory pamphlets as 'pathetic to read', Wilton continued this tradition by bombarding the government and the public with a long series of self-published pamphlets and petitions about Faulds. These curiosities ring with declamations of injustice and wrong, call for official recognition of Faulds and civil list pensions for his surviving daughters. They remorselessly recycle all the same material, in the same way that Faulds himself had done, but now with 'cheated Scotland' introduced (Faulds had kept his quest for retrospective glory strictly personal). In the end Wilton and his pamphlet campaign nearly outlived even the aged Faulds daughters themselves. The persistent barrister only died at age 101. Since then the issue has occasionally flapped around in the depleted air of Scottish Nationalism.[60]

Misinformation abounds in the history of the finger-furrows. Galton's role in fingerprinting was not as the first person to make *suggestions* about them. Many had done so, because that was easy to do once the patterns were noticed. Nor was Galton the first to collect a substantial number of prints, even if he had the largest collection in the world when he wrote his books on fingerprints. Instead he was the first to assemble a strong case, with carefully marshalled evidence, for their persistence, uniqueness and practicality of being classified and indexed in large numbers. This followed from both his extensive collection through the Anthropometric Laboratory and from his rigorous analysis of the probability of finding duplicate prints distinguished by minutiae.

E. R. Henry and Juan Vucetich were students of Galton's methods. To their credit neither man ever claimed to be more. The only biography of Alphonse Bertillon in English to date

56 Faulds 1912, 89.

57 K. Pearson 1911; Faulds to Pearson, 1911-11-23, PEARSON/598/F-O.

58 See, in addition to works already cited, Faulds 1915, 1917a, 1920, 1926 which is wearying enough for most.

59 Wilton 1938.

60 For Wilton's pamphlets one example should satiate non-masochists, e.g. Wilton 1962. See also Beavan 2001 for a retreaded version of this material.

has a dust jacket consisting entirely of a fingerprint, yet Bertillon was reluctant to adopt them and did not claim otherwise.[61] Faulds' spiteful publications by contrast did much to cloud the history of the subject. He simply did not understand the field that he claimed to have been instrumental in. His single claim to recognition was long thought to be his priority in print—a point endlessly, triumphantly, repeated by his promoters. But Tom Taylor anticipated him by three years, so even that must now be denied him. We can only hope that the ghost of George Wilton Wilton will be appeased by the fact that Taylor was a Scotsman too.

Figure 11.1: Galton in his seventies, taken in the 1890s by Eveleen Myers (née Tennant), National Portrait Gallery, London, UK.

[61] Rhodes 1956.

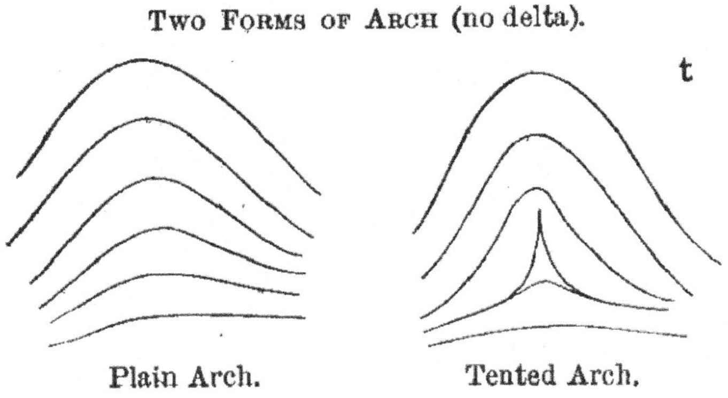

Figure 11.2: Fingerprint arches.

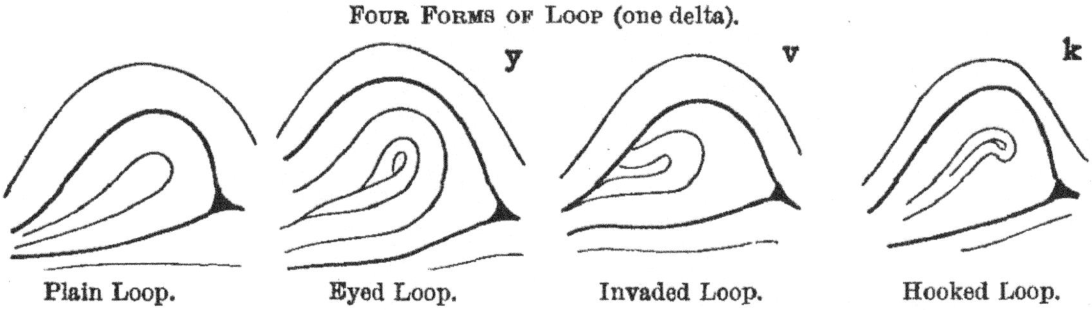

Figure 11.3: Fingerprint loops.

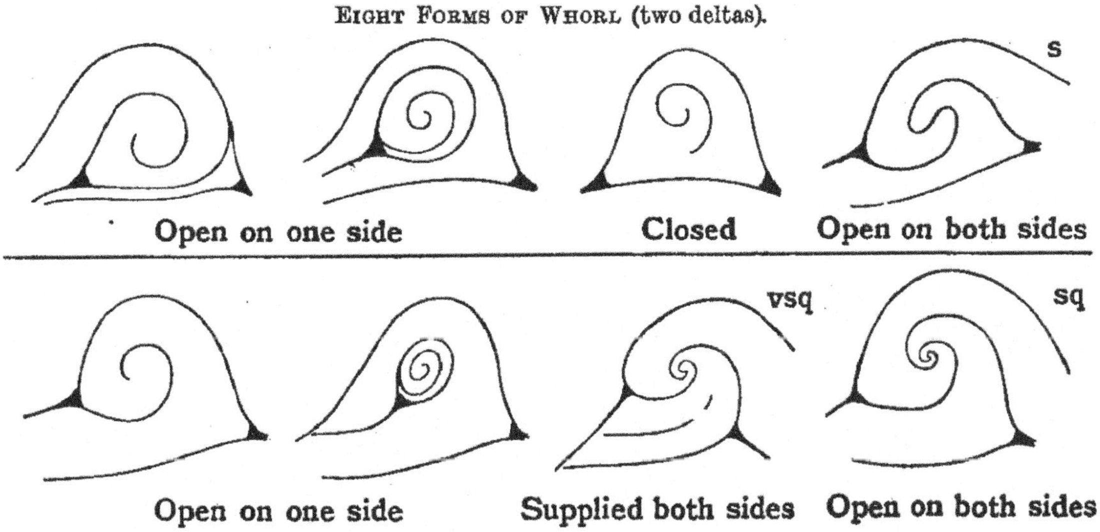

Figure 11.4: Fingerprint whorls.

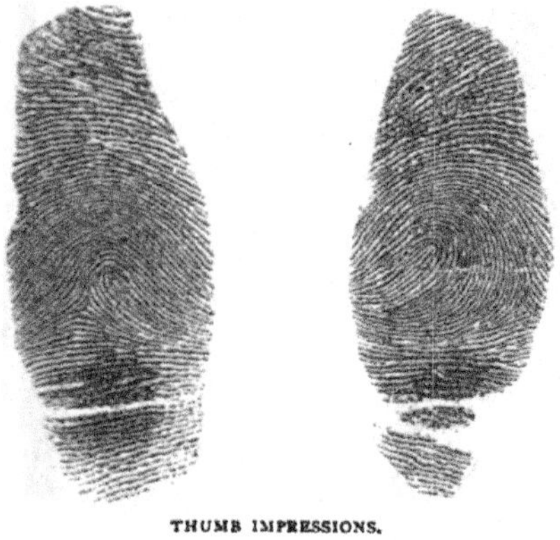

Figure 11.5: Thumb impressions from World of Wonders, *1883.*

Figure 11.6: William James Herschel circa 1865.

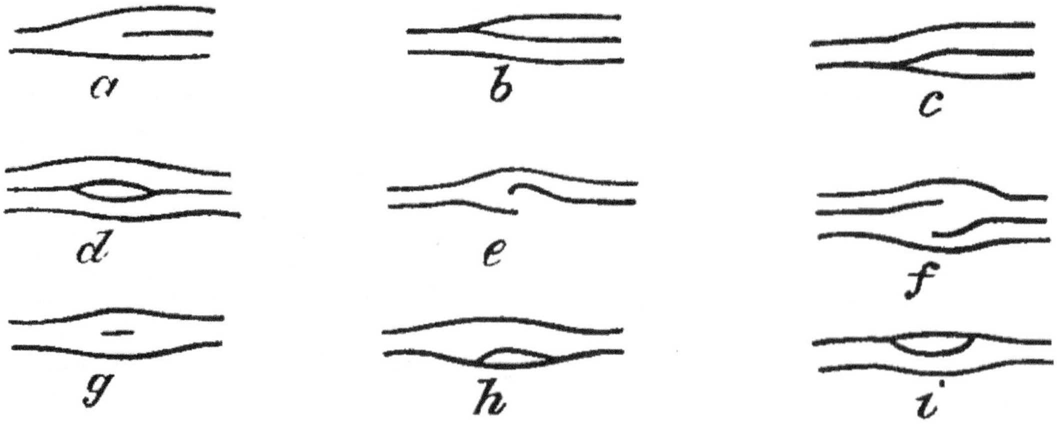

Figure 11.7: Fingerprint minutiae.

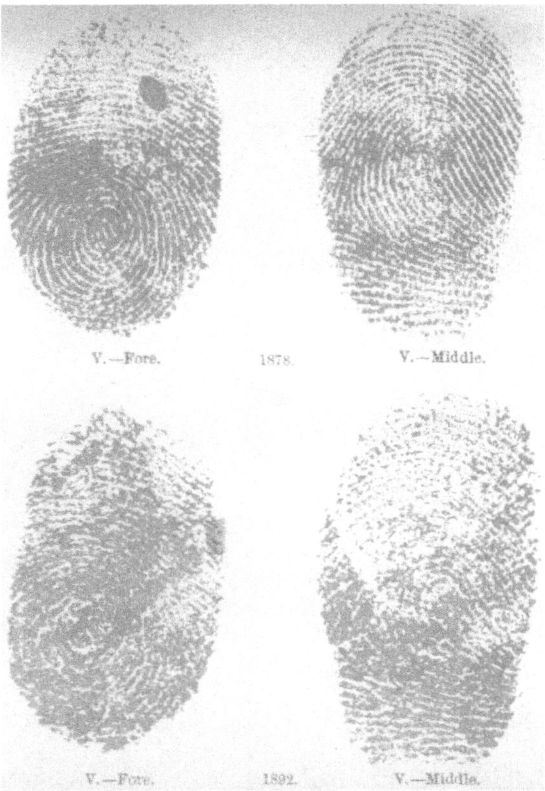

Figure 11.8: Blurred and imperfect fingerprints, Indian follow-up prints from 1892.

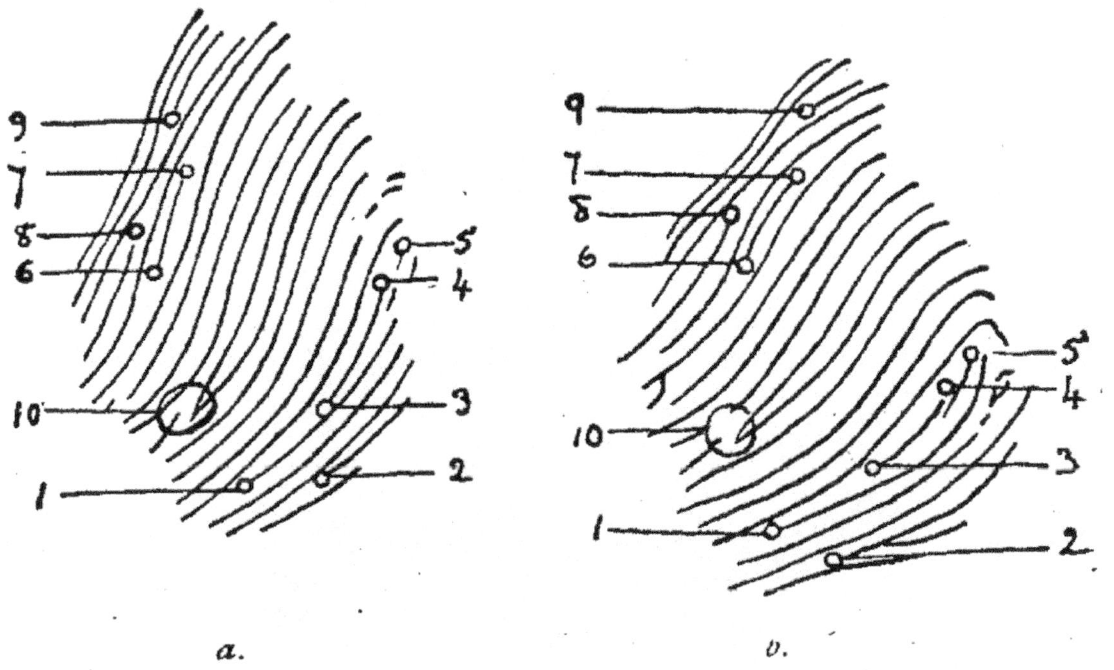

Figure 11.9: Comparison of forensic minutiae.

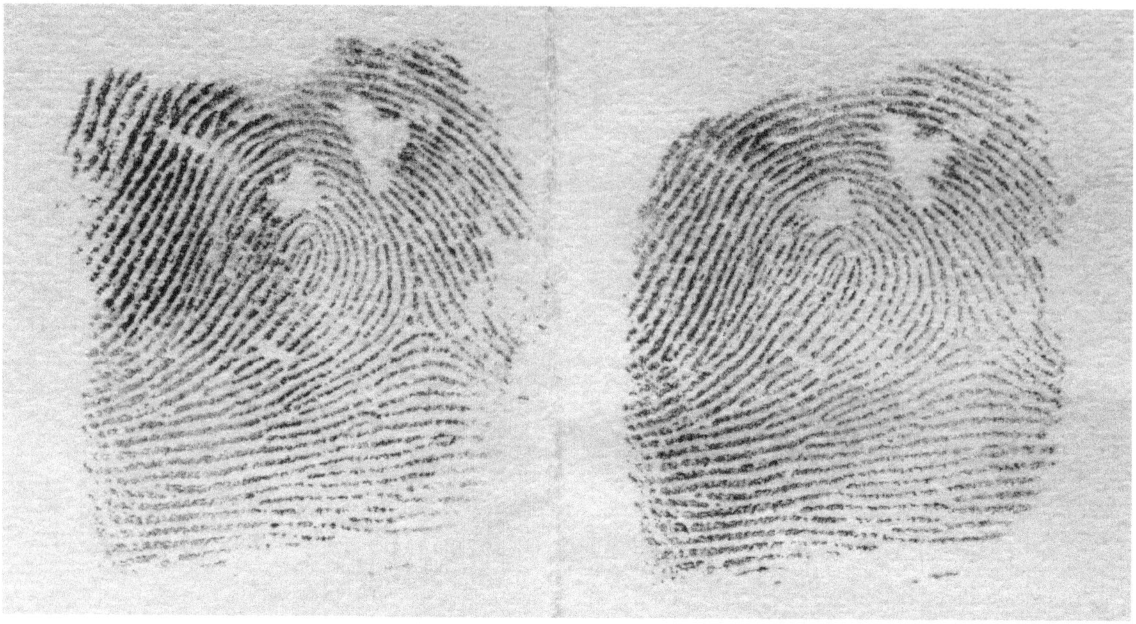

Figure 11.10: A repeated fingerprint of Francis Galton himself.

Figure 11.11: Alphonse Bertillon (1853–1914).

Figure 11.12: Dr. Henry Faulds (1843–1930) in 1920.

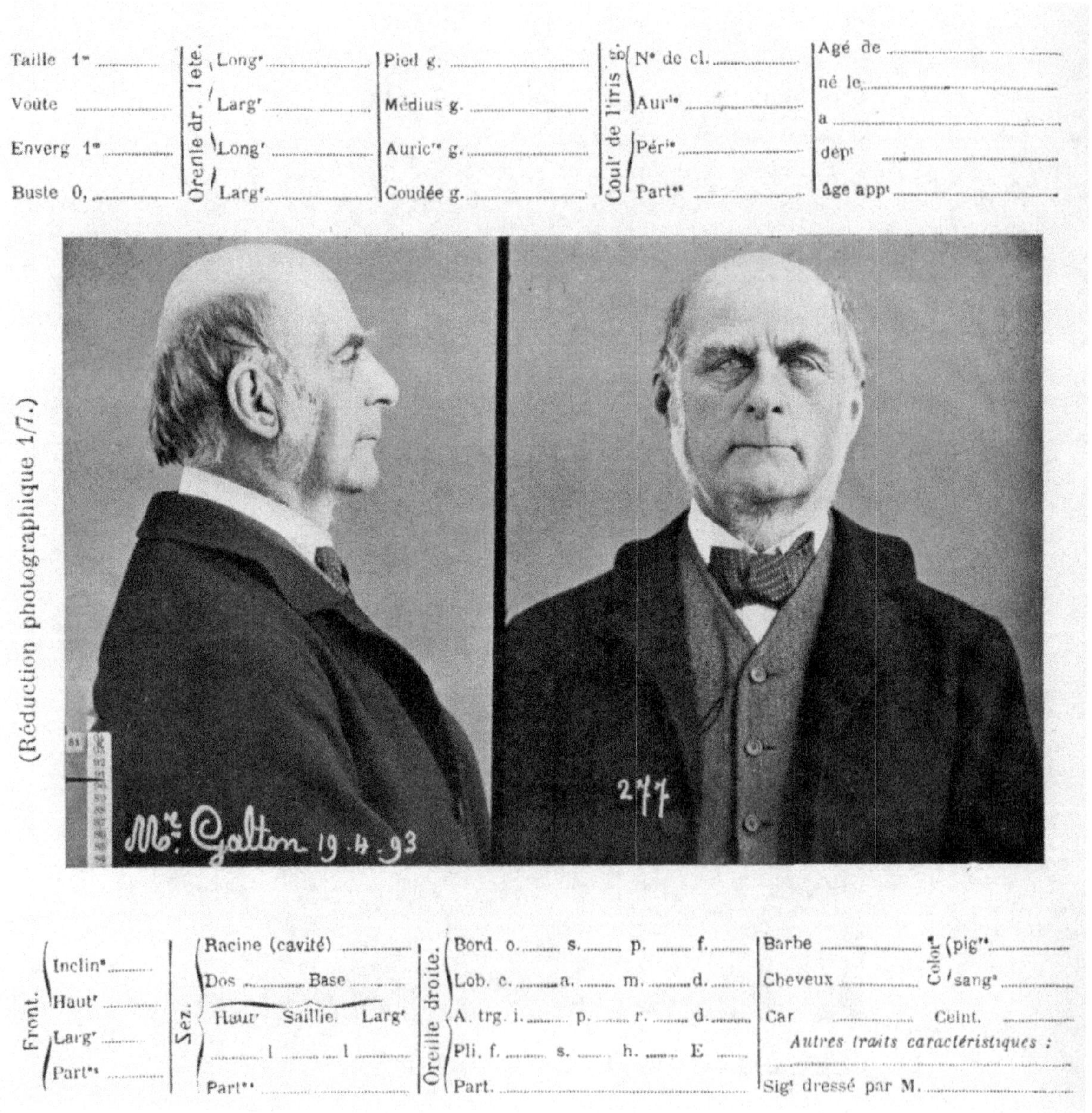

Figure 11.13: Bertillon identification card for Galton, 1893, K. Pearson 1930a.

Chapter 12

Measuring Intelligence

In 1820 the Prussian mathematician Friedrich Wilhelm Bessel (1784–1846), after whom 'Bessel functions' are now named, started working on an interesting problem. Astronomers seemed to make systematic errors when recording 'stellar transits'. Some tended to be slow, others fast. The Astronomer-Royal of England, Sir Nevil Maskelyne, had in 1796 dismissed his Assistant at Greenwich David Kinnebrook (1772?–1802?) for being slower than he was, interpreting this as a moral failing. Bessel was more forgiving, showing that all observers did this to a greater or lesser degree, giving rise to the idea of a 'personal equation' which, once measured accurately by chronograph, could be used to correct the systematic error per observer. This elevated long-dead Kinnebrook from infamy to martyrdom and greatly pleased astronomers, for whom timing is everything.

The personal equation had led to the study of mental chronometry by Frans Cornelius Donders (1818–1889), an ophthalmologist at Utrecht who recorded the time taken to react to stimuli. The Dutchman found that reacting to choices of stimuli took longer than individual stimuli, speculating that by subtracting the shorter time from the longer, one could measure the component of choice. At around the same time the Viennese physiologist Sigmund Exner (1846–1926) invented an instrument specifically to test 'reaction time' as he handily named it.

All this was familiar to Galton when he delivered his address to the Anthropological Section of the British Association at Plymouth in 1877.[1] He proposed that the 'personal equation' should not be treated merely as a nuisance but rather made a first-class object of study—the study of differences in psychology rather than universals. 'This is by no means the effect of inexperience or maladroitness, but it is a persistent characteristic of each individual, however practised in the art of making observations, or however attentive he may be'. It is 'a very fundamental peculiarity of his constitution'.

In general, quickness of mind varies and can be measured, 'the very foundations of the differences between the mental qualities of man and man admit of being gauged by a scale of inches and a clock' and while not known precisely yet are 'obviously a fundamental characteristic, and one that well deserves careful examination'. He proposed 'a simple inquiry that would throw much light on the degree in which its velocity varies in different persons'. Measuring this variation accurately, 'towards which I have already collected a few data', precedes his further speculation, 'how far it is correlated with temperament and external physical characteristics' (an idea which we have already considered at length in a previous chapter). The 'elementary judgment' involved in reaction time seemed promising, specifically the comparison proposed by Donders.

> If a simple signal be suddenly shown, and if the observer presses a stop as quickly as he can when he sees it, some little time will certainly be lost, owing to delay in nerve-transmission and to the sluggishness of the mechanical apparatus. In making experiments on the rate of judgment, the amount of this interval is first ascertained. Then the observer prepares himself for the exhibition of

[1] F. Galton 1877a.

> a signal that may be either black or white, but he is left ignorant which of the two it will be. He is to press a stop with his right hand in the first event, and another stop with his left hand in the second one. The trial is then made, and a much longer interval is found to have elapsed between the exhibition of the alternative signal and the record of it than had elapsed when a simple signal was used. There has been hesitation and delay: in short, the simplest act of judgment is found to consume a definite time. It is obvious that here, again, we have means of ascertaining differences in the rapidity of forming elementary judgments and of classifying individuals accordingly.

A compendium of such measurements of mental qualities could be collected, for example William Stanley Jevons' bean-guessing game. The economist had experimented in 1871 with instant-comprehension of the number of beans in a box, finding that he could accurately notice four at once but for larger numbers started making steadily increasing errors. Jevons had speculated that this was his own 'personal equation' and was probably different for others.[2] Galton supposed then that costly Physiological Laboratories would be needed in the future to make accurate measurements. Over the next twenty years he would actually find ways to do this relatively inexpensively.

The First Anthropometric Laboratory of 1884 seems like the natural vehicle for making reaction time measurements. In fact no such measurements were taken in that first iteration. There were only tests of fine sense discrimination, as described previously—of just-perceptible differences in weights, squareness of angles and divisions of lines. The catalogue of instruments does not mention reaction times, nor do the published descriptions.[3] Yet just two years earlier in his *Fortnightly Review* article on 'The Anthropometric Laboratory' Galton had described testing 'rate of response to stimuli' in both its simple and complex forms—only to omit it entirely from *Inquiries into Human Faculty* the next year.[4] Nearly 10,000 people were measured in 1884, but none for reaction time.

In September of the following year Galton lunched with an unusual young American, who would become one of his first major disciples, recommended to him by his friend John Shaw Billings, Surgeon General of the United States. James McKeen Cattell was finishing a PhD at Wilhelm Wundt's laboratory for experimental psychology in Leipzig, which had been founded in 1879. But the American was secretly unhappy and depressed in the unfamiliar environment there. He had been working on mental tests, following Wundt's own rather narrow interests in human universals. Now he was quickly converted to the study of differences by way of mental tests. Croom Robertson of *Mind* formed another link between the two men.[5]

After graduating in 1886, Cattell landed a 'Fellow Commoner' position (by then an awarded sinecure, rather than a paid-for privilege) at St John's in Cambridge. Galton was already referring at the Anthropological to the American's work on mental measurements, and writing to ask him about his plans.[6] Ostensibly Cattell was to study medicine at Cambridge, but was drawn instead to extending the program of mental tests that Galton had already sketched out, moulting Wundt's carapace. John Venn, Henry Sidgwick and James Ward took Cattell under their wing, and he kept in regular touch with Galton and Croom Robertson's Psychological Club when down in London.

By November 1887 an unofficial psychological laboratory had been started at Cambridge with some difficulty, allowing Cattell to start experimenting for a few months on small groups. However the American had left for a position at the University of Pennsylvania by the end of December. He would return the following year in April to give a course of lectures on psychology. There were plans to write a book on Experimental Psychology and he persuaded Galton to

2 Jevons 1871.
3 F. Galton 1884a,f,g, 1885b, 1889i. The presentation in F. Galton 1908, 248-9, misleads regarding the order of events here.
4 F. Galton 1882c, 336-7.
5 For Cattell see Sokal 1972.
6 F. Galton 1886e.

contribute a chapter to it on Anthropometry in lieu of co-authoring the text. Unfortunately the plan fell through, the first of many failed projects floated by the American. By January of 1889 he had returned to the US for good, newly married and intensely ambitious for the mental chronometry of individual differences.

Cattell's suite of extended tests would only appear in *Mind* in 1890, when he was back in America, first at the University of Pennsylvania, subsequently at Columbia. Meanwhile Galton did not have a fully satisfactory instrument for recording reaction times, and was glad as always to invent his own. He started with the design by Sigmund Exner, which involved a falling rod being arrested by pressing a button after a visual or sound signal. The distance fallen after the signal until the rod was arrested was converted to a time using a lookup table.

The reaction time test was introduced in January 1888 as part of the Second Anthropometric Laboratory (adjacent to the South Kensington Museum), which ran until February of 1891, when its land was taken over by the new Imperial Institute (later to become Imperial College). It was added to the measures of fine sense discrimination, which continued. Cattell's own measurements from that first year of the second lab survive (see Figure 12.2, p. 281). However in practice taking readings on the Exner took too long, so Galton devised a more convenient instrument using a substantial pendulum. An elastic thread attached to the pendulum was easily arrested by nippers when a button was pressed, again after a visual or sound signal. Distance travelled by the thread on a chord since that signal was converted to a time. This invention was exhibited at the Anthropological Institute in February of 1889, so one assumes that reaction times taken that year used it. The new instrument continued in use at the Third Anthropometric Laboratory (inside the South Kensington Museum, and often conflated with the Second), which ran from August 3, 1891 until its closure sometime in 1895.

While the Kensington laboratory accumulated data, Cattell was working on his instruments in Pennsylvania. An attempt to interest the public there in filing through a laboratory seems to have failed, and he would only be able to collect a large data set later, when he was at Columbia. When in 1890 *Mind* published the American's description of an expanded suite of instruments for 'Mental Tests and Measurements', Galton was disappointed.[7] The first ten tests described were not surprising. Many were already in Galton's own repertoire, as Cattell acknowledged.

I. Dynamometer Pressure.
II. Rate of Movement
III. Sensation-areas.
IV. Pressure causing Pain.
V. Least noticeable difference in Weight.
VI. Reaction-time for Sound.
VII. Time for naming Colours.
VIII. Bi-section of a 10 cm. line.
IX. Judgment of 10 seconds time.
X. Number of Letters remembered on once Hearing.

But Cattell went on to describe at least another 50 tests of discriminatory power, including 15 tests involving sight, 8 tests of hearing, 3 tests of taste and smell, 7 tests of touch and temperature, 4 tests involving 'sense of effort and movement', 7 tests of mental time, 2 tests of 'mental intensity' (including enigmatically, 'mental intensity as a function of mental time') and another 7 tests of 'mental extensity' (related to simultaneous perception). The taste and smell tests alone involved just-perceptible differences of cane-sugar, quinine, cooking salt and sulphuric acid, as well as the parts of the mouth they were tasted with. To 'effort and movement' he even suggested adding some more: 'Organic sensations and sensations of motion, equilibrium and dizziness'. There is no evidence that Cattell ever actually conducted these tests at any stage,

[7] Cattell 1890.

they seem to have been thought experiments in composition, part of his breadth-first approach to things.[8]

Some remarks by Galton follow Cattell's paper, demonstrating his dismay at the over-complications introduced by the American, returning the focus to basics and raising the issue of predictive validity through comparison with other measures. 'One of the most important objects of measurement is hardly if at all alluded to here and should be emphasised. It is to obtain a general knowledge of the capacities of a man by sinking shafts, as it were, at a few critical points. In order to ascertain the best points for the purpose, the sets of measures should be compared with an independent estimate of the man's powers. We thus may learn which of the measures are the most instructive'.[9] Galton goes on to emphasize the importance of getting the fundamental measurements right for the simple cases, describing his improvements to instruments, including the reaction time instrument put into use at South Kensington.

The Third Laboratory closed in 1895, as Frank's activity for the year in Louisa's Record explains. 'He published Finger Print Directories in May, & thus completed all he wished to say on the subject. Shortly after he closed his laboratory & gave the instruments to Oxford in charge of Prof. Thompson'. Something like 17,000 measurements had been collected, with over 7,000 containing reaction times and, one assumes, the tests of sense discrimination. No analysis of these mental measurements was published in Galton's lifetime. The ability to collect data had rapidly overtaken the ability to process it. In a retrospective, written a few years before the lab closed, Galton explained the difficulty.

> it never reached to my ideal. Besides the objects already named, I was almost equally desirous of establishing a place where the keenness of the senses and other faculties in any individual who applied, might be measured with all the accuracy and painstaking that is achieved by the few biologists who occupy themselves seriously in such pursuits. To effect this, it would be necessary to secure the occasional services of a skilled experimenter and to ensure at the same time that a sufficiency of persons should come to be measured. The time did not seem to have arrived for such an enlargement of the existing methods, though I hoped and still hope that it may not be far distant, as the utility of the laboratory becomes more widely appreciated. The measurements that have thus far been employed are of a comparatively rude but not ineffective character. It would give me pleasure at any time to receive suggestions as to new and useful special inquiries, such as might be carried on and brought to conclusion without a too serious expenditure of time and effort.

Perhaps Galton had hoped that Cattell would assume the role of a 'skilled experimenter', but with the American married and returned to the wilds of Pennsylvania, that fell through. It would be some years before the next follower came along, in the shape of Karl Pearson. Meanwhile the trove of data collected about mental abilities lay unused. Galton would be dead before the first substantial analysis of the reaction time data appeared.

After the First World War, Y. Koga and G. M. Morant obtained access to 9,000 Anthropometric Laboratory records, of which 3379 males matched their requirements and resources for analysis. They concentrated on age, acuteness of hearing and sight, and reaction time to sound and sight as measured by Galton's pendulum chronograph. The authors were under the impression that this data ranged over all of the labs, from 1884 onward, but they were mistaken as we have seen above. Their primary aim was to determine if sharper senses led to faster reaction times, which would suggest that the measure was confounded physically. Fitting cubics showed a small correlation between the senses and reaction times, while the kinds of reaction times correlated rather more closely to each other (see Figure 12.3, p. 282). Correcting for age had minor effect. 'We are forced to the conclusion that reaction times are a function of mental rather than sensory effectiveness, and must be taken as a measure rather of mental briskness than of sensory fitness'.[10]

[8] Following Cattell's biographer Sokal 1972.

[9] F. Galton 1890g.

[10] Koga and Morant 1923.

Although Koga and Morant had made good progress on validating that the measures used for reaction time conformed to basic expectations, they still did not have the independent validation criterion that Galton had himself seen the need for. Three more studies stretching into the 1930s covered other topics in more detail, like the effects of ageing, but it was not until 1985 that the full weight of modern statistical techniques was brought to bear on Galton's data.[11] The cards on which the data had been transcribed from journal entries were photocopied and digitized for computer-assisted analysis by a team of behaviour geneticists led by Ronald Johnson. They supposed that Koga and Morant had only analyzed data from the *first* laboratory, and that they were therefore the first to consider later data, but they were largely just folding the females back in from the same data set. This gave them 7,500 records in principle, but only 6,500 proved usable because of a change to the metric system after that. The split was 4,849 males versus 1,639 females.

Johnson et al were able to exploit over 400 records which were repeat measurements of the same person. Correlating these revealed the *item reliability* of the tests. The physical measures had high reliability, usually over 0.6. Due to unavoidable random errors, individual reaction times had modest test-retest correlations, in this case 0.22 and 0.26 for sight and sound of males, and 0.11 and 0.15 for females. If that seems low it is well to bear in mind that the test-retest reliability of individual items in modern IQ tests is not much higher, in the 0.2 range. Aggregation corrects this problem for IQ tests, which combine many such items. Random errors helpfully cancel out. To get a similar result for reaction-times one must repeatedly retest and average the results per individual to get a stable estimate for that person. Galton did not do that though, which is why it is only when his measurements are grouped to compare differences in means that they become well-behaved. As an example of this, one of the variables he had collected for his subjects was occupation.

Johnson et al created groupings of these into Professional, Semi-professional, Merchant-trades person, Clerical or semi-skilled, Unskilled, Gentleman, and Student-scholar categories. Age group means of reaction times for these categories, further divided into age-groups 14-25 and 26+, could be calculated. In this way analysis of variance can determine if there is a relationship between reaction time and socioeconomic class—an independent assessment of the validity of reaction time as a mental test, given that intelligence is known to be linked to class. The analysis of variance suggests that the group means for males appear to be different for 14-25 year-olds.[12] The mean reaction times to a light signal are given below for males, in 1/1000 secs.

Table 12.1: Reaction times.

Ages	N	14-25	N	26+
Gentleman	16	170	27	182
Professional	122	173	159	181
Student	1657	183	139	182
Semi-Professional	414	182	490	183
Merchant	58	190	103	183
Clerical	425	187	279	190
Unskilled	103	195	45	184
F		5.59		1.44

To get a feel for how large these differences are we need to consider the variability of the

[11] R. Johnson et al. 1985.

[12] F=5.59, $p < 0.005$ after Bonferroni-correcting 0.001 for two bites at the cherry over age groups. The smaller sample sizes for females meant that the results were inconclusive for them.

data. Incredibly, Johnson et al do not give standard deviations for their set. Nevertheless, since it is a superset of the earlier set analyzed by Koga and Morant, their standard deviation of 40.03/1000 secs for sight reaction time, which is for men only, must be close. Using this, the difference between the mean for gentlemen of ages 14-25 and unskilled workers, 25/1000 secs, is fully half a standard deviation. Which is a big advantage.

By 1895, McKeen Cattell had moved to Columbia, and began to accumulate his own data set of measurements. Like so many who would succeed him, he tested his own university students. They were to hand, cheap and uncomplaining. Each year 60 to 70 more were added to the set, yielding measurements for 'length and breadth of head, strength of hands, fatigue, eyesight, color vision, hearing, perception of pitch, perception of weight, sensation areas, sensitivity to pain, perception of size, color preference, reaction time, rate of perception, naming colors, rate of movement, accuracy of movement, perception of time, association, imagery, memory (auditory, visual, logical and retrospective)'. As a matter of course, all students at the college also had stature, weight, parentage, personal habits and health assessments recorded. Completing all the tests took 45 minutes per student. For reaction time, measured using a Hipp chronoscope of the kind Wundt had favoured, five repetitions were made for each student, but often only three were successful. By 1901 there were 256 reaction times for men, and 42 for women.

When Cattell's data set was analyzed in 1901, the technical work was outsourced to his assistant Clark Wissler, an anthropologist who had fallen under his influence at the American Association for Science and followed him to Columbia. The analysis became Wissler's PhD dissertation for the university. The anthropologist proceeded to calculate sets of correlations to see how the tests related to each other.[13] The key question he wanted to address was whether or not a *general* ability was being tapped by the mental speed (laboratory) tests, as opposed to specific abilities. Class standing—academic scale-of-five grade point average—was thrown into the mix as an independent measure of ability. The distribution of the grade point averages looked reasonably normal, as expected.

Wissler proceeded to calculate correlations between the measures, the most interesting of which proved to be that between reaction time and grade-point average. At -0.02 it was inconsequential, 'the tests of quickness seem to hold a chance relation to class standing'.[14] Tiny correlations dominated wherever the laboratory tests were involved. The conclusion seemed obvious to the investigator: tests of mental speed 'tell us nothing as to the general individual worth of college students or of adults'. Wissler also argued that the low correlations refuted the idea of general intelligence. Special abilities would have to sought as 'elements in a complex'.

Failure to find something is only interesting if it is expected to be there in the first place. Even Wissler showed some awareness that the quality of the measures used could attenuate the correlations found. 'It may be urged that in many cases failure to correlate is due to want of precision in the tests, so we must give some consideration to the validity of the methods employed'. Consideration took the form of looking at the probable error (equivalently the standard deviation) of the measures, which would drastically reduce the results if it was large.

For reaction times, 10 students with five measures were selected from the 1900 year. The average probable error was large at 12σ, but as that was 'within the range' of the overall probable error of 19σ, Wissler was not concerned.[15] An astounding conclusion to be sure! Simply retesting the same individual a few times produced nearly as much variance as picking different people from the entire population for each test. This is due to the low single item reliability of the reaction time, just as we have seen in the case of Galton's measurements, even with an average of 3 to 5 repetitions.

---

[13] Wissler 1901.

[14] Wissler naively interprets zero correlations as proof of independence throughout this paper. This is necessary but not sufficient.

[15] Here σ is the fraction of seconds unit.

In truth many more repetitions per individual must be averaged out. Wissler was at sea in the presence of error variance. 'The precision of a test may be estimated by correlating the successive trials; if the individual's own results do not correlate to a high degree, we have a test in which individual differences are relatively small as well as uncertain'. Using more repetitions did not occur to him. An additional source of error was the use of different operators over time to take the reaction times, each of whom would have introduced some variation. By contrast, every single one of Galton's 7,000 plus reaction time readings was taken by the same person, known only to posterity as Sergeant Randall.

Aside from low item reliability there are other problems here that Wissler did not understand. Cattell's expedience in studying his own students had condemned his data set from the start, since it *restricted the range* of ability. Columbia was then, as now, a highly selective institution whose pool of students had much lower internal variation of ability than the population as a whole. Selection truncated the data set to only admit very bright students, and the less capable simply did not apply. In a similar way, sprinters at the Olympic games do not include any fat people. In the general population, being overweight is a definite disadvantage for sprinting. Within the Olympic squad that relationship will be hidden, because whatever differences in body fat remain there are inconsequential. Between the restriction of range, which can be corrected for, the low item reliability of the response times, and the measurement errors undoubtedly introduced, it is not surprising that correlations between college standing and mental speed dwindled to nothingness. The only way to confirm this though is to test more representative groups, averaging the response time over many repetitions, and minimizing other measurement error as much as possible.

Cattell lacked the statistical *nous* to understand the shortcomings of Wissler's study. He wrote to Galton shortly before Christmas of 1901 to share the news. 'Please let me tell you how much I appreciate the amount of my own indebtedness and that of other psychologists to your pioneer work. My own work on Association, on individual differences and in other directions has been very directly based on yours. Have you seen, for example, Dr. Lay's Doctor's thesis on "Mental Imagery" and Dr. Wissler's on "The Correlation of Mental and Physical Traits"?'[16]

Galton's reply is not known, but it is unlikely that the Englishman was impressed. The American himself appears to have taken the negative results reported by Wissler to heart. He did no further work on mental chronometry. One could plead ignorance here in 1900, before techniques for correcting error variance and the drawbacks of restriction of range became widely known, as we shall see. By 1924 only lack of attention could explain why Cattell repeated Wissler's findings before the American Association for the Advancement of Science at Cincinnati, without any qualification: 'there was little or no correlation among the measurements or of these with the class standing of the student in different subjects'. Oddly he complained that reaction time and similar tests, which promised better information about the nature of intelligence than more complex tests, had been unfairly neglected in the interim. A few years later, when Cattell summarized the progress to date of psychology in America, Galton remained 'the greatest man whom I have known'.[17]

The shortcomings of Wissler's methodology were noticed at the time. Charles Edward Spearman (1863–1945) reviewed the study in 1904, as a prelude to his breakthrough technical formulation of general intelligence.[18] Spearman had pursued an unusual career path before becoming a psychologist. The grandson of a baronet in the Treasury, his mother endured reduced circumstances after his father died young. This steered young Charles from public school at Leamington College—obsessed with philosophy yet formidable on the playing fields—to the Royal Military College at Sandhurst instead of Oxbridge.

[16] Cattell to FG, 1901-12-21, GALTON/3/3/3/8. See Lay 1898.

[17] Cattell 1924, 1929.

[18] Spearman 1904a.

In the entrance examinations of June and July 1882, Spearman placed 5th out of 500 candidates, 79 of whom passed. His headmaster proudly wrote to the *Leamington Spa Courier* that 'Mr Spearman was a valuable member of the Football twelve and the Cricket eleven. He lived heartily the life of the school. And yet he has succeeded in gaining this brilliant position in the list, by school teaching only'. That is, without cramming. The curriculum at Sandhurst surely placed only moderate emphasis on academic attainments, since he qualified a year later in no more than the middle of the field (1979 marks versus the top scorer of 2394 and low scorer of 1550).

Joining the 2nd battalion of the Royal Munster Fusilier infantry as a cadet in October 1883, Spearman was posted to India and then Burma, where the Third Anglo Burmese War flared up in 1885. He was awarded a medal with two clasps (a special distinction), but the war would drag on fitfully for years. When not campaigning it was a life of polo, poker and gymkhanas. In 1889 he returned to England on 12 months of leave. In 1890 he qualified for the rank of Captain by examination.[19] Enrolling in the army Staff College in 1895, at some expense to himself, he graduated in January of 1897, only to resign his commission shortly afterwards. At age 34, realising that life was not long after all, he decided to forsake military sport and dabbling in philosophy for formal psychophysics. He enrolled, under Cattell's old mentor Wundt, at Leipzig—where paths could split or join more freely and philosophy was not sharply distinguished from science.

The Boer War interrupted Spearman's German studies in 1900, drawing him back in to the Army to serve at Guernsey in the English channel, to face a supposed domestic threat from France. He served as deputy-assistant adjutant-general from January to the end of April, when he resigned the position. There he met and married a Birnam native, Frances Aikman. In 1902 he returned to civilian life, moving to the Manor House near the village of Appleton, six miles from Oxford and four and half miles from Abingdon (see the photograph in Figure 12.4, p. 283, taken at this time).

The village was then in Berkshire, but is now in Oxfordshire, wandering as some villages are wont to do. 'One day, inspired by Galton's *Human Faculty*, I started experimenting with a little village school nearby'. 'This school was particularly favorable for my purpose, as it was within 100 yards of my own house; all the children and their families resided in the immediate neighborhood, so that I could easily obtain any information concerning them; the rector and schoolmaster most obligingly gave their valuable co-operation'. In truth he had already been collecting some test results on adults in Guernsey before he left there.

Spearman wrote to Galton, in search of previously unpublished material on mental imagery for a book he proposed to write on the subject, continuing and extending Galton's research.

Appleton Manor
Berkshire
28.10.02[20]

Dear Sir

I am writing a book which continues and extends your now classical researches into "mental imagery". I am therefore anxious to obtain all the works, articles, and returns published by you on this subject. I should be much obliged if you would have the great kindness to furnish me with a complete list of all such works as are not entirely included in your "Enquiries into the Human Faculty".[21]

I need hardly say that I should be especially grateful for any unpublished manuscripts. I of course bind myself to keep the names of the persons concerned quite confidential, while I shall make due acknowledgement in my book of any indebtedness to yourself (unless you prefer otherwise). I should be particularly thankful for a temporary loan of all the replies received by

[19] *Civil & Military Gazette* (Lahore) 1890-01-18, 7.

[20] Annotated in pencil: 'June 23 1903 I sent a circular'.

[21] Sic.

you in answer to your circulated "Questions on visualizing and other allied faculties", especially to the questions 11, 12, 12, & 14.

I may perhaps mention that two years ago I hoped to have the pleasure of making your acquaintance, having been invited to meet you by our common friend Sir Nicholas & Lady Elphinstone.

If I am not mistaken, you are also acquainted with Sir Richard & Lady Strachey, who are friends and indeed distant cousins of myself. I have not asked their permission to thus use their name, but do not think that they would have any objection.

yours faithfully
C. Spearman

Despite these introductions, nothing appears to have come of it. Disappointingly, Galton was away on the continent fleeing the English winter and only replied in June of 1903, annotating the letter 'Sent him a circular'.[22] A missed opportunity.

Meanwhile Spearman's attention had focused more closely. 'The aim was to find out whether, as Galton had indicated, the abilities commonly taken to be "intellectual" had any correlation either with each other or with sensory discrimination. The intellectual abilities I measured by the children's school marks in various subjects; the sensory discrimination, by a musical "dichord" of my own contrivance.[23] The reply of the experiment was prompt and decisive; all the mental powers measured did obviously correlate with each other in considerable degree'. The 'laboratory tests' for mental acuity used did *not* include reaction time. Altogether five series of data were collected by Spearman, the most important of which where the first and the fourth

I. The 24 oldest children from the village school in Appleton, without any further selection, were individually tested for discrimination of light, weight and sound. Tests were repeated many times per individual to converge to the correct personal threshold. Each sense was tested on a different day, for 15 minutes.

IV. Pupils from a preparatory school for Harrow, 33 pupils tested as a group for tone discriminations, 6 repetitions at 8 different levels. Spearman was confident about these measurements.

Spearman also obtained separate ratings for intelligence by using (i) class grades, (ii) an age-adjusted version of grades, or 'native capacity' (iii) individual ratings as 'bright', 'average' and 'dull' elicited from teachers and acquaintances (iv) separate 'common sense' evaluations by the two oldest children in the school.

At this stage Spearman was using the word 'correlation' in the non-precise conversational sense. So he worked hard at finding a numerical way to express the relationship. When he eventually read the prior literature, after a lot of solo work, he was dismayed to find out that he had long been pre-empted by Galton and others, 'and much better'. But there was another surprise, which turned out more favourably. 'Among those who had entered the field before me—although I knew it not—had been Cattell. This investigator had not only, unlike myself, made himself acquainted with the newly discovered correlational coefficients, but had even pushed so far as to apply these to mental abilities. But, to my amazement, I found that his results had been just the opposite to mine; for with him the different abilities did not appreciably correlate with one another. Had I seen his work earlier, I should certainly have thought the matter disposed of and should never have started my own work in this direction'.

That work had to continue back in Leipzig. Spearman found married life in England too expensive, so he auctioned off his household effects (including his rosewood piano) and returned

[22] Spearman to FG, 1902-10-02, GALTON/3/3/18/29.

[23] An instrument for measuring fine discrimination of tones.

to Wundt's laboratory.[24] Though he would submit his PhD thesis on 'Normal Illusions in Spatial Perception', graduating in 1906—producing papers like ' The Method of "Right and Wrong Cases" without Gauss's Formulae' and 'An "Economic" Theory of Spatial Perception'—his main interests lay in the mental testing work he had started in England.

Rumination suggested that at least one problem with Wissler's results was attenuation from random measurement error. Spearman promptly devised formulae for correcting attenuation, assuming that reliability can be estimated from repeated measurements in the same subject.[25] These methods are described in 'The Proof and Measurement of Association between Two Things', which also introduces correlation based on ranks only. Feeling that the Germans were overly parochial, and the British deficient in academic psychology, he published this and another landmark paper, "'General Intelligence," Objectively Determined and Measured' in the *American Journal of Psychology.*[26] Proofs of the formulae followed later, which showed that Spearman was repeatedly partialing out the unobserved true scores.[27]

It is clear from his description of his procedures that Spearman worked very hard to get good individual measures through deliberate repetition and convergence on thresholds, with close attention to the form of the tests and practice effects. He deliberately chose (mostly) school children to limit restriction of range, 'university students are not exactly the average from the schools but more or less a selection of the most able; hence they intellectively form a more homogeneous class, so that all their intellective correlations will be somewhat reduced in amount'. However he laid even more stress on item reliability. Wissler's weak results were, he pointed out, likely caused by 'alien elements' in the tests of discrimination, which were contaminated by the testing of three students at the same time, the use of different testers over time, and individual operator error from trying to capture too many details at the same time.

Spearman's correlations used rank-order, rather than raw values, because of the lower 'probable error' of that method. Where accuracy was critical he applied Pearson's (product-moment) method, already in common use, to the ranks; for less demanding work he used his own 'rank difference' method, which he had not yet corrected to be equivalent. He also corrected his data for age effects. For example, in data set IV he adjusted grades by taking differences between the rank of the grade and rank in age, a primitive kind of age norming.[28] In an appendix, he provided a complete listing of the data they were calculated over. This was 'open science' in 1904, but recalculation shows that there were some typos in the data given and the methods he used are ambiguous at times.[29] Hand-calculation inevitably produced some errors, but without affecting the crux of what emerged from the matrix of correlations. What intrigued the young PhD student was not just the fact that he had controverted Wissler's results for

[24] The auction notice advertised 'Brass-rail and Iron Bedsteads, Woven wire Spring, Hair and Wool Mattresses, heavy Witney Blankets, Bedding, Pointed Chests of Drawers, Toilet Glasses, Washstands, Sponge Baths, Stained Wardrobes, Toilet Sets, Four Excellent BEDROOM SUITES in Satin Walnut, American Ash, and Green Wood, Very Fine TURKEY. MIRZAPORE, BRUSSELS, and ART CARPETS, Oriental and other Rugs, Oak Kneehole Writing Table, Roll-top Writing Table, Oak Bookcases, Two Superior DIVAN CHESTERFIELD COUCHES. Spring stuffed Easy, Wicker and Occasional Chairs, Brass and Iron Fenders and Fire Implements, Brilliant-tone PIANOFORTE in Rosewood Case by the Brooklyn Piano Company, Handsome Ormolu Mounted Rosewood Curio Cabinet, Well-made Oak DRESSER SIDEBOARD. Set of 12 Oak Dining Chairs Leather. Oak Dinner Wagon. Large Oak-mounted Gong and Beater, Oak Umbrella Stand, Curtains and Window Draperies, Linen, Handsome Breakfast and Dinner Service, Dessert and Tea Services. China, Glass, Kitchen Furniture and Utensils, and other Miscellaneous Effects'. *Oxford Times* 1902-12-06, 2.

[25] If the observed correlation is r_{12} and the estimated errors are r_{11} and r_{22}, then the corrected correlation is $\frac{r_{12}}{\sqrt{r_{11}r_{22}}}$.

[26] Spearman 1904a,b.

[27] Philip 1995.

[28] Spearman 1904b. It is not always clear which method Spearman was using, complicating any reconstruction of his results from the published data.

[29] E.g. for data set I, rank order 6 appears twice because 16 was mutated into 6, and similar errors exist in data set IV.

sense-discrimination, but the additional pattern in them that stood out.

The matrix suggests that those who are good at one thing tend to be good at others (see Figure 12.5, p. 283, where 'Discrim' is discrimination of tones). Not perfectly so, but to a large degree. Those who did well at Classics did almost as well in French, and decently in Music. There is no *mathematical* reason for a pattern like this, it is perfectly conceivable that a universe exists in which those who are good at classics are no good at discrimination of tones, but it appears that these pupils do not come from that universe. The commonality in the correlations calls for explanation.

Spearman postulated that there is a latent factor, not directly measurable, that explains the commonality. He called this the latent *g Factor*, for 'General Intelligence', because (it is hypothesized) that factor shows up in all the tests used. Each one of those tests also shows something specific to its domain, since the correlations are not perfectly aligned. Which is to say that each has a *special factor* not shared by others. This came to be known as the 'two factor' theory of intelligence, but there are really *many* 'special factors' and just one 'general factor'. Although the latent general factor, which he would later call simply g, cannot be measured directly it can be revealed through its strength of association with each item in the matrix of correlations. Spearman calls that the 'saturation' of the item by the general factor (see Figure 12.6, p. 283). Some items are much more saturated (or 'loaded') than others.

Spearman had made an inspired leap here from his deattenuation formula for correlation. The true correlation cannot be directly measured, but it can be estimated using the repeated imperfect measures of it, correcting for the error. In the same way, a collection of sub-test items can be thought of as imperfect repeated attempts to measure the same underlying faculty. Superficially they look like different tests, but in practice they test the same thing. The correlation between a sub-test X and a collection representing 'general intelligence' (say the collection of school class ranks from data set IV) can be calculated by averaging its pairwise correlations involving X over that collection, and then correcting that for the measurement errors, estimated by internal pairwise correlations within the 'general intelligence' collection, and whatever error we can estimate for measuring X.

This was the first 'factor analysis' of a collection of tests, but the method used would hardly be recognizable to modern readers and practitioners. Although his paper is often referenced, it is seldom read closely. It would not be long before his methods were reworked to use much more general techniques from linear-algebra. The fact that he was able to obtain his results without those techniques shows how clearly he grasped the problem he was trying to solve.

In 1905 the British Association launched an Anthropometric Survey of the United Kingdom, following a suggestion by Galton. Part of this ambitious effort entailed a Psychological Subcommittee to survey mental characteristics, which was taken in hand by William McDougall. After distinguishing himself in the Natural Sciences Tripos at Cambridge in 1894, McDougall had accompanied the W. H. Rivers Torres Strait expedition of 1898, where he had collected anthropometric measurements using some of the equipment designed by Galton, whom he had come to know well. The psychologist became the Wilde Reader in Mental Philosophy at Oxford in 1904 after a spell at Göttingen under G. E. Müller, and at UCL under James Sully (where he retained a part-time post until 1907).

For the British Association survey McDougall enlisted the help of his most promising student, Cyril Lodowic Burt (1883–1971), who had enrolled at Jesus College in October of 1902. Burt was officially studying 'Greats', but keenly involved himself in the Physiological laboratory of J. S. Haldane (1860–1936).[30] He assumed Psychology as a 'special subject' under McDougall, who was impressed by the student's enthusiasm for Galton's ideas. As it happened Burt's physician-father, who had retreated from London to practice at Snitterfield in Warwickshire, knew the Galtons well through Francis' elder brother Darwin Galton. As Burt later

[30] Father of J. B. S. Haldane (1892–1964).

recalled:[31]

> When I was a boy, my father, who was a country doctor, used to take me on his rounds to visit his patients. Of these by far the most impressive was the squire of a neighbouring village called Claverdon, a small Warwickshire hamlet, lying between Stratford-on-Avon and Birmingham. The squire was an octogenarian named Darwin Galton. On his mother's side he was the grandson of Erasmus Darwin, physician, zoologist, botanist, and poet, and on his father's of a Birmingham Quaker who had made his money by manufacturing rifles. He himself was the eldest son in a family of nine. And, when he was fit, I was sometimes taken up to see him, and so met many other members of the same family, including the youngest, called Frank or Francis.
>
> Francis Galton was one of the most distinguished-looking people I have ever known—tall, slim, neatly dressed, with a forehead like the dome of St. Paul's. My father, who was an ardent Darwinian, used to try to inspire me with intellectual ambitions of my own by telling me of the remarkable investigations carried out and published by this exceptionally brilliant member of an exceptionally brilliant family. I met him again as an undergraduate at Oxford, and later in London shortly before his death.

Burt found his Oxford mentor had an 'almost Olympian' manner in public, and in private an 'air of lonely magnificence'. Though McDougall was no statistician, he could see the revolutionary significance of the new mathematical methods, so he invited Karl Pearson to address an Oxford student club on correlation. Burt was able to meet Galton's leading acolyte. The student was set to work on standardizing the psychological tests to be used. McDougall would describe him as one of his 'brands plucked from the burning and turned into the channels of productive research'.[32]

After Spearman returned to England from Germany in 1907, he was also pulled into the survey by the psychologist, who had been impressed by his work on intelligence. Moreover McDougall ensured that Spearman filled his own surrendered position as Sully's assistant at UCL. According to Burt, 'On McDougall's lawn at Boar's Hill Spearman and I used to meet and submit tentative ideas to him for criticism and suggestion'.[33]

Between 1907 and 1909 Burt worked on testing school children in an attempt to reproduce Spearman's earlier results, using a mixture of sense-discrimination tests with academic results for comparison. Subjects came from an elementary school and Lynam's Preparatory School, now known as the Oxford 'Dragon School'—the same school that J.B.S. Haldane attended.[34] One of the pupils tested was Karl Pearson's son Egon, later to become a distinguished statistician in his own right, which provoked interest from Galton's closest confidante. One of the more successful tests employed was a device invented by McDougall himself, a 'dot following' exercise. The matrix of correlations showed a similar pattern to that found by Spearman, with all tests showing moderate to high correspondence, suggesting a general factor.[35]

Burt detected that there were groups of moderate residual correlations in his data not explained by the general factor, proposing that there was a more complicated hierarchy than Spearman had supposed. 'The tendency for subordinate groups of allied tests to correlate together is discernible, but small'. There were 'distinct signs that the sensory tests are specially connected'. This suggested a hierarchy of factors, from the general factor through to group factors and down to special factors, a theme he and other authors would elaborate on extensively

---

[31] C. Burt 1962.

[32] C. Burt 1952; McDougall 1930.

[33] Burt suffers from some confusion here, stating that Spearman 'settled down for a while in a village near Oxford. He too had become interested in the Galton-McDougall scheme, and had started testing boys in the local school'. That was in 1902, before McDougall was at Oxford.

[34] Tredoux 2018.

[35] C. Burt 1909.

in coming years.[36]

James Ward, the editor of the *British Journal of Psychology*, was reluctant to publish Burt's paper, thinking that interest in mental tests would soon fade. 'A long and suggestive correspondence ensued, first with Pearson and then with Spearman, to both of whom Ward sent my manuscript. They, however, criticized each other even more vigorously than they criticized me'. One way to placate Ward was to leave out a detailed mathematical appendix, placed on file at the journal, trimming the paper down to just over 80 pages. Galton sent Burt 'an encouraging note'.[37]

Despite the promising results from Spearman and Burt using sense-discrimination tests, and Spearman's incisive diagnosis of Wissler's deficiencies, a more recent development soon starved discrimination and reaction-time tests of oxygen.

The French psychologist Alfred Binet (1857–1911)—like Galton, independently wealthy—had long been interested in cognitive abilities. Galton had been reading Binet since the mid 1880s. In an exchange with Max Müller on 'Thoughts without Words' he referred the German to Binet's study on mental imagery *La Psychologie Du Raisonnement*.[38] They shared an interest in calculating prodigies and the extent to which these rely on mental imagery for their operations.

In 1891, Binet had written to Galton to ask for copies of composite photographs, to use in the Physiological Psychology Laboratory of the Sorbonne, which Henri-Étienne Beaunis had entrusted to him for the winter.[39] Galton replied graciously in the new year, complimenting the Frenchman's book on mental imagery. Binet, by this time evidently in full charge of the laboratory, responded in kind, 'I am very happy that my little volume on reasoning meets with your approval—it would be very useful for us to keep in touch with you, our laboratory has much to gain from it, for I know that you have devised many ingenious methods of studying the sound man from the psychological point of view'.[40] Now the Frenchman was emboldened to ask for a photograph of Galton himself for the laboratory.

Galton had been testing 'mental arithmeticians' intermittently since the early 1880s. It was a family tradition of sorts. In 1817, his father Samuel Tertius had tested his servants for their powers during a national vogue for mental arithmetic. A 10 year old servant girl Harriet Bromley revealed an unsuspected talent for doing complex sums in her head, a story that young Francis had ingested as part of the family lore. The Bidder family of calculating prodigies had figured prominently in the mental imagery investigations incorporated into *Human Faculty*.[41]

In 1886 Galton had visited an 'Asylum for Idiots' at Earlswood, where he had toured a few years earlier with George Romanes, and another at Darenth, near Dartford. Now he was with Prof. Alexander Bain (1818–1903), editor of *Mind*. 'Span of prehension' (now called digit span) tests were given to a handful of ordinary female inmates, who were asked to recall strings of random numerals in the order read to them. Predictably they did rather poorly, though better at Darenth, where the inmates were 'merely imbecile'. But three male inmates at Earlswood had remarkable memories for historical dates, which they enjoyed memorizing from books, even though their digit span was no better than others.[42] These cases were unlike the otherwise-ordinary French calculating sensation Jacques Inaudi, who had been marvelling audiences since 1880, to the extent that he was formally investigated by a committee of French scientists led by Charcot in 1892.

---

[36] Among many other misrepresentations, Hearnshaw 1979 gets most of the history of this paper entirely wrong, even going so far as to claim that Burt did not examine residuals. Merely reading the original paper refutes this. See also Fletcher 1991; Joynson 1989.

[37] C. Burt 1952.

[38] F. Galton 1887e; Wolf 1973.

[39] Binet to FG, 1891-11-25, GALTON/3/3/2/16. None of Galton's replies seem to have survived.

[40] Binet to FG, 1892-03-19, GALTON/3/3/2/16.

[41] GALTON/1/1/7/1. See Tredoux 2023c and chapter 9, p. 159.

[42] F. Galton 1887d.

The following year Inaudi was publicly exhibited in England. His handler bore a letter of introduction from 'Professor' Binet to Galton, and the French committee's report. 'I recommend to your scientific curiosity, as well as to all your good will, Mr. Jacques Inaudi, a young calculating prodigy, on whom I published a study about a year ago. M. Jacques Inaudi, who is indeed of the family of Mondeux, Bidder and Colburn, shows a novel psychological character; in his operations of mental calculation, he does not see the figures, he hears them and pronounces them'.[43]

After discreetly assembling a local committee at a Royal Society *conversazione*, assuring the strictest confidence for all, Galton received Inaudi and his handler at Rutland Gate on the morning of May 11, for a *séance*. The scientists, including the biologist George Romanes and the physicist Oliver Lodge (1851-1940), examined the 'most marvellous' boy's mental powers.[44] They found that he had a 'prodigious memory for figures', that he could easily perform feats like calculating cube roots of seven figure numbers in his mind, but (somewhat obtusely) that his arithmetical abilities did not exceed those of an able bank clerk once the time saved writing things down was accounted for. In his mental operations he did not seem to operate visually, relying rather on sounds, with a special dislike for the number seven. 'He had that wearied look in the eyes & brow that I have seen in other boy calculators', Galton noted.

In 1894 Binet published *Psychologie des grands calculateurs et joueurs d'échecs* (Psychology of mental arithmeticians and blindfold chess-players), which covered the non-visualiser Inaudi and a Greek calculating prodigy who did visualise, Diamandi. Galton reviewed the book favourably in *Nature*.[45]

Interesting though the boy calculators were, the testing performed on them was not systematic and lacked standards of comparison. The prodigies were inherently very rare. By 1899 Binet had acquired a collaborator, Théodore Simon (1873–1961, sometimes given as 'Théophile ' for unclear reasons). They worked on more general tests of mental ability, motivated by the need to identify backward schoolchildren and the feeble-minded.

In 1893, Binet had already contacted Galton about mental tests. 'I venture to ask you if it would not be possible for you to point me to the precise list of your articles where you have dealt with the subject of Mental Tests. I believe that this question is very important, and I am currently studying it in the laboratory. You would give me great pleasure by telling me where I can find your written thoughts'.[46]

However the tests that Binet and Simon were investigating were now mostly composed of many tasks, each far more complex than Galton had used, though weight discrimination was also included. By 1905 the Frenchmen had realised that the performance of a child was relative to the age group to which it belonged—Spearman had made an equivalent correction for age the previous year, as we saw above.[47] Therefore the test items needed to be made increasingly complex by age group. A child of three would be asked to point to their eyes, nose and mouth. A five year old asked to order two weights, copy a square, count some small amount of money. A seven year old could be asked to complete the missing features in simple pictures of people, count fingers, copy a diamond figure, or repeat five numerals. An eleven year old could be asked to react to absurd sentences, define abstract terms, reorder a scrambled sentence and so on.

Binet and Simon implicitly adopted Galton's principle that a general ability underlay performance on disparate tasks. However the upper end of the ability spectrum was of minimal interest to them. Using a sample of some 200 Parisian schoolchildren, they established norms for ascending ages which allowed the level of performance (later called the 'mental age') of

[43] Binet to FG, 1893-04-20, GALTON/3/3/2/16. Galton's notes are wrong here, as Binet was not then a professor at any institution.
[44] Schwartz 2010.
[45] F. Galton 1894c.
[46] Binet to FG, 1893-02-10, GALTON/3/3/2/16.
[47] See Spearman 1904a.

a child to be determined. In 1908 they introduced a total score computed from the battery of tests to simplify this comparison. If a subject passed all but one of the test items for an age group, the mental level awarded is the highest such age group. In addition, for every five additional test items passed from even higher age groups, an additional year is added to the mental level (e.g. ten extra adds two more years). The method was extended to apply to adults by obtaining norms for them in a similar way.[48]

Neither Binet nor Simon appear to have fully grasped why their method really worked. By combining a large collection of sub-tests, each of which had modest individual reliability, an aggregate score with high reliability could be found. The *difference* between the age norm and the actual performance of the child gave a measure of relative backwardness. Application of the method quickly found that it now had decent predictive validity. It spread rapidly to the USA through Henry Herbert Goddard (1866–1957) and after that to England through Cyril Burt.

In 1912, a year after Binet succumbed to a stroke, William Stern (1871–1938) proposed *dividing* the mental age of a child by its chronological age to give an Intelligence Quotient (IQ). Multiplying the fraction by 100 simplified its use. Thus an on-par child would be scored 1x100=100, which is the origin of the modern IQ mean of 100. This was something that Simon himself was never comfortable with, but it gave a convenient way of comparing children and adults regardless of age and was widely adopted. Modern IQ scores dispense with this entirely, using pure intercomparison within the age distribution to get the scores we are now familiar with. The distribution is centred for continuity at 100, most commonly with a standard deviation of 15.

The success of the new IQ-style tests meant that tests of sense discrimination tended to recede into the background, though they continued to be used sporadically. Soon the lore of mental testing would maintain that Clark Wissler had consigned them to the dustbin, a claim that one can still find confidently asserted in the secondary literature.[49]

Wissler's name is often conjoined with another study by Stella Emily Sharp, published in 1899. Sharp considered seven (!) subjects for her Cornell PhD thesis. After a literature review she merely assumed what others have supposed she proved. 'The theory was provisionally accepted that the complex mental processes, rather than the elementary processes, are those the variations of which give most important information in regard to the mental characteristics whereby individuals are commonly classed. It is in the complex processes, we assumed, and in those alone, that individual differences are sufficiently great to enable us to differentiate one individual from others of the same class'. No correlations with external criteria were calculated. Few who reference Sharp could possibly have read her.[50]

Reaction-time and sensory discrimination studies of intelligence continued, but simply got far less attention. In 1933, Lester F. Beck (then at Brown) reviewed 28 studies relating reaction time to IQ tests. Of these 14 involved either simple or discriminative reaction time (requiring a decision) yielding a range of coefficients, —.32 to .90 with a median of .16. Another 14 concerned serial reaction time (requiring spatial location), yielding correlations between —.03 and .53, with a median of .18. However, meta-analysis, which takes sampling error and test reliability into account, would only be invented in 1975, so this review is weak.[51]

It was only in the late 1960s that reaction time was systematically revived using more modern methods. Hans Eysenck recalled noticing the deficiencies of Wissler and Sharp in the late 1960s and suggesting to Arthur Jensen that the subject could bear re-examination.[52] Jensen soon established that, provided enough repetitions of tasks like reaction time were made, say

48 Binet and Simon 1916.

49 Deary 2000.

50 Sharp 1899. For examples of this repetition see Sokal 1972, Sokal 1981, 338, Zenderland 1998, 91, Fancher 1985, 48, Wasserman 2012, 11. Many more could be given, see Deary 2000.

51 Beck 1933; F. L. Schmidt and Hunter 2004.

52 Eysenck 1998.

fifty, they yielded reliable averages. As long as the ability range was not restricted to (highly-selected) university students, the physical tests showed good predictive power, correlating substantially with traditional IQ tests at around 0.50.[53] They remain an active field of investigation today, as an attempt to *understand* the nature of intelligence at a more primitive level, in terms of physical efficiencies, as opposed to only using given IQ scores as predictors. This was a distinction already sharp in Spearman's early vision of 1904. But Spearman himself soon moved to the new tests. Factor analysis and understanding the general factor of intelligence g only required correlated subtests, and the Binet-style tests worked just as well for that.[54]

There are interesting parallels between these developments and Galton's investigation of physical correlates of intelligence. An Anthropometric Laboratory was established in 1885 at Cambridge, housed in the library of the Philosophical Society. It was run by John Venn along the lines that Galton had proposed in his Rede Lecture the previous year, using instruments that he had donated. The measurements were made by the librarian. Eyesight, height without shoes, weight, strength of pull and squeeze, and breathing capacity were included among these. Reaction time was not tested. Instead physical head measurements, which Galton had regretted omitting before, were added, as they were at the Second and Third Kensington laboratories. It was natural to suppose that more intelligent people had physically larger heads with larger brain volume. Galton had entertained this idea for many years. At meetings of the Royal Society it appeared to him that the more eminent men had more voluminous heads. As we have seen, he had collected hat sizes for his scientists in *English Men of Science*, and concluded that the scientists with smaller heads were more energetic. But he lacked an objective measure for energy and could make no links to intelligence beyond his personal impressions. He had never been able to obtain better data to analyze the hypothesis.

Now Venn was able to obtain head length, breadth and depth for over a thousand Cambridge students, mostly undergraduates, measured at the Lab between 1885 and 1887. Special instruments for the head measurements were manufactured by the Cambridge Scientific Instrument Company, known as 'Horace Darwin's Shop'. This had been founded in 1881 by yet another son of the naturalist, who shared Galton's fascination with precision instruments and ran the company for many years.[55] Head depth was tricky to measure, and Galton never included it in his Kensington Laboratory repertoire. Analysis of the Cambridge data started in late 1887. More than 25,000 calculations were required (rather than working out all the volumes individually and then averaging them, the averages were calculated first and then multiplied as an approximation). To get at differences in ability, the men were sorted at Venn's direction by those closest to the students at the colleges. 'I am getting the tutors to sort out for me into 3 classes: viz. High honours, Low honours and poll or pluck'.[56]

Venn published a general analysis of the Cambridge data in 1888, comparing the results with the equivalent data collected for the general public at the First Anthropometric Laboratory in 1884. The students were on average a full inch taller and ten pounds heavier, with stronger pull and squeeze. They also had markedly larger breathing capacity than the common man. Venn also compared the students by ability class for the physical measurements taken. The brightest seemed slightly more myopic, and they had weaker strength of squeeze. But the overall brawn advantage did not necessarily replace brains since the clever men had larger heads too. Other measurements showed no significant differences.[57] Galton appended a note analyzing the findings for head volume and ability in more detail. These were compared by age, 19 to 25 and up (see Figure 12.7, p. 284).

[53] Jensen 1998.

[54] Spearman 1927.

[55] Horace Darwin (1851–1928), Fellow of Trinity College. His daughter Nora Barlow became a noted chronicler of the family.

[56] Venn to FG, 1887-11-07, Verburgt 2022, 183.

[57] Venn 1888.

It appeared that at age 19 'high honour' men had roughly a 5% advantage in head volume over 'poll or pluck' men. By age 25 this advantage was reduced to about 2.5%. Throughout the age range 'high honour' men seemed 'more precocious and gifted'. Precocious because their heads appeared to have grown more quickly, after which their competitors narrowed the gap without erasing it. Galton drew graphs over the age range for each class, which he interpreted as good evidence for his conclusion (see Figure 12.8, p. 284). However he was well aware of the weaknesses in his data. There were far fewer cases for some age ranges, e.g. for ages 24 and over-25s.

These conclusions soon elicited a letter to the editor of *Nature* from somebody at Cambridge initialled 'FMT'. The author doubted if the small differences detected between classes of student could be reliable given the margin of error in individual measurements of the head. He estimated these to be '5 or 6%' from ten trials he had conducted at the laboratory by having two friends measured there repeatedly. Galton immediately responded (correctly) that the process of averaging over a large number of cases obliterated the individual measurement error in question.[58] Nevertheless he undertook to check the results using additional measurements accumulated since publication. The force of this response was lost on another correspondent, 'HJP' from Trinity College, who pressed exactly the same argument as 'FMT' a few months later, submitting his own variable measurements and offering the additional theory that hair in various stages of growth explained the differences.[59]

In 1890 Venn and Galton followed up with another analysis of the Cambridge data, which had doubled in size, with measurements for 2,000 individuals. The full data set confirmed the previous results, although it no longer appeared to be the case that more able men were weaker of sight. The average least able man was 4.6% stronger on the pull than the average most able man, but Venn was prepared to attribute this to more frequent exercise in the absence of academic interests. Galton confirmed that able heads were still larger earlier than the less able (see Figure 12.9, p. 285). Since more than a few students had been measured multiple times, the measurement error could be estimated—'one man, who seems to have had a morbid love of this physical inspection, has actually had his various dimensions and capacities tested no less than eighteen times during the course of some three years'. The 'probable error' of head volume was five cubic inches, which was reduced to less than one inch in the average, making the group A difference from C of seven inches highly unlikely to be a chance occurrence.[60]

At the Second and Third Anthropometric Laboratories in Kensington, between 1888 and 1895, Galton collected head breadth and length, but not height, of the general public. Since he had no direct ability measures, the only comparisons which can be computed are by occupational and social class. Using the data compiled in 1985 from the preserved records we see the following:

Table 12.2: Head Size by Class.

		N	L	B
Prof	14-25	122	7.66	6.01
Prof	26+	259	7.77	6.10
Semi	14-25	414	7.62	5.96
Semi	26+	490	7.72	6.06
Merch	14-25	58	7.62	5.99
Mech	26+	103	7.75	6.10
Clerical	14-25	425	7.58	5.93
Clerical	26+	279	7.69	6.05

[58] If there are n independent and identically distributed measurements included in an average, each with variance σ^2 then $\mathrm{Var}(\frac{1}{n}\sum x_i) = \frac{1}{n}\sigma^2$. This means that averages of large numbers of observations are highly resistant to individual measurement errors.

[59] Anonymous 1889b,c; F. Galton 1889a,b.

[60] F. Galton 1890c; Venn 1890.

Imputing male head volumes from length and breadth alone, given that height was not measured, the lowest socioeconomic grade has approximately $1,324\text{cm}^3$, while the highest grade has about $1,468\text{cm}^3$. However for women there is much less difference, $1,256\text{cm}^3$ compared to $1,264\text{cm}^3$. This assumes that lower classes did not on average have unmeasured *advantages* in head height, which is reasonable.[61]

Sheer head volume was not Galton's only interest. He had always been interested in physiognomy. A collection of portraits of Senior Wranglers at Cambridge suggested a distinctive pattern. In just over half of the portraits he noticed 'a form of forehead adapted to carry long, plain and strongly marked eyebrows'. But he was not dogmatic. 'It would be interesting to learn how far the independent judgment of others may confirm this conclusion'.[62] This speculation did not lead anywhere, but interest in head volume continued. Before the end of the decade, Galton's acolyte Karl Pearson picked up the topic again.

Pearson was skeptical from the beginning about the links that Galton had found between ability and head size. He encouraged Alice Lee, one of his research assistants at UCL, to investigate head volume using regression techniques. Living heads could not be filled with water or lead shot to determine their volume. However the measured volume of skulls could be regressed on head length, breadth, height, cephalic index and other predictors. The resulting formulae could be used to infer living brain volumes from feasible measurements. As part of this Pearson and Lee estimated volumes for leading anatomists and others. Ranking these by eminence, it was not obvious to them that there was any relationship to their brain volume, but they did not make any calculations about that, probably because it was hard to formalize eminence as a measurement.

Lee's work on skull measurements was submitted to UCL for a D.Sc. degree in 1899 but quickly ran into some obstacles. The examiners, who were mathematicians, shrewdly suspected that Pearson had supplied the chief ideas while Lee had taken on the labour of the calculations. The requirements of originality for the degree were not met. Galton, as a Royal Society man and an expert in the subject matter, was called on to prepare a report. This placed him in a tricky position. He did not sympathize with Lee's informal deductions about the link between head volume and ability, but that was just an aside within her thesis, not the main point. Pearson pushed strongly for the thesis to be passed anyway, not for its mathematics but for its science. He admitted that he had supplied ideas but maintained that Lee had done more than enough herself to deserve the degree.

Trusting his industrious collaborator fully on technical matters, Galton was bound to yield. He prevailed upon the UCL examiners after some adroit manoeuvres. Lee was granted the degree by late October. 'Sir H. Roscoe told me last night that Miss Alice Lee had got her degree. The mathematicians were however troublesome'.[63] The thesis was published less misleadingly as a joint paper with Pearson in 1901. It was soon followed by a more direct examination of the correlation between head volume and ability.

Pearson had already showed his hand about brain size in 1900 with the second edition of his positivist manifesto *The Grammar of Science*. He had theoretical reasons for doubting its importance. Here he was responding to Alfred Russel Wallace's skepticism about the natural selection of intelligence. 'His whole argument ... with regard to the brain turns upon its size, whereas it appears that it is the complexity of its convolutions and the variety and efficiency of its commissures rather than its actual size, which we should psychologically expect to have

[61] R. Johnson et al. 1985; Rushton and Ankney 1996.

[62] Note in a collection of portraits in the Wren Library Trinity College, dated 3 June 1894, Craik 2008, 132.

[63] FG to Pearson, 1899-10-27, K. Pearson 1930b, 515. There is a full account of this kerfuffle in GALTON/2/10/6 with further letters in GALTON/3/3/16/9/4. Pearson went to great trouble on Lee's behalf. He arranged for Galton to interview her, but behind the scenes prepared her personally on the best way to handle him.

grown with man's civilisation'.[64] Early in 1902 he published a preliminary report based on an examination of 1,000 students measured at the Anthropometric Laboratory at Cambridge and some data about a similar number of school children.[65]

The biometer had already broken the news to Galton the previous month. 'I have always felt we must go into the point more fully, since you laid stress on the view that ability was correlated with the size of the head in your criticism of Dr Lee's paper. There is still a chance that extreme genius may exhibit something abnormal in the size of head, but I think it is now pretty clear, if we are to look upon ability as normally distributed in the population, there is only a very small, practically negligible correlation between it and either the size or shape of the head'. Galton was not convinced.[66]

> The non-correlation of ability and size of head continues to puzzle me the more I recall my own measurements and observations of the most eminent men of the day. It was a treat to watch the great dome of Sylvester's head. William Spottiswoode was another of the 5 or 6 largest; so was that encyclopaedic physiologist Prof. Sharpey. That most accomplished & many-sided official, Sir John Lefevre (formerly a senior wrangler), was the largest of all. Gladstone's head, which I myself measured, was very large. Again, comparatively the other day, I was one of a deputation of physicists to the Treasury about the National Physical Laboratory and sitting behind the front row I marvelled at their skulls. Lord Rayleigh, Stokes, Lord Lister, Lord Kelvin were all remarkable partly perhaps owing to the powerful moulding of their heads, irrespective of size. A Frenchman collected the recorded weight of brains of many eminent people and published them in one of the French anthropological periodicals many years ago. They contained remarkable weights. However I can say nothing against the validity of your results. ... Another of my 5 or 6 large heads was Admiral Sherard Osborn. He was very broadset. Also he was considered generally to be the ablest man of his day in the Navy and the accepted mouth-piece of reform. (He died of heart spasm while still young.)

The Cambridge students were categorized by ability into the usual Honours versus Poll distinction. Handling these discrete categories required more complex methods than product-moment correlation, which assumes continuous measures. Pearson used an ingenious technique of his own devising that later became known as 'polychoric correlation'. It assumes that there are latent normally distributed measures underlying the observed discrete measures. The observed categorizations are used to infer the most likely parameters for the latent normal distributions, from which the correlation coefficient can then be calculated. The results were confirmed using the schoolchildren who were separated by sex and then classed more finely by their teachers as 'Quick-Intelligent, Intelligent, Slow-Intelligent, Slow, Slow-Dull, or Very Dull'. Height of head was also available for the children, who were now over 2,000 in number. Since ability did not increase with age, and boys were not more variable than girls, we may infer (though Pearson did not) that the rating was in practice relative to age and sex. In 1906 the full report with a slightly more elaborate analysis was published in *Biometrika*. The final correlations are shown in Figure 12.10, p. 285. Overall Pearson concluded that 'Very brilliant men may possibly have a very slightly larger head than their fellows, but taking the general population there is really a very insignificant association between size of head and ability'.

This analysis, with its large sample size, was thought by Pearson to be definitive, but there are in retrospect some obvious limitations. There is restriction of range in the Cambridge data. The students were at the top end of the ability distribution of the general population. The schoolchildren are somewhat more representative, but if the relationship between head size and ability firms up with age, using schoolchildren will obscure that. Head measurements (Galton had at least used their interaction, volume, as a predictor) are used as proxies for brain size, but were already known then to be only moderately associated with it. We now estimate the

[64] K. Pearson 1900, 539. In later editions Pearson removed this material from the book.

[65] K. Pearson 1902.

[66] K. Pearson 1930a, 248-250.

correlation at only 0.5. This introduces substantial measurement error. Similarly the ability measures are discretely categorized into a few compartments, and this loss of precision also attenuates any correlation found.

Spearman had already shown that measurement error attenuates correlations. However Pearson refused to entertain Spearman's correction for this, carrying on an acrimonious public controversy with the psychologist, in which he heaped scorn on his (unperturbed) opponent. Burt's pupil Charlotte Banks recorded one legend of this love not lost. 'There is an apocryphal story of Pearson and Spearman when they were both professors at University College. Dr. Philpott used to relate it with obvious enjoyment. The building which housed Statistics and the Galton laboratory was roughly diagonally across the quadrangle from the entrance to the psychology department. The story was that if, by accident, Pearson and Spearman emerged at the same time, each, on seeing the other, would hastily retreat and wait to come out only when the coast was clear'.[67]

Over the years a steady trickle of investigators returned to this question, using Binet-style IQ tests. A review in 2007 turned up fully 59 studies up until the year 2000, with varying sample sizes and IQ measures. The weighted mean correlation between IQ and head size was 0.2 (unweighted 0.21). These are values close to those Pearson would have obtained by correcting for obvious measurement error. However Magnetic Resonance Imaging (MRI) of brains provided a breakthrough. As we have stated above, the interest is in the size of brain, only imperfectly measured externally on the head, and intelligence, as measured imperfectly by IQ tests.

MRI gives a much more accurate measure of brain size. There is also an important age effect, which holds true for all associations involving intelligence. Up through adolescence, intelligence is more subject to variable influences than it is in middle age and later.[68] By measuring children, the correlation between (estimates of) brain size and intelligence is attenuated. When measuring adults, the correlation increases substantially. A meta-analysis of 37 studies concerning MRI-measured brain size and IQ, amounting to 1,530 subjects, suggests that for adult males and females the average correlation is 0.4. When children and adolescents are included, the correlation drops to 0.33.[69]

There is no doubt then that on modern evidence Galton was correct and Pearson was incorrect about the influence of brain size on ability, even though he had to make do with head size as a proxy. The fact that the correlation is not overwhelming confirms that other factors are critical. Size is not good enough on its own to measure intelligence. Nevertheless, it is a common misperception that moderate correlations of this kind are not useful. When aggregating people into groups they can become very useful indeed. Suppose for example that one is searching for a group of 10 very intelligent people. After they have been found the search can stop. It is expensive to test individuals. Even if we only have external head measurements and a correlation of 0.2 with intelligence, ordering them by head volume and then concentrating on an extreme subset of that group for individual testing will speed up the search and save time and money. The link to brain size also helps us to understand the physiological mechanism behind intelligence.

Was Galton just lucky here? It may be argued that the individual cases he observed in the places like the Royal Society cannot have been marked enough to support his inference from head size to ability. However he was observing extreme cases at the very tail of the ability distribution, far beyond the usual range of the studies we have cited. Out there, larger heads may well be detectable by eye (and, in the Anthropometric laboratory, by calipers) alone. Some people have far keener powers of observation than others. The Cambridge data he assembled

[67] Banks 1983. S. F. J. Philpott was one of Spearman's students.
[68] Haier 2017, 121.
[69] McDaniel 2005.

through Venn is in retrospect highly suggestive. The differences he was looking for showed up at the aggregate level as we would expect. Pearson's dismissal of the link shows a confusion between a direct test for intelligence and a useful criterion for it, not to mention an interesting means of understanding its mechanism. As usual, Galton was on the right track, just as he had been for reaction time.

Figure 12.1: A set of similar brass weights for testing fine sense discrimination. Manufactured in 1893 to Galton's specification by the Cambridge Scientific Instrument Co. run by Charles Darwin's son Horace. Science Museum Group, Creative Commons.

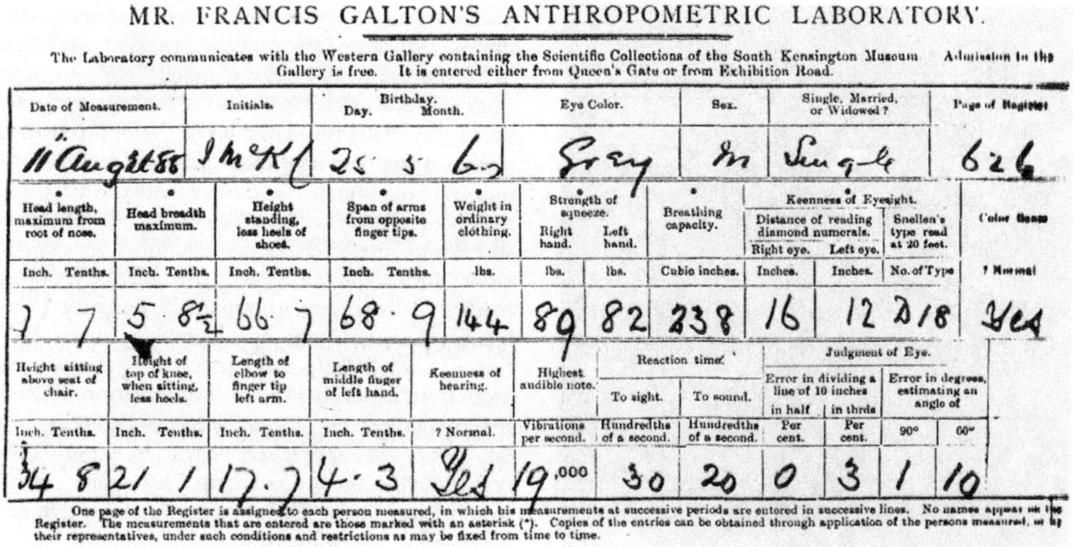

MR. FRANCIS GALTON'S ANTHROPOMETRIC LABORATORY.

The Laboratory communicates with the Western Gallery containing the Scientific Collections of the South Kensington Museum. Admission to the Gallery is free. It is entered either from Queen's Gate or from Exhibition Road.

Date of Measurement.	Initials.	Birthday. Day.	Month.	Eye Color.	Sex.	Single, Married, or Widowed?	Page of Register
11 August 88	JMcK C	25	5	60	Grey	M	Single	626

Head length, maximum from root of nose.	Head breadth maximum.	Height standing, less heels of shoes.	Span of arms from opposite finger tips.	Weight in ordinary clothing.	Strength of squeeze. Right hand.	Left hand.	Breathing capacity.	Keenness of Eyesight. Distance of reading diamond numerals. Right eye.	Left eye.	Snellen's type read at 20 feet.	Color Sense
Inch. Tenths.	Inch. Tenths.	Inch. Tenths.	Inch. Tenths.	lbs.	lbs.	lbs.	Cubic inches.	Inches.	Inches.	No. of Type	? Normal
7 7	5 8½	66 7	68 · 9	144	89	82	238	16	12	D/18	Yes

Height sitting above seat of chair.	Height of top of knee, when sitting, less heels.	Length of elbow to finger tip left arm.	Length of middle finger of left hand.	Keenness of hearing.	Highest audible note.	Reaction time. To sight.	To sound.	Judgment of Eye. Error in dividing a line of 10 inches in half	in thirds	Error in degrees, estimating an angle of 90°	60°
Inch. Tenths.	Inch. Tenths.	Inch. Tenths.	Inch. Tenths.	? Normal.	Vibrations per second.	Hundredths of a second.	Hundredths of a second.	Per cent.	Per cent.		
34 8	21 1	17.7	4 · 3	Yes	19,000	30	20	0	3	1	10

One page of the Register is assigned to each person measured, in which his measurements at successive periods are entered in successive lines. No names appear on the Register. The measurements that are entered are those marked with an asterisk (*). Copies of the entries can be obtained through application of the persons measured, or by their representatives, under such conditions and restrictions as may be fixed from time to time.

Figure 12.2: James McKeen Cattell's anthropometric measurements, 1888-08-11.

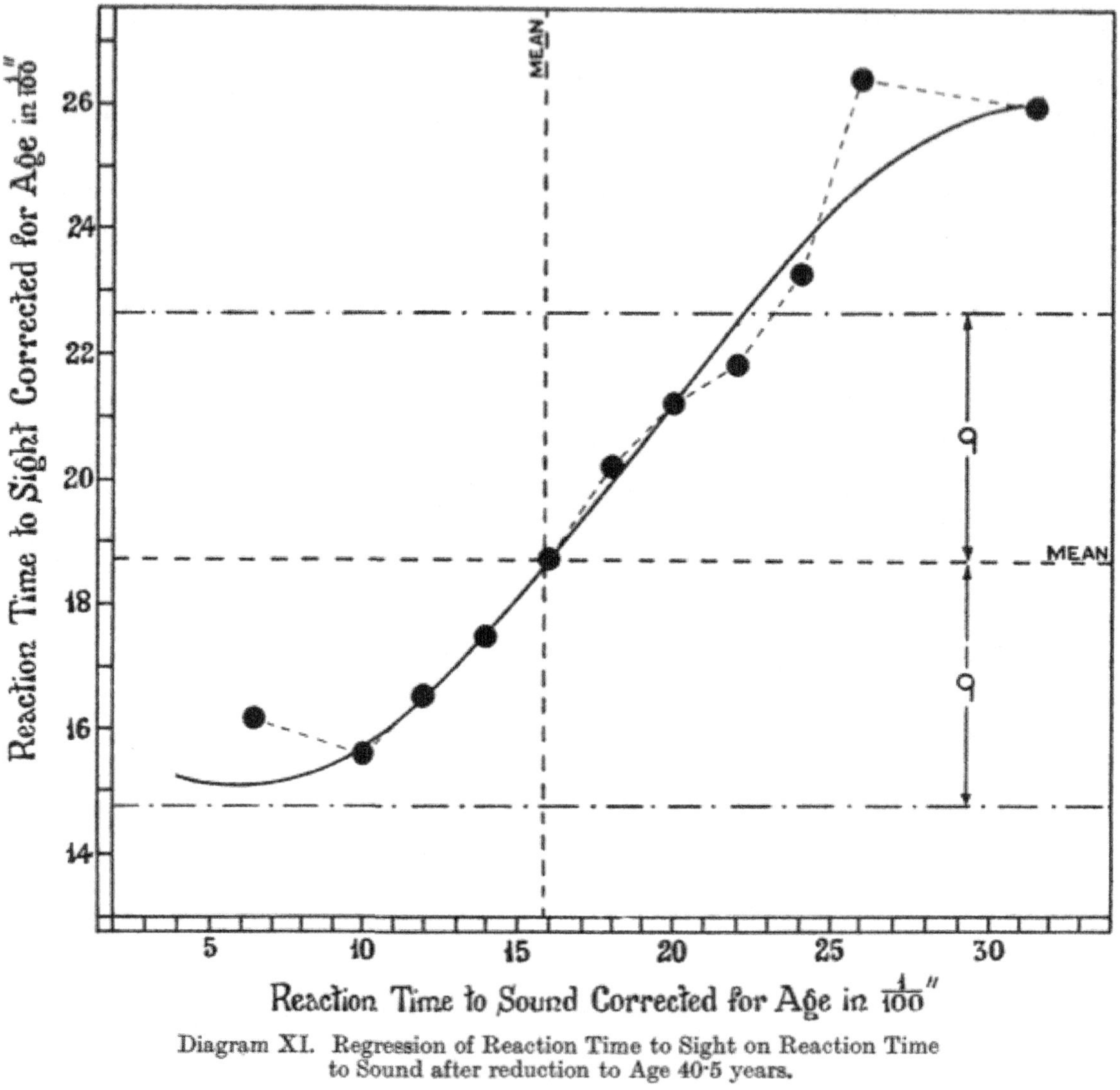

Figure 12.3: Koga and Morant relate sound and sight reaction times, from Koga and Morant 1923.

Figure 12.4: Charles Spearman (left) in 1902, with his wife Frances Aikman, daughter and parents in-law.

	Classics.	French.	English.	Mathem.	Discrim.	Music.
Classics,	*0.87*	0.83	0.78	0.70	0.66	0.63
French,	0.83	*0.84*	0.67	0.67	0.65	0.57
English,	0.78	0.67	*0.89*	0.64	0.54	0.51
Mathem.,	0.70	0.67	0.64	*0.88*	0.45	0.51
Discrim.,	0.66	0.65	0.54	0.45		0.40
Music,	0.63	0.57	0.51	0.51	0.40	

Figure 12.5: Spearman's matrix of correlations for data set IV, Spearman 1904a.

General Intelligence	Classics				+0.99
"	Pitch Discrimination				0.94
"	French				0.92
"	English				0.90
"	Mathematic				0.86
"	Music				0.70
Amal. School Place	Music and I	Av.	+0.72	0.02	
General Intelligence	*General Au*				*1.00*
					(p. e. ab 002)

Figure 12.6: General factor saturations by subtest for data set IV, Spearman 1904a.

Ages.	Class A. "High honour" men.	Number of measures.	Class B. The remaining "honour" men.	Number of measures.	Class C. "Poll" men.	Number of measures.
19	241·9	17	237·1	70	229·1	52
20	244·2	54	237·9	149	235·1	102
21	241·0	52	236·4	117	240·2	79
22	248·1	50	241·7	73	240·0	66
23	244·6	27	239·0	33	235·0	23
24	245·8	25	251·2	14	244·4	13
25 and upwards	248·9	33	239·1	20	243·5	26
		258		476		361

Figure 12.7: Cambridge head measures 1885–1887, F. Galton 1888d.

Relative Brain Capacity of Cambridge University Men according to their Proficiency and Age (from Dr. Venn's Tables).

The Numerals along the top of the diagrams signify the product of the three Head measures, viz. :— Length × Breadth × Height (in inches).

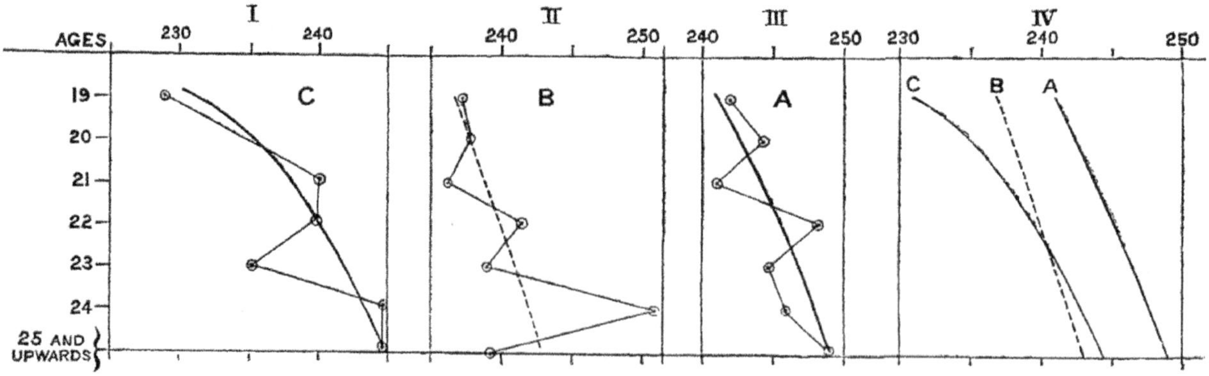

A, High honour Men; B, The remaining honour Men; C, Poll Men.

Figure 12.8: Head volume and growth in 1,000 Cambridge students by ability class, 1888, F. Galton 1888d.

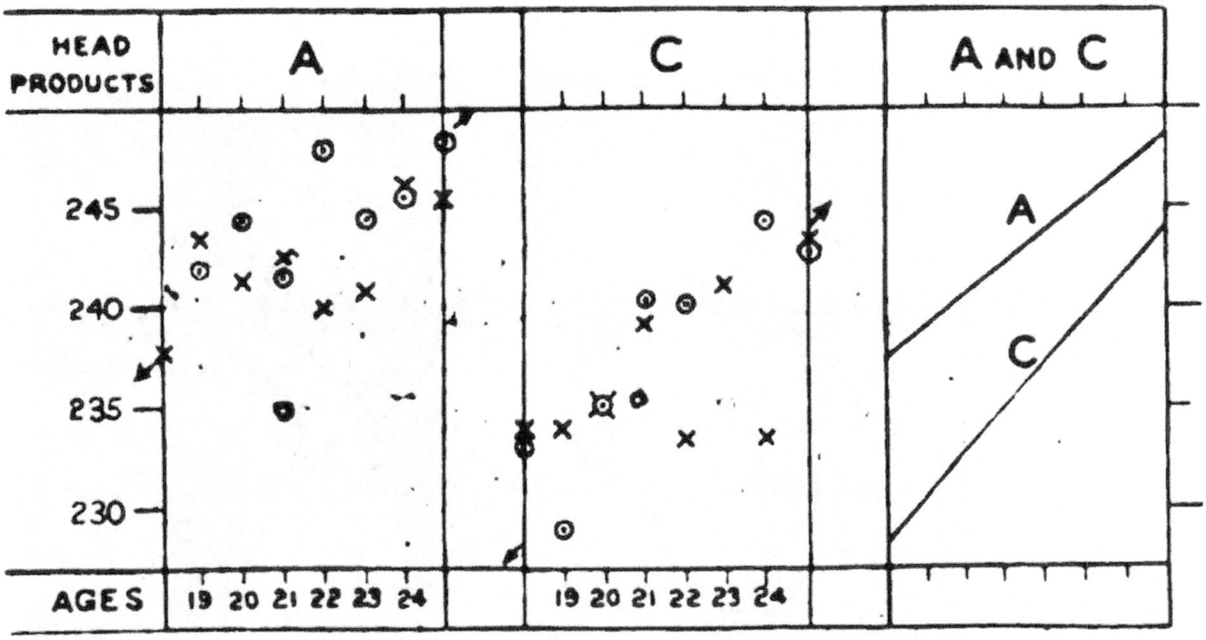

Figure 12.9: Head volume and growth in 2,000 Cambridge students by ability class, 1890. Circles=1888. Crosses = 1890, F. Galton 1890c.

Intelligence and	Cambridge Graduates	School boys at Twelve Years	School girls at Twelve Years
Cephalic Index ...	(1011) − ·061 ± ·021	(2345) − ·041 ± ·014	(2226) ·067 ± ·014
Length of Head ...	(1011) ·111 ± ·020	(2298) ·139 ± ·014	(2188) ·084 ± ·014
Breadth of Head ...	(1011) ·097 ± ·021	(2299) ·109 ± ·014	(2165) ·113 ± ·014
Auricular Height ...	—	(2290) ·073 ± ·014	(2166) ·055 ± ·014

Figure 12.10: Karl Pearson's head size by ability correlations, K. Pearson 1902.

Figure 12.11: Sir Cyril Lodowic Burt (1883-1971).

Chapter 13

Weightless Elephants on Frictionless Surfaces: Evolution

Does evolution proceed gradually, by accumulating small, almost imperceptible, nearly innumerable, changes—or must it *leap* to produce genuine novelty and so different species? Darwin and Alfred Russel Wallace were convinced gradualists, for which the *Origin of Species* is one long argument. But they always had opponents. Many doubted whether micro-evolution could ever differentiate species. Instead they supposed that evolution must be discontinuous—'evolution by jerks' as it would be caricatured many years later. T. H. Huxley, surely no jerk, entertained at least the occasional leap, as we will see.

In late 1889, the Royal Society sent Galton an intriguing new paper to referee. It was from a relatively unknown young biologist, Walter Frank Raphael Weldon, who had recently read *Natural Inheritance*. Metamorphosed by its new techniques, he had begun to apply them to the common shrimp, over a thousand of which he had already carefully measured and fitted distributions to. The referee had many comments. The paper was withdrawn, changes were made, and it was resubmitted.[1] And so an interaction was initiated which would have profound effects on evolutionary biology and Galton's own legacy. Soon he would be introduced through Weldon to Karl Pearson, an up-and-coming applied mathematician. The three would go on to found what came to be known as the biometric school of mathematical biology, with its own organ: *Biometrika*.

Weldon and Pearson would also do what Galton would not, which is engage in furious public controversies—most notably with the Cambridge biologist William Bateson and his followers, proponents of discontinuity who scorned 'actuarial methods' for subduing complex traits, in favour of clear-cut experiments on less agile prey like flower colours. The rediscovery of Mendel's work in 1900 would produce many bright flashes that one could almost read the fine-print by. Like the lightning that Galton witnessed near the Kalahari—but that also set large areas of the *veld* on fire, obscuring everything again. All sides would claim descent from elements of Galton's pioneering work. At the same time, all sides would discreetly claim to have transcended his limitations, alternately identified. However Galton's own position within these swirling arguments has been much misunderstood.

Raphael Weldon (1860–1906), the son of a struggling journalist who would later become a successful chemist and businessman, had shrugged off indifferent schooling to enroll at St John's College at Cambridge in 1878, by way of UCL and King's College. He intended for medicine. After winning an exhibition and a scholarship he switched to zoology, under the influence of Frank Maitland Balfour (1851–1882). Despite suffering a mental breakdown, Weldon was able to graduate with a first class degree in 1881, leading to a fellowship and a lecturing position. Only a year later, Balfour—a brilliant biologist well-known to both Darwin and Galton, having tried his hand at the blood transfusion experiments on rabbits—had perished in a mountaineering

[1] Weldon 1890.

accident on Mont Blanc.[2]

Throughout his life Weldon devoted his vacations to field work, first in locations like Naples and the Bahamas, later at the Marine Biological Association laboratory in Plymouth, where some of his 'common shrimp' measurements were first made (400 specimens). Others came from Southport (300) and Sheerness (another 300). He was trying to verify that measurements of animal bodies in the wild state really are normally distributed, as Galton had predicted and had already been shown for domesticated cases.

In the case of the shrimp (*crangon vulgaris*) Weldon worked with four different measurements of length, from sections of the front and rear, and these were certainly normally distributed, as his curves and some centile checks showed. Here he acknowledged considerable help from both Galton and Donald MacAlister, since his mathematical knowledge was still lacking. The mean values at the three locations were slightly different, as were the variations expressed as 'probable errors'—estimated, using Galton's handy method, from the quartiles of the ranked series. He supposed that these differences were probably due to local selection pressures, which would not be the same everywhere. A follow-up paper considered the correlations between organs, which appeared to the same regardless of location. This was published a few years later, and used Galton & Dickson's handy ratio method for approximating the correlations.[3]

When Weldon moved to UCL in 1891, after being elected to the Royal Society the previous year, taking up the Jodrell professorship of Zoology, he soon encountered Karl Pearson (1857–1936). The applied mathematician had been interested by Weldon's biometrical papers, and they found common cause in a then-current movement to reform the university. The reforms ultimately came to naught, but by then the two had become increasingly close.

Pearson, actually christened Carl, was a multi-talented scholar, barrelling along with energy and determination, loath to leave off once his bite had locked. His father William was a successful Inner Temple Chancery lawyer, originally from Yorkshire. From the University College School, Carl went up to Cambridge, where he was coached in mathematics by the renowned Edward Routh (1831–1907)—himself a product of William Hopkins, who had coached Galton in the early 1840s. Most of the noted wranglers of that century were produced by the Hopkins–Routh dynasty.

Indeed young Carl went up early, working for a year under Routh as a non-enrolled 'brute'. He tried for Trinity, which in earlier years had dominated the wrangler lists, but he did not yet have the broad knowledge required. So he had to settle for King's, a college formerly reserved for Old Etonians and not exactly known for mathematics. If his recollections are to be believed, the blunt Yorkshireman refused to attend compulsory divinity lectures, which were dropped after he would not back down. Likewise compulsory chapel attendance, which he then made a show of attending *voluntarily*, when the mood took him.

All this did not detract from Tripos preparations, which were intensely interesting nationally. The Victorian craze for betting encouraged one punter to approach young Pearson while out walking, to ask him directly what his chances were. 'What the devil business is it of yours?'—which completely fails to account for bets being very good business indeed.[4] In February of 1879 he placed as Third Wrangler, a very strong showing but a bitter disappointment

---

[2] Balfour's philosopher brother Arthur (1848–1930) would later become Prime Minister of England (1902–1905) and an early President of the Sociological Society, which Galton helped to found and was closely associated with the Eugenics Education Society. See chapter 14, p. 330. For Balfour's help with the rabbits, see chapter 5, p. 84.

[3] Weldon 1892. If the measures of organs a and b for individual i are x_i and y_i, with means x_m and y_m, then for any individual

$$r_{ab} \approx \frac{x_m/Q_b}{y_i/Q_a} \approx \frac{y_m/Q_a}{x_i/Q_b}.$$

[4] The exact source for this anecdote is now lost among my mental notes, so *caveat emptor*.

to him at the time. The undistinguished careers of the first and second wranglers suggest that the gruelling week-long, eight hours per day, problem-solving marathon tested a lot of other qualities beside real mathematical talent. In 1898, the superlative pure mathematician G. H. Hardy would only place fourth, then have the Tripos reformed beyond recognition a decade later.[5]

Pearson certainly transcended narrow specialism, a source of pride. 'In Cambridge I studied Mathematics under Routh, Stokes, Cayley and Clerk Maxwell—but wrote papers on Spinoza'. After the Tripos results of February he could plunge to the depths of all things Germanic, spending the next twenty months abroad, sustained by a Fellowship in April of 1880. As an honours man, he had not 'gone out' as one of the *hoi polloi*, but nevertheless he (rather implausibly) claimed to be 'a democrat and in sentiments and language one of the *hoi polloi*', dissolving (older and wiser) Oscar Browning's advice to mix freely abroad.[6] 'In Heidelberg I studied Physics under Quincke, but also Metaphysics under Kuna Fischer. In Berlin I studied Roman Law under Bruns and Mommsen, but attended the lectures of Du Bois Reymond on Darwinism'.[7]

At least, that was the view looking back. At the time, Darwinism disappointed. He had been stirred up by *Das Capital*, writing of Reymond's lecture, 'He has an audience of perhaps 400 students, who think that some solution of their difficulties is to be obtained from the theories of evolution. Poor fellows they go & listen attentively to the possibility of producing a permanent race of mules, as if that could be any cure for tea at 6p a lb., and no marmalade at 1/4 a pot!'.[8]

Thus Carl became Karl, author as 'Loki' of *The New Werther* (1880)—really, the travails of young Karl, agape at the endless vista of knowledge. He versified in *The Trinity: a Nineteenth Century passion Play* (1882). Then he turned back toward his father's profession, law, reading at the Inner Temple in November 1880, qualifying as a barrister within a year. But vacations would see the *wandervögel* regularly back in Bavaria and the Tyrol. He suffered from fluctuating Teutonomania through the rest of his life.

At home in late 1880, it was more than law, for Karl was burning candles at all ends in multiple dimensions. 'Back at Cambridge I worked in the engineering shops but drew up the schedule in *Mittel-* and *Althochdeutsch* for the Medieval Languages Tripos'. Beside 'lecturing on Heat at Barnes, on Martin Luther at Hampstead and on Lassalle and Marx on Sundays at revolutionary clubs round Soho. Indeed, I contributed to the Socialist Song Book hymns which I believe are still chanted' (see Figure 13.3, p. 310).[9] The 'hymns' were not sentimental. 'At last our soul with hate shall burn / Too Long has love been lord!'. He even corresponded with Karl Marx. But an offer, plus sample, to translate *Das Kapital* was sarcastically rejected by the bearded Basilisk of the British Museum Reading Room. So ended the dalliance between Karl M. and Karl P.

Enthusiasm for law soon waned and Pearson applied for several professorships of mathematics, eventually landing a chair at UCL in June 1884. The previous occupant had been no less than William Kingdon Clifford (1845–1879), an enthusiastic evolutionist who had been intro-

---

5 All three wranglers were coached by Routh. The Senior was Andrew James Campbell Allen (St. Peters College, 1856–1923), son of a Belfast banker. He soon took up religious education, finally becoming a country vicar on £50 a year, and apparently produced no mathematics of any note. The Second was George Francis Walker (Queen's, 1855–1883) from Durham, who went out in 1883 to Auckland in New Zealand as Professor of Mathematics, but drowned in the harbour—beware of new colleagues who take you yachting—just a few days after arriving. Though his career was cut short, emigrating to the periphery is suggestive.

6 Karl Pearson to Oscar Browning, circa 1879, Porter 2004, 74. Browning was an enthusiastic socializer, like that other Oscar, with a taste that extended to sailors.

7 Filon et al. 1934, 20. Kuna Fischer was lecturing on Kant.

8 Karl Pearson to Fanny Pearson, 1879-11-24, Porter 2004, 74.

9 Filon et al. 1934, 20. See also Porter 2004, 106 fn 31. Examples include 'Songs of the Proletariat, I' and 'II,' *Cambridge Review*, 2 (1881-05-18), 335; 1881-06-01, 367. Signed 'K.P.'.

duced to Galton by Charles Darwin while still a student. Clifford had died young, his *Common Sense of the Exact Sciences* incomplete. Pearson busied himself with applied mathematics, completing both Clifford's book and a Miltonian multi-volume treatise on the *History of the Theory of Elasticity* (no man ever wished it stretched longer) started by Isaac Todhunter before he too had perished, perhaps from the effect.

At the same time, Karl P.'s interest in radical politics continued to engorge: communism, 'free-thought', women's rights, lectures for working men. Olive Schreiner—unhappy author of *Story of an African Farm*, 1883—and many other fringe figures like Annie Besant were drawn into his Men and Women's Club. So was the woman he would actually marry in 1890, Maria Sharpe (1853–1928). His communism apparently involved nationalizing all land and capital in the name of mutual aid, invoking evolutionary 'for the good of the group' arguments. He would also consort with the Fabians for a time, before (inevitably) going his own way.

Under Weldon's influence, Pearson took up mathematical statistics, until then not one of his main interests. He had read Galton's *Natural Inheritance* in 1889 when it came out, and would later credit the book with liberating his thinking from the fetters of causality, revealing to it purely descriptive correlation. But Pearson seems to have done little about that right away—a single lecture that year had reservations about Galton's methods—until he met Weldon.[10] After 1891, the biologist—while trying manfully to expand his neglected mathematical skills—moved on from the common shrimp to the shore crab. At Plymouth Sound and the Bay of Naples Weldon collected a thousand adult females and industriously took eleven separate measurements from each body (thus 11,000 measurements).

One dimension, frontal breadth, was puzzling for the Naples shore crabs. Unlike the others it was *not* normally distributed, but skewed to the left. It was surely Galton who suggested to Weldon (though he does not say so) that two distinct races, with different frontal breadth distributions, had probably been mixed up. The race with lower mean bulged out the left hand side. The challenge was to fit those separate curves, which Pearson attacked with relish, inventing the 'method of moments' (later beloved of economists) for fitting the parameters.[11]

Correlations between pairs of measures—using Galton's approximation formula as before—were roughly the same, given the modest sample, in the Naples and Plymouth crabs. Weldon was now in love with the beauty of his weapons: 'the problem of animal evolution is essentially a statistical problem'. To detect natural selection in operation is to detect changes in the distributions, shifts in means and variances. The degree of change measures the speed of the evolution underway. But that is 'a question of arithmetic'. It would soon become apparent that it is not so easy as all that.

Skew distributions were hardly new. Galton himself had published a paper in 1879 noting instances—response to stimuli, growth of population and capital, etc.—which compounded multiplicatively, with a geometric rather than an arithmetic mean, and a lopsided distribution. An accompanying paper by Donald MacAlister described the mathematical form.[12] Later this came to be known as the *log-normal* distribution. Taking the logs of all the data elements—a simple transformation—produces the familiar, convenient, normal distribution, which can be analyzed as usual; then results can be mapped back to multiplicative space by exponentiating. Galton pointed these papers out to Pearson, who seems to have digested them. But the mathematician was not satisfied; he would exhaustively characterize skewness in all its aspects, then systematically root out the stoats who ignored its subtleties.

The (abstracted) paper explaining Pearson's new method for solving Weldon's mixture problem followed immediately after the abstract by the biologist in the *Proceedings of the Royal*

[10] Magnello 1993.
[11] Weldon 1893.
[12] F. Galton 1879c.

Society for 1893. The full version, published the next year, ran to over fifty dense pages.[13] It was the first of many 'Contributions to the Mathematical Theory of Evolution' by Pearson in the coming decade. He was easily able to find many other published examples of skew measurements, once curved, from Bateson's earwigs to Thomson's prawns, Roberts' eyesight of schoolboys, and even F. Y. Edgeworth's prices. There were others. He distinguished a number of ways in which skewness could arise, noting that Weldon's crab data was just one special case of a general theory of skew distributions.

The 'moments' of curves (ideas inherited from the field of mechanics) can be used to find the best parameter estimates to fit them, since taken together these describe their shape. The more moments you specify, the closer you control the shape. The mean is exactly the first moment or 'center of mass', the second is the variance, the third sets the skewness or lopsidedness, the fourth sets how weighty the tails are (kurtosis). Higher-order moments like the fifth and sixth are more subtle, relating the tail weight to skewness and so on. Using just two moments, and the area under the curve, a single normal distribution can easily be fitted to a data set.[14]

To fit a mixture of two curves Pearson set the first *five* moments equal to those calculated from the observed measurements, and minimized the sixth moment. Old Tripos days had trained him relentlessly for the feat of endurance in applied mathematics that followed. The poison was in the dose, as he laboured to calculate the moments (not so easy before mechanical aids), and then force-marched simultaneous higher-order equations over reams of paper, eventually strong-arming them into a more familiar quadratic in the areas under the curves. This led to a test, based on the existence of real roots for that quadratic, determining whether or not mixtures made sense for the data at hand. Using that test on Weldon's data, the Third Wrangler was happy to report that 'it is shown that the Naples crabs are breaking up into two different sized families, owing to evolution in their foreheads' (that is, frontal width).

Galton was asked by the Royal Society to referee Pearson's paper. The elderly scientist approved it, with the disclaimer that he was no expert on the advanced mathematical methods used, but expected that they would stimulate further work and soon be found out by other experts if they were faulty. (Of course, they proved accurate.) Writing privately to Pearson he was skeptical that the data set was good enough to stand up to the long-and-exceeding-fine interrogation sessions that Pearson put it through. He had himself tried mixing male and female distributions, known to be rather different individually, but had been perplexed to find how imperceptibly the mixture deviated from a standard normal.[15]

Soon Pearson generalized his technique further. He noticed that the skew distributions he had obtained could be expressed as the solution of a simple differential equation. When that equation was generalized by adding a few more parameters, a rich family of solutions was teased out. He proceeded to investigate them in great detail, laying the foundations for much of the early statistical theory that the new century inherited. Two members of the family turned out to be ubiquitous: the beta and gamma distributions. With time several others proved to be either trivial or just not very useful in practice.

None of this enthused either Galton or Weldon, who didn't see the relevance of this extravaganza of skewness in most practice. Weldon complained privately to Galton that Pearson had allowed himself to be carried away, seduced by plasticity, losing touch with the needs of a working biologist like himself. In real data, skewness was often at the tails only, a product of outliers and other unusual observations in fallible data, not the vast bulk of it. Likely it was an illusion that skew curves usually made a better fit, once the small amount of anomalous data was ignored, as Weldon was happy to do. The choice of which skew curve to use, and the

---

13 K. Pearson 1893, 1894.

14 The normal is the only distribution in which the mean and variance are independent.

15 The referee report is dated 1893-11-22, and is transcribed in Stigler 1986, 332. FG to Pearson, 1893-11-25, GALTON/3/3/7/20.

implications of that fit when happenstance was so important, meant that interpretation was no longer clear-cut: too many other curves were close fits too.[16] (Even today this remains a central problem in all applied statistics based on observational data; it is all too easy to chase random noise in data when given sufficient flexibility.)

Perhaps the unadorned normal would do as an approximation for practical work with the sort of observed data one was likely to get? Pearson was rooted: only the best fit would do. Weldon wished there was another mathematician to serve as a counter-weight to the immovable Third Wrangler. Galton was privately in agreement, though he preferred not to debate the issue in public. Perhaps by then he knew the tenacious Wrangler all too well. He continued over the years to press others, again in private, to find simpler alternatives. In effect—anticipating David Hilbert—he would not have Karl P. drive working scientists from the paradise that Quetelet and he had created.[17]

At this stage an old friend of Weldon entered the debate, with relish. The world would hear a great deal more from him. His name was William Bateson, and he had been another pupil of Frank Balfour at Cambridge, just one year behind Weldon.

Bateson came from a scholarly background. His father William Henry had been Master of St John's College in Cambridge, and had married a Scottish woman with distinguished collaterals. The Master's son had gone up to St John's after surviving Rugby, but did not enter in there for mathematics, taking the natural sciences Tripos instead. After obtaining a first class degree—absolutely essential he later advised all—he had taken up invertebrate morphology under the ill-fated Balfour.

For two years Bateson studied the acorn worm in the United States, where it is plentiful. Despite all appearances and common belief, he concluded that the worm is in most ways far removed from the invertebrates. A fellowship followed on his return in 1885, and he went off to the Siberian Steppes, in search of evidence that animals really did adapt to their environment through natural selection, by looking at salt lakes. He found little evidence of that after two years, from which he concluded that they did not. It proved to be an irreversible leap in his thinking, corresponding curiously to the leaps he now supposed that evolution required. Heretical ideas like that had to be kept hidden—or at least so he thought—in an atmosphere completely dominated by the Darwin-Wallace consensus.[18] In their account, natural selection on negligible differences drove evolution with infinite patience, and no leaps were contemplated.

It is clear that Bateson fastened on to discontinuity in evolution at this early, concealed, stage and then spent the rest of his life trying to locate evidence for it. The first step was to gather all the instances he could find in the literature of anything like a sudden departure from a biological norm. This would supplement Darwin's own massive compendium of variation, already published in 1868, but also allow Bateson to pursue interpretations and place emphasis where Darwin did not.[19] *Materials for the study of variation* struggled from the press in 1894. Generally it met with a cool reception and poor sales, but the idea that coming out for leaps would harm his career proved false.

Yet there was one distinctly positive commentary on Bateson's *Materials*. It appeared in the journal *Mind*, somewhat out of the way for evolutionary matters. It gave the commentator 'utmost pleasure' to discover a fellow spirit. He too had argued that leaps may appear. Yet 'all along I seemed to have spoken to empty air'. His suggestions had been ignored. 'I never heard nor have I read any criticism of them, and I believed they had passed unheeded and that my

[16] Weldon to FG, 1895-01-23, 1895-01-27, 1895-02-11, 1895-03-03, GALTON/3/3/7/28. Transcribed in Stigler 1986, 337.

[17] In time, Pearson's methods would be superseded by R.A. Fisher's maximum likelihood technique, which wastes less of the observational data when fitting curves. But Pearson would not abandon his offspring at any price, and went to the grave arguing otherwise.

[18] B. Bateson 1928.

[19] W. Bateson 1894; C. Darwin 1868.

opinion was in a minority of one'. But he was careful to add that leaps were not *required*, they were just *possible*, so that evolution might be fast or slow. Careful as well to qualify the leap interval, which could be 'at least as large as it has been in any hitherto observed instance, and it may be smaller in any less degree'. (Rather a lot of flexibility.) He even gave instances from his own research where discontinuity seemed undeniable. The instances—including previously unnoticed finger print pattern genera, and calculating boys—singled out the author, Francis Galton.[20]

Encouraging Bateson was not the primary aim, for Galton wanted to refine ideas that he had floated previously about the relative stability of types and their relationship to 'sticking points' in evolution. As discussed previously, Galton's views on this have been much misunderstood, even by his follower Karl Pearson. In part this is because his understanding evolved over time and was not always clearly explained. As a result regression to the mean came to the forefront rather than the real issue, stability of type, of which regression is no more than a necessary mathematical consequence.[21] Types were critical to Galton's views on evolution, in both its natural form and its possible eugenic form, so it is important to grasp them firmly.

For an example of the confusing way that Galton sometimes explained his thinking, consider this assertion at the Aberdeen meeting of the British Association in 1885. 'The appearance of a new and useful family peculiarity is a boon to breeders, who by selection in mating gradually reduce the preponderance of those ancestral elements that endanger reversion. The appearance of a new type is due to causes that lie beyond our reach, so we ought to welcome every useful one as a happy chance, and do our best to domicile and perpetuate it'. (In the third sentence, the 'new type' is actually the family peculiarity of the previous two sentences.) It is not necessarily easy to permanently shift the distribution defining the type. 'The limits of deviation beyond which there is no regression, but a new condition of equilibrium is entered into, and a new type comes into existence, have still to be explored'.[22]

A particular scenario must be called out here and clearly understood. Galton argued, entirely correctly, that the descendants of a gifted couple will inexorably regress to the population mean *if those descendants mate at random with other members of the population.* After enough generations of random mating, any descendant will be indistinguishable from a member of the population drawn at hazard, and the best prediction or estimate of that person's mean trait will be the population mean, since we know nothing more about them. What selection does is actively *alter* the breeding population in an attempt to alter the type. But how easy is this?

Galton had already suggested as early as 1877 that shifting the distribution by selection might be harder than some might suppose. It was an idea that he would develop considerably in coming years but really amounted to no more than pointing out that certain kinds of selection are ineffective, in that they do not really alter the race—the pool of hereditary particles that defines the population.

There were in Galton's view two very different *kinds* of variation that one could try to select on. As he put it in his preface to the second edition of *Hereditary Genius* (1892), 'the word variation is used indiscriminately to express two wholly distinct conceptions: sports, and variations properly so called'.[23] He also called the latter '*mere* variation', reserving it to describe those changes that do not result in a shift of the type. In his review of Bateson, he improved these terms, using *transilient* variation for shifts in the type, and *divergent* variation where the type is not changed.[24]

Trivially, selecting on variation that is random or an environmental effect (or as we would

20 F. Galton 1894b.
21 See page 215 and following
22 F. Galton 1885a.
23 F. Galton 1892b.
24 F. Galton 1894b.

say, not heritable) will not shift the type, since it is 'mere variation'. For traits driven to fixation in the population, all variation will be of this 'mere' kind. Another source of resistance to change in type derives from Galton's prescient but faulty theory of heredity. Recall that this theory entails the inheritance of ancestral characters from the 'stirp', not necessarily manifest in the parents but transmitted by them. We might think of this as a kind of portable side-channel gene pool, handed down from person to person, from which fresh draws are made when breeding.

The stirp (in Galton's understanding) tends to swamp other characters, somewhat like mating within a larger breeding group would do. This is not insuperable, since with enough selection over a long time, the mean of the stirp itself will *eventually* be shifted. Regression to the (same) mean is involved in this precisely to the extent that the distribution does or does not change.

Galton did not believe, or argue, that it is *impossible* to alter a type, i.e. the distribution, but it might be very slow. 'It must be borne in mind, that there is no stability in a breed improved under the supposed conditions; but that, as soon as selection ceases[25] it will regress to the level of the rest of the population through stages in which the deviation at starting, sinks successively to $w, w_2 ... w_n$ of its value. It may, however, happen that a stable form will arise during the process of high breeding, that shall afford a secondary focus of regression, and become the dominant one, if the ancestral qualities that interfere with it be eliminated by sustained isolation and selection. Then a new variety would, as I conceive, arise'.[26]

There were other remedies to hand, side-stepping pure selection from a subset of the population. As he argues in *Human Faculty*, one way to accelerate changes in type is to introduce new racial elements (which we would call new genes) through immigration. He preferred introducing talented populations like the Huguenots. The immigrants will already be at a different 'sticking point' and therefore shift the combined population in proportion to a new overall 'sticking point'. But underlying this is a shift in the pool of 'hereditary particles' associated with the population, which shows that Galton clearly understood the crux of the problem.

All these considerations led Galton, in the second edition of *Hereditary Genius* (1892), to preface his earlier enthusiasm for selection with qualifications derived from his more recent work on heredity. But only to point out that it would be much more *difficult* than he previously thought, not *impossible*.

For the process of evolution which resulted from the 'sticking points' of types, Galton used the analogy of a faceted sphere rolling abruptly, under impetus, from one stable position to another. Characteristically, he had a wooden model constructed and used it to demonstrate the idea to people. The distance between each position of equilibrium need not be large—very many positions might exist close to each other, through which the population could 'click' as it evolved. It was an idea that went all the way back to the rough stone rolling analogy in the closing speculations to *Hereditary Genius* in 1869.[27] The calculating boys like Jacques Inaudi that he had examined struck him as possible examples of sudden discrete leaps in a trait.[28]

Galton also argued that positions of 'organic stability' could even arise wholly independent of selective forces. He used an example he had gleaned from his work on fingerprints. The pattern genera in the prints seemed to have no possible functional or adaptive significance in themselves. They could not have been sexually selected for either, since people could not see them. Yet there were only a few dominant sets of patterns, which appeared to be stable. This

[25] By 'selection ceases' Galton probably means that the sub-population resumes random breeding with the broader population, since that is what a breeder would do in practice.

[26] F. Galton 1897a.

[27] F. Galton 1869a, 369.

[28] F. Galton 1869a, 368. There is no point in quibbling with this argument today, since it is well-known that mutations in single genes *may* have large effects, though non-fatal ones typically do not. When they do, they are either driven to fixation rapidly, if positive, or selected out of the population if negative.

suggested to him that some features of the animal world arose for other (perhaps internal) reasons wholly outside the framework of natural selection.

The flexibility of Galton's ideas about all this means that they are hard to grapple with. Arguing as he does mostly by analogy and suggestion does not help. There is ultimately no real distinction between continuous change and discrete change if we are prepared to let the distances between discrete positions become as small as we like. 'The interval leapt over in a transilience may be at least as large as it has been in any hitherto observed instance, and it may be smaller in any less degree'. If all states between positions of equilibrium are allowed, when the population is merely *in motion* as a wooden sphere would be, then that even permits continuous evolution with spurts of rapid change and extended periods of stasis. It is a theory with enough flexibility to account for a wide range of possibilities, but it closely anticipates later developments, no less flexible.

Weldon for one was confused by Galton's flexible leaps and novel positions of stability. The biologist wrongly supposed that regression to the mean would simply eliminate any such discontinuity! 'About "regression," I will say only this, that Galton was himself a good deal mixed, at least in his exposition, when he wrote *Natural Inheritance*: and that I cannot conceive that characters "which do not mix" are thereby rendered independent of the phenomenon of regression'.[29]

All this annoyed the world's leading gradualist, who paid no regard to Galton's nuances. From Dorset, Alfred Russel Wallace let off super-heated steam to E. B. Poulton. 'For nearly a month I have been wading through Bateson's book & writing a criticism of it — & of Galton who backs him up with his idea of "organic stability". Privately, I think Bateson's book one of the most pretentious, and most worthless — "as a contribution to the study of the problem of Species"— I have ever met with'. Worse followed. 'I met Bateson when I was at Cambridge in the spring & it struck me at once that he had a high opinion of himself. Neither he nor Galton appear to have any adequate conception of what Natural Selection is, or how impossible it is to escape from it. They seem to think that given a stable variation & natural selection must hide its diminished head! Bateson's preface, concluding reflexions &c. are often quite amusing in their cool bumptiousness! He is so cock-sure he has made a great discovery—which is the most palpable of mare's-nests!'[30]

The 'mare's-nest' was the idea (first formed by Bateson in Siberia at the salt lakes) that since environmental variation was itself continuous, it could not produce discontinuous variation in organisms. Indeed, Bateson could not see how even tiny advantages could be of selective value. A row with Weldon was in the making. Soon his old friend was active in the pages of *Nature*, asking for precise definitions of 'discontinuous' and suggesting that while small variations were, by definition, hard to notice and therefore curate in museums, they might exist for all that. To find out whether or not (say) tooth size varied in real animals, large numbers of those should be collected and measured—to find shifts in distributions, though Weldon did not spell this out. Bateson might well have turned this around to ask for a precise definition of 'small', and how many selection steps made up 'continuous', but did not. It is a central point, but neither Weldon nor Bateson would ever understand it.

Yet Bateson found some (private) support from T. H. Huxley, who had a 'hasty glance' at the copy of *Materials* sent to him by the author. 'I see you are inclined to advocate the possibility of considerable "saltus" on the part of Dame Nature in her variations. I always took the same view, much to Mr. Darwin's disgust, and we used often to debate it. If you should come across my article in the *Westminster Review* (1860) you will find a paragraph on that question near the end'. That paragraph finishes with 'Nature does make jumps now and then,

29 Weldon to Bateson, 1894-02-15, quoted in Provine 2001, 44.

30 Wallace to Poulton, 1894-09-08, Hope Entomological Library, Oxford University Museum of Natural History, ARW 50. Wallace Correspondence Project, WCP4397.4659.

and a recognition of the fact is of no small importance in disposing of many minor objections to the doctrine of transmutation'.[31] But the former surgeon of the *Rattlesnake* mistakes the dose. Bateson is not adding *saltus* to taste, it is the main dish—it is all *saltus* all the time.

Galton had already allowed for arbitrarily small but still discrete leaps, but would no longer play a public role in the controversies that followed. Weldon and Bateson would soon form opposing schools, with Pearson drawn to Weldon. In turn, Bateson would gather his own disciples. Early in 1895, the fault-lines opened more clearly in public. There was an unenlightening tea-cup tempest in *Nature* over the origin of domesticated Cineraria flowers—selected gradually or hybridized quickly? Thistleton-Dyer and Weldon versus Bateson. Weldon suggested that Bateson had misread the history of *Cineraria* domestication. Bateson took that as a slight and fumed by letter. Weldon shrugged. Then Weldon's crabs resurfaced, in a Royal Society report.[32]

The idea of obtaining solid experimental evidence for evolution had long been one of Galton's pet projects. In a series of long letters, Wallace had suggested creating a test farm for that purpose in the early 1890s, but Galton thought the idea poorly specified. He wanted a crisp idea of how the evidence gathered would be decisive.[33] By 1894 the Royal Society had been persuaded to found and fund a committee—no more crisp—tasked with getting data. Galton was in the chair, with Weldon, Francis Darwin, Raphael Meldola and E. B. Poulton on the committee. Karl Pearson was added later.

The first report issued by the committee was not based on experimental data, but observational. Weldon had compared the crabs in Plymouth Sound at different ages, and thought he could detect a decline in variability (probable error), of the frontal breadth of the crabs, with age, suggesting that active selection reducing extremes was underway. There was no decline of variability in an unrelated character, right dentary margin.[34]

Bateson sprang into action almost at once. The methods used did not convince him biologically. Were the crabs measured at the same stage of moulting? It was not an especially good argument—two large random samples probably would not differ for that. Nevertheless, he repeatedly badgered Galton, as chairman of Royal Society committee, with complaints. Galton conciliated by expanding the committee, inviting more biologists, including Bateson, to rebalance Weldon and Pearson's mathematical skew. Pearson would later claim that there was a deliberate plot hatched by Bateson to take over the committee,[35] but this approaches paranoia: the biologist *turned down* Galton's initial invitation to join, and only changed his mind when asked to reconsider. It was Galton's idea to expand it, apparently to give the other side a fair hearing. But that shifted the committee from whatever position of equilibrium it had found, into an unstable state, rolling forward toward another facet.

Weldon soon found a better natural experiment to demonstrate natural selection in action. A mile-long artificial breakwater built in Plymouth harbour had disturbed the local environment of the shore crabs, trapping sewage (increased by growing towns) and fine moorland clay silt deposited into the bay by drainage. Collecting measurements of male crabs, and comparing to those obtained in earlier years—thus 1893, 95, 98—it seemed that narrower-width crabs were being selected for over time by the enriched diet of sewage and silt. That is, the mean frontal width was reduced, which Weldon speculated made the crabs more efficient at filtering out the silt.

These results were checked by experimentally confining crabs to laboratory tanks in which

[31] Huxley to Bateson, 1894-02-20, L. Huxley 1902, 394; Thomas Henry Huxley 1893, 77.

[32] See the able summary in Provine 2001, 45-9, for article references and letter excerpts.

[33] Transcriptions appear in K. Pearson 1930a, 128-33.

[34] Bulmer 2003, 305-6, points out that Weldon failed to take into account the *shape* of the crab carapace, which changes with age. Bulmer's calculations suggest that doing so eliminates the observed difference in widths.

[35] K. Pearson 1930a, 287. The new boys included F. DuCane Godman, Walter Heape, E. Ray Lankester, Maxwell Masters, Osbert Salvin.

they could be progressively poisoned by gradations of silt and putrid sewage. Again it seemed that narrower crabs survived fine silt and potent sewage, but at lower levels the pollutants made no difference to width. Presenting his results to the British Association of Science, held at Bristol in 1898, Weldon aimed squarely at Bateson: 'numerical knowledge of this kind is the only ultimate test of the theory of Natural Selection, or of any other theory of any natural process whatever'.[36]

It was strong evidence, and it is inherently easier to show a shift in means than it is to show a convincing change in variation (for that, much more data is needed). But with enough motivation, objections could always be found. Perhaps the crabs from earlier years benefited from unusually warm conditions then, increasing their width and robustness? Maybe the crabs dying off in the laboratory were just the younger ones, which were more vulnerable and just coincidentally narrower in frontal breadth? Weldon did not have definitive data—in which age is clearly separated from width through complete lifecycles—to counter this objection. And his abandonment of his former hypothesis, reduction in variation, eroded some trust.[37]

Meanwhile, Bateson was in pursuit of (what he thought would be) contrary evidence, for leaps. In 1897, he had persuaded the Royal Society committee to award him grants for experimental breeding studies, crossing varieties of plants. His stated intention was to examine the resulting characters statistically. It is doubtful whether he really had the required skills to do this, though surely he could have acquired collaborators who did.

Tensions within the committee spiked. Pearson and Weldon, now firm friends, were dissatisfied with the influence exerted by Bateson and the more traditional biologists, who they considered old-fashioned and a brake on the new methods they favoured. Galton had also become close to the two friends, but still maintained a cordial and professional relationship with Bateson, who had broken with Weldon. He continued to press for more data, advocating that photography be used to make measurements of livestock at one remove. By standardizing distances at which the shots are taken, it is possible to measure characteristics relatively within the photograph itself (an early example of photogrammetry). Some experimentation showed that this could be very accurate. Composite photographs of, for example, horse heads could also be used to characterize prize bloodlines, an echo of their application to anthropology a decade earlier. Accumulation of data from ever-popular livestock and thoroughbred competitions could permit inference about heritability of characteristics. After proposing this at the British Association, a Royal Commission was appointed to look into the matter, and was used effectively by Galton, who drove its favourable report. The effort appears to have petered out afterward though: one can photograph horse-fanciers at water, but fail to make them drink.[38]

In January of 1900, when he was abroad in Egypt, Galton resigned from the committee by letter, joined by Weldon, Pearson, Thistleton-Dyer and Meldola. Bateson and the rest, now in control of the funds, cheerfully soldiered on. As it turned out, Bateson, like all other investigators pursuing the laws of breeding experimentally, had been beaten to the punch by an obscure Augustinian monk he called 'Mendl' in Brnö (then part of the Austro-Hungarian empire, now in the Czech Republic) fully 35 years earlier. In August 1900, Bateson informed Galton of this remarkable rediscovery, which had been brought to his attention by another advocate of discontinuity, Hugo De Vries (they had met in England the previous year at the Horticultural Society Conference on Hybridization).

9 Aug
1900

Dear Mr. Galton

[36] Weldon 1898. Weldon withdrew his previous paper on the subject, Weldon 1895, which he conceded had 'neglected several important facts' (left unstated).

[37] Joseph Thomas Cunningham (1859–1935) in Cunningham 1899.

[38] F. Galton 1898b,c,d,e.

de Vries no doubt sends you his papers and so you will have seen his new work on the splitting up of the characters of hybrids compte rendu, 26 Mar, 1900 more fully given in Ber. Deut. Bot. Ges. 1900 p. 83.[39] But I am writing to ask you to look up the paper G. Mendl, Verb. naturf ver. Brünn, 1865.IV, p. 3 to which he there refers, in case you may miss it.

Mendl's work seems to me one of the most remarkable investigations yet made on heredity and it is extraordinary that it should have got forgotten.

In the current Bot. Cblt LXXXIII. No. 3, p. 84 is a review of a new paper by Tschermak[40] confirming Mendl's work and adding to it. Tschermak's paper I have not yet received.

This group of papers, if the work is all reliable, carries things on a long way.

I don't like de Vries' introduction of squares of D and R which is very misleading, but that is a tricky point as yet.

yours truly
W. Bateson

Mendel—well-educated and then earning a living as a physics teacher, shortly to be elected a salaried Abbot—had hybridized, or bastardized as he put it, large numbers of garden peas. Some of these he found tasty. He crossed varieties with alternative characters, for example two different (discrete) colours of seeds (yellow and green), or alternate shapes (smooth versus wrinkled). He noticed that in each case the character emerged in a nearly constant 3:1 proportion in the set of second-generation offspring and beyond. He deduced that an unobserved factor with alternate forms (say, A and a, which any plant may have in combination in four possible ways: aa, aA, Aa or AA) determined the character. He supposed that these alternate forms split apart (segregated) during crossing and then recombined in the offspring, one from each parent, in one of the four possible ways. The monk proposed that A dominates a in garden peas, so that if the offspring gets Aa or aA, that is the same as AA, which yields one character in three different ways, while aa produces the other character in just one way. He also tested more complicated combinations of alternate characters, which follow a binomial law.

This inspired exercise in hypothetical model fitting got rather limited attention—we can only speculate why—while Mendel was still alive. In early 1865 he had given a two-part talk on his results to the local scientific society, which was published a year later in the *Verhandlungen des naturforschenden Vereines in Brünn* [Transactions of the nature research association of Brünn]. Four years after that he published another paper on less interpretable results from hybrids of hawkweed (Hieracium), completing his scientific output.[41]

A general survey of hybridization by W. O. Focke noticed Mendel's papers (inaccurately) in 1880/1, and this would lead others back to them decades later, by different paths.[42] There were various mentions of Mendel in passing by other authors, but none recognized the significance of his conclusions or engaged with the substance of his experiments. Curiously Charles Darwin owned a copy of Focke, but we know that he did not read the relevant Mendel reference in it, since those pages were uncut. The naturalist lent it to G. J. Romanes—who also failed to cut those pages, but nevertheless did include Mendel in an article he wrote on Hybridization for the *Encyclopedia Britannica.* Mendel's fragile personality may have had something to do with this neglect—novel theories benefit from brash advocates.[43]

[39] Comptes rendus de l'Académie des Sciences.

[40] Erich Tschermak, 1871–1962.

[41] Mendel 1866, 1870. Translated along with Mendel's correspondence and critical commentary in Stern and Sherwood 1966.

[42] Focke 1881.

[43] There is now a small industry organized around Mendel, with much product-differentiation among scholars. For a clear summary see Hartl and Orel 1992.

Now, in 1900, the atmosphere had changed. Bateson, well-primed to receive its ideas by his expectations of discontinuity—colours changed in one step—had known of Mendel's work since that May, alerted by Hugo de Vries. He had read the original paper in German on a train to a Royal Horticultural Society meeting and immediately incorporated it into the address he delivered there, the first to introduce it to the English scene. But Galton had been abroad in Egypt and Greece at the time.

Galton certainly knew Hugo de Vries (1848–1935)—an enthusiast for the evolutionary significance of discontinuous mutations of hereditary particles, which he called 'iterations in pangens'—well enough. The Dutchman had written to Galton in September 1894, noting that he had 'been working during several years on the introduction of your great principles of research into the study of heredity in plants' and to express 'gratitude for all I learned by the study of your books and papers'.[44]

However there is a mystery here: De Vries appears to have known of Mendel's overlooked paper since 1895, from a paper he had been sent by an American breeder,[45] without fully appreciating its implications. He was then reminded of it in 1899 when one of the few people who had an original copy forwarded it to him. Perhaps he noticed then how closely it accorded with his own results. It is possible that De Vries had by then discovered the 3:1 Mendelian ratios himself, through his own 'monster-producing' hybridization experiments. But this is disputed—he may have nudged his estimates to match Mendel's, once he fully comprehended them.[46]

Galton seems to have familiarized himself with the rediscovered paper as soon as he learned of it from Bateson. As he wrote to Weldon in 1905, 'I don't believe anybody would have appreciated your work more than Mendel himself had he been alive. Dear old man; my heart always warms at the thought of him, so painstaking, so unappreciated, so scientifically solitary in his monastery. And his face is so nice. I can't give you any useful hints. I wish I could. I am just a learner, and bad at that now'.[47]

The neglected Abbot would be praised in Galton's *Memories*. 'I must stop for a moment to pay a tribute to the memory of Mendel, with whom I sentimentally feel myself connected, owing to our having been born in the same year 1822'. (There is another curious parallel between the two men, since Mendel also reacted adversely to stress—recall Galton's 'regular breakdown'—with severe and lasting physical symptoms. In one early incident, it took Mendel a full year to recover.) 'Mendel clearly showed that there were such things as alternative atomic characters of equal potency in descent. How far characters generally may be due to simple, or to molecular characters more or less correlated together, has yet to be discovered'.[48]

Unlike their mentor, Weldon and Pearson remained unconvinced by Mendelism to the end. In 1900, Pearson was promoting his own theory of 'homotyposis' which equated variation between individuals to variation *within* individuals, in their 'undifferentiated like organs'. Homotyposis is now forgotten, but it fuelled another dispute with Bateson, who poured biological scorn on it. One curious consequence of the theory would be that sexually reproducing organisms would be no more variable than asexual ones! Bateson refereed the resulting paper for the Royal Society and circulated his criticism before it was published. Pearson took extreme unction at this irregular proceeding, correctly extracting an apology from Bateson, but (again, correctly) no biological concessions. However it was enough to persuade Pearson that the Royal Society would not be a welcoming forum for his ideas. Soon Pearson and Weldon were pregnant with a new journal for their own use.

[44] De Vries to FG, 1894-09-27, GALTON/3/1/2.

[45] Bailey 1892, which he later misremembered as Bailey 1895 (this does not book Mendel's study but does mention him once by name).

[46] Kottler 2006.

[47] FG to Weldon, 1905-03-31, K. Pearson 1930b, 542.

[48] F. Galton 1908, 308.

It was Pearson's idea to call the new journal *Biometrika*. It was pitched to Galton as a forum for applications of mathematics to biology, which Pearson was persuaded, perhaps correctly, no other forum would then welcome. With Galton on board as a nominal editor, lending his prestige and depositing his cash in a guarantee fund, Weldon and Pearson could assume day-to-day control of their own organ. In other hands this might not have worked, but both men had prodigious energy. And so the 'biometric' movement was founded, which only strengthened over time, simply because it made perfect scientific sense, regardless of the specific quarrels attending its birth.

The first issue of *Biometrika* appeared in June 1901, with a foreword on 'Biometry' by Galton. He declared that 'a new science cannot depend on a welcome from the followers of older ones'. This new science could feasibly detect evolution in progress. 'Evolutionary changes are exceedingly slow as a rule, but supposing that a thousand years or thirty generations of mankind, would suffice in some particular case for some conspicuous alteration in a species, exact measurements ought to discover its progress well within the limits of a human lifetime'. It is not clear if he has revised his position of 1894 that evolutionary change *need* not be slow, given the qualifier 'as a rule'. But then the 'organic world as a whole is a perpetual flux of changing types', implying less stasis than might have been inferred in 1894. He envisaged a journal 'for those who are interested in the application to biology of the modem methods of statistics', who would marshal mobs of data into orderly arrays. The misguided could be saved from misconceived labour by better methodological advice.

One of Galton's *en passant* ideas is startling in its foresight. 'I hope moreover that some means may be found, through its efforts, of forming a manuscript library of original data. Experience has shown the advantage of occasionally rediscussing statistical conclusions, by starting from the same documents as their author. I have begun to think that no one ought to publish biometric results, without lodging a well arranged and well bound manuscript copy of all his data, in some place where it should be accessible, under reasonable restrictions, to those who desire to verify his work'.[49] If closer attention had been paid to this, the 'replication crisis' of modern times might have been moderated.[50] That repository has still not been realized.

The new journal had an American presence too, with Charles Benedict Davenport (1866–1944) acting as local editor there. Davenport, then a Zoologist at Harvard, had been in touch with Galton since 1897, drawing a warm response. 'What perhaps gratified me most was that you perceive a unity in my work although there is much variety in the subjects. I quite feel it myself, but have often feared that others might think more concentration needed'.[51] The scientist discreetly pressed his new convert to look for simplifications to Pearson's skew curves. 'Cannot you stir up some good mathematician at Harvard, to interest himself in these matters & to master Karl Pearson's papers. It would considerably help the course of statistical inquiry, if the entirely fresh minds of good mathematicians were diverted to it. The want now seems to be of approximation formulae that can be used in place of Karl Pearson's accurate but very troublesome ones, whether in normal or in skew systems, and to find out their properties. It will give me great pleasure to hear of anything going on in that direction in America'.[52]

However it was not to be simplifications of skew curves that proved the sticking point for Pearson, but rather Mendelism. Davenport soon absorbed the emerging experimental literature and became a convert, submitting a paper to *Biometrika* on those lines. When Pearson tried to squelch it, the American baulked. The mathematician would not accept more than a few hypothetical Mendelian factors—too many degrees of freedom—and Davenport proposed many. The American had not expected the journal to be run 'in the interest of one idea', he wrote to

49 F. Galton 1901a.
50 Ioannidis 2005.
51 FG to Davenport, 1897-04-06, APS.
52 FG to Davenport, 1897-07-29, APS.

Pearson in 1903. 'I gather from your last letter that a mistake may have been made in inviting me to be a co-editor of *Biometrika*—considering my heretical views'. 'To relieve you from any embarrassment, I resign'.[53] (The Harvard man would soon go on to found an influential laboratory at Cold Spring Harbor and play an important role in the American eugenics movement. His name continued to appear on the masthead of *Biometrika* for several years.)

It had been questionable from the start whether or not Mendelians would be given a friendly hearing in *Biometrika*. In the second part of the very first volume, Weldon had lodged objections in a minute analysis. 'Mendel's Laws of Alternative Inheritance in Peas' gives Mendel muted credit but doubts whether his theory is universally applicable: 'segregation of seed-characters is not of universal occurrence among cross-bred Peas, and that when it does occur, it may or may not follow Mendel's law. The law of segregation, like the law of dominance, appears therefore to hold only for races of particular ancestry'. For general applicability, Weldon thought that ancestry, not just the characters exhibited by the immediate parents, had to be taken into account. More general Mendelian laws seemed unlikely to succeed. In a second paper in the next volume, Weldon went on to suggest that Mendel had coarsely classified colours with intermediate shades, which would be better handled by biometric methods.[54]

Tabulating the data, Weldon noticed that all but two of Mendel's seven results for the second hybrid generation were well within the expected 'probable error'. Even the two deviations were not much greater than expected. He did not state explicitly that this was suspicious, but the implication is clear enough, 'if they were repeated a second time, under similar conditions and on a similar scale, the chance that the agreement between observation and hypothesis would be worse than that actually obtained is about 16 to 1'. He then applied the χ^2 test, recently invented by Pearson, on the correlated combinations of colours that Mendel had found, and determined that 'if the experiment were repeated a hundred times, we should expect to get a worse result about 95 times, or the odds against a result as good as this or better are 20 to 1'.[55] He found similar better-than-expected concordance for the third generation data. Privately he wondered in a letter to Pearson whether it was conceivable that the tonsured monk had fiddled his data, 'I cannot help wondering if they were not too good'.[56] Later, the eminent statistician R. A. Fisher would take up this lead.

Weldon's cool, somewhat obtuse, dismissal of Mendelism—safe in doubting universal application, but failing to really exploit its new insights—with the implication that its adherents were a little naive, bothered Bateson to the point of frenzy. The biologist thought at first that Pearson might be detached from Weldon, and made an ally of in this dispute, completely misreading the mathematician in a patronizing letter. 'I respect you as an honest man and perhaps the ablest and hardest worker I have met, and I am determined not to take up a quarrel with you if I can help it .... There has probably been no discovery made in theoretical biology that we can remember which approaches Mendel's in magnitude, and the consequences it leads to. This is not a matter of opinion but certain. You have worked well in the same field and if through any fault of mine you were to be permanently alienated from the work that is coming, I should always regret it. With Weldon it is different. He is a naturalist. He goes in with his eyes open. Besides, as between him and me it is too late'.[57]

The reply from Pearson—never known to forgive those who crossed him—was swift and brutal. 'I think sometimes you cannot be aware that Weldon has been for many years past one of my closest and most valued friends; that I do not readily make friends, and that when I say a man is my friend I mean that I have tested the strength of his affection in the graver matters

[53] Davenport to Pearson, 1903-06-05 and 1903-07-07, cited in Witkowski 2008, 40.

[54] Weldon 1902a,b.

[55] W. Palin Elderton had included a handy χ^2 table a few pages previously.

[56] Weldon to Pearson, circa 1901-11, cited in Magnello 2004. Photographs of Mendel suggest that he was not tonsured.

[57] Bateson to Pearson, 1902-02-13, cited in Provine 2001, 63.

of life, and am prepared to do for him and to accept from him anything that one human being can or will do for another. I think, as I say, that you have not known this, or possibly your references to him,—only three or four, but my memory is very jealous in such matters—would have been more guarded. As to the scientific side of the present controversies, I am perfectly ready to hear both sides, and will willingly reserve space in Part III of Biometrika for your defence of Mendel, if you think our Journal a suitable locus for your paper'.[58]

Bateson, no doubt stung, perhaps wary of editorial intervention, did not take up Pearson's pro-forma offer. Just two months later he rushed out—'at once to defend Mendel from Professor Weldon'—*Mendel's Principles of Heredity: a Defence*, including a (slightly) revised translation of the monk's key paper, and the later less substantial follow-up by the Abbot.[59] The preface contained a lot of spluttering, 'exactness is not always attainable by numerical precision: there have been students of Nature, untrained in statistical nicety, whose instinct for truth yet saved them from perverse inference, from slovenly argument, and from misuse of authorities, reiterated and grotesque'.

Weldon is named as 'a vehement preacher of precision' who dispenses with it in order to criticize, and who is by implication 'worthy of scant respect'. Although his fellow editors of *Biometrika*, Galton and Pearson, are praised as 'great intellects', they are unfortunately 'not trained in the profession of the naturalist'. More extreme statements had been filtered out by the publisher. Bateson also pressed Mendel retrospectively for an anti-Darwinian, a notion not supported by the monk's text—he would surely not have taken Bateson's shilling.[60] Similar arguments were made in a *Report of the Evolution Committee* which listed the results from experimental breeding obtained by Bateson.

It is curious that Bateson and Pearson held comparable views about hereditary particles, which, as we have seen, Galton had proposed early in the 1870s but neither of the two men entertained. Pearson, an early positivist, saw no need to invent unobserved causal entities and preferred to stop at describing the data. Anything more was metaphysics. Bateson thought that transmission between generations took place not through an exchange of particles, even if these were present, but rather through a kind of vibratory process between parents and offspring, which he was never able to satisfactorily describe. It is possible to improve his thinking here, to speak of *information* transmission during reproduction, but that goes far beyond his own writing and is a mathematical abstraction he would likely have scorned.[61] Opposition to particles would cause Bateson a lot of trouble later on.

In his more substantial comments, without adding any new data, Bateson denied that ancestral characters had any relevance at all once it was known that a parent had some particular character. The factors for characters were passed down entire, or 'pure'. There was no side-channel for passing ancestral influence, no Galtonian 'stirp'. This exposed the biologist to obvious criticism. In the *New Phytologist* for 1902, George Udny Yule—who had been a 'demonstrator' for Pearson at UCL in the early 1890s, then assistant professor of mathematics there—noticed Bateson's latest work. The biologist's 'turgid and bombastic' preface was deplorable: 'one cannot help feeling that his speculations would have had more value had he kept his emotions under better control; the style and method of the religious revivalist are ill-suited to scientific controversy'.

Yule saw great potential in Mendel's 'ingenious and remarkable' work, but calmly pointed out that, because of environmental influences and dominance, one might not *know* what (cryptic) factors the parents actually carried, based solely on the (somatic) characters they displayed. Knowledge of ancestral characters could help to disambiguate that. Then Galton's Law, inter-

[58] Pearson to Bateson, 1902-02-15, cited in Provine 2001, 63-4.

[59] W. Bateson 1902.

[60] R. A. Fisher would later dismiss this claim with prejudice, Fisher 1936, which is borne out by Mendel's wording.

[61] Here I respectfully disagree with Cock and Forsdyke 2008, without scorn.

preted strictly as a predictive model, sans stirp and not as a mechanism, would be applicable and improve on Mendel. The latter is a special case of the former, the two systems are not contradictory.[62] Yule was right. But it was a far-sighted first step toward intellectual and political reconciliation of a kind that Weldon and Pearson did not necessarily welcome—nor Bateson.

By now Weldon and Pearson were exceptionally close, taking vacations in the country together—in reality, a near-frenzy of work on their joint child *Biometrika*. So it was that after Easter of 1903, 'a biometric camp was formed at Peppard on the Chilterns; here the "Consulting Editor" and one of the co-editors had established themselves, and the Weldons took a week-end cottage. The three Oxford members of the party arrived partly on cycles and partly on four feet, and were often met *en route* by the residents in the uplands, the numbers being swelled by the addition of biometricians from the London or Oxford schools. Hence arose a series of Friday "biometric teas," for the discussion of the week's work and plans for the next two days. Saturday and Sunday morning were given to steady calculating and reducing work, and much was got through'.[63]

The conflict between Bateson and the biometricians found a new local maximum in August 1904, at the British Association meeting camping at Cambridge that year. Bateson was president of the Zoology section (D), and delivered an elaborately rehearsed attack on the 'Ancestrians' as he now called them. From Galton he appropriated the idea that variation is divided into the true kind, leaps, and the merely apparent kind—not-leaps, subject to regression. The real mechanism of evolution is not Darwinian selection on minute differences (ignoring Galton's flexibility there), an idea which he claims had stymied further progress since 1859. Rather it is selection of much larger atomic leaps, as explained by Mendel's principles. Specific types are fixed. Mutations generate integral leaps between types. Repeated selection on these leaps is downplayed, extending to a short period and a few steps at most. (He is forced into this position, since those would amount to no more than a change in step size in the Darwinian account, from tiny, though unspecified, to large, but also unspecified.)

A footnote to the published address—'when the unit of segregation is small, something mistakably like continuous Evolution must surely exist'—suggests an unusual talent for simultaneously stating the solution to a problem while failing to recognize that it is, in fact, the solution.[64] No doubt Pearson and Weldon, who were both in the audience, winced at the statistical howlers. 'If individuals from necessity are treated collectively, it must be proved that their composition is identical'. Surely they grimaced at: 'In direct contradiction to the methods of current statistics, Mendel saw by sure penetration that masses must be avoided'.[65]

The day after the address, an open discussion was held. A crowd had packed the hall, anticipating 'demonstrations of partisanship so new to the serene abstraction of section D', as *The Times* drily put it.[66] They were not to be disappointed. Weldon—who had cycled over all the way from Oxford—opened by questioning Mendel's colour classifications again, finer shades of which might be affected by ancestors. Likewise, 'hairy' versus 'glabrous' classifications of plants surely had intermediates. He argued that Pearson's predictions of the colour of hybrid mice (bred by Cuenot in Paris) were even closer than Mendel's laws provided, with the added advantage that they did not require a 'cumbrous and undemonstrable gametic mechanism'.

It was not especially strong stuff, but it set Bateson off on a furious rehearsal of now familiar arguments, ending with a straight comparison between the 'Ancestrians' and the 'Flat Earth' reactionaries of days past, all of which 'culminated in the President dramatically holding aloft

[62] Yule 1902. Galton had known the statistician's uncle Sir Henry Yule (1820–1889), a distinguished geographer, from the RGS, and had also collected material from him for *Human Faculty*.

[63] K. Pearson 1906, 42-3.

[64] B. Bateson 1928, 238.

[65] B. Bateson 1928, 244.

[66] *The Times*, Saturday August 20, 1904, 8. This is the most complete published account.

the volumes of this Journal [Biometrika] as patent evidence of the folly of the school', as Pearson told it.[67] The mathematician then rose celestially to defend statistical methods, clarifying that the 'introduction of methods of precision had nothing to do with Mendelism or Ancestral law'. The Mendelian theories were too vague and in flux to test rigorously, and Pearson called for a public ceasefire for some years until the details could be worked out without personal feeling. Against this the Rev. Thomas Roscoe Stebbing, an amiable Darwinian Zoologist, bounced up to state that he rather hoped they would keep having it out in public instead, 'from which the world could only gain'.

It was all very entertaining. Galton, who was not there due to his deafness, hastened to meet Weldon and Pearson at the Bull Inn, Burford, to learn all about the 'rather savage phrases' by Bateson that the *Times* had shielded its readers from.[68]

The dispute between the Biometers and the Mendelians, in search of a new local maximum, next moved to the question of the coat colours of horses. A large source of data was available from *Weatherby's General Stud Book of Race Horses*, and had already been used by Pearson for other purposes—as an aside, the mathematician had noted that coat colour did not seem to be Mendelian at all. Inspired by previous experimental results from rabbits, the Batesonian Charles Chamberlain Hurst (1870–1947) countered that the stud book showed that bay and brown were dominant to chestnut. Weldon refereed the resulting paper that Hurst submitted to the Royal Society and pored obsessively over the studbook to find counterexamples. He dramatically pointed these out when Hurst read his paper late in 1905, and followed with a written note a month later. Bateson, who had communicated the paper for Hurst, promptly withdrew it.

Always tenacious, Hurst was able to argue that the anomalies were just errors of entry in the studbook, and the paper was resubmitted, then published, in the new year. A note he added, countering the anomalies claimed by Weldon, seems to have incensed the biologist—perhaps because Hurst signalled an intention to treat any anomaly as either a data error, or something to be expected occasionally, irrefutable protection from criticism.

Weldon then embarked on a massive study of the studbook, while ostensibly on holiday in Rome, even when jointly holidaying with Pearson near Woolstone in Oxfordshire. An entire memoir on horse colours was planned. The overworked biologist was still supposedly on holiday in April of 1906 when he caught pneumonia after several outings in the countryside; he died in a nursing home on Good Friday. Pearson, who had been cycling with him a few days before, took this as a bitter blow. As he had told Bateson years earlier, he did not make friends easily, and his loyalty to his companion in the 'biometric teas' of earlier times was nearly boundless.

After Weldon's death, the calmer, less obsessive, head of George Udny Yule continued to work at resolving the ongoing dispute through better mathematical models. In 1906 (it would be published a year later) he presented one to the Third International Conference on Genetics, put on by the Royal Horticultural Society in London. An earlier result by Pearson purported to show that Mendelian factors could not account for observed parent-offspring correlations, such as height.[69] Theory predicted 0.3, but 0.5 was observed in the extensive data sets collected by Pearson. Yule pointed out that this was a consequence of an assumption that dominance was total. If this was relaxed to allow for varying levels of dominance then a parent-offspring correlation of $1/2$ would result when dominance was absent. Likewise $1/4$ for grandparents, $1/8$ for great-grandparents and so on.

Thus Mendelism—'the theory of the pure gamete' sans ancestors—could approximate the biometric approach as more factors contributed to a character without complete dominance.

---

[67] K. Pearson 1906, 44. Newspaper reports do not mention this detail.

[68] FG to Pearson, 1904-08-23, GALTON/3/3/7/20. FG to Millicent Lethbridge, 1904-09-06, K. Pearson 1930b, 528. An earlier letter from Weldon about those 'savage phrases' has gone astray.

[69] K. Pearson 1904.

Yule also argued that non-heritable factors would also affect 'compound characters' (which we call polygenic phenotypes). These were confounded in observations with the presence of dominance, but a complete theory of heredity had to account for them.

This paper has many of the core ideas needed for a complete reconciliation of Mendelism and biometry, but, once again, it had little influence at the time. Possibly because Yule's approach was terse (it runs to three pages only) and not a few mathematical details were left for the reader to fill in.[70] In 1909 Pearson accepted its conclusions—for gametes and characters affected by multiple factors—without mentioning Yule. 'That ancestry does not matter if we know the gametic constitution of the parents, that it does matter if we only know the somatic character of the parents appears to be the solution of one of the difficulties which some have found between the Mendelian and biometric methods of approaching the subject'. If absolute dominance was discarded, 'there would be little left of contradiction between biometric and Mendelian results as judged by populations sensibly mating at random'.[71] However the way in which he interpreted this going forward was that Mendelism was just a simplistic nuisance.

In the meanwhile, Bateson—who had held only minor appointments at Cambridge until he was appointed Reader in Zoology in 1907, operating largely through the Royal Society—had landed a major coup. In 1908 a donor had funded a chair at the university intended for the novel subject of genetics. Nevertheless it was named for biology, with Bateson as the first occupant—later was it changed to genetics. The following year he brought out *Mendel's Principles of Heredity*. The title is confusingly close his to 1902 fusillade, which he had allowed to go out of print, but drops 'a defence'. Like its namesake it contains translations of the two Mendel papers, but those are now secondary to a substantial description by Bateson of the advances made since then. In the same year he noticed what he called attraction and repulsion between genes, but was soon known as linkage, but failed to notice the role played by the chromosomes in this, which he did not think important and resisted for years.

Unlike Pearson, Bateson maintains that the difference between the biometric and Mendelian approach is 'absolute and irreconcilable'. The biometric school has only professed the two theories compatible after 'every possible means of discrediting the truth of Mendelian facts had been exhausted'.[72] Which is to say, the biometricians were not *true and sincere believers through the exercise of their own free will.* Underlying Bateson's antagonism is a religious dimension which explains some of his heat. Pearson was following along, 'I have just been looking at Bateson's book. I take the place of Weldon as the butt for his contempt'.[73]

Galton had asked Bateson for a copy of the new book, but had to remind the biologist through his niece Eva that none had been received. Bateson seems to have felt a little bashful, since it contained not a few unflattering remarks about the older man too. He hastened to say that he would certainly send the book, sugaring the pill. 'I don't think many people admire, or can admire, Mr Galton more than I do. The novelty of his thoughts and the freshness of his outlook on nature are not to be found in any other living writer, so far as I know. I often remember the thrill of pleasure with which I first read *Hereditary Genius* and the earlier chapters of *Natural Inheritance*, and every year when I read aloud pieces of those books to my class, as I always do, I can see what excitement they have the power to cause'.[74] The book appears to have been the source of the three pages of detailed notes related to Mendel—made by Galton on July 20, 1909—which survive in the Galton papers.[75]

Mendelism was co-opted by Bateson because it offered the entire parts and discontinuities he had hankered after since the 1880s. Whole units, segments and changes in higher-level body

[70] See Bulmer 2003, 266-272, for a detailed reworking in modern terms.
[71] K. Pearson 1909.
[72] W. Bateson 1909, 127,130
[73] Pearson to FG, 1909-05-07, K. Pearson 1930a, 383.
[74] Bateson to Eva Biggs, 1909-07-07, K. Pearson 1930a, 288.
[75] GALTON/2/4/19/11.

plans always dominated his thinking. His Yale lectures of 1907, published eventually as *The Problems of Genetics* in 1913, explain that he thinks species are distinct and immutable, yet have somehow evolved, in a way that he cannot explain. Adaptation is neutered—an inversion of Fleeming Jenkin's blending conundrum, since now nothing will ever transition. He can see mutations sometimes giving rise to new species but is not so rash as to believe that all extant species can be generated in that way, entire. The most important problems in biology seem hopeless as a result. All in all, 'an unsatisfying conclusion'.[76] In the meanwhile the biologist glibly advocates multiplying the classification of species indefinitely—as many as one can get away with. It is all as thoroughly anti-Darwinian as anybody could have conceived of at the time, and a dead-end road.

As Galton told Leonard Darwin, Pearson could understand Bateson, but Bateson could not understand Pearson. Darwin relayed this to R. A. Fisher, who improved it by having Bateson know nothing of mathematics, and Pearson know nothing of biology (this ignores Weldon).[77] However it is misleading to blame Bateson's lack of mathematical sophistication here. That does seem to have blinded him to the accumulation over time of modest but persistent advantages, forming the very substantial ones he was looking for at the species level—the kind of steady accretion that Galton, by contrast, fully appreciated.[78] But it is like stating that a man has not walked to Tartary because he has not the shoes, whereas he actually has no desire to go there in the first place. Bateson liked his evolutionary porridge lumpy and could not see how any amount of mathematical pulverizing would improve it.

The dispute would drag on grimly, in the attenuated form of sarcastic asides, until the participants, increasingly irrelevant, finally exited from the scene. Neither side would concede defeat, though it was (necessarily) the biometric school who laid most of the mathematical foundations for a unification of the two theories and not Bateson's Mendelians. The distribution of blame is distinctly skewed toward the Batesonians. The geneticist gave up his Cambridge chair in 1910 to head the John Innes Horticultural Institution, dying prematurely in 1926. Pearson outlived him by ten years only, after which his statistical son Egon steered *Biometrika* to even greater success.

The personalities of the participants were strong determinants of this outcome. Both Pearson and Bateson were energetic controversialists. In her collection of Bateson's 'Essays and Addresses', his wife Beatrice takes up three pages with a *rough* list of the controversies that her husband engaged in. She reproduces part of a letter he wrote to a protégé, cheerfully defining 'controversial' as 'a term used by those who are wrong, to apply to the persons who correct them'.[79] Pearson, too, had many disagreements which turned into intractable feuds—with Charles Spearman (as we have seen), C.B. Davenport, Major Leonard Darwin, and (as we will see) R. A. Fisher. There were more. Still, even given all the rancour of the above dispute, one wonders whether Gregor Mendel would not have benefited from at least a modicum of the dogmatism and public aggression on display. Being insulted in public is surely preferable to merely being ignored—a well-liked Abbot remembered only for an unusual talent for gardening, a fallen tree mouldering in Moravia.

Galton himself played no direct part in this fracas after 1901. He was a bemused spectator by then, as he had told Weldon, and always his own man. Not merely because he was being diplomatic—or because he avoided controversy, though he did—but because he really had sympathies for both points of view. He had after all allowed for *some* leaps in evolution, but of an arbitrary size, without committing to how prevalent 'large' ones were, or what 'large' even meant. Yet he endorsed evolutionary gradualism as a rule, as he had written in the very first

[76] W. Bateson 1913, 249.
[77] Bennett 1982; Fisher 1956.
[78] F. Galton 1910c.
[79] B. Bateson 1928, 464.

number of *Biometrika*. Galton could comprehend Mendel without kicking out the cornerstone of his personal world-view, always pragmatic. His own theory of heredity, the ancestral law, was a rough take based on an improvised notion of how inheritance worked, but pointed the way forward, suggesting which techniques might be fruitful. He was not wedded to it.

——oOo——

In 1918, the statistical geneticist R. A. Fisher, then a hard-up country schoolmaster, would build on Yule and Pearson to show that a trait influenced by multiple genes of small individual effect has an approximately normal distribution, with closer and closer agreement as the number of genes increases toward infinity.

It makes perfect sense, given the history of this dispute, that Fisher's ideas would not be received sympathetically when, in 1916, he first tried to get the results published by the Royal Society. Neither by the Mendelians—in the form of R. C. Punnett, a protégé of Bateson at Cambridge—nor by the biometers, now assuming the form of Pearson. Reg Punnett could not follow the mathematics, and said so—besides he thought it of little practical relevance to biologists, like 'weightless elephants upon frictionless surfaces'. But if the mathematicians liked it he would not object. Pearson found it 'in some respects suggestive' (!) but rather doubted a profusion of hypothetical entities (genes) toward infinity, which he thought would take things beyond the range of practical confirmation by experiment. Still, if the biologists liked it, he would not object.[80]

It is true that the Royal Society referees did not refuse publication *outright*, as Pearson's son Egon protested many decades later. Each put the burden on the other, equivocating. But Fisher correctly saw that responding to these sorts of comments was hopeless. The frames of reference were incommensurate. He fumed in the avuncular ear of Major Leonard Darwin. It was Fisher's turn to seek an outlet that was not blocked by two generations of the old guard. With Darwin's help he had the paper published by the Royal Society of Edinburgh, which wanted extra money for a paper of that length. The Eugenics Education Society gladly supplied the necessary, even after the canny Scots annoyed Darwin by upping their price on hearing the news.[81]

The incident was the first of many disputes between Pearson and the man who would eventually succeed him as Galton Professor at UCL, then bathe his memory in tepid praise cut with scorn. Fisher would lay the foundations for the modern theory of evolution in his *Genetical Theory of Natural Selection* (1930). The 'actuarial methods' which had baffled Bateson showed that selection acts by shifting gene frequencies in populations, with even very small amounts of selective pressure leading inexorably to substantial mean shifts over time.[82] When combined with the work of Sewall Wright and J.B.S. Haldane, the biometric/Mendelian debate was consigned to the past.

Fisher had been an (outstanding) undergraduate at Cambridge when Galton was still alive, but they do not seem to have met or corresponded. By that time the scientist was largely withdrawn from the world, though he did receive the young Harold Laski at Rutland Gate (see below). In 1911 the newly-formed Eugenics Society at Cambridge heard an astute talk from the student on 'Mendelism and Biometry'. Fisher was a founding stalwart of the society, along with John Maynard Keynes. He mentioned in passing the odd accordance found by Weldon (left uncredited) between the expectations of chance and Mendel's data, while wondering if a doubtful plant here or there had been nudged to the expected column.[83]

[80] Norton and E. S. Pearson 1976.
[81] Bennett 1982; Box 1978.
[82] Fisher 1930.
[83] Fisher 1911.

In 1936 Fisher returned to the subject of Mendel's data, in 'Has Mendel's work been rediscovered?' The article was little-noticed for a long time but ultimately prompted many duelling analyses. When closely examined, Mendel's data showed distinctly depressed variability that was highly unlikely to occur without intervention. Fisher concluded that the data had been adjusted to fit the theoretically expected ratios—again, he did not mention Weldon, who got there first. It is a conclusion that has never been refuted. No innocent procedural explanation is compatible with the descriptions given by Mendel of how the data was produced. Fisher himself suspected that the monk may have been misled by a research assistant. However that may be, the data is too good to be true. The theory—formed early in the process, perhaps even before it—elicited the data, rather than the other way around.[84]

Figure 13.1: Karl Pearson (1857-1936) as a young man, National Portrait Gallery.

[84] Fisher 1936; Weldon 1902a. The controversy is authoritatively summed up in Franklin et al. 2008.

Figure 13.2: Karl Pearson in the 1880s.

POETRY.

SONGS OF THE PROLETARIAT.

I.

Johann Most's Workmen's Reveille*

Ye, men of toil, arise !
The sun of freedom hies,
  To end black night.
Clasp, Proletariat band !
With faith the brother's hand,
Around in ev'ry land,
  The foe to spite !

Why still dost hesitate ?
Why thus inert await
  What cometh not ?
Halt'st thou 'fore horrors pale ?
With coward fear dost wail ?
For thee can never fail
  Thy heavy lot ?

Hope even to the grave,
Yet none will come to save
  From heaven's height.
Will'st thou a strong free race,
No longer servile place ?
Then for thy good right face
  Thyself the fight.

See, how so many yon
Luxurious live on,
  Upon thy toil !
Leaving thee nought to gain,
But crown of thorns' domain ;
Thy wage is only pain,
  Till death doth foil.

Then man by man unite
To join the ranks of fight,
  United be!
Men do the hardest thing,
Who boldly forward spring,
As sure to goal will bring
  Our unity !

Hail, labour ! hail it, folk !
It shall 'neath heavy yoke
  No longer be.
Haste, Proletariat band,
Hand truly clasped in hand,
Strive forward, one aim scann'd,
  Thyself to free.

K. P.

*Most's Proletariat Liederbuch.

Figure 13.3: Songs of the Proletariat I, by Karl Pearson, Cambridge Review *2, 1881-05-18, 335.*

SONGS OF THE PROLETARIAT.

II.

*G. Herwegh's Hymn to Hatred.**

Come on! come on! o'er hill and dale,
T'wards red of dawn with zeal,
To faithful wife last kiss not fail,
Then grasp the faithful steel!
Until our hand to dust doth turn,
It shall not drop the sword;
At last our soul with hate shall burn,
Too long has love been lord!

Love cannot help us on the way,
Love cannot end our pain;
Summon, O Hate, thy judgment day,
Break thou, O Hate, our chain!
And where disdainful tyrants spurn,
Lay hold of them unawed;
At last our soul with hate shall burn,
Too long has love been lord!

Let each who has a heart then try
The growth of hate to speed;
On every side is timber dry,
Our fiery flame to feed.
Children of freedom, ye must learn,
To sing with one accord;
"Now let our souls with hatred burn
Too long has love been lord!"

Then struggle early, struggle late,
'Gainst tyranny on earth,
More holiness will gain our hate,
Than all our love was worth.
Until our hand to dust doth turn
It shall not drop the sword;
At last our soul with hate shall burn
Too long has love been lord.

K. P.

* At a recent Sunday concert given at the German Social Democratic Club, Soho, in aid of the "Freiheit" Defence Fund, one of the German workmen remarked to a friend that this song was the best of the "Sozialdemokratische Lieder."

Figure 13.4: Songs of the Proletariat II, by Karl Pearson, Cambridge Review *2, 1881-06-01, 367.*

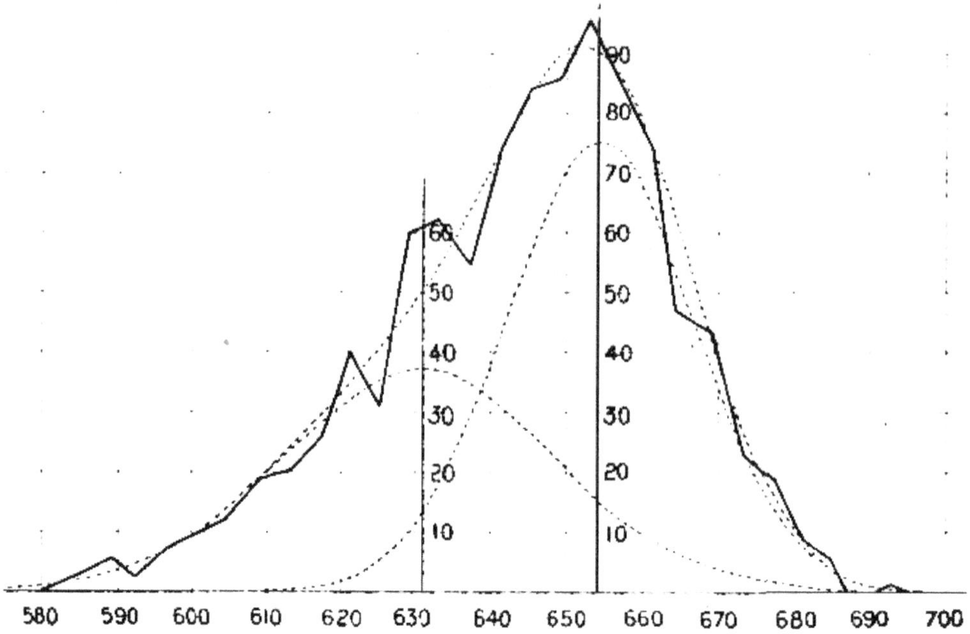

Figure 13.5: Weldon's skew distribution: frontal width of shore crabs from the bay of Naples.

Figure 13.6: Walter Frank Raphael Weldon (1860-1906), from the Biometrika *obituary.*

Figure 13.7: William Bateson (1861–1926) in 1905.

Figure 13.8: Ronald Aylmer Fisher the unifier, in 1924.

Figure 13.9: Francis Galton, from a magazine woodcut.

Figure 13.10: Francis Galton, by an Unknown Artist.

Chapter 14

Wellborn

In March 1874, Beatrice Potter—wealthy, serious-minded, beautiful but only on the verge of womanhood—recently returned from America and now bound for London—diarised the 13th. 'Now if I am a wise girl I shall go away at Easter, especially if F.G. comes—ten to one if I see him again I shan't be able to resist making a lot of silly castles in the air about him; and that is what I want to avoid'. She channelled her guardian angel. 'You are really getting into a nasty and what I should call an indecent way of thinking about men, and love, and unless you take care you will lose all your purity of thought, and become a silly vain self-conscious little goose'.[1]

'F. G.' was Francis Galton. Decades later Miss Potter would become Beatrice Webb, after turning down Joseph Chamberlain for Sidney Webb and abandoning herself to socialism. They had just two children: the Fabian Society, and the London School of Economics. In her memoirs Beatrice would be more guarded when recalling 'F. G.'. She saw the scientist often at 25 Rutland Gate, the home of her sister Georgie Meinertzhagen, but it may have been that old family friend of the Potters, Herbert Spencer, who introduced them first:[2]

> the one who stays in my mind as the ideal man of science is, not Huxley or Tyndall, Hooker or Lubbock, still less my friend, philosopher and guide Herbert Spencer, but Francis Galton, whom I used to observe and listen to—I regret to add, without the least reciprocity—with rapt attention. Even to-day I can conjure up, from memory's misty deep, that tall figure with its attitude of perfect physical and mental poise; the clean-shaven face, the thin, compressed mouth with its enigmatical smile; the long upper lip and firm chin, and, as if presiding over the whole personality of the man, the prominent dark eyebrows from beneath which gleamed, with penetrating humour, contemplative grey eyes. Fascinating to me was Francis Galton's all-embracing but apparently impersonal beneficence. But, to a recent and enthusiastic convert to the scientific method, the most relevant of Galton's many gifts was the unique contribution of three separate and distinct processes of the intellect; a continuous curiosity about, and rapid apprehension of individual facts, whether common or uncommon; the faculty for ingenious trains of reasoning; and, more admirable than either of these, because the talent was wholly beyond my reach, the capacity for correcting and verifying his own hypotheses, by the statistical handling of masses of data, whether collected by himself or supplied by other students of the problem.

Over a quarter of a century later, on the 14th of September 1901, 'F.G.' is perched in the Royal Albert Hall, peering intently at the stage through opera glasses. The venue is sold out, but he has secured one of the 15,000 seats, crammed with admirers of the male form, spilling dangerously into the aisles. Perhaps he has walked over from nearby Rutland Gate, to observe the throngs along the way. No cabs are to be had. For once he is not just people-watching but intently scanning the platform, where the impresario of 'physical culture', the great Sandow, is to select the three finest physical specimens from 60 competitors. From the 'physical culture' academies mushrooming all over Britain, all of the entrants have been preselected by Sandow, a marvellously muscled, well-proportioned but not-especially veracious

[1] Mackenzie and Jeanne 1982.

[2] B. Webb 1926.

strongman, who reinvented himself on the circus and Vaudeville circuit after fleeing Königsberg and the Prussian draft.[3]

A thousand guineas in prize-money is on offer, along with solid gold silver and bronze statuettes, based on a miniature by William Pomeroy of none other than Sandow himself. The Prussian is said to 'sculpt' his body through exercise alone, using classical Greek statuary as models. Two judges will assist him. One is Sir Charles Lawes (1859–1930), once a renowned Cambridge oarsman, subsequently a mile runner and cyclist, but now better known as a sculptor. The other is Sir Arthur Conan Doyle, the internationally-known creator of 'Sherlock Holmes' and author of historical romances—if Galton does not already know Conan Doyle, he will get to know him well a few years later when the author visits Rutland Gate to discuss fingerprints.[4] But now the Band of the Irish Guards is striking up a tune of Sandow's own composition as the competitors troop onto the platform. The Boer War is still dragging on, and not going all that well for the world's largest empire—some noses have been bloodied. All proceeds will be donated to the 'Transvaal War Relief Fund', or so Sandow says.

Galton was robust enough himself to survive the rigours of two years of rough African exploration in malarious latitudes. His family also boasted the well-known Regency strong man Robert Barclay Allardyce.[5] But now, at last, the on-stage exhibitions of chest expanders, wrestling, fencing and gymnastics are over. It is time to see the 60 hopeful competitors exhibit their proportions and tone on pedestals in choreographed poses. Black tights done up with black jockey belts bulge under innovative electric lights, leopard skin shoulder-pieces discarded. The field is thinned to twelve.

Soon the audience is hushed. Conan Doyle and Lawes circle, taking notes. When deadlocked they consult Sandow, who is crouching up close to inspect muscle tone and form minutely. Early on, Galton fixes his opera glasses on two of the competitors—D. Cooper and A. C. Smythe—and is pleased when these make the cut of twelve. Perhaps he keeps score using one of the pocket data 'registrators' that he habitually carries, so as not to attract attention when taking notes (these will be lost afterward). But the competitors are too muscular and not well-balanced overall, far from Sandow himself and the Greek ideal. Whatever their other merits, the British are not the best stock to start from in search of physical beauty. He had long concluded that the bathers he often saw in the Serpentine were less shapely even than the Ovambos.

When the prizes are announced, William L. Murray wins the Gold Statuette, Cooper takes the silver, and Smythe the bronze. As the Irish Guards play 'See the conquering hero', Galton is disappointed that he did not select the winner, but he must concede on reflection that the judges have chosen fairly. This reinforces his conviction that men can be classified accurately through careful inspection alone.

A few years later, when talk became rife in the newspapers of national deterioration, 'F.G.' would be confident in the latent potential he had seen, if sheer beauty is not an object: 'the whole of a race which was able to furnish the large supply that is produced in Great Britain of men who are sound in body, capable in mind, energetic and of high character, has the capacity (speaking as a rearer of stock) of being raised to at least that same high level. How to do this is a question of both Nature and Nurture'. For Nature he would propose that 'if a strong and intelligent public opinion should ever become aroused in favour of improving the breed of our race, influences are in existence which even now operate in a small way in entire conformity with sentiment and law, and which are capable by development of producing great results. Individually their power is small, but in cooperation they would wield a practically

---

3 Sandow was born Friedrich Wilhelm Müller, in Königsberg, then in East Prussia, to mixed German and Russian parentage. He seems to have taken his Russian mother's maiden name, anglicized from Sandov. For much more on this flamboyant character see Chapman 1994.

4 According to Galton's great great nephew Hesketh Pearson (no relation to Karl), H. Pearson 1938, 52-3, who used to visit 'Uncle Frank' at Rutland Gate between 1905 and 1910.

5 See Tredoux 2023c.

overwhelming force'.

But the promise comes with a warning: 'we have a great deal yet to learn on matters bearing upon race-improvement, which lies within the plain province of anthropology, before it would be justifiable to attempt a crusade. Otherwise great mistakes would arise, efforts would be wasted, the results would not agree with anticipations, and the movement would become discredited and collapse. My attitude, which has usually been misrepresented, is to urge serious inquiry into specific matters which still require investigation in the well-justified hope that a material improvement in our British breed is not so Utopian an object as it may seem, but is probably quite feasible'.

Improvement of mankind has been on Galton's mind for a long time, at least since 1864. At first he concentrated on mental ability, but expanded later to cover physical qualities. However it is only after 1900 and the rediscovery of Mendel that the idea catches on broadly. This will catalyse a social movement and field of study, *eugenics*, catching the imagination of the whole world, before it eventually fizzles out sometime in the 1950s. The Fabians and the Fabian cast of mind that Beatrice Webb helped to form will play a critical role within this first epicentre of Eugenics. Curiously, Sandow's first name is Eugen. But the 'solid gold statuette', however much it resembles the Prussian, will prove to be only gilded bronze.

It was at Cambridge that Galton first started thinking seriously about the inheritance of abilities. As he told a meeting of enthusiasts in 1908, at the house of his Rutland Gate neighbour Montagu Crackanthorpe, 'I date my first opinion from my days at college, as far back as 1840, where competitions of all kinds showed most clearly to an unprejudiced eye that men are not equal in their natural powers, but most diverse in mind as well as body. It was also noticeable that high gifts of both of these tended to run in families'.[6]

Darwin had drawn the analogy between *artificial* selection, as practised by animal domesticators, and what he called *natural* selection, practised in the abstract by nature. Galton turned this back around: if mankind is subject to the same forces as animals are, and the human race was formed by natural selection, could the race not be evolved consciously and deliberately through selection of favoured families and individuals? There were three separate questions to be considered.

1. Are human traits really inherited and therefore a feasible target for selection? How should we measure those traits? What natural laws are involved?
2. Should we try to consciously seize control of human evolution? What are the pros and cons?
3. In practice, which traits should be selected for? How long will it take before we see any effect?

These are distinct questions, and one could say yes to the first, a purely factual question, but no to the second, a matter for ethical judgement, in which case the third doesn't arise.

The long articles on 'Hereditary Talent and Character' of June and August 1865 had broached these questions and provided some initial data.[7] They included wry speculations about marriage by organized competition—perhaps in some future Utopia where, based on rigorous examinations, state-provided stipends would incentivize the winners, with their status boosted by awards at Olympian ceremonies. 'What an extraordinary effect might be produced on our race, if its object was to unite in marriage those who possessed the finest and most suitable natures, mental, moral, and physical!' All the more urgent given the rate at which

[6] In fact this speech was read by Crackanthorpe on Galton's behalf, due to his deafness, though the scientist was present.

[7] F. Galton 1865b,c. See Appendix C, p. 399, where these papers are reproduced with editorial clarifications, for the full context of their creation and reception.

technological progress was outstripping human abilities. However it was not that mankind was degenerating in absolute terms—Galton made no mention of that there.

The idea of degeneration was already widely present in the journals, popular magazines and books of the 19th century, but each account used its own definitions and laundry-list of supposed causes—diet, drunkenness, too much sex, masturbation, homosexuality, clothing, relaxed natural selection, fecund criminals and imbeciles, Godlessness, inept marriages, city life, the decline of corporal punishment, racial admixture, disturbed states of mind during copulation. Usually with little evidence except anecdotes, operating on eccentric theories of hereditary transmission—or in some case with clear signs of monomania.[8]

In Galton's account, mankind is instead falling behind relative to the new demands placed on it. Enthusiasm erupts: 'what a galaxy of genius might we not create! We might introduce prophets and high priests of civilization into the world'. The rest of his long life was devoted to conducting the research programme sketched above, to address this world rather than some other world. That programme tried to obtain more definite answers, pursuing multiple secondary but related questions at all levels, from the foundation of heredity, through the differential psychology of traits, to the specification of desirable outcomes. Throughout, Galton returned periodically to the theme of conscious direction of human evolution. And in private he would construct more than one whimsical utopia in which the reins of evolution had been seized, as we shall see from the notebooks that went unpublished in his lifetime.

In 1869, *Hereditary Genius*—particularly the penultimate chapter on 'Influences that affect the natural ability of nations'—resumed and greatly amplified the argument for controlled evolution, without altering the gist of the original formulation (see chapter 4, p. 51). Though the mischievous effects of clerical celibacy and persecution of freethinkers were added now, the principal focus was *not* on the spectre of steady demographic degeneration. It is an important point, often overlooked, quite distinct from later doomsayers. Galton's primary motivation was positive liberation, not fear of degeneration. As he put in his introductory chapter, 'each generation has enormous power over the natural gifts of those that follow', so that 'it is a duty we owe to humanity to investigate the range of that power, and to exercise it in a way that, without being unwise towards ourselves, shall be most advantageous to future inhabitants of the earth'.

This is not merely counteracting whatever degeneration exists, only treading water, but much more ambitious: a scheme for completely breaking the current shackles of heredity. That is reflected in Galton's suggestions for taking control of human evolution: lowering the effective marriage age of the talented and raising that of the less talented, reducing the cost of society to families, supplying incomes from professional sources rather than inheritance, opening up opportunities for talent to flourish through liberal provision of scholarships and exhibitions, raising the social status of advantageous marriages, voluntary accommodation of the weak in nunneries and monasteries, and the introduction of new talent through encouragement of selective immigration.

An important follow up, 'Hereditary Improvement', appeared in January of 1873. There was some trouble getting this piece published. It was originally titled 'Is Viriculture Possible'—by then Galton was using the term 'viriculture' for the directed evolution he would later call 'eugenics'. At the close of 1872 Galton had approached, among others, James Thomas Knowles of the *Contemporary Review*, with what he described as an 'audacious' proposal.[9] The editor was still smarting from the heat of the 'efficacy of prayer' pyrotechnics[10] earlier that year, and turned it down. 'You will think that editors are a "feeble folk" & so perhaps they are—but it is

[8] See the early bibliography in Holmes 1924 for many now-forgotten references to this 19th century melange of ideas, not all of which are nonsense.

[9] James Thomas Knowles, 1831–1901, was the son of a distinguished architect, also named James Thomas Knowles, 1806–1884. He also practised as an architect. Father and son are easily confused.

[10] See page 109 and following.

certain that our constituents (who are largely clergymen) must not be tried much further just now by proposals following Tyndall's "friends" on prayer—and of a similarly bold or as you yourself say "audacious' character"'.[11] It was an early sign of the prevailing source of opposition to the idea of 'viriculture' for many decades to come.

So Galton tried his audacity on William Allingham at the more adventurous *Fraser's Magazine*, which had double the circulation of the *Contemporary* but tended to a middlebrow rather than a highbrow audience.[12] He reassured the editor that even though it is on a 'somewhat audacious' topic, the article 'contains not a word or allusion to indelicacy'. Indeed 'I should be very sorry unnecessarily to offend people, as Malthus contrived to do'. Allingham accepted it gladly with some softening and the piece appeared in the new year. It applied a word for the first time which would stick to the subject forever after. 'One word you pencilled, I have retained, "nurture"; it seems reasonable enough to me—and more expressive than any other I can think of, in respect to influences mainly occurring in the earliest years of life'.[13]

'Hereditary Improvement', as it was now titled, was a detailed response to those critics who had argued on the lines that 'philosophise as you will, men and women will continue to marry as they have hitherto done, according to their personal likings; that any prospect of improving the race of man is absurd and chimerical, and that though enquiries into the laws of human heredity may be pursued for the satisfaction of a curious disposition, they can be of no real importance'. Galton countered the last point first, reiterating the profound effects of shifting the entire population's ability distribution to the right a few grades, through modest selection repeated over a long period, 'sweeping away a legion of defectives' and 'introducing in very much greater proportions the number of men of independent and original thought', 'who are the salt of the earth, and who make nations what they are'. He reiterated his earlier arguments that human abilities had evolved in a much simpler environment, but the increasingly extreme demands of cooperative modern civilization called for a corresponding increase of the native mental and moral abilities required to successfully operate it.

An argument based on compassion, which would be reiterated often in the future, is also introduced. Natural selection operates blindly and wastefully, even recklessly, without remorse. There is a *moral* obligation, Galton argued, to reduce suffering by consciously directing the population toward a fitter state, reducing the vagaries of natural selection (the opposite of what is now called 'Social Darwinism'). This is really an obligation to what he called, somewhat vaguely, the Whole, but we may interpret to mean the gene pool or mankind itself. Society must evolve toward 'those higher forms of civilisation, vainly tried as yet, of which the notion of personal property is not the foundation, but which are, in honest truth, republican and co-operative, the good of the community being literally a more vivid desire than, that of self-aggrandisement or any other motive whatever'.

Contemporary well-being matters less than our (what we would call genetic) endowment to posterity. This is not just to counteract the 'deleterious influences of modern civilisation' on that endowment, but goes beyond that to 'raise it higher still' by at least the same degree. To get there, more than mere changes in living conditions would be required, since even the wealthiest among us still suffer from myriad diseases. Which is to say, *socialism* = *viriculture*, an idea that would subside in Galton's own writings but feature strongly in the thinking of his followers many decades after.

The practical workability of viriculture is addressed by pointing to the power of education founded on rigorous scientific demonstration, reiterated and inculcated by an elite to the point of acceptance similar to religious belief, producing an internalized conviction of moral duty. People

[11] Knowles to Galton, 1872-11, GALTON/3/3/11/8.

[12] Ellegård 1958.

[13] Letters from FG to William Allingham, 1872-11-28, 1872-12-06, Allingham and Williams 1911, 185-186. As this little-known piece, F. Galton 1873b, is not widely available it is reproduced in full in Tredoux 2023a.

would have no option but to flock to systematic knowledge patiently established cooperatively over time, 'if observation broadly confirms the truth of the present doctrines of heredity, quite as many social influences as are necessary will become directed to obtain the desired end'. No 'anticipatory action' beyond this is needed.

Galton explicitly disavows coercion in marriage in contemporary society, 'such an idea would be scouted now-a-days almost as much as that of polygamy, or of infanticide'. Castes might spontaneously arise which improve themselves inexorably using sure knowledge of heredity, and would simply outcompete competitors choosing to remain ignorant, replacing them in the population. But not all is sweetness and light. Perhaps in some state of society in the far future, things might develop to the extent that the enlightened caste loses patience with the benighted and proscribes them as enemies of the state.

The objection that the rewards of viriculture will be too slow and uncertain to attract the individual—what has posterity ever done for me?—is met by pointing out that forest conservation measures demonstrate that people certainly can be persuaded to care about longer-term effects which they will not personally benefit from (a modern example here would be the global concern for the far-future effects of 'climate change'). By leveraging these sorts of sentiments, the population might be steered into a 'perfect enthusiasm' for viriculture, aided by official registers of abilities and traits.

Along the way Galton anticipates an idea which will only be appreciated fully a century or more later. He contrasts 'race'—which he will soon come to call 'nature', and we call genes—to nurture, or the 'various circumstances' to which individuals are 'perforce subjected, especially in early life'. But then he goes further. 'Race has a double effect, it creates better and more intelligent individuals, and these become more competent than their predecessors to make laws and customs, whose effects shall favourably react on their own health and on the nurture of their children'.

Cyril Darlington would make the same point in his *Facts of Life* (1953): 'the environment with which a man's genotype reacts is itself, in its most profound character, the creation of men of like genotype'.[14] In modern terms, genes may transmit their effects through the environment they construct as 'genetic nurture'—using recently developed methods this can be detected from genes which are not, by random chance, transmitted to offspring, yet still have a measured effect on their outcomes.

Because of this double action of race/nature, Galton argued that while the 'sterling value of nurture' should not be underrated, race (or nature) dominates it. Thus, it is race, or nature, which must be improved to make really significant strides in well-being.[15]

Far from starting a firestorm of controversy, 'Hereditary Improvement' raised little stir. Galton sent a copy to Alphonse de Candolle, once again describing it as 'audacious', but what remains today of their long and intensely cordial disagreement, then just starting, does not mention it again.[16] His cousin Charles Darwin, to whom he had also sent a copy of the magazine, was sympathetic with the stated aims, and *The Descent of Man* had warned at length of deterioration due to relaxation of natural selection, as through vaccination and charity. 'Thus the weak members of civilized societies propagate their kind. No one who has attended to the breeding of domestic animals will doubt that this must be highly injurious to the race of man',

[14] Darlington 1953, 285. The author's copy of this book, widely-read in its day and revised as Darlington 1964, was once owned by the distinguished anthropologist Carleton S. Coon, who annotated this passage in pencil 'man makes his environment'.

[15] It is curious that Galton also anticipated, while discussing the degenerating effects mankind is subject to, the later discovery of the evolutionary trade-offs made by genes for sickle-cell anaemia, 'in every malarious country, the traveller is pained by the sight of the miserable individuals who inhabit it. These have the pre-eminent gift of being able to survive fever, and therefore, by the law of economy of structure, are apt to be deficient in every quality less useful to the exceptional circumstances of their life'.

[16] Tredoux 2023b.

'hardly anyone is so ignorant as to allow his worst animals to breed'. The only remedy Darwin could see for this, given mankind's greater revulsion to present suffering than future calamities, was arranging 'that the weaker and inferior members of society do not marry so freely as the sound' voluntarily. Regretfully 'this is more to be hoped for than expected'.[17]

Darwin was skeptical about the practicability of implementing Galton's proposals for improvement, which went much further than his own hopes.[18]

> Very many thanks for Fraser: I have been greatly interested by your article. The idea of castes being spontaneously formed & leading to intermarriage is quite new to me, & I sh^{d} suppose to others. I am not, however, so hopeful as you. Your proposed Socy.[19] would have awfully laborious work, & I doubt whether you could ever get efficient workers. As it is there is much concealment of insanity & wickedness in families; & there w^{d} be more if there was a register. But the greatest difficulty, I think, w^{d} be in deciding who deserved to be on the register. How few are above mediocrity in health, strength, morals & intellect; & how difficult to judge on these latter heads. As far as I see within the same large superior family, only a few of the children w^{d} deserve to be on the register; & these w^{d} naturally stick to their own families, so that the superior children of distinct families would have no good chance of associating much & forming a caste. Though I see so much difficulty, the object seems a grand one; & you have pointed out the sole feasible, yet I fear utopian, plan of procedure in improving the human race. I sh^{d} be inclined to trust more (& this is part of your plan) to disseminating & insisting on the importance of the all-important principle of Inheritance. I will make one or two minor criticisms. ... surely nature does not more carefully regard races than individuals, as (I believe I have misunderstood what you mean) evidenced by the multitude of races & species which have become extinct. Would it not be truer to say that nature cares only for the superior individuals & thus makes her new & better races. ... I doubt whether you have made clear how the families on the Register are to be kept pure or superior, & how they are to be in course of time to be still further improved.

Galton's reply to these points has not been preserved. Stronger criticism was expressed privately by the botanist Joseph Dalton Hooker who had been alerted to the article by Darwin the following day. 'Have you read Galton's Essay in Fraser? The idea of castes forming spontaneously & the members inter marrying seems quite new; but the more I think of the subject the less possible does it appear to me that the plan could ever be carried out'. Hooker was not especially interested, given that he only got around to it over a year later. 'I read for the first time F. Galton's article in Frazer (Jany 1873) on the advisability of securing a race of intellectual & physical athletes who were to marry inter se—I thought it the weakest thing of his that I had seen'.[20]

Notices in the press were mixed. One syndicated wag supposed in his 'London Letter' that the author must have been inspired by the Smithfield Cattle Show, and noted a 'good deal of amusement among our male gossips'. However *The Scotsman* thought it a 'remarkable article', and that 'a sneer will not dispose of it' since its schemes might just work if only a beginning could be made, though this seemed unlikely. Likewise the *Newry Reporter* detected 'a considerable grain of truth' in the argument, and suggested extensions to diminish the criminal classes, using arguments which revealed a firm belief on the part of the reporter in the inheritance of acquired characteristics. But then the salty *Aberdeen Press and Journal* dismissed the whole scheme as 'Utopian'.[21]

Viriculture of a kind extended to the geographical level too. In June a letter from Galton to the *Times* outlined a whimsical scheme to encourage the Chinese–whose native abilities he

---

17 C. Darwin 1871b, 127.

18 Charles Darwin to FG, 1873-01-04, GALTON/1/1/9/5/7/14. See also DARWIN, DCP-LETT-8724.

19 For collecting registers about abilities and traits.

20 Darwin to Hooker, 01/05/1873, DARWIN, DCP-LETT-8726; Hooker to Darwin, 1874-03-24, DARWIN, DCP-LETT-9371.

21 London Letter, *Cardiff and Merthyr Guardian, Glamorgan, Monmouth, and Brecon Gazette*, 1873-01-18, 5; *Scotsman*, 1873-01-07, 2; *Newry Reporter*, 1873-01-30, 4; *Aberdeen Press and Journal*, 1873-01-15, 6.

praised and declared to be only temporarily obscured by an 'effete system of classical education' which 'made originality a crime'—to populate Africa. He expected them to greatly bolster the indigenous Negro population. He rejected the idea that Negroes were inherently incapable of advancement, though they were certainly 'lazy palavering savages', since they had produced many individuals who demonstrated otherwise, but he thought they required a large measure of external guidance and support—their curve of ability was well below the Chinese, who would thrive anywhere and possibly replace the Negroes in time. He proposed that this sort of competitive emigration and replacement was the rule in human history.[22]

The most remarkable support for viriculture was forthcoming in August from Darwin's mathematical son George (see Figure 14.3, p. 356), soon to give up law due to a breakdown and return to a second fellowship at Trinity College. By then Knowles of the *Contemporary Review* had overcome his fear of controversy enough to publish George's blunt 'On Beneficial Restrictions to Liberty of Marriage'.[23] Though George found Galton's paper 'very curious and interesting', the creation of wellborn castes could lead to social fratricide and the means for inducing them in the first place seemed weak. To bolster Galton's 'build it and they will come' scheme, he proposed the more direct route of marriage prohibition, gradually introduced, including examinations for mental and physical disease and the production of untainted pedigrees. This could be justified by appealing to the greater good, as with compulsory education, vaccination, sanitation and the prohibition of various forms of incest. The burgeoning study of comparative anthropology provided many examples of the successful formation of marriage prohibitions by custom and law. Belief in the truth of heredity would grow from a mere germ to a body of sure knowledge that the population would embrace. Much of this reworked arguments that Galton had already made more positively. However Darwin emphasized degeneration in a way that Galton had not: growing hereditary insanity and the general deterioration entailed by relaxation of natural selection.

'Beneficial Restrictions' like 'Hereditary Improvement' elicited no more than mild attention. The following year the biologist St. George Jackson Mivart (1827–1900)—a Catholic convert who had proven to be one of the most tenacious opponents of natural selection—mentions it in passing while anonymously reviewing recent anthropology in the *Quarterly Review*. Mivart suggests that Darwin 'speaks in an approving strain of the most oppressive laws, and of the encouragement of vice to check population'.[24] This enraged the X Club, alive to slights of this kind. They closed ranks to 'cut' Mivart, and for a while George pondered going to law over the sneer.[25]

Thinking better of the courts, George followed up instead with research on the dangers of first-cousin marriages. His own father had married his first cousin Emma Wedgwood, and had been interested for some time in the possible ill-effects of inbreeding. The result looked disappointing for believers in deterioration, but set his father's mind at ease. George had to concede that he could find little clear evidence of ill-effects, based on comparing frequencies, where he was able to look—lunatic asylums, circulars, registries, doctors, Cambridge lists and the like—except possibly for sporting ability.[26] But he conceded that his methods were rough and subject to confounding, and personally felt that 'there are nevertheless grounds for asserting that various maladies take an easy hold of the offspring of consanguineous marriages'.

It all would have been easier to study if the census collected this information, as John

---

[22] F. Galton 1873a, see Tredoux 2023a. As noted in chapter 4, p. 51, Galton's ranking of racial abilities was not then controversial and was shared by Darwin, T. H. Huxley and most scientists of his era. The replacement of 'barbarous' races by civilized ones was mentioned, in a matter-of-fact manner, in C. Darwin 1871b, 122.

[23] G. H. Darwin 1873. For Darwin, George Howard, see page 451. For an alternative perspective note C. Pritchard 2018.

[24] Mivart 1874, 38.

[25] See the letter from Charles Darwin to George, 1874-08-01, DARWIN, DCP-LETT-9580.

[26] G. H. Darwin 1875.

Lubbock and others had urged, but the legislators had turned that down, and 'the tone taken by many members of the House shows how little they are permeated with the idea of the importance of inheritance to the human race'! Galton wrote to congratulate and caution George at the same time. 'You certainly have exploded most effectively a popular scare. Would it be possible to make any "probable error" sort of estimate of your results'.[27] George was, of course, right in his *personal* view, but his data and methodology were wholly inadequate. Much more advanced methods, not invented then, are needed to pick up the deleterious effects of cousin marriage, which are serious when the practice is kept up over multiple generations.

'Viriculture' did not catch on yet, with one or two exceptions over the Atlantic. Since the 1840s John Humphrey Noyes, charismatic leader of the small Oneida religious commune in Upstate New York, had been advocating and then actually practising selective breeding of humans under the name 'stirpiculture'. Noyes had read Galton along the way—he seems to have had the idea before Galton even published 'Hereditary Talent and Character' in June 1865—but he had his own fringe ideas about reformation of marriage and breeding. Noyes divided these into (i) polygamous free-love for recreation only, based on the withdrawal method, and (ii) procreational mating under expert supervision by committee. In the decade beginning with 1869 at least sixty children were produced in this way.

Another version of the practice had also been advocated by the leading Congregationalist preacher Henry Ward Beecher, a Darwinian brother of the enormously popular authoress Harriet Beecher Stowe. Yet even though 'stirpiculture' appeared in the American press and also in the speeches of politicians in the 1870s, it was little noticed in Britain. Galton seems to have been completely unaware of American stirpiculture and Noyes, let alone Beecher, until much later.[28] The Oneida experiment fizzled out in the late 1870s as Noyes declined physically. The cult members rapidly reverted back to traditional monogamy and rebranded themselves as highly successful joint-stock cutlery manufacturers, somewhat embarrassed in later years by their colourful past.

Patience and persistence would prevail in the long run. Viriculture was always in the background of *English Men of Science* (1874) but only broke into the foreground every now and then.[29] Readers were left in no doubt that a priestly class of scientists was desirable and ought to be encouraged in every way, as we have seen in chapter 6, p. 99. There was however the question of sexual selection (only recently introduced by Darwin) which could be 'repugnant to an improvement in the human breed' if the scientifically gifted preferred not to marry each other. Galton's data suggested that the parents of scientific men in his data set 'tended to produce differentiation and purity of race', implying that the parents were more harmonious for scientific traits than contrasted. But he had no data 'showing how far the selective sexual instincts of the population generally would thwart, be indifferent to, or co-operate with the influences of future social restrictions on unsuitable marriages, or encouragement of suitable ones'.

It was 1883 before Galton returned at length to the subject of directed evolution. Many sections of his *Inquiries into Human Faculty* (see chapter 9, p. 159) were devoted to the subject, which was treated from multiple angles.[30] Here the term 'eugenics' was substituted, and would finally stick (as it happened, the completely unrelated compound 'eugenic acid' was already in wide use among chemists). The word had broad application from the start, encompassing all 'questions bearing on what is termed in Greek, *eugenes* namely, good in stock, hereditarily endowed with noble qualities. This, and the allied words, *eugeneia*, etc., are equally applicable to men, brutes, and plants. We greatly want a brief word to express the science of improving

[27] FG to G.H. Darwin, 1875-11-10, GALTON/3/3/7/8.
[28] See for example *The Nation*, 1876-08-10, 92.
[29] F. Galton 1874c.
[30] F. Galton 1883d.

stock, which is by no means confined to questions of judicious mating, but which, especially in the case of man, takes cognisance of all influences that tend in however remote a degree to give to the more suitable races or strains of blood a better chance of prevailing speedily over the less suitable than they otherwise would have had'.

The eugenic argument that sprawls all over *Human Faculty* needs some reconstruction. To study the problem scientifically, data first needs to be collected from pedigrees, assisted by family registers and twin studies. A preliminary data set is described and analyzed. Statistical methods are required to use that data, and the central role of the normal distribution is described. The mechanism of hereditary improvement is selection, by analogy to stock breeding and the domestication of animals. This works if it has heritable variation to act on. Selection can operate at multiple levels, from the individual, through early marriages of those with desirable traits, to the group level. The fluctuation of demographic groupings geographically shows that the latter is feasible even if it proceeds slowly by replacement.

Traits to select for include physical health and vigour, energy, good moral character and intellect, as judged by common repute and the usual signs of professional success. Traits to select against include criminality and slavishness. People should be persuaded that this sort of selection is the right thing to do, as an internalized moral code. Nurture from an early age into a creed is required for that, and early associations are analyzed accordingly. Psychometric experiments reveal the basic nature of such mental associations, and exposes the role of the unconscious mind in thought. Other incentives may be added: pride in ancestry through official awards for 'family merit' and marriage-supporting payments from endowments, to remove economic obstacles. Eugenics would generate its own demand by discovering and transmitting knowledge about heredity. *Thus Galton proposes to use nurture to alter nature by fostering selection for desired traits.*

This version of the eugenic doctrine was deliberately restrained. 'It is one thing to check the course of a huge steam vessel by the shock of a sudden encounter when she is going at full speed in the wrong direction, and another to cause her to change her course slowly and gently by a slight turn of the helm. Nay, a ship may be made to describe a half circle, and to end by following a course exactly opposite to the first, without attracting the notice of the passengers'. He happily conceded that his list of traits to favour (e.g. in the awards for ancestral family merit) was illustrative only.

There is a subtle combination of logically distinct ideas in play here: (i) the scientific study of *mechanisms*, and (ii) the prescriptive practice of improvement using knowledge of those mechanisms. The natural analogy is to hygiene, where no distinction is drawn in practice between studying the nature of germs and processes of infection, and the practice of eliminating or avoiding them where they are harmful. There is no constituency for encouraging harm from infection, everyone wants to eliminate that. For many decades Galton and his subsequent followers would think the same way about eugenics. It is simply taken for granted that one would *want* to improve the human race if that is at all possible, and that there would be broad agreement on what improvement meant, exactly.

Many other questions were (necessarily) left unposed and unanswered for the moment. What are the trade-offs between traits? If you select for, say, intelligence, would that entail sacrificing some other desirable trait? Does intelligence come with a handicap, such as ill-health or madness? Is it really true that having more and more of the traits listed is necessarily a good outcome, or are there upper limits of the population frequency beyond which they may become less desirable or even harmful?

To some extent Galton had anticipated these questions in *Hereditary Genius*, where he had argued that greater mental ability was associated with better than average mental stability, physical health and strength. This is repeated in *English Men of Science*, where his subjects are described in terms of all-round magnificence. Now he has come to realize that this association will not be enough. So he adds health itself and other bodily characteristics as criteria in their own right for marking—that is, for selection. But then would it be best to have the smartest of the physically fit, or the most physically fit of the smartest? Or some other combination of the two criteria?

The idea of testing for physical fitness was pursued energetically over the coming decades. The successive Anthropometric Laboratories run by Galton incorporated, among their expanding set of measures, various tests of strength, for both muscles and lungs. He publicly advocated that schools ought to award marks for physical condition, and that the civil service should examine for it, not just narrowly focus on mental ability. Physical prowess would best be measured using his method of rank percentiles. Some experiments showed that committees agreed fairly when making evaluations by inspection alone.[31]

Recall that when travelling in the Sudan in 1845, fresh from Cambridge, Galton had been amused by the chance to have himself 'evaluated' on the spot by a slave-trader. 'I was curious to know my worth as an animal. He took evident pains, and I think was fairly honest, though with a bias towards flattery. Having regard to the then high state of the market, he estimated my worth on the spot, at a number of piastres that was about equal to £20'.[32] A tidy sum in those days, but far short of a fortune. Regrettably we do not know Galton's *rank* among his friends, who included Montagu Boulton, grandson of Watt's partner Matthew.

It was just as well that the perennially sour *Spectator* had not been to Khartoum. The magazine was deaf to these eugenic concerns, seeing no need for more than a minimum standard of health. Surplus 'health' might even be a 'positive drawback'.[33]

It is easily forgotten now how poor the state of health of the average Victorian was. Contemporaries lacked a good frame of reference, so what seemed expected to them appears miserable to us. It showed up starkly in the puny volunteers for the army during Queen Victoria's numerous 'little wars'. Life expectancy was only 48 in 1900, the average height was 169cm (or under 5'6"), with weight under 140 pounds. Most people used more calories than they could get. Tuberculosis was rife, causing a substantial fraction of all deaths. Natural differences in robustness were accentuated by primitive public health measures. The water and milk supplies teemed with bacteria, smog from coal and wood fires choked the cities, horse dung fouled the roads and walkways. With antibiotics not yet discovered, and infection ever-present, natural constitutional resistance was very valuable. The Galtons themselves were wealthy yet constantly ill, as shown by the diaries of Louisa. They squelched their way through the spas of continental Europe, where monied victims congregated in misery.

Alfred Russel Wallace (see Figure 14.4, p. 356) had praised *Hereditary Genius* when it was published, and had always shown a great deal of sympathy for protecting the quality of human stock. In September 1890 he invoked a higher authority. 'In one of my latest conversations with Darwin he expressed himself very gloomily on the future of humanity, on the ground that in our modern civilisation natural selection had no play, and the fittest did not survive. Those who succeed in the race for wealth are by no means the best or the most intelligent, and it is notorious that our population is more largely renewed in each generation from the lower than from the middle and upper classes'.[34] But by then his views about the best practical way to address the problem had shifted radically.

Wallace had nothing against plans to encourage superior intellects to breed, he even welcomed them, but doubted if this would change the average much even if it gained some traction. It was more important to persuade the below-average to reproduce less, though he saw no morally acceptable way to compel this (but see below for what he did find feasible to compel).

---

[31] F. Galton 1889d,e,f,g, 1890a,e, 1905b.
[32] F. Galton 1890f. See Tredoux 2023c for a fuller account.
[33] F. Galton 1890b.
[34] 'Human Selection', *Fortnightly Review*, Wallace 1890.

Here his mathematics let him down, since the changes are equivalent as far as the average is concerned.[35] 'Exceptionally great and good men are always produced in sufficient numbers, and have always been so produced in every phase of civilisation'(!) But be that as it may, he thought that he had found a much more general solution to not just dysgenics, but all of mankind's other problems too: *socialism.* He had become a believer: 'our present phase of social development is not only extremely imperfect but vicious and rotten at the core'.

Under full socialism, 'we shall find that a system of selection will come spontaneously into action which will steadily tend to eliminate the lower and more degraded types of man, and thus continuously raise the average standard of the race'. Thus Wallace actually believes in *eugenics through socialism*, which is on his view both necessary and sufficient. His idea of socialism, 'just' and 'rational', is taken not from real life—who ever heard of a spontaneous event under the elaborate choreography of socialism?—but unironically from the science fiction of the novelist Edward Bellamy, in *Looking Backward* (1888).

Once Wallace has 'cleansed the Augean stable of our existing social organization', 'education' will do most of the work in transforming conditions to a state of material equality, allied (ominously) with state-awarded grades for good conduct, industry and intelligence (left unexplained). Indeed everyone will work for the state, it is just a question of finding the job—overseers, superintendents, general managers—that they are best suited to do for that solemn body. (No menial positions are mentioned, suggesting some endless Möbius strip of supervision.)

Just *how* society is to be transformed into this blissful situation of plenty-for-all in 'one big family' is not elaborated, but elsewhere Wallace talks of confiscating all private property and reallocating it. (Without resistance?) Then all material restraints on marriage will be lifted, and sexual selection (by women) will inexorably improve the race, since nobody will want to marry until they are mature, nor would women want to marry the idle and vicious. (Somewhat contrary to present experience.) It seems that education and indoctrination will again do the necessary. 'The most careful and deliberate choice of partners for life will be inculcated as the highest social duty'. Doing the opposite will be considered an offence against society. Prison camps are not explicitly called for, but they lurk nearby.

It may be relevant to all this that Wallace had also become a believer in the stage trickery of spiritualism, but neither that folly nor his sinister 'social cleansing' scheme seem to have drawn any comment from Galton, though he surely noticed both. He would perhaps have asked how inborn inequalities of talent would simply dissolve after confiscation and reallocation of property. More to the point, the natural abilities needed to thrive under a totalitarian state are not evenly distributed either.

In 1894, eugenics bobbed to the surface again. A long memoir by Galton in the *National Review*—ostensibly a review of the recently published Fabian treatise *Social Evolution* by Benjamin Kidd—considers the role of religion, repurposed to spread the eugenic word.[36] Kidd argues that the state will evolve to inexorably take over all social functions, mercifully replacing the wasteful and cruel 'rivalry' of natural selection. Channelling competition through the state, rather than free enterprise, will result in more efficient use of that energy. Altruism derived from ultra-rational (that is non-rational) religious motivations will replace cruel natural selection, through the salvation of socialism.

Kidd's characterization of religion as non-rational strikes Galton as wrong-headed. The scientist suggests instead that religion has persisted because it serves to unite groups, giving them a selective advantage over their fragmented rivals—which is to say, religion itself has a rational evolutionary basis. That makes the supposed non-rational altruism rational at one remove. In the next century this argument would be reformulated by William D. Hamilton in

[35] Since addition is associative i.e. $(x + i) + y = x + (i + y)$.

[36] F. Galton 1894d; Kidd 1894.

terms of selection for shared genes underlying altruism, rather than selection of groups—but the level of selection concentrated on is partly a matter of taste, given the inter-relatedness of group members and the hereditary nature of the trait that Galton automatically assumed.[37] (An argument that seems to go completely over Kidd's head when he has the opportunity to respond at the end of the review.)

Galton expects that socialist experiments of Kidd's kind will be tried at first, to address mounting social problems such as those caused by overpopulation, but will disintegrate into anarchy, owing to the 'moral and intellectual incompetence' of mankind, which is still in a 'low state' and unable to handle more advanced civilization (man's real 'original sin', as he described it in 1865).[38] From that state of anarchy only the uniting force of a new eugenic religion, with its accompanying improvement of human nature can offer safety. Though the spread of knowledge will inexorably discredit the old superstitious religions, the new scientifically-based eugenic religion will improve human nature and co-evolve positively with it, enabling basic utopian experiments to succeed, e.g. by bringing calamitous population growth under control. Society will strive to make every individual efficient, both through nature and through nurture, as a religious duty. The same sentiments *underlying* religion will be appealed to, but their objects will be redirected. The fact that religious sentiment is universal but varies dramatically by society and over time shows that its concrete objects are changeable, almost arbitrary.

Until 1896 eugenics lay dormant again and few traces remain in Galton's private or public writing. In July of that year he was in Germany at Wildbad in the Black Forest, recuperating under doctor's orders from 'gastric catarrh'. It rained heavily for weeks, putting paid to sight-seeing, so he whiled away the time in his notebooks. He took up a theme he had first explored in 1892 when Mars was passing close to earth. He had written then to the *Times* about the possibility of communicating with Martians, if there were any, by means of sun signals flashed from large mirror arrays. In the 1850s he had invented a successful hand-held heliograph, with a built-in aiming mechanism, which had come into wide use in the navy, so he had long experience in the general principles of sun signalling. The trick was to 'bootstrap' communication with the Martians by first flashing an encoding scheme, which we now recognize as digital, with sequences of long and short states, then using that encoding to send mathematical and astronomical facts. Predictably, the *Spectator* saw no point.[39]

In 1893 Galton elaborated on his information-encoding scheme as part of another Friday Evening Lecture to the Royal Institution, describing a vector-graphics protocol for drawing pixelated pictures by moving in arbitrary directions, with pen up or down.[40] He didn't mention Mars, referring instead to Earthly telegraphy. But Mars kept bubbling to the surface, and a note from Galton to Norman Lockyer—his old colleague from the *Reader* and the founding editor of *Nature*—envisaged a scheme of digital information trading, with either side acting as examiner for the other. Across the cosmos parties could exchange valuable facts about advanced mathematics despite the great distance between them.

The Wildbad notebook took up these ideas but quickly transformed into a comic science fiction tale, in which the Martian observatory notices flashing signals from Earth, then scrambles to decode the content and construct apparatus to signal in reply, despite domestic opposition from the more frugal citizens. The Martians turn out to be highly intelligent and warlike creatures, resembling centaur-sized egg-laying ants, with faces like diving helmets. They are keen-sighted but deaf, communicating using motions like London deaf-mutes, through twitching antennae, which Galton wryly notes must be covered with headgear at diplomatic conferences

[37] Hamilton 1964a,b.

[38] This anticipates E. O. Wilson's entomological quip about communism: nice idea, wrong species. See F. Galton 1865b,c.

[39] Anonymous 1892; F. Galton 1892d.

[40] F. Galton 1893d.

to avoid betraying information. They are specialized into three castes: fertile females, neuter worker females and warrior males. Intriguingly these adumbrated ants have taken charge of their own evolution using unsentimental eugenic techniques.

> Life is begun in the form of an egg the size of that of a hen or a crocodile which is laid about a fortnight after the close of the mating season. Each fertile female lays a large number of them in rather quick succession with small strain to her health & strength. Afterwards she cares no more about the eggs, but returns to her usual & queenly & housewifely life, while they are looked after by the neuter females. In another week's time, the larvae are hatched which are about 6 inches long at first. They are tended with great care by the neuters who assiduously note & record [the] physical & instinctive peculiarities of each larva, with the view of determining which should ultimately preserved and which destroyed. Perhaps 20 times as many larvae [leave] the egg as are allowed to pupate. The attachment of the neuter nurses to the larvae is of a curious business-like kind, there is no love in it, for these neuter females seem incapable of true affection. Anyhow they nurse, watch & observe the larvae sedulously, and from time to time a formal consultation takes place between the queen & males, after which the neuters kill quite coolly those of their growing charges who it has been judged undesirable to preserve.
>
> Some few unpromising larvae are always kept for scientific investigation, to see how they will turn out in after life, it being of obvious importance to the community that the selected larvae should really be the fittest and it is only by careful test experiments that the value of the rules under which the selection takes place can be confirmed & the rules themselves gradually improved. The larvae lead a purely animal existence feeding & fattening enormously; their numbers are from time to time reduced as above described until at the time of pupation 12 months after they began to live; there remain but only nine that are intended to live, each being a huge grub some 4 feet long & six or 7 inches thick. When the larvae show signs of approaching pupation they are placed into separate cells where they weave a cocoon about themselves. There they lie, with little help, outwardly impassive but inwardly undergoing rapid & continuous changes, until the time comes that they burst their cases, aided by anxious neuters, and emerge feeble & limp but as perfectly good M[ars] folk, so far as physical form is concerned. There follows a long period of hardening and what may be called school & college life, after which certain probationary tests with accompanying ceremonies are gone through, and the probationaries receive the full rights of citizenship.

The story develops amusing newspaper controversies (particularly disdained by Galton) in the Martian daily *The Bellona Gazette*, as the sequences of messages are progressively decoded and responded to by the Martian astronomers, who marvel at the idea of intelligent vertebrates. The most interesting parts remained unpublished in Galton's lifetime, languishing in a spider's web of heavily revised, barely-legible, notebook pages. When Galton got back to England that year he sent off a denatured version to the *Fortnightly Review*, retold from the point of view of Earth. The Martians make first contact, and we hear little about them, with nothing about their eugenic practices.[41]

At the close of October 1901, Galton was awarded the Huxley Medal (in platinised silver) of the Anthropological Institute, and so gave the second annual Huxley Memorial lecture. it was delivered at the Society of Arts Meeting Room. John Lubbock (Lord Avebury), who had given the first Huxley lecture the previous year, was in the chair, with Karl Pearson and many other

[41] F. Galton 1896. For a complete transcription and reconstruction of the Mars story, along with supplementary material, see Tredoux 2023d.

fellows of the Royal Society prominent. H. G. Wells and many other Fabians were present in the crowded audience.

Galton takes as his subject 'the possible improvement of the human race under the existing conditions of law and sentiment' which, despite being the 'grandest of all objects', 'has not hitherto been approached along the ways that recent knowledge has laid open'. Here he means expanding knowledge about the laws of heredity, ignorance of which entails that the subject has 'a less dignified position in scientific estimation than it might'. To date his own periodic advocacy of viriculture and eugenics has been 'smiled at as most desirable in itself and possibly worthy of academic discussion, but absolutely out of the question as a practical problem'.

The normal law of error is recapitulated: good enough for illustrating the distribution of 'social worth', without depreciating Karl Pearson's more exact skew curves. A simple diagram shows how this worth regresses to the 'genetic centre', not just the 'parental centre'.[42] Nevertheless, the talented have the best chance of producing talented offspring.[43] Diplomas for civic worth may be awarded by juries of peers, to foster pride. Favorable marriages, made early, should be encouraged by awards and grants, thereby crowding out less desirable marriages through greater fecundity. Benevolent charities and patrons should take up deserving protégés from less fortunate classes, to raise them up. Conversely, criminals are best kept locked up so that they do not reproduce, abolishing a recurring source of suffering and misery.

Above all, accurate records must be kept, tracking how promising people actually turn out. We want a society for the promotion of the study of inheritance, 'publishing intelligence and memoirs, holding yearly elections, administering large funds, establishing personal relations like a missionary society with its missionaries, keeping elaborate registers and discussing them statistically with honest precision'. But sure-footed caution is necessary. 'More is wanted in the way of unbiased scientific inquiry along the many roads I have hurried over, to make every stepping-stone safe and secure, and to make it certain that the game is really worth the candle'.

The audience is receptive. They laugh when Galton recounts, as a regressive whim, a lady who had fallen in love with her father's undertaker. They laugh again when he reports the remarks of an official at a women's college: one third of graduates profit from the education, another third gain little, and the remaining failures marry.[44] They applaud when he declares that it is 'imperatively necessary that England should lead the way in this matter, for we planted our stock all over the world'.[45] He does not need to mention the elongated embarrassments of the Boer War, still in progress.

The press and periodicals treat the lecture with respect, but hardly any more seriously than previous airings of eugenics. Exceptions are the *Yorkshire Evening Post*, which anticipates a 'minor shock' at the proposals, and the *St James's Gazette* which (a little sardonically) calls them 'unnatural selection'.[46] By contrast, the *Times* editorial notices the gentle pressure proposed to encourage favourable marriages, but rather suspects this would produce the fat, the lazy, the conceited and the pampered instead, with the youth pairing off as they please regardless of requirements cooked up in laboratories. Bribes will only produce misery and disappointment. For all his statistical skill, Galton's opinion on who should marry whom carries no more weight than any other man.[47]

Sometime in 1901 Galton began to sketch another whimsical eugenic utopia in a notebook. 'The Donoghues of Dunno Weir' (puns intended) would never be published, but gives interesting

42 Here Galton is wrong. Regression is toward the parental centre, as the best estimate here of the breeding pool centre, but this is a consequence of his mistaken notion of ancestral influences through 'stirps', which act like portable gene pools carried by each individual.

43 An argument that Galton had elaborated on in *Nature*, see F. Galton 1901b.

44 *Westminster Gazette*, Wednesday 30 October 1901, 5.

45 *London Evening Standard*, Wednesday 30 October 1901, 8.

46 Wednesday 30 October 1901, 2; Wednesday 30 October 1901, 6.

47 Wednesday, Oct. 30, 1901, 7.

insight into his thinking and affectionate sense of humour.[48] The story must be glued back together from the fragments that remain. 'John Donoghue' is a family patriarch, of Quaker descent, grown hugely wealthy from an industrial concern he started. His father 'Jacob' had been an amateur stock breeder, inspiring 'John' to pursue this more scientifically, from an early age. The interest deepens after the publication of the *Origin of Species*. As a result, 'John' educates his sons William and Eustace in science at university. (For unclear reasons, Eustace is made into a cripple with poetic fancies. We are told that the 'sympathetic turn of his mind & his brilliant conversation introduced a delightful element into the household'.)

To promote better human breeding, 'John' uses his surplus wealth charitably, not in the 'namby pamby sense' but rather to build 'a hundred or even a thousand' houses at a place called 'Dunno Weir'. It is a more productive advertisement of surplus wealth than the usual showy mansion. The model town is located on waste heath bordering Surrey, and was originally founded by his father Jacob (dead at 50, alas, in a riding accident). Stables, paddocks and other amenities are provided and the river is dammed. It is a 'School for Heredity', a 'eugenic experiment'. (In an alternative sketch, the wife of 'John' has hereditary weaknesses and dies young, persuading him that eugenic improvement is critical for the race.) The roomy houses in the semi-rural town are 'exceptionally sanitary', forming hubs for rearing large and healthy families. Some guest-houses for shorter term stays attract people to the settlement. Tenants are charged nominal rent but are selected rigorously to identify 'good character', 'fine pedigree' and inborn talent. Occasionally, the more successful of the residents endow the settlement generously.

By producing more children than average, the picked tenants of 'Dunno Weir' boost the general health and intellectual quality of the human stock, crowding out others. Hired help assists in the raising of these large but capable families, freeing parents from drudgery. Continuous selection at the rate of 1 in 10, repeated over generations, maintains standards. Wealthy Godfathers from outside 'Dunno Weir' act as patrons for especially promising offspring. The town, mostly populated by artisans, is organized into trades union. The aim is to produce 'sensible vigorous men with principles and desire to live' rather than geniuses. Galton hopes—after all, it is a utopia—that there will be 'no over-drudged persons, no low life, no feckless characters, no termagant persons'. Residents would be mutually helpful. Infant mortality would be reduced.

This is all indubitably a form of socialism, but may be excused (Galton notes) since there is so much of that 'in the air' anyway, and at Dunno Weir the emphasis is not on levelling down, but rather lifting up. Other settlements spring up spontaneously to emulate the successful example of 'Dunno Weir', but ordinary English artisans are 'gifted with too small individual doses of civic capacity' to do so successfully *without* such an example to follow.

The story was prematurely put aside by year end, destined to lie unread in his notebooks until taken up again in 1910. It had been directly inspired by a newspaper clipping from March 1901, about the wealthy Quaker chocolate baron George Cadbury (1839–1922) and his Peabody Trust.[49] Cadbury's Bournville experiment, a £170,000 investment, aimed to create a 330 acre town in Worcestershire, with a view of the Severn, for housing 1,000 families of Birmingham-area factory workers in improved conditions at low rents. This resembled 'Dunno Weir' in most aspects, but it completely ignored eugenic concerns, naively assuming that improved environments would be inherited, presumably as 'acquired characters', by descendants. Unless this misconception was corrected, Galton believed that such experiments, though well-intended, would perversely have *dysgenic* effects by encouraging proliferation of unfavourable stock. 'Dunno Weir' was his playful eugenicization of Cadbury.

If there is an inflection point in the early history of eugenics, it is surely the first meeting of the newly formed Sociological Society, held on May 16th 1904, at the 'School of Economics

[48] GALTON/2/4/19/5.

[49] *The Standard*, March 10, 1901.

and Political Science' in the University of London. It is significant that the school had been started by the Fabians in 1895, when they had been bequeathed £20,000 by the Derby lawyer, Henry Hunt Hutchinson (1822–1894). 'That they may apply the same at once, gradually, and at all events within ten years to the propaganda and other purposes of the said Society and its Socialism, and towards advancing its objects in any way they deem advisable'. Hutchinson was a sometime Fabian, at times disaffected, who had committed suicide the previous year. Fabians—present, former and future—made up a influential proportion of the audience and speakers that Monday evening: Karl Pearson (chair), H. G. Wells, Mrs Dr. Drysdale Vickery, Benjamin Kidd. Those who could not attend in person but submitted written comments included George Bernard Shaw and the liberal journalist John Mackinnon Robertson.

The Sociological Society was the first important organization to promote eugenics. It had been formed in late 1903, concocted in the rooms of the Royal Statistical Society that May. (As noted before, statistics was not then a theoretical or even a mathematical subject—instead it involved reams of accumulated social data, which it might at best average.) Leading forces were the businessman Victor Branford (1863–1930), the botanist Patrick Geddes (1854–1932), the wealthy liberal businessman James Martin White, the former Corpus Christi fellow Leslie T. Hobhouse, then a liberal journalist, the King's College statistician Charles Stewart Loch (1849–1923), the Liberal MP James Bryce (1838–1922) and congeners. Galton himself was not at the founding meeting, but sent in a letter of support. He cooperated with the Provisional Committee led by Victor Branford, tasked with examining the formation of the society.

Through Victoria Lady Welby,[50] Branford solicited a paper from Galton for the second session of the society in May 1904. 'Several leading members of the Society have expressed a desire that very considerable attention should be paid to the subject of "Eugenics"'.[51] This paper became 'Eugenics: Its Definition, Scope and aims', and it was delivered at 5PM on May 16 in the Great Hall of the School of Economics. But there was more to it than that.

The biometric-Mendelian dispute lurked in the background (see chapter 13, p. 287). Branford promised Galton that he would invite Pearson, Weldon and Bateson to the Sociological, evidently in the hope that the discussion would prod the feuding factions into cooperation on heredity in man. Though Pearson chaired the meeting in person, and Weldon attended, Bateson was prominently absent, sending in written remarks instead. Branford appears to have hoped for even more—that the Sociological Society would facilitate a grand synthesis of knowledge, from biological foundations up through to the structure and dynamics of society.

In his paper, Galton carefully restates the essentials of his eugenic ideas, purposely adding nothing beyond principles he had already gone over multiple times in the past. He repeats his cautions about the dangers of over-zeal and the need to let public acceptance gradually grow. But that cat was already out of the bag and scalding off down the road.

That sometime student of T. H. Huxley, leading Fabian and best-selling author, H. G. Wells, rose to dissent from mild schemes of positive encouragement. 'It is in the sterilisation of failures, and not in the selection of successes for breeding, that the possibility of an improvement of the human stock lies'. But he thought that more subtypes for matching should be identified.

Mrs Dr. Drysdale Vickery (1844–1929), otherwise known as Alice Vickery, rose to share her conviction that Britain was in no state to absorb its excess population production (from inferior stocks) of half a million a year. Vickery is little remembered today, but was somewhat notorious then as a free love campaigner, women's rights advocate, contraception promoter and abortion enthusiast. She had qualified as a chemist and then as a doctor, one of the first women to do so legally. She lived openly in an informal arrangement with Dr Charles Robert Drysdale (1829–1907), who shared her radical political and social views. Naturally, Vickery told the audience that better treatment of women was necessary if the race was to be improved.

[50] For Welby, Victoria Lady, see page 459.
[51] Branford to FG, 1904-04-11, GALTON/2/4/19/9.

By contrast, that other Fabian Benjamin Kidd—whose book on *Social Evolution* had been reviewed by Galton in 1893—argued that eugenic aims were laudable but should be pursued with caution, as there was 'much danger of incomplete knowledge, and even of downright quackery, clothing itself with the mantle and authority of science'.

George Bernard Shaw, absurdly Bohemian playwright and founding Fabian, was not there, but had sent in provocative written comments. He was in no doubt that 'nothing but a eugenic religion can save our civilisation from the fate that has overtaken all previous civilisations'. Even though 'we never hesitate to carry out the negative side of eugenics with considerable zest, both on the scaffold and on the battle field', people 'select their wives and husbands far less carefully than they select their cashiers and cooks'. Rather than attacking marriage, it should be sidestepped. 'What we must fight for is freedom to breed the race without being hampered by the mass of irrelevant conditions implied in the institution of marriage'. These breeding experiments would be run 'concurrently with the promiscuity of ordinary marriage', which would be preserved for entertainment purposes.

There were several less entertaining speakers who preferred environmental remedies for social ills. Dr Robert Hutchinson (1871–1960), a Scottish physician, spoke with confidence that feeding alone could solve many of the supposedly racial problems of the day. L. T. Hobhouse (soon to become an influential sociologist) denied any knowledge of biology and undertook to remain silent about heredity until more definite results were available from the biologists themselves. Although he suspected that both stock and environment played roles, it was unclear how stock could be altered in desirable ways.

Some of those present were far out of their depth. The Consulting Physician for Mental Diseases at Charing Cross Hospital, C. A. Mercier (1852–1919), interpreted probability as the notion that it is 'quite impossible to predict from the constitution of the parents what the constitution of the offspring is going to be, even in the remotest degree'. The psychiatrist Henry Maudsley (1835–1918) thought that infectious biological agents determined relative success or failure within families, without giving evidence. He was loath to compare men to animals, and inclined to just let things take their own course as they had always done.

Others turned the discussion to their own ends. William Bateson's written remarks merely prosecuted his dispute with Weldon and Pearson. He explicitly depreciated 'actuarial methods', which is to say biometry, as old fashioned, pointing to Mendel's results from experimental breeding. Any results derived in the meantime from 'actuarial methods' would be a poor approximation to the truth and a perilous basis for eugenic programs. However his rival W. F. Weldon was there in person, pointing out contra Mercier that 'actuarial techniques' allowed good predictions of traits without pretending to be causal explanations. Weldon invoked Karl Pearson—present in the Chair, despite considerable personal misgivings that sociology would ever be a numerate science of the kind he could recognize—for support. Bateson's depreciation of 'actuarial methods' was premature, for they would still be needed to analyze and so verify the results of breeding experiments, where those were possible. Many human traits did not seem amenable to Mendelian analysis so far. 'Actuarial methods' based on a mass of data yielded better results for man so far.

In response, Galton, who usually abhorred public controversy, was pointed. (His remarks would be reported accurately in the *American Journal of Sociology*, but toned down drastically when they appeared in the domestic *Sociological Papers* the following year.)[52] 'When this debate began, I was extremely unhappy at the quality of it. The two first speakers [Maudsley and Mercier] really seemed to me to be living forty years ago; they displayed so little knowledge of what has been done since. More than one of the later speakers were really not acquainted with the facts, and they ought not to have spoken at all. We are much indebted to Professor

[52] F. Galton 1904c, 1905c. A summary without any of the debate at all appeared in *Nature*, see F. Galton 1904d.

Weldon for raising the debate to a higher level'. Evidently the extremism from the likes of H. G. Wells and Bernard Shaw had not impressed him. 'I do feel that if the society is to do any good work in this direction, it must attack it in a much better way than the majority of speakers seem to have done tonight'.

That hard-hitting conclusion was completely omitted from the *Sociological Papers* version. In a letter to his sister Emma, Galton described the discussion as 'wishy-washy', and sardonically noted that the offprint he was sending her was 'well-printed, anyhow'.[53] He took up Bateson's criticisms directly with the biologist.

42 Rutland Gate SW
May 18, 1904

Dear W. Bateson

Your printed communication re eugenics, of course interests me, but I am not at all sure that I rightly catch your full meaning.

It is certain that the ordinary data concerning family status enable us to foretell with fair average accuracy, and with a pretty exact knowledge of the probable error of a prediction. What I wish to know is whether you think that additional information (of a procurable kind), treated in any particular way, could reduce the probable error, making the prediction approach nearer to certainty. If you have any such idea I sh[d] be grateful to learn it, for I want to arrange a questionnaire that shall be approved by all who can be considered experts in heredity. It must of course be one that is likely [to] be answered truly.

yours faithfully
Francis Galton

Bateson's obstacle was that he wanted experimental evidence, not observational. But he also believed that deliberate experiments involving polygenic traits (those caused by multiple 'factors' as he put it) were too 'complex' to analyze and therefore use as a basis for eugenics.

Grantchester, Cambridge.
28 May. 1904

Dear Galton

Various interruptions have prevented me from writing sooner. I am sorry that what I wrote for the Eugenics meeting was obscure. As you tell me my note has been printed, I will see if I can emend it when the proof is sent. Meanwhile I have no verbatim copy.

The meaning I intended to convey was not so much that alterations in the "questionnaire" would be likely to make a sensible improvement in the results, as that, in the light of what we now know of heredity, this method of investigation can scarcely be expected to contribute much more than it has already done.

It is true, as you write, that the ordinary data have in several cases provided predictions which hold good in the cases with fair average accuracy, a result of indisputable value. The question is whether similar materials, however extensive, or well collected, will carry the inquiry further? If the data you receive are minutely analysed, instead of being compounded together (as such materials have hitherto been), it is I admit, within the bounds of possibility that the true laws of inheritance of some very simple and well chosen character may appear. But having regard to the classes of human characters hitherto selected for statistical study such as those of your own R.F.F., or those recorded at the anthropometric stations, I see no hope of results corresponding more closely with physiological fact being attained by the same means.

[53] FG to Emma Galton, 1904-07-05, K. Pearson 1930b, 527.

Until experimental breeding with controllable cases has got some way further, it seems to me unlikely that these complex phenomena will be unravelled. On the other hand, when from the results of experiment we know what to look for, this analysis shall not be impossible.

I do not know if you have found time to examine the results of the recent experimental work on heredity. It is evident that none of the discoveries of Tschermak, Cuenot &c. could have been made without an analytical method of investigation applied to materials originally homogeneous. In illustration of a much simpler case I send you two cobs of maize, grown by R. H. Lock at Peradeniya. That numbered 3 is an ordinary E1xR. (The ink marks represent an imperfect attempt to trace phyllotaxis).

That marked 9 (see back of label) has a rather more complicated history, but from what has now been learnt by analytical experiment the significance of such a case is perfectly clear.

I have not as yet heard of other methods which would have in any way contributed to the solution even of this comparatively easy problem. Human families are "cobs" with about four seeds each! The original materials are heterogeneous and particulars of even three generations are seldom complete.

Still more at fault most non-experimental methods be in dealing with more difficult cases—such as the Stocks, where the character of the offspring depends on the order in which the parental forms are introduced; or the comb characters of Malays, where there are five types of males and two types of female each with distinct powers of transmission, though outwardly indistinguishable—as yet, not to speak of colour characters which are determined—probably in many species—by two factors at least, each separately transmitted.

The human characters to which your questions have hitherto related must be at least as difficult as these, and are probably complicated by the disturbing effects of external conditions. Renewed study of such facts by comprehensive "actuarial" methods seems likely to defer real progress in the science of heredity, by habituating the mind to unnatural conceptions of the phenomena. Adherence to these methods can only be justified on the ground that the facts are assumed to be of such an order of complexity that a proper physiological analysis must be indefinitely postponed—an assumption which from all analogy seems to the last degree improbable.

I am
yours truly
W. Bateson.

P.S. I hope shortly to return your circular as to "Noteworthy Kinsfolk". I have not found it easy to form a clear idea of note-worthiness, especially in females, but I suppose if I send a full return you and your assessors will eliminate as they think proper.

Galton's reply shows a desire to conciliate, rather than any real basis for agreement with Bateson. Observational data is all that is ever likely to be available.[54]

June 12 1904

Dear Mr. Bateson

[54] Daniel Kevles alters the meaning of this letter, stating that in it Galton 'declared that he had never intended to condone the mating of men and women "as we please, like cocks and hens"'. As the full exchange given here shows, 'he had never intended to condone' is inserted by Kevles himself, see Kevles 1985.

Your letter of May 28 sh^{d} have been answered earlier, had I not delayed in hope of receiving your promised answers to my 'Ability in Families' circular, & reply to both at once.

I quite understand now (I think) your point, and to a great extent agree with it. But what are we humans to do, if any 'eugenic' progress is attempted? We can't mate men & women as we please, like cocks & hens, but we could I think gradually evolve some plan by which there would be a steady, though slow amelioration of the human breed. The aim being to increase the contributions of the more valuable classes of the population & to diminish the converse. We now want better criteria than we have of which is which.

Do what we can (within reasonable limits as regards mankind), fraternal variability will never be much lessened, but I do think that the fraternal means might on the whole be raised. That is the ~~practical~~ problem as it seems to me, to be held in view; also that an exact knowledge of the true principles of heredity would hardly help us in its practical solution.

I do indeed fervently hope that that exact knowledge may be gradually attained and established beyond question, and I wish you & your collaborators all success in your attempts to obtain it.

Yours faithfully
Francis Galton

Do you want the cobs of maize back?

But however muddled this debate was, forces supporting eugenics were clearly emerging now in late 1904, and in coming years Bateson too would periodically come out in support. Writing to his niece Milly, Galton enthused. 'I have been full of my "Eugenic" plans, so full that I have not even written up my little pocket journal for a week. I stayed last Sunday with Lady Welby at Harrow, who is a most zealous friend, and have consulted with Sidney Lee, the Editor of the *Dictionary of National Biography*, with Barron, the Editor of the *Ancestor*, with Professor Weldon, with Branford, the Secretary of the Sociological and with Lionel Robinson, a general littérateur, and see no less than seven different ways of making the first effort, between which a wise choice has to be made. Every one of them seems hopeful at present. The iron is kept red hot on the anvil'.[55]

In October of 1904, Galton donated £1,500 to the University of London, at £500 per annum over three years, for a Galton Research Fellowship in the study of eugenics, after talking the idea over with his friend the Principal Arthur Rücker. It was rapidly accepted by the University Senate on the recommendation of a committee (including Galton himself) which described it as a 'matter of urgency'. He shared the pleasing news with Milly, 'I feel the seed is planted in good soil'.[56] Now 'everything is done in the name of the University' and 'the word "Eugenics" is officially recognised. I am very glad of all this as it gives a status to the inquiry, so that people cannot now say it is only a private fad'.[57]

By the end of the year Galton had finally joined the Sociological as a dues-paying member (one guinea). Branford realised that this was a major coup. The new member was immediately pressed for advice about a submission from Dr. Robert Reid Rentoul (1856–1925), an eccentric Irish-born surgeon who practised in Liverpool and was notorious for bombarding professional bodies with caustic motions on any topic remotely controversial.[58] In mid-December Rentoul

[55] FG to Millicent Lethbridge, 1904-10-08, K. Pearson 1930b, 529.
[56] FG to Millicent Lethbridge, 1904-10-15, K. Pearson 1930b, 530.
[57] FG to Millicent Lethbridge, 1904-10-30, K. Pearson 1930b, 531.
[58] For Rentoul, Robert Reid, see page 458.

had sent Victor Branford a prickly outline of ideas, 'The Proposed Sterilization of Certain Degenerates', as a topic for discussion at the Sociological.

The doctor asked rhetorically if the number of degenerates in the United Kingdom was increasing (yes), whether they could be cured (no), and briskly proceeded to the question of which methods of sterilization might be preferable: orchotomy (removal of the testicles), oophorectomy (removal of the ovaries), vasectomy, or salpingotomy (ligature and excision of the fallopian tubes). These measures he would bolster with: immigration restrictions on lunatics, criminals, paupers and prostitutes; anti-miscegenation laws; prohibition of child and female factory labour; and education about procreation and venereal diseases. He left abortion and suicide as open questions. Galton was taken aback by the proposal and came out strongly against having anything to do with Rentoul.[59]

Dec 20 1904

Dear Mr Branford

Lady Welby writes to me enclosing yr letter (wh: I send back to you as you may want to use it again), and saying that you w^{d} be glad of some opinion of my own about the advisability of accepting Dr. Rentoul's paper.

I have an opinion, but it is more decidedly in the negative than I can easily justify. Besides your reasons, with wh: I agree, a society sh^{d} do nothing likely to shock many of its members. This is more especially the case with a young society that has not yet found its bearings. There is plenty of good work for the society to do, without running amuck against prejudices far stronger than its own powers can batter. It would get a momentary notoriety, coupled with an abiding bad name. I could mention not a few other subjects for discussion as suitable in the one sense and as grossly unsuitable in another.

Supposing that many experts agreed with Dr. Rentoul, their natural pulpit would be the medical papers, at first; then in reviews. The time might come when the question had grown familiar & the disputed points were few & well defined; then, & then only as I think, would discussion be proper between so mixed & "popular" a society as the Sociological, as suitable virginibus puerisque.[60]

But, as I began by saying, this imperfectly expresses all that I feel in disfavor of accepting the paper for reading & discussion, or even for publication.

Very faithfully
Francis Galton

As I am writing, I may as well go on about a quite different subject. Dr Saleeby was here yesterday & in the course of conversation it occurred to me to suggest as a possible field of activity for the Society, the discussion & adoption (partial & with or without amendments) of the Report of the National Deterioration Cttee, and of memoralising Govt thereon. I sent him last night, an open letter of introduction to the Scty of the Cttee, Mr. Grant Pooley, Scotter, Well Walk, Hampstead, N. W., to use or burn after consultation with you. Probably "the more systematic feeding of school children" Report p. 69 would be thought to be well within the purview of the Society, & on this Sir John [Gorse] has lately expressed himself forcibly. But I have my own hands full, so will say no more on this, which, while it is cognate to, is not identical with my present chief subject of interest.

Very sincerely
Francis Galton

[59] Keele, GB172 LP/1/1/2/7.
[60] For girls and boys, i.e. family friendly.

In truth Rentoul had already publicly raised these topics in the *Medico Legal Journal* earlier that year,[61] and even earlier in 1903. He would re-emerge a few years later.

More help was needed. One of Weldon's biometrical research assistants from Oxford, Edgar Hermann Joseph Schuster (1879–1969) was appointed as the Galton Research Fellow at UCL in January 1905, beating five other candidates, though he was not Galton's favoured contender. Ethel M. Elderton, who had already been assisting Galton with fingerprint data, was made his assistant.[62] Under Galton's control, with assistance from Weldon, they made up the 'Eugenics Record Office' in the university, at 50 Gower Street—near the house where Charles Darwin had first lived after returning from the *Beagle* voyage, and Galton had visited more than once as a medical student. Later that year a Lady Secretary would be added; now Schuster was promptly dispatched to join the Sociological Society, where he would be introduced to Branford.

At the 'So-so', as insiders had started to refer to the Sociological, more momentum was gathering. On February the 14th of 1905 Galton submitted a paper—it was read for him, as he was in Italy—on 'Restrictions in Marriage', arguing that the public was more pliable than many had supposed, once the scientific basis of eugenics had permeated popular consciousness. He gave an extensive list of prior restrictions on marriage from the study of world anthropology, in the same way that George Darwin had in 1874.

Another paper followed immediately on 'Studies in National Eugenics'. This laid out a program of study suitable for the new Eugenics Record Office: estimation of the average quality of the offspring of married couples, from their personal and ancestral data; effects of action by the state and by public institutions on eugenics; study of other influences that further or restrain particular classes of marriage; collection of accurate facts about heredity; catalogues of existing data relevant to eugenics; cooperation with other influential bodies; and the construction of a scheme for issuing (voluntary) eugenic certificates, to testify to the worth of marriage partners.

The ensuing discussion (this time published only in the *Sociological Papers* and a year later) showed that opinion had shifted to mostly positive. Almost all agreed on the desirability of eugenic ends. A few, such as the Zionist intellectual and physician Max Nordau, the Fabian sexologist Havelock Ellis, the anthropologist Sir Richard Carnac Temple and the zoologist Charles A. Witchell, were still highly skeptical about the practicability of the policies suggested. Those only mildly skeptical about methods included the psychiatrist Frederick Walker Mott, the sociologist Prof. Ferdinand Tönnies (Kiel), and the anthropologists Prof. Giuseppe Sergi (Rome) and Alfred Henry Huth.

Strongly positive responses were received from influential figures like the anthropologists Alfred Cort Haddon, A. E. Crawley, and Edvard Westermarck; the zoologist Yves Delage (France), the constitutional lawyer A. Posada (Spain), the ethnologist R. S. Steinmetz (Leyden), the cell biologist August Weismann, the historian B. Altamira (Spain), the Scottish physiologist John Gray McKendrick, the philosopher John Henry Muirhead, and the evolutionary biologist Edward B. Poulton.

The logician and public intellectual Bertrand Russell—who would prove notoriously receptive in the future to ideas about non-traditional marriage, at considerable cost to his own personal relationships—went so far as to declare that 'I have read Mr. Galton's two papers in abstract with much interest, and agree entirely with the view that marriage customs might be modified in a eugenic direction'. Ideologues who would bend the discussion to their own ends were still there: Dr. Alice Drysdale Vickery (accompanied by her common-law husband Dr. Drysdale), was pleased to talk once more about women's issues and birth control.

No exasperated remarks were called for now. Instead Galton politely rebutted and tacked on another piece as his main response, 'Eugenics as a Factor in Religion' (which he thought not substantial enough to be published as a memoir in its own right). Now he felt more freedom to

[61] Jan. 1, 1904.

[62] W. Palin Elderton, Ethel's brother, had spoken at the May 1904 meeting of the Sociological Society.

openly advocate his replacement 'virile creed', which is 'full of hopefulness'. It forbids sentiment harmful to the gene pool, while encouraging the ties of kinship and broader altruism, so as to improve that reservoir.

So encouraged, the Eugenics Record Office busied itself from 1904 until late 1906 with a study of the relatives of living Royal Society fellows that Galton had started in 1904. It was a detailed follow-on to his pedigree studies from the 1860s, but concentrated on the families themselves, not the individuals.[63] Galton had surveyed all 454 of the fellows himself for details about their relatives. About half of the fellows had replied. As he had expected, ability tended to cluster by family, as verified from independent sources, but not invariably. But he had only built a preliminary index.

By the middle of 1905 Galton was writing to his niece Milly, 'I must now "boss" these Eugenic matters a good deal, to make them "hum" as the Yankees are pleased to say'.[64] Thus an 'Advisory Committee' was also launched that October. 'Mr Galton would be glad to utilise the room of the Eugenics Record Office, after office hours, for the occasional meeting of a few invited persons who seriously desire to promote Inductive Research in matters connected with Eugenics'. Schuster, Victor Branford, and John Willis Slaughter were roped in, with Miss Elderton as secretary. Karl Pearson, who detested committees in general, detected squeaking rather than humming. 'It really only hampered the Research Fellow'. But both Branford and Slaughter would contribute to founding and running the Eugenics Education Society a few years later—which Pearson would also detest.[65]

Schuster industriously dug up biographical details for the (male) members of 66 families of living fellows (those which contained at least 3 noteworthy members, defined as appearing in sources like *Who's Who*, *The Dictionary of National Biography*, encyclopedias and similar reference works). Results were cross-checked by the fellows themselves. The results were summarized in *Noteworthy Families* which was completed by September 1905 but only published in early 1906. It was chiefly Schuster's work in the details, but not the analysis.[66] Galton argued that these were men of great ability, but not the most outstanding ability, who are much harder to study because they are enigmatic and vanishingly rare. There is a poignant recollection of 'my own private knowledge of the somewhat abnormal frequency with which eccentricity, or other mental unsoundness, occurs in the families of very able scientific men'. (This is somewhat at odds with his assertions in *Hereditary Genius* to the contrary, see chapter 4, p. 51.)

The data could be used to determine if being an FRS sizeably increased the probability that relatives were noteworthy. Only males were considered because females had almost no opportunity to distinguish themselves. Calculating the necessary frequencies was not so straightforward, since it was challenging to estimate how many total relatives each family had. Half the returns were missing this information. Galton had been vexed by this problem—estimating the total number of relatives or offspring—ever since working on *Hereditary Genius* in 1868/9. It was also relevant to fertility concerns. In 1873 he had posed it as a challenge in the mathematical *Educational Times*, in terms of the extinction of family surnames, offering a prize for precise solutions. With no plausible solution forthcoming, a partial analysis, subsequently improved, was elicited from his old friend, the Rev. H.W. Watson, which Galton read to the Anthropological Institute. It turns out that Watson's analysis was not entirely correct, mistakenly concluding that *all* surnames must go extinct—yet the population grows!—a mathematical error which was missed by Karl Pearson. Today this question, which Galton was not satisfied with and Watson evidently lost interest in, has been properly solved and generalized to apply to all sorts of related problems as the study of 'Galton-Watson Branching Processes'.[67] Galton resolved the

[63] F. Galton 1904a,b.
[64] FG to Millicent Lethbridge, 1905-06-25, K. Pearson 1930b, 545.
[65] K. Pearson 1930a, 438.
[66] F. Galton and Schuster 1906.
[67] F. Galton 1873d, 1874a; K. Pearson 1924; Watson 1873. Watson overlooked (or for some reason ignored) an

question here by just using as a rough estimate the figures obtained from those fellows who had given full details of their families. He had no choice but to accept whatever bias was introduced by the fact that fellows who had notable relatives were surely more likely to respond than those who did not have them.

In the returns, there was roughly one fertile male relative for each fully specified kind of relationship (which 'grandfather' is not unless the line of descent is given). By assuming that these surviving relatives had sufficient opportunities to become notable, he could divide to get the frequency. Here he only counted relatives who were at least as old as the fellow himself, and of these only fathers, brothers, grandfathers, uncles, great-uncles, great-grandfathers, and male first-cousins. Complicated weighting systems which took into account degrees of eminence were found to make little difference overall.

There remained the hard problem of determining the total sample size given that just 207 out of 467 surveyed had responded. Pragmatically Galton took the mean of 207 and 467 as the total for calculating overall frequencies, or 337. Getting the ratios would cancel the total out anyway. The noteworthy were then 3.37 of 337, or 1% of the total set. But fathers of fellows made up 81 of 337, or 24.1 times more. Of first cousins of fellows, 99 of 337 were notable, or 24.75 times more. This suggested strongly that being a fellow conferred a much higher chance of having notable relatives, even in relatively remote degrees.

Schuster and Elderton promptly moved on, working on the extension of the *Hereditary Genius* data that was published as *The Inheritance of Ability, Being a Statistical Study of the Oxford Class Lists and of the School Lists of Harrow and Charterhouse.* Schuster also worked up the investigation fulsomely titled *The Promise of Youth: a statistical inquiry into the question whether success in the examination for the B.A. degree at Oxford is followed by success in professional life.* To which the data suggested 'yes'. These would both be published in 1907.

Schuster, no doubt wearied by countless hours of mining biographical information about relatives of eminent men, abruptly resigned his fellowship in October of 1906, taking up an unpaid fellowship at Oxford to resume biological work. His mentor Weldon had suddenly died earlier that year, which may have impressed his former pupil with a more definite sense of time. As he wrote to Galton, 'my own interests lean entirely in the biological side of Eugenics; the sociological & economic investigations which form so important a part of the subject are, I am afraid, neither in accordance with my aptitudes nor my inclinations'. He suggested that 'experimental breeding of animals & plants' was a more promising line of investigation into heredity.[68] Still, Schuster remained interested in the field, publishing his own book on *Eugenics* in 1912.[69] He would open his own Anthropometric Laboratory at Oxford. Now Galton would turn to Karl Pearson, until then only marginally involved with the Eugenics Record Office.

The Sociological Society had more or less run out of steam by mid 1906. Enthusiasts within the 'So-so' for the cause of eugenics were rubbing up against the Patrick Geddes type of mostly-innumerate sociologist who had little interest in biology and preferred to study higher level topics like 'civics'. So at the end of the year Galton appealed to his Rutland Gate neighbour, the socially well-connected barrister Montague Crackanthorpe QC. Oxford-educated, with a double first in classics and mathematics, 'Crack' shared his population concerns. 'Dear Crack, May I consult with you on the following? Is not the time ripe for some association of capable men who are really interested in Eugenics & to whom the existing Eugenic Office of the London University might serve as a centre? If you think so & cared to suggest the outline of a working plan & a few good names, I shall be grateful. I am too much of an invalid to push forward

additional solution to the fixed-point problem involved. See Kendall 1966, 1975 for a comprehensive history including earlier anticipations of the analysis and solution.

[68] Schuster to FG, 1906-10-22, GALTON/3/3/18/4.

[69] Schuster 1912.

[anything] except by letter. Still I think something useful might be done even in [that way]. I do not yet see it very clearly & am desirous of fresh ideas?'[70] It is unclear exactly when Galton first met 'Crack' but they were already close friends by 1905.

Dr. Robert Reid Rentoul—whose radical paper advocating sterilization had been rejected on Galton's advice by the 'So So' in 1904—was back in public view by now, having published *Race Culture or Race Suicide: a Plea for the Unborn* that year.[71] Galton read it, writing to Karl Pearson, 'It is full of shortcomings but very suggestive, and shows how much has been written and done here and in America. There is a massive movement going on out of the public sight. He mentions a fact that may be new to you, as it was to me, that a certain "Malthusian League" sends pamphlets "Why have any more children?" to the parents mentioned in the Birth Columns of Newspapers. Have you seen any of these? I recollect something of the sort long ago by (?Annie Besant)'.[72] It would be a while before Pearson got around to reading Rentoul himself, with less generous results: 'his exaggerations and fallacious statistics spoil an otherwise strong case. He is wretchedly careless also in expression—rather a medical Bernard Shaw. In fact he displeases me'.[73]

Early in 1907 the Eugenics Record Office was renamed 'The Francis Galton Laboratory for the Study of National Eugenics' and placed under Pearson's control. Galton was now too old to involve himself directly in its operations, and acted after that only as a kind of elder statesman and benefactor. Pearson also continued to run his biometric laboratory at the university, bearing a heavy combined workload, made worse by his determination to collect very large datasets. Nobody could ever accuse Pearson of insufficient dedication to the pursuit of his ideas, though many would later argue that these were often in mistaken directions.

Pearson replaced Schuster with David Heron, a young Scottish mathematician who had graduated from St. Andrew's in 1905 and appears to have joined the Biometric laboratory at UCL around that time (recollections were somewhat vague). Ethel Elderton was finally recognized, at Pearson's suggestion, as 'Galton Research Scholar'. The lab immediately set about tackling what were considered to be the major social problems of the era: alcoholism, tuberculosis, insanity and mental defect. The 'So So' and its allied circles had struck the mathematician as a haven of pseudo-scientists and cranks. He disliked Bernard Shaw and H. G. Wells. He had flirted with the Fabians early on, but had taken his own path, despite agreeing with them broadly about most political matters.

What set Pearson apart was an abundance of scientific caution, and an abhorrence for flashy claims and publicity. With that went a prickly temperament and a tendency to feud with opponents. Assembling extensive data sets for careful analysis, to understand the precise nature of hereditary transmission, mostly of complex traits, would occupy the Galton Laboratory for many years. More room would be needed at Gower street (the office was now at number 88), a library, extra assistants, and a continuous stream of publications to issue to the public. Much was doing. The education committee of the London County council offered the lab data for over 8,000 schoolchildren, which was gratefully accepted.

While Pearson was getting going in the lab, Galton had the opportunity for further public promotion of eugenics, after being invited (through E. B. Poulton) to give the 1907 Herbert Spencer Memorial Lecture at Oxford. At first he turned the invitation down but then gave in to repeated requests from the Vice Chancellor. 'Probability the foundation of Eugenics' was delivered on June 5th in the Sheldonian Theatre by his cousin Arthur Galton, as the aged scientist was still recovering from sciatica after a fall.[74] Galton took a great deal of trouble over

[70] Only the draft of this note survives, GALTON/3/3/7/22. For Crackanthorpe, Montague, see page 451.

[71] Rentoul 1906.

[72] FG to Pearson, 1906-12-20, K. Pearson 1930a, 304.

[73] Pearson to FG, 1907-04-08, K. Pearson 1930a, 312.

[74] He had rested on an insecure crockery table in the dark, and was only found lying there some hours later in the morning.

the lecture, which is written in the comfortable & confiding style that he had honed for many years; it draws on Keats to establish common ground with his predominantly literary rather than scientific audience.

Galton means 'probability' in the sense of population distributions, and defends the use of biometry, which Bateson had dismissed a few years earlier as 'actuarial methods', for studying the inheritance of human traits through those distributions. Variation, which is not 'something indefinite and capricious', has been neglected in the past but matters as much as the means. The concepts can be explained in simple terms to any lay audience—teasing Pearson—as could intercomparison of ranks, and correlation. It is by no means clear how many of *that* audience really followed the detailed explanations he gave, but it is likely that no simpler explication of this material has ever been given (it was worked up a few years later by Miss Elderton and her actuarial brother William into a *Primer of Statistics*).[75] After summarizing prior research regarding eugenics, he goes on to argue once more that the 'tyranny' of public opinion, once shaped, can overcome the admitted myopia of individuals to the consequences of injudicious marriages. A 'jehad' or holy war for eugenics is needed in due course, but *only when the scientific facts are known*, and not before then.

Even a few years earlier the newspapers would have been bemused by all this. Now they were scrupulous in their coverage. More specific details had already been dealt with a few weeks earlier, on May 17th, by Karl Pearson himself. Coincidentally, the world's leading 'biometer' had been invited to give the Boyle lecture to the Oxford University Junior Science Club on 'The Scope and Importance to the State of the Science of National Eugenics', and had given an extensive summary there of established correlations for human characters.[76]

By the middle of 1907 Crackanthorpe would publish his eugenically-minded book *Population and Progress*, containing a dedication 'to my friend' Galton.[77] That friend appreciated the effort, writing to Karl Pearson that June, 'Crackanthorpe's "Population and Progress" interests me much. His last chapter (VI) opens out quite a new horizon to me, and suggests a subject for discussion at some future Hague Conference—viz. limitation of populations! The pullulating nations have ever been the primum mobile of invasions'.[78]

A few months earlier, the Fabians had issued Sidney Webb's *Decline in the Birth Rate*, an interim result of nearly two years of investigation. It finds that birth rates appeared to have declined in both urban and rural areas, that the decline is tied to the inconvenience of having children, that it applies also to those who keep servants, and is due to some new voluntary cause. Crucially, it 'appears to be much greater in those sections of the population which give proofs of thrift and foresight than among the population at large'.

Webb looks for a comprehensive solution through incentives and liberation along the same lines as Galton proposed. 'Once the production of healthy, moral and intelligent citizens is revered as a social service and made the subject of deliberate praise and encouragement on the part of the government, it will, we may be sure, attract the best and most patriotic of the citizens. Once set free from the overwhelming economic penalties with which among four-fifths of the population it is at present visited, the rearing of a family may gradually be rendered part of the code of the ordinary citizen's morality. The natural repulsion to interference in marital relations will have free play. The mystic obligations of which the religious-minded feel the force will no longer be confronted by the dead wall of economic necessity. To the present writer it seems that only by some such "sharp turn" in our way of dealing with these problems can we avoid degeneration of type—that is, race deterioration, if not race suicide'.[79]

75 W. P. Elderton and E. M. Elderton 1909.

76 This was the talk by Pearson attended by the young Cyril Burt, see chapter 12, p. 261.

77 Crackanthorpe 1907.

78 FG to Pearson, 1907-06-11, K. Pearson 1930a, 375.

79 S. Webb 1907.

This alignment of forces led to the formation of a dedicated Eugenics Education Society at a meeting held in the Caxton Hall, Westminster, on November 15, 1907. The new society formed a home for 'So so' members like John Willis Slaughter, Havelock Ellis, and Dr. Caleb Williams Saleeby. Other impetus came from disaffected members of the 'Moral Instruction League', which described itself as 'a non-sectarian and non-party association to urge the vital importance of efficient moral and civic education in schools'. It was a humanist movement—that is deliberately non-religious, as a competitor to religion—with improvement of the working class as its declared aim. Note though that both this and the Sociological would continue in parallel with the Eugenics Education Society, with their own focus.

By December 9th a Provisional Committee for Eugenics Education was in place, chaired by Slaughter (1878–1964), an American originally from Alabama. Slaughter would end up at Rice University in Houston but was then lecturing in sociology at the London School of Economics. The organizing force was not really Slaughter, who would continue as secretary of the 'So so', but rather the redoubtable organizer Mrs. Sybil Gotto (1885–1955, see Figure 14.7, p. 358).

Mrs. Gotto was born Sybil Katherine Burney, but had married Lieutenant Arthur Corry Gotto RN in 1905.[80] Widowed in 1906, she occupied her time and considerable energy with the cause of birth control and a crusade to eradicate prostitution and venereal disease. After reading Crackanthorpe's eugenics-redolent book, she had met John Willis Slaughter at the Sociological, where he was secretary (Victor Branford had moved on to Honorary Secretary). In turn she had been introduced by the American to Crackanthorpe. A meeting with Galton, whose books Gotto had steeped herself in, followed. She became secretary to the Eugenics Education Society (the *Eugenics* hereafter), under its first president the eminent Scottish alienist Sir James Crichton Browne.[81]

The first General Meeting of the society was held on Feb. 14th, 1908, at Denison House in London. Initially Galton was not directly involved, though he supported the creation of a non-scientific body to stir up public consciousness—as he had suggested to Crackanthorpe over a year earlier—and had offered help when Slaughter first wrote to ask. Karl Pearson had turned down an invitation to join the society, as he preferred to concentrate on the science, and he was wary of those he thought might harm the eugenic cause through scientific naivete and rash public pronouncements. Still, Galton politely kept his own counsel. A crisis soon presented itself and caused him to withdraw his offer in what must have been a testy letter (now vanished) to Slaughter. Only the icy response from the American survives.

> February 11th, 1908
>
> Dear Mr. Galton
>
> I am exceedingly sorry for the attitude you feel necessary to take towards the Eugenics Education Society; however, there is a mistake somewhere as the Society in question has not as yet held any lectures, and, indeed, I cannot think what your friend reported upon unless it was the lecture of a certain individual whose vagaries we can hardly be held responsible for, and whom nobody with an ounce of scientific sense ever takes seriously.
>
> In spite of this being a mistake, I will not ask you to change your attitude and renew your kind offer of help, as it is possible that we may, even with the greatest care, admit something that would give offence. It is very difficult to find competent lecturers, as those who have had the best opportunities to acquaint themselves with Eugenics, prefer to keep the whole matter within the limits of a small academic circle. On the other hand, general interest in the subject is now thoroughly awake and the

[80] She was later known after her second marriage in 1917 as Sybil Neville-Rolfe, and was ultimately awarded the O.B.E..

[81] There is an autobiographical prefix in Neville-Rolfe 1949 with several factual errors.

demand for instruction will be met somehow, if not by the competent, then by the incompetent. The movement will take place anyway, and the question is whether it is better to guide it, or to treat it with contempt. I have no doubt that the discussions will, again and again, show an appalling ignorance, but it seems to me that ignorance is someone's opportunity.

I hope this letter does not savour of impertinence; please do not trouble to reply to it.

Yours faithfully
J. W. Slaughter

Apparently Galton had sent a ticket to Miss Elderton for a public meeting. He believed it was held under the auspices of the *Eugenics*. Elderton had written in early February to complain that sex matters were discussed at the meeting.[82] The scientist was mortified. 'The tidings in your letter about the Eugenics Education Society pain me much. Thank you greatly for sending them. I have written to Dr Slaughter withdrawing an offer of help that I made in response to an exceedingly sober and well-written letter from him, and said that I cannot consent to be connected with it at present. It is very sad'.[83] But Galton was (comically) mistaken, as Slaughter (uncomically) pointed out.

The *Eugenics* got around soon enough to hosting their own public meetings, and a ready cause was close at hand. Showing the hand of the Moral Instruction League, Dr. Caleb Williams Saleeby was incensed by the decision of the London County Council to close down its home for chronically inebriated women, since they were incurable—letting them loose on the streets, there to cavort and breed. The society issued a condemnation of the action—cures were optional, fertility was dangerous—and generated some rousing newspaper controversy in the august *Times*. The Home Office co-opted the society onto a committee of enquiry. It was a start.

In March there was an embarrassing setback when Dr. Stanton Coit became involved in a bizarre incident, in evening dress, with a bus conductor, one Alfred Ernest Roper. Coit was an American who had naturalized in Britain and involved himself in 'Humanist' and 'Ethical' movements. He was then loosely associated with the *Eugenics* and slated to deliver a lecture series. On the lower deck, hands mysteriously touched—too long. On the upper deck, to which Coit had followed the conductor, bodies brushed—too closely. Alarum! and the lecturer ran off down the road, still in evening dress, where he was tackled and pinned by the infuriated conductor. A charge of indecent assault was pressed when a policeman promptly arrived (a most unlikely contingency today).

Coit was duly sentenced to a month in jail by an indignant magistrate. Pearson glumly relayed the news to Galton, calling it 'a bad blow'. His mentor was more sanguine. 'Dr. Coit! what an unfortunate name that man has. It is enough in itself to make ridicule out of eugenics'.[84] *Coitus omnibus interruptus.* Fortunately for him, Coit managed to have his conviction fully interrupted on appeal before the 'sessions'. His counsel argued (in retrospect, rather unconvincingly) that the body contact was only an accident, and adroitly shepherded the flustered conductor into the folds of contradiction.[85]

Galton was not yet officially involved in the education society, but wrote to Pearson (ever skeptical) that it 'promises better than I could have hoped. Crackanthorpe is serious about it, and Professor Inge has joined it! I can't find that Crichton-Browne has as yet done much. Saleeby acts as a restrainer, but is very eager, & they have got a particularly bright lady

[82] K. Pearson 1930b, 628.
[83] K. Pearson 1930b, 585.
[84] Pearson to FG, 1908-03-26, K. Pearson 1930a.
[85] The case was fully covered in the newspapers, e.g. *South London Press* 1908-04-17, 3. Forrest 1974, 275, badly mangles this incident, incorrectly assuming that the name redacted by Pearson was John Willis Slaughter.

secretary "Mrs. Gotto" who acts and works hard for the love of the thing. I have not yet ventured to join it, but as soon as I am assured it is in safe management, shall do so'.[86] He was asked to serve as Honorary President by the end of the month, and accepted.

Hesketh Pearson later recalled that his great-uncle Frank was explicitly set on using glacial displacement rather than shrill controversy to promote eugenics. 'It should be spread by Fabian methods, a gradual inculcation of its needs; it would do more harm than good if forcibly pressed forward in its entirety—we would become cranks in the public eye'.[87]

At a cosy local meeting in late June, held at Crackanthorpe's Rutland Gate house, Galton had a lucid lecture on 'Eugenics' read by the host on his behalf—he was too deaf to speak himself—while he scanned the audience to count telltale fidgets, or so 'Crack' later recalled.[88] What is advocated is a 'living and growing science' with 'high and practical aims', the reverse of fanatical. 'Statesmanship is concerned with the nation; Charity with the individual; Eugenics is concerned with and cares for both'.

Nurture matters too. 'Families which are likely to produce valuable citizens deserve at the very least the care that a gardener takes of plants of promise. They should be helped when help is needed to procure a larger measure of sanitation, of food, and of all else that falls under the comprehensive title of "Nurture" than would otherwise have been within their power. I do not, of course, propose to neglect the sick, the feeble, or the unfortunate. I would do all that available means permit for their comfort and happiness, but I would exact an equivalent for the charitable assistance they receive, namely, that by means of isolation, or some other less drastic yet adequate measure, a stop should be put to the production of families of children likely to include degenerates'.

Misapplied, charity can be contra-eugenic. Properly applied, it helps the 'desirables'. A large though gentle eugenic effect can be carried by public opinion alone. Accurate knowledge of heredity is crucial, towards which Karl Pearson and the University of London are working and already issuing memoirs with results.

The audience was select, but newspaper reporters had been shrewdly invited by 'Crack' and proceeded to publish extensive excerpts. Most of the content may have been familiar to long-time followers, but it was new to the broader public, and pitched at just the right level for them. A fresh wave of members washed into the *Eugenics*. Soon Galton enthused to (a privately skeptical) Pearson. 'That Eugenics Education Society seems really promising. The prospectus has been re-worded and members are coming in. Mrs Gotto is marvellous in her energy'.[89] When Crichton-Browne chose not to serve as President for a second term, Crackanthorpe would replace him.

Membership of the *Eugenics* was split into two tiers: full and associate. The former had voting rights, and paid a guinea. Those who joined were by definition social reformers, and dominated by liberals and even radicals. In 1909 there were 112 full members and 229 associates, growing to 155 and 300 the following year, then rising again to 196 and 331 in 1911. Membership peaked in 1913 at 406 full members and 307 associates, after which it gradually declined over the years. But these numbers do not convey the full influence of the society, which counted well over fifty people eminent enough to make the *Dictionary of National Biography*.

The modern reader will still recognize some of these names: Arthur J. Balfour (a former prime minister), Neville Chamberlain (another future prime minister); the economist and polymath John Maynard Keynes; the botanist Sir Francis Darwin; his brother, the precision instrument manufacturer Horace Darwin, who counted Galton among his clients; the statistician R. A. Fisher; the Bohemian bluestocking Lady Ottoline Morrell; the radical socialist Harold

[86] FG to Pearson, 1908-05-18, GALTON/3/3/7/20.

[87] H. Pearson 1921, 73.

[88] See Crackanthorpe 1911, reproduced in Tredoux 2023a. The fidget theme appears rather often in stories about Galton.

[89] FG to Pearson, 1908-07-10, K. Pearson 1930a, 342.

Laski; the pioneering psychologists William McDougall, Charles E. Spearman and Cyril Burt; the sexologist and crank Havelock Ellis. Then there was Major Leonard Darwin, who was perhaps the first man to become known chiefly for simply being a resolute eugenicist, aside from being a son of the naturalist. Others were major figures in their day but are less well-known now: Rev. Edward Lyttleton (headmaster of Eton), Patrick Geddes (botanist and sociologist), Archibald Geikie (geologist), Sir Frederick W. Mott (neuro-pathologist); W. R. Inge (Dean of St Paul's and radio regular); and Sir John Arthur Thomson, the Scottish naturalist. There were politicians and poets; scientists, mathematicians and medicos; university administrators and school headmasters (Harrow was not to be left out); philosophers and philanthropists; economists and eccentrics; Fabians, socialists and holy men.

A journal, the *Eugenics Review*, was established in 1909 and flourished, moving steadily from advocacy and popularization to more serious analysis over time. Galton contributed a foreword and two papers to the first volume,[90] which also featured 'Crack', Saleeby, Slaughter, Havelock Ellis, W. R. Inge and J. Arthur Thomson. The second volume included Sidney Webb, Saleeby and the philosopher F. C. S. Schiller.

Glaringly absent from all this was Karl Pearson, who had little sympathy with the *Eugenics* popularizers. In January of 1909 he shared his forebodings. 'If our youthful efforts were mixed up in any way with the work of Havelock Ellis, Slaughter or Saleeby, we should kill all chance of founding Eugenics as an academic discipline. Please don't think I am narrow, or that I do not admit that these men have done or may do good work. All I say is that I could not get the help we are getting from the medical profession, from pathologists or physiologists, if we were supposed to be specially linked up with these names. Rightly or wrongly it would kill Eugenics as an academic study. All I want is to stand apart doing our scientific work, not in any way hostile to the Eugenics Education Society, giving it any facts we can or an occasional lecture, but not being specially linked to it in any manner. For this reason I am rather sorry that Heron has gone on to its Council, because it makes a link, which I think it is better for Laboratory and Society not to forge—it will hamper the freedom of both. My policy, however, with my young people is to show them my own standpoint, but in no way to control their action'.[91]

To some extent Pearson's fears were justified by the George Bernard Shaw incident in March of 1910. Galton had forebodings too. 'Bernard Shaw is about to give a lecture to the Eugenics Education Society. It is to be hoped that he will be under self-control and not be too extravagant'.[92] On the 3rd of March Shaw got up at Caxton Hall and put Galton's fears in deep shade, calling for the abolition of private property and marriage. Eugenics required widening the scope of sexual selection, a broadening of the field. Doing so might produce more of the best people, like himself (laughter). Income would be provided for all by the state, and the idea that human life was sacred would have to be revised. Some murderers should be praised, depending on the victims, and a certain *number* of murders per person, like bites per dog, could be tolerated. The state could not leave marriage to free choice since it might want to experimentally cross-breed wholly incompatible tempers. Motherhood did not require the mothers to live domestically with the fathers. Then he sat down abruptly (more laughter).[93]

Usually the *Eugenics* asked for the text of addresses in advance, but Shaw had not provided one, claiming that he did not know what he would say—which, as Crackanthorpe later put it, may even have been true. Galton took it in stride, 'I don't mind good jokes, but Bernard Shaw makes such bad ones'.[94] The press ran with the idea that socialism-a-la-Shaw and eugenics were identical. G. K Chesterton had a field day in the *Illustrated London News* arguing that

---

[90] F. Galton 1909, 1910b,c.

[91] Pearson to FG, 1909-02-07, GALTON/3/3/16/9.

[92] FG to Pearson, 1910-02-26, K. Pearson 1930a, 426-7.

[93] *London Evening Standard*, 1910-03-04, 8, and other newspapers.

[94] K. Pearson 1930a, 427.

the cracked ideas on offer were really mainstream to the eugenics movement—'cartloads of Eugenical nonsense'—and nothing new. He did agree though, that the world would be 'a jollier place if there were fewer quacks in it' (like Shaw).[95]

A more serious rift had threatened in the meanwhile. In May, Dr. C. W. Saleeby published an article on 'The Progress of Eugenics' in the supplement to the *New Age*, edited by Patrick Geddes of the 'So So'.[96] Geddes had solicited a paper from Galton himself, but had to be content with what he called, to the dismay of the ailing scientist, a 'note of encouragement'. Mortification followed when Saleeby, who was tolerated as useful but widely disliked, was represented by Geddes as Galton's 'foremost disciple'.

In his article, Saleeby promoted Bateson's Mendelian genetics, at the expense of Pearson's biometrics, though 'if and when it can serve eugenics, the eugenist will of course avail himself of it'. He vaguely identified 'racial poisons—those substances, inorganic, organic, microbic, or other, which produce blastophthoria or damaging of the germ-plasm', which had to be eliminated. He enthused over the possibilities of sterilization, which most of the *Eugenics* rejected. But some of his effusions here proved lasting, such as his distinction between positive (encouraging good breeding through rewards) and negative (preventing bad breeding through controls) eugenics.

Galton placated his disciple, writing 'I had seen the *New Age* and was vexed at Dr. S.'s remarks,' and 'The Council of the Eugenics Education Society have, I learn, extruded Dr. S[aleeby] by not putting his name on the candidate list. As I am told, certain members of the Council strongly objected to serving longer with him, and Mrs. G[otto] undertook to tell him so, which she did, doubtless with all practicable tact, but I have reason to know his feelings are much wounded. He however spoke very nicely to Mrs. G[otto]'.[97]

The rift gaped further in public view when Ethel Elderton of the Eugenics Laboratory completed 'A First Study of the Influence of Parental Alcoholism on the Physique and Ability of the Offspring' that same month.[98] The hand of Karl Pearson shows strongly in the text. This was issued as a Eugenics Laboratory Memoir and drew attention in the *Times*, since it concluded that, contrary to the expectations of the temperance movement, it could find little evidence of negative constitutional effects of parental alcoholism on children of those alcoholics.[99] It was based on data from the Edinburgh Charity Organization Society Report (over 400 families) and a manuscript account of the children in the special schools of Manchester (over 600 families).[100] The parents were classed by observation into alcoholic categories.

Elderton found that children of alcoholic parents had a higher death rate than those of the non-alcoholic, but had slightly higher weight and height.[101] The alcoholic parents earned marginally lower wages. Overall health of the offspring of the alcoholic was slightly better than the non-alcoholic, and there was no sign that alcoholic offspring were more prone to be mentally defective (again, classed by observation), nor that they had worse vision.[102]

A substantial portion of the *Eugenics* were convinced temperance advocates, being drawn from sources like the Moral Education League, for instance Saleeby and Montague Crackanthorpe. They refused to believe these results, at least partly because, like Saleeby quoted above, they clung to beliefs about acquired characters, resulting in poisoning of the germ-line.

[95] *Illustrated London News*, 1910-03-12, 2.

[96] Saleeby 1910.

[97] FG to Pearson, 1910-05-07, K. Pearson 1930a, 428. Alfred Ploetz (1860–1940) edited the periodical *Archiv für Rassen- und Gesellschafts- biologie*, which had been founded in 1904.

[98] E. M. Elderton and K. Pearson 1910.

[99] 'Alcoholism And Offspring', *Times*, 1910-05-21 (39278), 14.

[100] The sample totals have to be inferred from the tables in the appendix to the report.

[101] Note though that 'there is a connection between the mother's alcoholism and her child's physique, and this connection appears to be greater for girls than boys'.

[102] Elderton was using Pearson's 'polychoric' correlation coefficients, inferred from differences in means, since categorical variables were involved.

Crackanthorpe took to the *Times* a few months later, to argue that the results were 'contrary to general experience' and that 'biometrical methods' did not provide 'a safe rule of conduct' when dealing with so 'complex' a problem, based as they were on 'the law of averages' and 'the theory of probability'.[103]

Many cogent criticisms could have been offered by a technically adept critic—measurement error from induced categories based on observation could easily overwhelm the observed correlations, as could sampling error. Failure to find something impresses only when it should probably have been found, and the observational data presented was questionable. Skepticism about results based only on limited numbers of noisy observations, not yet replicated, is entirely healthy.[104] Instead Crackanthorpe impugned statistical methods in general, without offering an alternative beyond prior convictions.

Galton—still more or less the patron saint of the *Eugenics* and thus embarrassed—wrote a pointed response: refuting Crackanthorpe's description of biometrics; perplexed at the reference to averages; offering a précis of its methods; denying that the effects of alcoholism were beyond the reach of statistical methods.[105] In reply, Crackanthorpe clarified that he meant that considerable variation in *individual* results should be expected, which the correlations (inherently averaged over observations) would mask.[106]

Pearson, confirmed in his skepticism of the laity and constitutionally predisposed to take any criticism personally, was more severe. He wondered if Crackanthorpe had even read the report in question, or had relied on newspaper reports instead, and so was disinclined to take him seriously. Talk of 'general experience' was bluster, to conceal lack of real data, which Elderton had provided along with the relevant statistics. He went on to wonder if other reports that the *Eugenics* were submitting to government would show similar disregard for evidence and rigorous methods.[107]

Crackanthorpe soon asked Galton to send him a copy of the Elderton report to read. Some peacemaking by their indulgent patron was needed. 'Yesterday I got together to tea, Miss Elderton, Crackanthorpe and Ploetz. C. made himself very agreeable in a long tete-a-tete with Miss E., but I fear was insufficiently penitent to receive full forgiveness'.[108]

Curiously William Bateson had much in common with his sworn foe Pearson here. The Cambridge geneticist had come out in favour of eugenics periodically, at least for clear-cut cases of imbecility—sometimes habitual criminality too, but here he vacillated—and he scorned egalitarianism. Inheritance of abilities was obvious, he argued, and unequal distribution of them equally so. There was great and genuine inherited heterogeneity in society, but this was exactly the sticking point: who would decide *which* traits to select for, and what were the unintended consequences of doing so? Remixing them might, in the future, produce desirable, perhaps essential, combinations.[109]

However a more fundamental factor giving Bateson pause was the fear that getting science mixed up in politics would retard the science without making any difference to the politicians, who were not known to be guided by facts so much as by emotions and popular perceptions, or more base motives. Eventually this persuaded him to altogether leave off discussing eugenics

103 'Alcoholism And Offspring', *Times*, 1910-06-02 (39288), 6.

104 Modern evidence shows that severe drinking by *mothers* during pregnancy carries substantial risks for offspring. A great deal of Eldertons' analysis is confounded by fathers, perhaps because it was not as clear then as it ought to have been that only mothers could offer a direct physiological pathway to children through drinking, and then most logically while pregnant.

105 F. Galton 1910a.

106 *Times*, 1910-06-07 (39292), 9. This is true enough, but absent further data the individual errors could just as well be positive (good effects) as negative (bad effects).

107 *Times*, 1910-06-10 (39295), 9.

108 FG to Pearson, 1910-06-10, and FG to Pearson, 1910-07-04, K. Pearson 1930a, 429.

109 W. Bateson 1905.

in public—he would turn down an offer to lecture at Cambridge on eugenics, 'taboo waters'.[110]

In July of 1910, an article in the *Westminster Review* on the 'The Scope of Eugenics' caught Galton's eye in a sheaf of press cuttings sent to him. It contained an assured and comprehensive summary, in eleven full pages, of all that he had been advocating. It neatly balanced the promise of Mendelian methods with the readily available results of biometrics. It quoted the very latest research from Pearson. It denied that acquired characters were transmitted. It quoted Sidney Webb's result, that the average offspring of English intellectuals only numbered 1.5, and that the talented were having fewer and fewer children in general. It called for an urgent solution, by improving the breed. 'Upon the framework, with which genetics provides us, we must build a strong political superstructure. We see the necessity of a radical reform in the basis of our life'. But he was puzzled by the author, 'H.J. Laski'. 'The name was unknown to me, so I wrote to him "Care of the Editor," and hear from Laski this morning in a very nice, modest letter that he is a school-boy at Manchester, aged 17!! It is long since I have been so much astonished. The lad has probably a great future before him and he will make a mark if he sticks to Eugenics, which he says has been his passion for two years. I as yet know nothing more about him, but hope to learn'.[111]

Perilously frail, Galton seldom received visitors. But now young Harold Laski and his brother Neville were invited round to Rutland Gate for inspection. 'My wonderful boy Jew, Laski by name, came here with his brother to tea. Eva was out, but Miss Savile fortunately called and did the necessary. The boy is simply beautiful. She is an artist and quite agreed. He is perfectly nice and quiet in his manners. Many prodigies fail, but this one seems to have stamina and purpose, and is not excitable, so he ought to make a mark. The two boys are grandsons of a famous Russian Rabbi, a mystic and a great Kabbalist. They told me much about the Kabbala; how only the initiated in it know the proper pronunciation of Jahveh. I told them about Professor Robertson Smith, who knew it, and on pronouncing it before a great Rabbi visitor at Cambridge was cursed by him from the crown of his head to the soles of his feet, and withered and died within three months!'.[112]

These were heady days for eugenics. Laski, from an accomplished Jewish family ensconced in Manchester, had been introduced to the cause the previous year by his soon-to-be girlfriend and future wife Winifred Mary 'Frida' Kerry. Miss Frida—a Fabian and a suffragette, from a Suffolk farming family, scandalously non-Jewish and eight years older than Laski—had already been proselytizing for the movement since 1908. The couple had met at one of Frida's lectures on eugenics, when he was only 16 and still at Manchester Grammar School. Later they would mortify his family by eloping. Harold's article on eugenics had been published just as he turned 17.

In short order Galton was corresponding with the wonderful boy: 'it would be excellent eugenic policy to favour the marriage of those who are somewhat but not necessarily much more likely than the rest to produce capable citizens. The average level might thus be raised a grade or two, with little difficulty, and sports from that level, two or three grades higher than it, would be common and would produce very able men. Whereas an equally high deviation from the lower level would be very rare'.[113]

Karl Pearson was in touch too, and invited Laski up to the Eugenics Lab to work there before he went up to Oxford the following year. Pearson thought him a most promising find. A *Biometrika* article followed, and a Eugenics Lab memoir.[114] However, once at Oxford the boy wonder—who joined the Galton Club, addressed the Union on eugenics, and cockily planned

[110] W. Bateson 1912, 1919, B. Bateson 1928, 388. This was much later, in 1925. He meant a religious taboo from the established churches.

[111] FG to Millicent Lethbridge, 1910-07-11, K. Pearson 1930b, 606.

[112] FG to Millicent Lethbridge, 1910-08-07, K. Pearson 1930b, 607.

[113] FG to Harold Laski, 1910-08-26, K. Pearson 1930b, 608-9.

[114] Laski 1912; Laski et al. 1913.

to write the definitive book on the subject—soon discovered that his natural field was not biology, or anything scientific, but rather history and politics, so this association would not last. Nevertheless there is every indication that Laski, already a radical Marxist, and his wife Frida both remained committed eugenicists thereafter, even though neither would work directly in the field.

In the interim a curious package has arrived at Galton's desk. It contains the journal of Professor I. Donoghue—perhaps of the Dunno Weir branch? we don't know. A letter, which strikes the scientist as quite forward, asks him to look over the manuscript, perhaps to help publicize it. We only have Galton's very fragmentary summary now, and an old notebook, from which the story must be reconstructed.

Prof. Donoghue describes himself as a man of 'strong statistical bent' (indeed he is a 'Professor of Vital Statistics'). He has stumbled—while taking a sabbatical?—on an English colony, Kantsaywhere—presumably offshore, though again we do not know for certain. It was founded around 1820, after the Napoleonic Wars, as a private venture. Under covenant, it has its own endowment, considerable now, from a Mr. Neverwas. From a founding population of 200 the colony has grown to 10,000. It has a single town of medium size—situated among farms which sell timber to nearby mines—and a college, which dominates its life, with major holdings in farms, houses, hostels, besides its endowment. Heavy taxes support a sort of provident socialism based on native inequality and carefully assessed merit.

The residents—direct and sweet, alert, frank, courteous, progressive, good looking, high toned and friendly without taking liberties, prone to breaking out in song to relieve excess energy—receive the wanderer coolly. They keep a watchful eye on him, quietly assessing his unimproved, raw, genetic worth—his 'prairie value'. The fragments describing how the professor came to meet the Allfancy-Ambruster family are missing now, but we know that they take him under their wing. A good thing too, since the ways of Kantsaywhere are different. Homes are hung with many composite photographs of their families and ancestors. Soft ghostly images of family types peer out from these, sometimes approvingly, other times sternly. Local visionaries sometimes see these ruling figures directly, without the aid of photographs, and contemplate their shades (perhaps as in Figure 14.8, p. 358). Differences of opinion exist about the power of the evanescent forms. Donoghue suspects the inhabitants of ancestor worship. To many of them life is just 'the narrow crest of the line of breakers of a never-resting and infinite and inert continent'—the potential far exceeds the actual.

The Allfancy-Ambrusters have a son Tom and a daughter Augusta. The wandering professor becomes intrigued by the vital statistics of the latter, but the family are hesitant. In Kantsaywhere marriage is *de rigueur*, and a serious business, administered through the benevolent fund. The College awards certificates of 'genetic worth', and encourages good matches only. Worth is based a combination of factors: the eminence of forebears, a series of examinations, and whatever accomplishments may be added later. Immigrants are encouraged—a founding population of 200 would have been noticeably inbred without 'fresh blood'—but must pass similar examinations if they wish to stay.

Miss Augusta has just taken her examinations and has scored respectably—honours, a member of the Gold Button Order!—but not stratospherically. Donoghue is persuaded to try for the coveted Gold Button and Miss Augusta's hand, by taking the examinations—first a 'Pass' then an 'Honours'. If he succeeds then the fund will set them up and foster their children if needed, with stipends and rough immigrant labour from Boers.

The exams are thorough, physical and mental. First, the professor's ancestry must be scored. His forebears are problematic, since Donoghue only has an imperfect grasp of them—to the consternation of the locals, who take an extra-keen interest in theirs, and have all the records, like their composite photographs, at hand. The qualities of ancestors are combined in a weighted sum using their decreasing fractions of relationship. (The close resemblance of this formula to Galton's ancestral law suggests that *Natural Inheritance* had reached their libraries.) Happily, despite his deficient records, the professor is awarded the 'Passed in Genetics' degree.

Next, the 'Honours' process. An Anthropometric Laboratory—rather like the kind that the professor had surely seen himself at the South Kensington Museum in London—measures a number of physical qualities: stature, arm-span, weight, breathing capacity, pulling and striking strength of arm, reaction time, sense discrimination, acuteness of vision and so on. They take his fingerprints as a matter of course. He is quizzed about his sporting accomplishments—rowing at Cambridge comes in handy. His aesthetic and literary faculties are weighed. He is asked to read aloud, even to sing briefly. Marching demonstrates his elegance and posture. Medical exams are conducted, and he is quizzed about his medical history—luckily he is in good shape. Four hours have elapsed, not counting breaks.

Allowances are made for rank outsiders, but overall, when the measures are combined via a formula, the itinerant professor's family-record disadvantage means that he falls a few points short of Miss Augusta's Gold Button level. What to do? The family likes him, but they remain wary. Miss Augusta is whisked off in the meanwhile as a debutante, matched to those of her grade. She needs a good match to overcome her respectable but not stellar placing and secure a College Marriage. Awards are made based on the combined merit of the partners. There are only a limited number of eligible summands. As for the professor, he is introduced into society, perhaps to get second opinions as to his real worth.

The College Registrar keeps extensive records of all inhabitants in a vast archive, to be constantly mulled over, for signs of promise fulfilled or surprise disappointments. Family and race are prized. Scores are periodically recalculated. The professor is shown over these stacks and notices—that 'strong statistical bent'—that the scoring system leaves a lot to be desired. Formulae should be revisited. A bright idea: why not assign Donoghue to this grinding task—he might even enjoy it—and award him some extra points for donating his time? A bonus. 'The data was magnificent, much fuller, more trustworthy and covering more generations than any other in existence and afforded splendid material for discussing the statistical & possibly other laws of heredity'. Donoghue's 'special mathematical knowledge', not widely available in Kantsaywhere—it is, after all, a small place—is indispensable.

The College rules forbid the use of funds for ordinary charity, which is repugnant to the covenanters and left to other bodies in the outside world. Those who do not receive a PG (Passed in Genetics) degree are sponsored for emigration, and placed under surveillance if they do not depart. If they breed they are expelled or segregated for degrading the communal stock—this is a covenant worshipping posterity as much as provenance. No blame attaches to those passed 'PGs' whose offspring fall short naturally through no fault of their own. The offspring of College-sanctioned marriages are well-known by all to be better than the average of Kantsaywhere. Boiling off the less able keeps up the standards, perhaps even raises them further.

In Kantsaywhere, we are told, lovemaking is 'as elsewhere', but 'manly' and within-marriage. The 'mammalian' women dress their hair in a simple knot behind and are solidly built—like the figures in Guido Reni's painting 'Aurora and the Hours' (see Figure 14.9, p. 359). The men are firm and resolute. The sexes are where one expects them to be. Women are encouraged to marry young, but not too young—22 is supposed to be a good age for childbirth—while men are encouraged to delay until 30.

The Balliol manner is absent—they are smart but not overly so. Scandal and gossip are disapproved of. The prevailing accent is 'pure English—simple, sound & hearty'. Their observations are accurate and to the point, including specifics while leaving out the unimportant, and avoiding exaggeration.[115] Bohemianism is unknown. They treat intellectual competition as the English do sports—with intense interest. A simple silence is kept before commencing

[115] A positive trait that Galton associated with the Darwins.

dinner, in Quaker fashion. Civic religion centres on the unity of present with past & future, and pride of descent—with a day of introspection and forgiveness. Memorial ceremonies feature a song of happiness that 'so & so' lived, with a chapter of that life.

The Kantsaywhere version of *Debrett's Peerage* is more like a racing form book, bulging with recorded scores and marks of scientifically assessed merit. It is just as well then, as we learn from the notebook, that the professor and Miss Augusta succeed in summing their scores, and join the family trading business.

In late 1910 Galton diverted himself for a few months with this mildly dystopian joke at his own expense. Occasionally he toyed with 'Kantsawha' as the title. It follows on from his earlier eugenic tongue-in-cheek utopias, which can be traced back to the Mars notebook of 1896, and the 'Donoghues of Dunno Weir' fragment of 1901. It was completed as a novella, finally titled *The Eugenic College of Kantsaywhere.* 'I have at last finished Kantsaywhere, if that be the best orthography. Miss Jones has been staying here during Eva's absence and copied it in fair writing. But I must keep it by for a little while and add and alter before sending it to be typed. Then I must ask you and other friends to kindly read and criticise. It would now fill about 20 to 30 pages of *Nineteenth Century* size and type'.[116]

Help was available nearby in Haslemere, where Galton had retreated for the winter, in the form of Henry Graham Dakyns (1838–1911). The retired classical scholar and translator, a rugby enthusiast who had once been tutor to Tennyson's sons Hallam and Lionel, proof-read it and offered corrections.[117] So Galton submitted the novella to the publisher Algernon Methuen that December. 'This morning's post brought the neatly typed Kantsaywhere revised and done up in book form. Methuen comes here if he can on Sunday or Monday afternoon, so I must keep it for him. This is, I expect, just the most awkward time for new publications, politics and Xmas both in front. I shall soon know more about all this'.[118]

The publisher was not enthusiastic. Galton told Milly that Methuen 'takes no interest in Eugenics'.[119] After Christmas Galton gave up on it. '*Kantsaywhere* must be smothered or be superseded. It has been an amusement and has cleared my thoughts to write it. So now let it go to "Won't-say-where"'.[120]

After the death of her uncle, Milly wasted no time in burning most of the book, including a complete copy that Edward Wheler-Galton appears to have possessed. Parts of the family never were reconciled to eugenics. She left only the more technical sections for Karl Pearson to consider. Nearly all of the comical story is lost, though some pieces can be filled in from a barely-legible notebook. This destructive act shocked and annoyed Pearson, who pointed out, from the safety of 1930 when the final volume of his biography of Galton was published, that his subject had not left any such instructions or even appointed a literary executor.

As it turned out, Karl Pearson was never to be reconciled to the *Eugenics.* A letter from niece Eva Biggs to Karl Pearson shows conservative Catholic opposition within a branch of the family to the *Eugenics* and their aims.[121]

> Dec. 31, 1910
>
> Dear Professor Pearson
>
> Uncle Frank has given me y^{r} letter. I don't believe the devil leaves you protestants & agnostics alone; but he doesn't torture you as he does the Catholic countries. You don't stick a knife into Professor Bateson or Whitaker, Saleeby, etc. as I sh^{d} probably try to in y^{r} place—but who but the devil prevented you doing what I asked—

[116] FG to Millicent Lethbridge, 1910-10-23, K. Pearson 1930b, 613.
[117] Dakyns would only outlive Galton by a few months, dying of a heart attack at age 72.
[118] FG to Millicent Lethbridge, 1910-11-29, K. Pearson 1930b, 615.
[119] FG to Millicent Lethbridge, 1910-12-06, K. Pearson 1930b, 615.
[120] FG To Millicent Lethbridge, 1910-12-28, K. Pearson 1930b, 616.
[121] GALTON/3/3/5/8.

viz.—persuade Uncle Frank off that tiresome Eugenic Education Society—you & your own pupils do not let y^{r} names appear among that riff raff, so why sh^{d} Uncle Frank's name be put at the head of them!

The doctor has been & keeps Uncle Frank in bed all day, to rest as it has become the rule now once a week. He did so enjoy y^{r} visit. I wish you came oftener as it cheers him up.

Ever sincerely yrs
L. E. Biggs

And so Pearson resolutely kept his prickly distance, concentrating instead on amassing datasets, to be skew-curved and correlated. His methods would inevitably be superseded by emerging techniques. Meanwhile the eugenics movement grew from strength to strength, establishing itself not just over the British Empire but internationally, across the European continent and Russia, over to the Americas and beyond. In each of these locations it took on its own local character, assimilating earlier movements with related concerns. But Galton would see none this himself, since he had little time left to live.

——oOo——

It should be clear from the above that eugenics was not a politicized topic in Galton's lifetime, and this is true up until about the mid 1930s. Most of those associated with the movement were radicals, simply because proposing it supposed that far-reaching reforms were necessary to remake society, but parties did not divide over the issue. Conservatism in the traditional sense is naturally opposed to social engineering, and it is an egregious error to suppose that any of those involved were there *because* of, rather than *despite*, the fact that they were conservative. In the intellectual milieu that Galton in particular came from, the great divide was between science, or Darwinism in particular, and received religion—which in England meant the established state.

As far as personal politics went, Galton was a Liberal Unionist, just as his father Tertius had been a thorough Whig. Most of the Darwinians were.[122] They opposed Home Rule in Ireland because of the control that would have given the Catholic church over education and thus the teaching of evolution in schools and universities. In all of his published and unpublished writings Galton says nothing of any consequence about party politics, which seem to have bored him.

Hesketh Pearson recalled that when, barely twenty, he had told Galton that he was transferring loyalties from the Conservatives to the Liberals, his venerable great-uncle was taken aback. 'Fancy being a Conservative at your age!' 'Most men begin life as red-hot Republicans and end life as stiff-necked Tories. Why, I thought all the young men in the country nowadays were Socialists. All thinking men change their politics, and the majority change them for worse. If you are a Conservative now, I shan't envy your fireside acquaintances when you are seventy. You'll be a blood-red revolutionist at an age when most men are content with the opinions of their grand-parents!'.[123] This suggests that Galton was thoroughly detached from political feelings at the close of his life.

The idea that Galton was a Conservative can be traced to Ruth Cowan. Her PhD thesis on Galton (Johns Hopkins, 1969) referenced recollections by a son of his brother-in-law H. Montagu Butler, who claimed that Galton always voted Tory (actually, this appears to be true of H. Montagu Butler himself). This factoid was repeated without examination in the first popular biography by Derek Forrest, and thereafter.[124] For those who make the claim it is

[122] See Tredoux 2023c for more on Galton's politics.

[123] H. Pearson 1921, 74. Any anecdote from Hesketh Pearson should be treated with caution.

[124] Cowan 1969; Forrest 1974.

invariably pejorative. It is a bad error, gauchely projecting contemporary political divides back in time, where they do not make sense.

Along with the Fabians, every major geneticist of the era was a eugenicist in principle, though not necessarily a member of an advocacy group like the *Eugenics*, even if many like R. A. Fisher and Reginald Punnet were. Some of those were also communists or their fellow-travellers. Harold Laski was hardly the only Marxist to join the eugenic cause. The Nobelist Herman J. Muller is a good example. Muller had been converted to communism while a student at the University of Texas in the 1920s. He relocated to the Soviet Union in the 1930s to do genetic research there, hoping he could persuade Stalin to implement a eugenics program, and so further enrich his worker's paradise. He sent the manuscript of his tract *Out of the Night* (1936), advocating this and more, to Stalin. But far from receiving a favorable welcome—Stalin had it translated so that he could read it—he found that he had to discreetly exfiltrate back to the west, via Spain and the civil war there, to avoid the deadly consequences that had befallen the geneticist Solomon Levit and many others. It is significant that his book was the first book published by the Left Book Club founded in 1936 by Victor Gollancz.

Gollancz had not yet received the memo. Highly attentive to the party line, as he would confess years later after falling out with it, he would publish nothing like *Out of the Night* ever again. Stalin had turned against genetics in general and eugenics in particular in the early 1930s—perhaps as early as 1930—and had started to persecute scientists who worked in or around the field, having them shot or dispatched to the Gulag to be ground into 'camp dust'.[125] The politicization of eugenics can be traced to this period. In the West it was still a nascent development, and it is no coincidence that those geneticists and biologists who started then to distance themselves, however tentatively, from eugenics were fellow-travellers of the Soviet Union.

Lancelot Hogben is an interesting case study. Descended from Plymouth Brethren, with whom he broke to become an atheist and then a communist, Hogben had trained as a biologist at Cambridge. Afterwards he taught in Canada and South Africa before returning to UCL in 1930 as professor of Social Biology. He would make a career on the side as an especially successful science popularizer. Early on he had sympathies for the Soviet Union and its cause. Though he never appears to have formally joined the party, he was a dependable fellow-traveller, often seen with Soviet delegations to the west. By 1933 he was critical of eugenic proposals to encourage the reproduction of the fitter classes, arguing that this was an attack on the working class. Few eugenicists conceded this point, since identifying and accelerating talented cohorts from the working class was a central plank of their program, given the sheer numerical size of that class in comparison to others. However Hogben was careful to qualify his criticism, insisting that he was in favour of eugenics in principle, and only opposed to certain policies he thought disreputable.[126]

Hogben would soon be joined in this loyal-opposition role by J.B.S. Haldane and other left-wing scientists who had until recently been staunch advocates of eugenics and hereditarianism. Loyalty to science itself would become increasingly difficult for this loose grouping, as Stalin's persecution of genetics and promotion of Lysenkoism deepened, and spread abroad, becoming a matter of party discipline. Haldane in particular was forced to retreat into glum silence.[127]

It was not until the 1960s that eugenics began to acquire the distinctly unfavourable reputation that it has today. Even in 1962, the noted biologist Julian Huxley was still openly advocating the policy, as chairman of the *Eugenics*. As was Herman J. Muller, whose views had not changed. The society continued to operate until 1989, when it changed its name to the Galton Institute. Now it has relinquished even that name, after falling into the hands of

[125] For more on this turn, see Tredoux 2018.
[126] Hogben 1933.
[127] Again, for more Lysenkoism and the role of Haldane, along with his career in espionage, see Tredoux 2018.

anti-hereditarians, a poignant decline.

The now-prevailing notion that eugenics was ever a 'conservative'—let alone a 'right wing' or even a 'Nazi' ideology, a term of straightforward abuse—is badly mistaken. The subject has become politicized in retrospect, precisely when its principled adherents have more or less vanished. To that extent, the contemporary anti-eugenics posture resembles nothing so much as a football jersey. Wearing a certain colour signals that you adhere to a party line. Ascribing contrary allegiance to those who wear other colours, simply for the sake of wearing them, is to act out an artificial conflict that bears no genuine relation to the colours themselves. None of this has any substantial connection to Galton or his concerns, since he would not even have recognized the ideas or allegiances involved.

Figure 14.1: Beatrice Webb née Potter (1858–1943).

Figure 14.2: Punch 1880-03-20. Happy thought! Let us all have a voice in the matter. Noble breeder of shorthorns: Well, you are a splendid fellow and no mistake! Prize Bull: So would you be, my Lord, if you would have chosen your pa and ma as carefully and judiciously as you chose mine! By George du Maurier (1834–1896), grandfather to Daphne.

Figure 14.3: George Howard Darwin (1845–1912).

Figure 14.4: Alfred Russel Wallace (1823–1913), by John William Beaufort, 1923.

Figure 14.5: Francis Galton at age 80 in 1902, K. Pearson 1930a.

Figure 14.6: Sir Francis Galton at age 81 in 1903, by Charles Wellington Furse. About 20,000 brush strokes.

Figure 14.7: Sybil Gotto née Burney. Later Neville-Rolfe.

Figure 14.8: Peering Progenitors from Kantsaywhere?

Figure 14.9: Guido Reni, 'Aurora and the Hours', whose figures the substantial women of Kantsaywhere are said to resemble.

Chapter 15

Knight of the Bath Chair

> I find old age thus far to be a very happy time, on the condition of submitting frankly to its many limitations.[1]

Though illness, often serious, was seldom further than arm's length, Galton lived to a remarkable age for his times. He was at middle age before he mastered his mysterious physio-psychological challenges. Malaria may have lain behind some of the recurring physical illnesses—'ague' and 'fever'—that he suffered from after he returned from the Middle East in late 1846.[2] He may have caught a more serious strain of malaria in Africa. Other infections seem to have struck both of the Galtons with metronomic regularity.

Louisa in particular acquired the reputation among her Butler relatives for hypochondria, on top of miserliness.[3] Her honeymoon trip to France was interrupted by serious illness and frailness draped itself around her. She always seemed to be at death's door, forever preparing for the end, only to make a miraculous recovery. This may be unfair. Diaries show that *both* Galtons suffered from a recurrent series of infections, for which no effective remedies were really available. Internal medicine before the 1930s consisted mostly of moral support and encouragement, where it was not positively harmful.

In 1897, while they were travelling abroad at Royat in France, Louisa Galton died at age 75 of an internal complaint, which was probably chronic and gastrointestinal. Later, Frank recalled that her vigour had been distinctly declining. 'Examination showed the cause of Louisa's long ill-health and final death to be an extremely small stomach and an extremely constricted outlet due to her illness 19 years ago. The stomach was barely one third the natural size, and the outlet leading out of it no larger than would just contain a common lead pencil'. A letter to Leonard Darwin has further details.[4]

> Dear Leonard Darwin
>
> You and your wife will both be sorry to hear my tidings namely that my dear wife died here (Grand Hotel Royat) at 2¾AM this morning. It was a violent form of a bilious attack that is now prevalent here. She was well on Sunday, and had packed up to start on a mountain tour the next morning, when the attack came on. The doctor thought little of it on Monday, & not much on Tuesday morning, but it increased in strength as she became feebler. She was semi-conscious only, I think for the last day or more, & she passed away so suddenly that I was doubtful for many moments whether she had ceased to live.

1 F. Galton 1908, 7.

2 Contrary to speculation by an early biographer, Forrest 1974, there is no reason to believe that Galton contracted venereal disease in the Levant, since he did not progress through the well-known stages of a disease like syphilis.

3 H. Pearson 1964, 148, quoting Ralph Butler, who recalled Uncle Frank's generous doling out of Xmas sovereigns when pinched Aunt Louisa was not looking. Dorothy Middleton, also a Butler relative, had similar recollections of family legends about Louisa's meanness with money

4 FG to Leonard Darwin, 1897-08-13, GALTON/3/3/7/10.

> It is early as I write, and I do not yet know about French formalities & time of burial; when all is finally over I shall go back to my desolate home, at least for a while.
> With affect. & love to you both
> ever sincerely
> Francis Galton.

Traditional ceremonies and stoicism reverted. 'Dear Louisa was buried with simple decorum yesterday in the cemetery of Clermont-Ferrand. The day was lovely, the mountains looked singularly imposing, the English Chaplain, Mr Wilcox (of Battersea Park Road) officiated, and a most kindly and tactful clergyman ... A feeling allusion was made in the sermon of to-day, and appropriate hymns were sung ... I have very much to be grateful for, but our long-continued wedded life must anyhow have come to an end before long. We have had our day, but I did not expect to be the survivor'.[5]

There is every indication of affection in the marriage, despite its failure to produce children. Neither of Galton's brothers produced children. Darwin married multiple times but his only child died in infancy. Erasmus, a noted eccentric who had enrolled in the navy and volunteered for the militia, never married. Perhaps infertility ran in the male line. In earlier days Louisa had interested herself as much as she could in her husband's work, but this dropped off in later years. He would often interrupt their European trips to dash home alone to the annual British Association gathering, while she proceeded on her round of relatives.

At least two hereditary ailments plagued Galton himself, both of which worsened with age: asthma and deafness. His father had suffered from both. Asthma drove the carpets out of 42 Rutland Gate, which acquired spartan furniture to keep the dust down. Progressive deafness was more serious. As it got worse, Galton was forced to scale back his activity in scientific societies, where he had previously played leading roles. By 1900 he had more less ceased all participation. That January he was abroad anyway, travelling in Egypt.

The Boer war was underway when Galton set out for Egypt in December 1899 at the age of 79, trying to lose a severe cough that had persisted all year. An earlier trip to Spain had failed to completely clear it up. He was frail enough to require the company of his niece Eva Biggs (in London his Swiss manservant of then over 30 years service, Albert Gifi, was always present to assist).[6] The British Army was already suffering serious setbacks and news of their blunders followed him as he travelled. He had been to Egypt before, in 1845, when he was bewildered by personal events. Then he had sailed down the Nile in a small native *dahabieh*, past the cataracts, taking the desert short-cut to Khartoum. After returning to Cairo he had gone on to Damascus, living there in native *mufti* until late 1846.[7]

Now—after being invited by the Geographical Society of Cairo to address them about those experiences early in the new year—Eva and her beloved uncle Frank were steaming down the Nile in the comfort afforded by the 'Anglo-American' Company, at the cost of native intimacy. They were in Luxor at the Hotel Karnak by the 15th of December, sightseeing on seven hour donkey rides. Archaeological diggings had exposed much that was still hidden in 1845, 'the clearings of the very few last years have added immensely to what was to be seen before, especially to the many bright-coloured wall paintings and hieroglyphics. The unearthed bases of many columns have made them much more stately'. 'The wonders are just unspeakable'.[8] A few days later they waited at Aswan for a steamer to take them between cataracts, returning to Luxor and the Hotel Karnak by early January.

When booking his passage in London, Galton had a stroke of luck bumping into Flinders Petrie, who was going out to continue his Egyptian excavations, moving to about half-way

5 K. Pearson 1930b, 503.
6 For Gifi, Albert, see page 455.
7 See Tredoux 2023c for a full description of this Middle Eastern sabbatical.
8 FG to Bessy Galton, 1899-12-01, K. Pearson 1930b, 515.

between Aswan and Luxor.[9] The Egyptologist invited Galton to stay with him at the desert camp near Abydos where his systematic digging took place. Petrie was long-familiar to Galton, having corresponded with him since the late 1870s, astounding the scientist with his ability to visualize a slide-rule by parts and make calculations using the mental image; he had also offered numerous suggestions for improvements to anthropometric measurements and instruments. A series of photographs sponsored by the British Association followed, at the suggestion of Galton, to record depictions by the Egyptians of different races. For greater clarity, Petrie made plaster casts of the originals where possible, and photographed those (see Figure 15.5, p. 378).[10]

The archaeologist had no formal schooling, having been educated at home by his father, an engineer, who taught him surveying, and his mother, who had taught herself Latin and Greek and published scholarship on mythology. Matthew Flinders (1774–1814), the explorer of Australia, was his maternal grandfather. Petrie had steeped himself as a teenager in the collections of the British Museum. The pyramids of Egypt had attracted him when his father became intrigued by the occult theories of the astronomer Piazzi Smyth, who was thoroughly familiar to Galton. Careful surveying disproved the idea, initially believed by Petrie, that the Pyramids had been built using a divine 'pyramid inch' measurement, handed down since the time of Noah—Piazzi Smyth thought that the pyramids encoded celestial distances and even divine prophecies in their construction. But not a little good came from this self-paid fool's errand: observing the pillaging of Egypt for artefacts, Petrie resolved to preserve what remained. By 1892 he was the first Amelia Edwards Professor of Egyptology at UCL, and the dominant figure in the field.

After running into Galton at the booking office, the Petries had left London on November 17th and were at their camp seven miles outside Abydos (el-Araba) by the 30th. They hoped to receive Galton's party in mid-December, reassuring him that 'we have here a lady doctor, Miss Johnson, out for her own health. So you can get any advice you might want while here'.[11] But instead Galton and his niece made plans to join Petrie on January 5th, after returning down the Nile from the cataracts. They were forbidden to bring any outside servants of any kind with them. The real reason for this secrecy would become apparent later, and was only hinted at for now.[12]

> 23 December 1899
> Arabah, Baliana
>
> My dear Sir,
>
> We shall be delighted to meet you & your niece on the 5 Jan. at Baliana & bring you safe over here. But we still hope that you will see your way to doing without a Luxor boy. We see the cloven hoof in all such, & avoid having any such outsiders about the place or the work. My wife will attend to anything about the premises that you want, & our house boy, a vivacious little negro, will be at your service all day in & out.
>
> We find the need of keeping our work & discoveries as dark as possible in this country. The dealers—especially from Luxor—stick at nothing if they see a chance of robbery. Such are the difficulties that all our workers are drawn from a village 50 miles away, & no man of the plunderers & dealers here is allowed about the place. If a Luxor spy was found prowling in the desert our men would catch & thrash him.
>
> In this state of siege, which generally prevails about our excavations, you will perhaps see in what light a Luxor boy would be looked on about our premises. So I

[9] F. Galton 1908, 97, where he incorrectly dates this trip to the Winter of 1900/1901—it was 1899/1900. For Petrie, William Matthew Flinders, see page 457.

[10] Petrie 1887, GALTON/2/7/2/4/31, GALTON/3/3/16/20.

[11] Petrie to FG, 1899-12-09, GALTON/3/3/16/20.

[12] Petrie to FG, 1899-12-23, GALTON/3/3/16/20.

hope he is not indispensable to you. If he took care of himself he would live in the village & tell people of our work and finds. All our workmen live behind our house, & are forbidden to go to the village on pain of dismissal, so needful is it to keep our business to ourselves.

This system which we have had to make is, of course, so unusual that you could not anticipate it, or see beforehand what antipathy we have here to the sophisticated native. We will undertake that you shall have as many good boys to run about as you can use; & my wife or Miss Johnson will always be at hand. So the wily Luxorian would be needless as well as dangerous.

Hoping to see you all well on Jany. 5
Believe me
yours very sincerely,
W. M. Flinders Petrie

At the Baliana railway station they were impressed by the mounted figure of Miss Alice Neville Vowe Johnson, the bustling thirty-year old lady medical student (not yet a doctor) who was over in Egypt helping Petrie to excavate. It was not a placid province. The authorities had advised visitors not to travel at night unaccompanied. Miss Johnson, who never tired of riding into the village despite having only a smattering of Arabic, used her horse, tongue and hippo-hide *kurbash* to extricate them from the rabble of importunate station natives, then guided them over the seven mile moonlit donkey ride to the diggings.[13]

At the camp they met Dr. David Randall-MacIver, who would later excavate the Great Zimbabwe ruins but was now testing sight and colour sense, anthropometry that Galton had himself pioneered; Arthur Cruttenden Mace (1874–1928) a cousin of Petrie who would join Howard Carter in the Tutankhamen discovery twenty years later; and John Garstang, (1876–1956), who would excavate at Meroe in the Sudan, where Galton had passed through in 1845-6. The camp structure, with stout theft-proof storerooms, had been built by Mace before the Petries moved in. 'We had each a room with mud walls, nine feet long, seven feet wide and eight feet high, and a bedstead and empty packing cases for furniture. There was no regular door, but a mat, hung in front of the doorway, kept out the prowling dogs. It was on the desert sand, 150 yards from the palm trees, etc., and the floor of the hut was made by that sand. Every one had to throw away their own slops. A well was dug close by, to supply water'. Recent finds were arrayed in front of the mud huts. The workmen lived behind one corner of the camp, where once a week they danced and howled like dervishes. 'Their nature seemed to require occasional doses of these ebullitions'.[14]

The whole operation was run with invincible industry by Petrie and his wife of two years, Hilda (née Urlin), supervising 130 hired diggers split into three teams. They were racing against time to mine the rich heaps of debris left behind by the previous excavator of the Royal Tombs, Émile Amélineau (1850–1915). Over his five year concession the Frenchman had, according to Petrie, 'reduced to chips the pieces of stone vases which he did not care to remove, and burnt up the remains of the woodwork of the 1st dynasty in his kitchen'.[15] The site had been widely considered to be exhausted as a result. Since the concession given to Amélineau still ran in theory to June 1, Petrie disguised activities, leaving the one previously untouched tomb for later in the year, feinting that he was only looking through scraps, just in case the greedy Frenchman should turn up again.[16]

Galton saw that 'everything found was assessed and paid for to the workmen', 'bits had to be pieced together and every day some interesting "finds" took place'. The 'fellahs' who

[13] For Johnson, Alice Neville Vowe, see page 456.
[14] F. Galton 1908, 100.
[15] Petrie 1931, 172-3.
[16] Petrie's diary entries, in Drower 2004, reveal this strategy.

wanted to sell their own finds to the Petries lined up outside the camp every morning to submit themselves to the firm valuation of the master. Petrie noted in his journal that Galton was 'much interested in all our proceedings and especially in the skeletons'.[17]

The 'feast of the full moon' (Zikr) fell in that week and was punctuated by solemn chants. The clear desert nights were brutally cold, as they had been in the Namib. But a pyjama suit, drawers, jersey, thick socks, hot water bottle, three doubled blankets, thick Jaeger rug, sheet, thick morning coat, dressing gown, thick Jaeger shawl, Ulster coat, overcoat and pillow piled on top of the feet, though they helped, were less effective than two animal skins had been on the Skeleton Coast. By contrast, Hilda and Flinders seemed 'peculiarly independent of ordinary comfort', but 'the consumption of marmalade at their table was enormous' (in fact Hilda's journal shows that she needed five blankets on at least some nights).[18]

The level of energy that the preserves powered was impressive, given the all-consuming nature of the work, 'the originality and fertility of the ideas of Professor Petrie and the ingenuity of his explanations were marvellous'. 'The mere digging can be delegated, but the rest seemed to occupy every faculty of our hosts at full stretch from early morning to late evening every day. There was drawing, copying, photographing, recording, comparison of specimens, piecing of them together, discussing them and planning new work, besides attending to the discipline of many men not concentrated at one spot, but dispersed among different diggings'.[19] Writing home to his sisters, he was agog. 'It has been to both of us one of the most interesting experiences of our lives. I am more than ever taken with admiration of Petrie, and his wife is as nice as possible'.[20]

After a week of this 'delightful visit', as Petrie would recall, the Galton party was safely dispatched back to Luxor and the Hotel Karnak once more. Staying on for another season in the desert, that most capable woman Alice Johnson would qualify as a doctor at Brussels and successfully practice at Carmarthen in Wales, then bustle at the 'Bird in Bush' Mothers and Infants Welfare Centre in the Old Kent Road, London. She lived dutifully until her death with a senior Welsh suffragette, Alice Abadam (1856–1940), keeping in touch with Galton while he was alive, dropping by for tea and corresponding about the mental qualities of the children she had charge of.[21]

At Luxor Galton had another reminder of his fifty-years younger self when he met, first, the South African big game hunter and collector Frederick Hugh Barber (1847–1919); then, second, another noted hunter, the Englishman Henry Seton-Karr (1853–1914). Karr was returning from Omdurman, recent scene of the ferocious revenge over the followers of the Mahdi, witnessed by a young reporter named Winston Churchill in his just-published *River War* (1899). The Irish archaeologist Professor R. A. Stewart Macalister (1870–1950) showed up too, en route to the Petrie camp, where he would study the innovative stratigraphical techniques practised there.

There was no hurry to return to smog-bound London. The desert chill had aggravated that persistent cough. Once he had steamed back down the now dramatically-depleted Nile to Cairo, Galton addressed the city's Geographical Society on the 24th of February, reminiscing divertingly about the Egypt he had known in 1845. Of D'Arnaud, the Bey at Korosko who had inspired him to travel purposefully—'the road to Khartoum is open'. Of the changes he noticed since then—'At that time the crocodiles were basking on the sand-banks and positively dangerous in Assiout'. He introduced them to his most recent scientific progeny, fingerprints, rapidly propagating internationally and already in operation at Cairo.[22] Some desert botanizing

[17] Drower 2004, 158.

[18] Drower 2004, 164.

[19] F. Galton 1908, 98–99.

[20] FG to Bessy and Emma Galton, 1900-01-14, K. Pearson 1930b, 516-517.

[21] GALTON/3/3/10/8.

[22] Published as F. Galton 1901c. For a full translation of the French text, see Tredoux 2023a. In F. Galton 1908, 97, the death of Arnaud is misremembered as 'not many months previously' but it was really as far

and geologizing, with the German explorer Georg August Schweinfurth (1836–1925) and other companions, was arranged in early March. The Bedouin were still a constant danger to desert excursions. *Khamsin* winds blew hot.

At the city, Galton was able to inspect in person the fingerprinting system put in place by Colonel Harvey Pasha. He found the central Identification Office a 'model of method and efficacy'. It revivified verisimilitude, 'while the natives of India and of Egypt have beautiful traits of character and some virtues in an exceptional degree, their warmest admirers would not rank veracity among them'. England was not free of fraud but nevertheless 'if a map of the world were tinted with gradations of colour to show the percentage of false testimony in courts of law, whether in different nations or communities, England would be tinted rather lightly and both Bengal and Egypt very darkly'. The new system also transfixed the bewildering variety of *legitimate* names available there according to taste. This could prevent the release of the wrong 'Hasan Mohamed' from prison, which had certainly happened in the past. Now new recruits to organizations could be authoritatively screened for past convictions. The office made him up an identification card with a photograph, dated 1900-03-22.[23]

By March 24, the Petries had returned to London—to attend to UCL duties, publish their finds with the usual expedition, stock up on marmalade and sit for G. F. Watts (not a good likeness, the family thought). In November, with Amélineau safely out of the way, they would return to Abydos to excavate the held-over tomb. An English resident and his wife would be murdered in Baliana, and Hilda would narrowly escape a pot-shot taken at her as she exited her bedroom at the camp.

Galton's party returned home by way of Greece, which he had last visited in 1840, in mind of Byron.[24] They left Alexandria for Athens in early April, and sister Emma was glad to hear soon after that 'everything is so strikingly beautiful'. She had advised her brother not to rush home, since it was brutally cold in England and influenza was rampant.[25] Then it was on to Tripoli in the Peloponnese and Kalamata by mule, accompanied by a Dragoman, in search of more excavations. Only by late May were they back in Leamington Spa, by way of Patras and Venice, much to the relief of sister Emma and her neighbour Bessy.

A few years later Galton forwarded Petrie's name for election to the Royal Society. Karl Pearson would live in the house next door to the archaeologists in Well Road, Hampstead, and become a firm friend. They sent the biometer thousands of crania from the Egyptian diggings for measurement and statistical sorting into distinguishable races, using his 'coefficient of racial likeness'.[26] Flinders—yet another remarkable man from an age which churned out remarkable men—would finally expire at Jerusalem in 1942, covered with honours and over a hundred books to his name.

After the Egyptian trip, Eva stayed on with her increasingly frail uncle to manage his household in London, where his world progressively shrank back to the propagation of Eugenics, the measurement of resemblance, a stream of bright ideas and book reviews for *Nature*, an active international correspondence and a devoted circle of friends—including Karl Pearson, Raphael Weldon and Lady Welby. Winters were evaded in the warmer parts of the continent: the Italian and French Riviera, and finally Spain.

He lived largely through correspondence now. For instance, he maintained a close friendship with Victoria Lady Welby, at one time a Maid of Honour to Queen Victoria, who also happened to be her Godmother.[27] She was an autodidact, assembling an expanding library of the latest books and journals that she crammed cover-to-cover with notes. The friendship dates back to at

back as 1884.

23 F. Galton 1900c.

24 See Tredoux 2023c for a full description of this trip and its family precursors.

25 Emma Galton to FG, 1900-04-18, GALTON/1/4/2.

26 This work would be extended by G.M Morant, e.g. in Morant 1925.

27 For Welby, Victoria Lady, see page 459.

least 1890, when Galton offered some sympathetic remarks on Welby's early paper 'An apparent paradox in mental evolution'. This had first been delivered before the British Association—where it was titled 'Is there a break in Mental Evolution', and was read for her by E. B. Poulton, though she was present—and then the Anthropological Society, where it was read by the secretary, though again she was present.

Welby argued that animals (and therefore at some early stage, man's ancestors) must be fully rational about their environment in order to survive. Thus the presence among primitive man of elaborate superstitions—ghosts, spirits, fetishes, ancestor worship—implies that either some dramatic break from rational to irrational had occurred at a certain stage of intelligence, or else that man had been going steadily backward as intelligence evolved. Either alternative appears paradoxical. She took it as given that evolution progresses, an idea that few of her contemporaries would have disagreed with. Moreover, at some stage intelligence further develops to the point where magical thinking is rejected. How to explain that? And why were the superstitious not simply outcompeted by the less superstitious among them?

In the ensuing discussion Galton made two points. Firstly, the development of reasoning requires the power of imagination, not possessed by 'brutes' but developed by more evolved 'savages'. On its own, without additional constraints, this imagination blurs the distinction between fantasy and reality. Secondly, the reason why that tendency is not simply selected against is because the united action that it enables through myth and superstition may confer competitive advantages on those who possess the required credulity. (This is essentially the same argument that Galton would make a few years later in response to Benjamin Kidd's contention that religion is irrational, see page 326.)

Lady Welby does not appear to have paid much attention to these remarks, and soon broadened her scope to what she called 'Significs', by which she meant something like the study of the 'meaning of meaning' or what others would call semantics.[28] Her scope is much broader than that though. Her analytical distinctions come—like sorrows, crosses and omnibuses—in threes. There is for example a difference between sense, meaning and significance. Roughly, this is what one intends by a statement, versus what others take the statement to mean, versus what it really means, given its context, in profound totality. These are incremental levels of indirection, since the second and third senses cannot be got at directly. She thought that an astronomical metaphor was useful: sense is 'planetary', meaning is 'solar', but significance is 'cosmic'. Then again—something that later commenters have usually passed over silently—she found this threefold scheme applied in some way to mental evolution too, from the brute stage (sense, or planetary) to savage (meaning or solar) and on to modern man (significance or cosmic).

A dominant ambition was to clarify the confusions created by imprecise use of language, by moving discourse into the significance realm. She offered prizes to those who could contribute to clearing up confusions, in order to get mankind there, liberating psychology, anthropology and other fields. There is plenty of discussion in her works about the marvellous end state in which true mastery of 'significance' has been attained in the use of language, with its attendant benefits in clarifying thought, but the prizes had yielded only more discussion and no plausible methods for getting there.

Her thinking here, which produced a number of books and articles, is fresh and highly suggestive—many more ideas were roped in—but hard to pin down. It was really not Galton's personal cup of metaphysics. As he told Karl Pearson, 'Lady Welby is irrepressible in her inquiries. She was with us at Lyndhurst for more than a week, full of mystical triads, etc. and much else. Socially she is very charming and good'.[29] A few days before, Pearson too had received an enquiry from Lady Welby, asking why threes seemed to turn up all the time, but

[28] Welby 1903, 1911.

[29] FG to Karl Pearson, 1909-04-08, K. Pearson 1930a, 379-80.

had thought better of a sarcastic reply.[30]

Nevertheless Galton was always full of encouragement. In 1903 Lady Welby communicated her excitement at reading the tea leaves she divined in *The Principles of Mathematics* by Bertrand Russell. 'I had been told by everyone that it would be useless for me to get Bertrand Russell's big book as even some of the mathematical experts find him difficult to follow, and he must needs be quite unintelligible to the "outsider". Now I of course don't know, so to speak, even the "a" much less the "b" and "c" of mathematics. But when (in consequence of one review) I did send for the "Principles of Mathematics Vol. 1." I find to my amazement that here, set forth in intricate & subtle schemes of reasoning often too technical for me to follow are some of the very ideas I can find nowhere else and which in my crude & feeble & too often I fear incoherent fashion I am trying to express', 'I feel I must somehow get into touch with Mr. Russell'.[31] And no doubt she did—one longs to know what she made of the successor, *Principia Mathematica* (1910 onward).

We have seen her involvement in facilitating the foundation of the Sociological Society that year,[32] but soon Lady Welby was dipping into Mendelian genetics too. 'I have been carefully studying another book of engrossing interest: Hugo de Vries on the experimental evidence for his mutation theory. It seems to me that the bearing of the facts he brings forward on our ideas of human heredity may prove to be of far-reaching importance'.[33] Her readings were heterodox. 'You will know about Mr G. Smith's very important studies on sex in crustacea & will have seen the notice of them in *Nature* which is the first I have seen on the subject. They give me (as so often happens) precisely the right which I had despaired of, the scientific right that is, to tell the Man that my appeal was to the dormant mother sense in him as well as in us, since in the male also there is an organic capacity for motherhood'.[34]

Towards the end, Welby expressed gratitude and affection for this long friendship, 'the work of my life will bear some fruit. Never can I forget that you were one of the very first to welcome and encourage it (e.g. in the Anthropological paper & to me at the Brit. Assoc.'. And a few days before his death, 'I can never forget the practical sympathy you showed many years ago with my crude efforts to promote the "significal" mode of approaching all studies and promoting all true advance'.[35]

By turns, Frank's elderly siblings died, taking their shared memories, their unresolved conundrums, with them. Provincial Darwin in January 1903, thrice-married but heirless at age 89 (one child died in infancy). Intelligent practical Emma in August 1904 at age 93, unmarried and childless. Another sister, Elizabeth Anne ('Bessy', Mrs. Wheler), in January 1906 at age 98, leaving a son Edward and a daughter Lucy. His eccentric brother Erasmus, sailor then inventive farmer, in 1909, never-married and childless at age 94 but still experimenting. Not all the nine siblings—not counting two sisters, Agnes and Violetta, who had died in infancy—were this long-lived or infertile. Sister Lucy Harriot (Mrs. Moilliet) died in 1848 at age 39, leaving six children: Lucy-Amelia (Biggs), Frances Anne Adèle, James Keir, John Lewis, Tertius and Emma Sophia. Another sister Millicent Adèle (Mrs Bunbury), who had a daughter Millie (Mrs Lethbridge), died in December 1883 at age 73. Frank too was childless, and would die just short of 89. But nine children from nine couples is a fertility rate of 1, well below replacement.

The last trip abroad was a return in 1905-6, by way of Pau and Biarritz, to St-Jean-de-Luz and San Sebastian—the Basque country that he had last seen with Louisa in 1860. Back then he had been swept up by a mania for climbing, ascending the Pyrenees to prop himself upright in a sheepskin sleeping bag and watch a tremendous lightning storm through the night. Now he was

[30] Karl Pearson to FG, 1909-04-06, K. Pearson 1930a, 378-9.
[31] Victoria Welby to FG, 1903-10-27, GALTON/3/3/22/16.
[32] See chapter 14, p. 330.
[33] V. Welby to FG, 1905?-10-02, GALTON/3/3/22/16.
[34] V. Welby to FG, 1907-09-22, GALTON/3/3/22/16.
[35] V. Welby to FG, 1909-07-13 and 1911-01-11, GALTON/3/3/22/16.

confined to the hotels and country inns by bad weather and old age, while *Noteworthy Families* went through corrections to press. 'I want vicarious exercise, like being tossed in a blanket (of course, not occasionally bumped on the floor as in school-boy days). Some mechanism ought to be devised for shaking elderly people in a healthful way, and in many directions'.[36] He day-dreamed of a crane on a tripod designed to hoist his recalcitrant old frame onto a donkey once again. In reality, he was reduced to discussing philosophy in an armchair with the invalid Lionel de Tollemache at Biarritz. But in the vicinity of St-Jean-de-Luz cheerfulness would break in again—when he was not being distracted by a wild boar reared at the Inn—with the mountaineer Henry Russell (1834–1909), steeped in archaic Basque lore. Come early April he was back at Rutland Gate, marvelling at the new military teapots with inserted gauze baskets as strainers. In England for good.

Nature continued to receive a stream of Galton's bright ideas. Throughout his career he sent these in to journals. Many of his communications remain unidentified because they were not always signed. Sometimes initials give a clue. They ranged broadly.

- Stereoscopic maps for depicting mountainous country in 3-D, which he had actually implemented in collaboration with a cousin.[37]
- Accelerated 'passage charts' based on wind speed patterns, unfortunately bypassed by steam ships.[38]
- Calculations, based on the inclination of strata, justifying reports of a current in the recently-opened Suez Canal.[39]
- The results of his experiments in Regent's Park with Herbert Spencer's recipe for extreme giddiness—pronating while bent over a walking stick and then raising the head quickly and attempting to walk straight as normal, producing 'frightful giddiness and feeling of sickness'.[40]
- Ruminations in *Mind* on 'free will', prompted by his data on the correspondences between the life-outcomes of twins—he thought free will an illusion.[41]
- Measures of the imagination, reckoned from the amount of information that the mind could fill in when hearing was hard, showed how inventively he was prepared to push his methods.[42]
- Self-conditioning suggested that smell could be used for arithmetic: two whiffs of peppermint could be associated with one of camphor, though he found the effort exhausting and retention temporary.[43]
- Weber's Law meant that corporal punishment of criminals should be increased exponentially rather than linearly, given that the worst were the least sensitive to pain—here he would bashfully 'shrink from subscribing my name', though he did provide initials.[44]
- Hearing that strawberries aggravated gout, Galton cheerfully produced refuting evidence from the 'dog-Latin' autobiography of Linnaeus, who reported that strawberries progressively cleared up his own case of gout.[45]
- A derivative of composite portraiture could be used to extract those features *unique* to a particular individual by a complex process (of his own design) that amounted to *differencing* the portrait of the individual from the generic composite or 'type'.[46]

[36] FG to Millicent Lethbridge, 1905-12-17, K. Pearson 1930b, 555.
[37] F. Galton 1865e.
[38] F. Galton 1866d.
[39] F. Galton 1870d.
[40] F. Galton 1878b.
[41] F. Galton 1884d.
[42] F. Galton 1893c.
[43] F. Galton 1894a.
[44] F. Galton 1898a.
[45] F. Galton 1899.
[46] F. Galton 1900a,b.

— A scheme for planting fruit-trees at maximum density in hexagonal spacing was scotched before publication by his practical-farmer brother Erasmus, who pointed out that straight lines were needed for access by farm equipment.
— While Charles Furse was painting his *very-like* portrait in 1905, Galton whiled away the posing hours by counting the brush-strokes: at a fair estimate, a portrait required about 20,000—about the same number, Galton noted with interest, as he had counted in the portrait by Graeff in 1882. The portrait remained unfinished.[47]
— Even tea-cakes—hard to keep fresh once cut—did not escape scientific attention. Removing a whole wedge with a straight chord, after which the pieces could be pushed together, served just as well as more difficult symmetric excisions.[48]

Elections, juries and other group decisions became grist to the mensurating mill too. Galton argued that the *median* rather than the mean estimate of opinions is usually most appropriate. Cranks might skew the result of simple averaging by polluting the pool with their absurd over- or under-estimates. The middle-most answer is more stable, less sensitive to outliers—an early foray into the field that later became known as *robust statistics*. Some empirical data came along in the form of the voting cards used at the West of England Fat Stock and Poultry Exhibition held at Plymouth in December 1906. After paying sixpence, contestants estimated the dressed (butchered) weight of a fat ox purely by inspection, for the chance of prizes given to the most accurate estimates. Galton could array their named and numbered cards without leaving the comfort of Kensington. The median estimate of 787 legible entries was 1207 pounds, 0.8% over the true value of 1198, a result 'more creditable to the trustworthiness of a democratic judgment than might have been expected'. Curiously the distribution of estimates was not normal, being systematically lower than expected below the middle range and not large enough above it. Galton could only speculate that contestants 'corrected' their initial normal errors, magnifying negative differences and minifying positive ones.[49]

The last of these forays would come in 1909 when Galton tackled the question of 'cursed' church property seized by Henry VIII, which his niece Eva appears to have believed in, along with other Roman Catholics. First sons of the inheritors were rumoured to be doomed. Statistical investigation commenced. An antiquarian provided data about the ownership, between 1800 and 1909, of 245 properties, some of them originally sequestrated from the church. There was hardly any difference, Galton calculated, between the lifespans and tenure of the owners of former church property versus non-church property. It seemed that former Church property did come onto the market more often, but Galton was willing to ascribe that to the discomfort of former church buildings and the expense required to update them.

In-between, Galton had published a successful autobiography, *Memories of My Life*. A literary adviser to Methuen planted the suggestion in late 1907. It was taken up as a handy way to fill increasingly bleak hours. There were enough of those to complete the manuscript in just over 6 months—it was published within the year. But he was already in his late 80s, and knew that his memory now for dates was questionable. Immediate family were for the most part dead, though Emma and Bessy had kept enormous scrapbooks of miscellaneous family history. Louisa's old diaries helped, but few of his own had survived. Even his records of his own publications were fragmentary.

Literary style had always interested Galton, and he considered ways in which scientists could write better prose—if not poetry, as his grandfather Erasmus had done. He was enlisted into the Royal Society of Literature, where he repeated what he had already said before the Royal Society (of scientists), that 'memoirs published by scientific societies are blamed with justice

[47] F. Galton 1905d.
[48] F. Galton 1906.
[49] F. Galton 1907b,c,d.

for being more difficult of comprehension than need be, owing to a want of simplicity in their language, of clearness of expression, and of logical arrangement'. He did not pose as a literary critic, but 'a man need not be a cobbler in order to know when his shoe pinches'.

Obvious faults included 'superfluous use of technical expressions', which ought to be minimised. Papers should start with clear abstracts, to explain why they were written. Neologisms should be discouraged (he objected to the recent 'dominant' and 'recessive'). Long words, especially portmanteau constructions, as in chemistry, were an obstacle to comprehension. Unfortunately literary style was prized more in France than in England, where education of exposition among scientists was defective. He called on the councils and referees of learned societies to demand higher literary standards, especially for clarity. Regular anthologies of bad science writing might also help. The audience was receptive. Nothing happened. And so it remains today, though abstracts at least are now standard.

Over a decade earlier Galton had described his own writing technique to George Bainton.[50]

> I have a singular difficulty in writing, due, I believe, to a natural habit of not thinking in articulate words, when thinking hard, but using gesture thought largely instead of language: so that when I want to express myself, I often find that I have, as it were, to translate. As a matter of fact, I re-write my MSS many times, and correct each time very much indeed. For example, at this moment I am revising the proofs of a forthcoming book, 'Natural Inheritance.' In this, besides frequent re-writings of the MS, I have had it type-written almost throughout, twice in succession, and yet have had largely to correct the proofs. The only tendencies that enable me to write intelligibly, are a great desire to be clear in thought and distinct in expression, and an inclination to take much pains. Also I have great appreciation of good and clear writing by others, and a love of getting at the exact meaning of words. I constantly consult good dictionaries, finding a large Dr. Johnson, and the handy *Skeat's Dictionary*, published by the Oxford Clarendon Press, the most useful. Also I find *Roget's Thesaurus* of the greatest help in disentangling the different meanings of a word. It rarely has enabled me to find one, but constantly has enabled me to observe a want of clearness. This very day I have spent a good half-hour over a word *process*—'the processes of heredity'—which as yet I cannot better, but which does not explain exactly what I want. It is easy enough to write off-hand, like this letter, but it is difficult to write a book. In what the difference consists is a little difficult to say, and I will not attempt to express it now. I occasionally see *The Educational Times*, and when I do, I marvel at the beauty of the prize translations in it. Good writers have the art of building their sentences in the simplest way, with the important parts first, and of placing what follows in the most easy-going order.

The prose of *Memories of My Life* is remarkably clear and disarming, free from all pretension and vanity. It is full of warmth and affection for family, friends and colleagues. Inherited family tendencies to measure, count and organize are traced. Quaker temperament likewise. The Birmingham childhood recalled is happy, with resilient family support at every turn. The schooldays are robust, though frustrated by an antiquated pedagogical system ill-suited to young Frank's talents. The descriptions of his medical studies provide insights into his temperament and the genesis of his thinking about human differences. The Danube-to-Istanbul adventure of 1840, before going up to Cambridge, and the trip home through quarantine, is given in fond detail—it is the only published source, though diaries survive confirming most of it. The early personal crises are openly described: the nervous breakdown at Cambridge, with its subsequent flight, the unsettled years—few manuscript sources remain now—the slow recovery through exploration and the techniques of travel, and the great personal crisis of 1865-8.

However, the description given of Galton's rise through Victorian science will puzzle most readers by its choice of emphasis. An entire chapter is given over to the Kew Observatory in Richmond, where Frank spent many happy hours calibrating instruments—it also lingers lovingly over the Meteorological Society and his long devotion there to inventing new instruments for automatically capturing data. Respects to collaborators are given dutifully throughout. An

[50] Bainton 1891.

entire chapter on 'social life', fascinating to biographers for its cryptic hints, contains more grateful acknowledgments. The Royal Geographical Society and its expeditions, which Galton was administratively involved in, also gets an entire chapter, heavy with reminiscences of Burton, Speke and the Geographical bugbear Henry Morton Stanley. The work on the anthropometric labs, composite portraits, human faculty, heredity, and eugenics is strangely arranged, appearing at the end only and taking up just of over 60 pages out of over 300. Consequently, the unifying themes of Galton's scientific endeavours and their importance are lost sight of.

And as Galton expected, many of the dates and other minor details in his *Memories* are wrong, and it is a rash reader who depends on them. He was elected to the Royal Society in 1860, not in 1856. He was not joint secretary of the Ethnological society with Lawrence Oliphant, but rather with Thomas Hodgkin. He could not have spoken to General Whish 'years later' about the death of Montagu Boulton because the general died in 1853. He did not visit Bertillon before but rather after his lecture to the Royal Institution on personal identification in 1888. He stayed with Flinders Petrie in the winter of 1899/1900 not 1900/1901. The list of his writings given, aside from being far from complete, has manifold errors in it. Many more examples could be given.

On June 14, 1909, Galton received what he at first mistook for a catalogue from a wine merchant, but turned out to be an official notice from the Liberal Prime Minister, Herbert Henry Asquith: he was to be knighted on November the 9th, the King's birthday—traditionally known as the Birthday Honours List. When the date of the ceremony was fixed, he replied with a 'piteous appeal' to the Home Office that he could not possibly attend in person, given his continually declining health, so they agreed to mail him the patent instead. His neighbour Montague Crackanthorpe was quick to offer congratulations. 'I found him just going out for his morning "trundle." "Why, they should have made you a K.C.B.!" were my first words. With a twinkling eye he replied, "Well, I am a sort of KGB. I am a Knight of the Bath Chair."'[51] In a letter to Eva, Galton suggested that his knightly mission would be to 'succour distressed perambulaters'.[52]

Why did the award come so late in Galton's career? One can only speculate. Perhaps because the Darwinians were seldom in good favour then with the non-scientific establishment? Charles Darwin was never knighted, and was only buried in Westminster Abbey after a campaign in which Galton himself played a prominent role. T. H. Huxley was not knighted either. Darwin's son George was made a K.C.B. in the same honours list of 1909, but had won his reputation as an astronomer, not as an evolutionist. It would be 1913 before Darwin's other son Francis, a biologist, was honoured.

There may have been contributing reasons for this delay in official recognition. In earlier days Galton annoyed many of the great and good with his sallies against prayer and the intense public controversy that followed.[53] As we have seen, when *Inquiries into Human Faculty* was reissued by Everyman in 1907, the publisher insisted on omitting the three chapters with anti-religious themes, originally published in 1883.

Not everything was going downhill. The following year Galton's neighbour Crackanthorpe was introduced to the new electric ear trumpet that the scientist was trying out. 'His first sure trial was his deafness, which cut him off from scientific gatherings where at one time he was a familiar figure. This defect he remedied with the help of an electrical instrument very much in the form of a camera. I well remember going to see him a day or two after this new acquisition. Pointing to it as it stood on the table by his side, he said:— "That is my ear. If you will speak to it without raising your voice I shall hear all you say." The experiment was successful, and

[51] Crackanthorpe 1911. The more prestigious 'Knight Commander' (KCB) is the Most Honourable Order of Bath. A 'bath chair' is an old name for a wheelchair.

[52] FG to Eva Biggs, 1909-06-21, GALTON/3/3/7/11.

[53] See chapter 6, p. 99, above.

he talked gaily on De Quetelet's letters and digressions from curve of frequency'.[54]

Though Galton was pleased by the knighthood, he told his grandnephew Hesketh that it had drawbacks. While it added prestige to the cause of eugenics, it also trebled the correspondence he had to deal with.[55] A more significant award followed in November, recognizing his 'researches into heredity', with which he was less ambiguously pleased. This was the Copley Medal of the Royal Society, by far their most prestigious, with only one awarded each year since 1731. It was not limited to the United Kingdom, and ranged over all science. 'It is the "blue ribbon" of the scientific world, and I am of course deeply gratified'.[56] Two years previously A. R. Wallace had won it, the following year George Howard Darwin would win it, and before them a pantheon—Sylvester, Huxley, Cayley, Charles Darwin, Joseph Hooker, Karl Weierstrass, Virchow, Pasteur, Stokes, Mendeleev, ... 'Another Gold Medal! How glad I am – I feel a golden glow, no?', his old friend Lady Amy Pelly wrote.[57] It is an impressive collection.

1. 1853. Royal Geographical Society. Gold Medal.
2. 1886. Royal Society. Gold Medal.
3. 1901. Anthropological Institute. Huxley Medal.
4. 1902. Royal Society. Darwin Medal.
5. 1908. Linnaean Society. Darwin-Wallace Medal.
6. 1910. Royal Society. Copley Medal.

George Darwin agreed to collect the Copley on behalf of Galton, who would not stand up to the rigours of a ceremony, especially one in early November.

Ancient medicinal substances that he knew well from the Levant and the Kalahari eased his chest as the winter of 1910/11 approached. 'Asthma has plagued me, but I stave off the worst bouts now, by smoking a cigarette of *bhang* (Indian hemp-*hashish*). It is curious to perceive the spreading of the narcotic effect over the lungs and everywhere'.[58] When winter came Galton had retreated once again to the country—this time to Grayshott House in Hampshire, just four miles from the Surrey town of Haslemere, near where he had wintered the previous year. He received few visitors—the geologist Archibald Geikie (who lived nearby), the publisher Methuen, his nephew Edward Wheler-Galton—attended to as always by niece Eva and his long-serving manservant Albert Gifi, now assisted by a male nurse, Joseph Alfred Charman—their patient needed carrying up and down stairs. Karl Pearson came for dinner on the 28th, and enjoyed arguing about the Devil once again with the Catholic convert, niece Eva.

Two weeks later, on Sunday January the 15th, Eva notified Pearson that Galton had suffered a bronchial attack, picked up a few days previously, and had a temperature. 'Uncle Frank is one degree better to-day but still in danger'. 'He is quite easy in his mind and very clear when he speaks, but too weak to speak more than a word or two. My cousin Edward Wheler, a very dear nephew of his, is here—we never leave him a minute'. For some reason Pearson saw no great cause for alarm, expecting the old traveller to pull through as he had always done before.

The next day, Eva reported a further decline to Milly. 'I am sorry to have no good news. Dr Lyndon[59] considered Uncle Frank worse this afternoon his breath is so difficult to get, he is in great discomfort and very weak, but so sweet and cheerful, always saying something witty if he can speak a few words'.

Then on the evening of Tuesday the 17th, 1911, Galton died of apparent heart failure

[54] Crackanthorpe 1911.
[55] H. Pearson 1921.
[56] FG to Millicent Lethbridge, 1910-10-08, K. Pearson 1930b, 611.
[57] Amy Pelly to FG, 1910-11-08, GALTON/3/3/16/12.
[58] FG to Karl Pearson, 1910-08-4, K. Pearson 1930a, 430. This is surely the 'cigarette of peace and quiet' he referred to a few years earlier in FG to Millicent Lethbridge, 1906-06-16, K. Pearson 1930b, 573.
[59] Dr. Charles Lyndon.

brought on by the bronchial attack. 'Just before he died, when almost too feeble to speak, he was given a prick of strychnine in the wrist; this interested him intensely and though we didn't want him to exhaust himself talking, he wouldn't let the doctor alone without having it clearly explained what the strychnine would do for him. He was most excited about the oxygen they gave him and wanted Edward Wheler to tell Dr Lyndon all about his experiments with it—this an hour or so before death', 'he became unconscious about 3/4 of an hour before death; Gifi looked in and Uncle Frank opened his eyes and smiled at him, and then never opened them again; he seemed in a sort of torpor'.[60]

The scientist had died, full of honours, just one month short of his 89th birthday, mentally sound until the end. On Saturday the 21st of January, he was buried in the Galton family vault at St. Michael's Churchyard, Claverdon, near Stratford-on-Avon in Warwickshire. Not far from Leamington where his parents had retired to, and where his brother Darwin had farmed at Loxton. The grave-site was neglected for many years, but restored in 2011.

——oOo——

The will was proved on March 7. The gross estate came to £115,932, with a net personalty of £113,371 (in 2017 money, about £10 million) Galton left £65,000 in personal bequests: £19,500 to his niece, Millicent Lethbridge; £15,000 to his nephew, Edward Wheler-Galton; £3,700 to Arthur Francis Butler; £2,000 to Amy Pearson; and other smaller amounts, including £200 for his manservant Albert Gifi (about £15,000 in 2017 money, or perhaps 2 years of wages for a skilled tradesman).

Curiously, his niece Eva Biggs, who had looked after him for so many years, had insisted in writing that she was not to receive any money—her reasons are not clear, and it was against Galton's express wish—but she was nevertheless allowed to choose whatever personal effects she liked to the value of £25.

All the rest of the money, which exceeded £60,000 (or about £6.5 million in 2017 money) went to the University of London to establish a professorship of Eugenics, 'to pursue the study and further the knowledge of National Eugenics, that is of the agencies under social control that may improve or impair the racial faculties of future generations physically and mentally'. The first Galton Professor was to be Karl Pearson, who was named in a codicil to the will as the favoured candidate, though without binding the university.

[60] Eva Biggs to Karl Pearson, 1911-01-25, K. Pearson 1930b, 618.

Figure 15.1: Francis Galton near the turn of the century.

Figure 15.2: Francis Galton at age 75 taken in 1897 at Royat, K. Pearson 1930b. Restored.

Figure 15.3: Flinders Petrie in 1886.

Figure 15.4: Hilda Petrie in 1897.

Figure 15.5: From Petrie's Racial Photographs: Mam and Uaata Pay Tribute to the Egyptians.

Figure 15.6: Francis Galton at age 85 in 1907, K. Pearson 1924.

Figure 15.7: Francis Galton at age 87 in 1909, with Karl Pearson, K. Pearson 1930a.

Figure 15.8: Francis Galton at age 88 in 1910, by Eva Biggs, K. Pearson 1924.

Figure 15.9: Francis Galton going on 89 in late 1910, by Frank Carter, K. Pearson 1930a.

Appendix A

Ethnography Notebook, 1864

Page f.1

a phrase in Herbert Spencer "First Principles" p. 15 "... prompts imagination that it afterwards projects into the external world and bye & bye mistakes for realities." ([maybe] thinks religion is not so [...])

Page f.2

Farrar see Reader Oct. 1864[1]
Dutch at the Cape
Spaniards & America
French in Canada
Negroes in America
Danes & Norwegians in Greenland
? Iceland
America books
Herera General History of the doings of the Spaniards in America translated by Stephen 8vo 6 vols 1740[2]
Casas las[3] — 8vo London 1693
In Greenland Crantz Hist of Greenland
1—242 244.[4] Prévost Hist. Genl. des Voyages tom XV 152 not. (96) [5]

Page f.3

Buffon Hist. Nat. III 484—IX, 103, 114[6]
[Histoire naturelle] recherches Philosophiques sur les Americains[7]
Rousseau
Ae up to p. 17
Ulloa[8]

---

1 Report of British Association talk by F. W. Farrar 'On Fixity of Type' *Reader* Oct. 29, 1864, 550–551. 'There was at one time a universal impression that the diversities of type and complexion observable in the human race might easily be accounted for from the effects of climate, custom, food, and manner of life. The opinion is now entirely abandoned by the majority of scientific men'.

2 Antonio de Herrera y Tordesillas (1559–1625). *The general history of the vast continent and islands of America: commonly call'd the West-Indies, from the first discovery thereof: with the best accounts the people could give of their antiquities.* Translated by John Stevens (not Stephen). London: Printed for J. Batley, 1725-26.

3 Bartolomé de las Casas (1484?–1566), a Spanish historian.

4 Crantz, David (1723–1777) *History of Greenland* (1767).

5 Abbé Prévost (1697–1763) *Histoire générale des voyages.*

6 Georges-Louis Leclerc, Comte de Buffon (1707–1788).

7 Incorrectly pencilled in under Buffon. Cornelius de Pauw, *Recherches philosophiques sur les Americains* 1770. See also his *Defense des Recherches philosophiques sur les Americains* 1768.

8 Antonio de Ulloa (1716–1795) who had governed Louisiana for Spain.

Humboldt[9]
Robertson[10] II 62 feebleness of constitution 63. Smallness of appetite, 65 Absence of love insensibility to beauty the natives treat their women with coldness & indifference, neither tender attachment nor ardent desire. The missionaries express astonishment at their coldness (see quotations see also p. 103) 71 Difference in certain localities 71 Hardly any were deformed mutilated or defective in their senses, but when taken care of they are so see Ulloa I, 222

Page f.4

Albinos among the rare & monstrous animals that Montezuma had collected—see Cortes [in] Ramusio III, p. 241 F[11] 108. Parental love is strong 110 Filial love small
... "render the union between husband and wife unequal. They shorten the duration & weaken the force, of the connexion between parents & children["]
399 Difference of temper between [an] Amer. Indian & a negro quoted from Tertre ii, 490[12] (ask Mr. Douglas — Mauritius).[13]
The many independent communities in America having common qualities shows that these are due to nature not tradition.
140 Among the Natchez Indians despotism & suttee.

Page f.5

141. In Hispaniola Cuba [...] the Caziques had extensive power gave rain & sunshine
Bogotá ranked in civilization next to Peru & Mexico
150 Love of vengeance (vengeance is the principle instilled into the minds of children)
159 Cruelty to prisoners
161 Indifference to death, fortitude under torture
165 proceedings in S. American [towards] captives
167 Passive fortitude—patience not valour
177 Attentions to dress & ornament had more reference to war than to gallantry
188. The cold phlegmatic ~~operating~~ temper renders all their operations languid—"the work of an Indian" is

Page f.6

Robertson America II
anything in the execution of which an immense time has been employed & much labour wasted
195 Religion consists in averting wrath of evil spirits
199 Religion of Natchez—sun
200 " Bogotá–sun & moon
human victims
207 Incantations soothsayers auguries
210 Dancing–a serious exercise have no relation to love & gallantry as those of the African & Asiatics
213 Immoderate love of gaming
214 Drunken
220 Murder parents
226 Non gregarious
228 No sympathy—the Spanish

---

[9] Alexander von Humboldt (1769–1859) the Prussian polymath and explorer.
[10] William Robertson (1721–1793), *The History of America* (1777).
[11] Giovanni Battista Ramusio (1485–1557), *Navigationi et Viaggi.*
[12] Jean-Baptiste Du Tertre (1610–1687), *Histoire naturelle et morale des iles Antilles de l'Amérique. Enrichie de plusieurs belles figures des raretez qui y sont décrites. Avec un vocabulaire Caraïbe.* First edition (Rotterdam, 1658).
[13] Sholto James Douglas (1820–1868).

Page f.7r

sometimes found it necessary to enforce the common duties of humanity by positive laws
229 never care for their gods
230 is a serious melancholy taciturn animal—will sit whole days in one posture without opening the lips marching Indian file—rowing canoe all in silence
232 Secrecy—dissembling—An insurrection at [Fera] was hatching for 30 years unsuspected
233 Perseverance & dignity
234. Partisanship (to community) zeal for honour of their tribe
236 A man can conceive no other sort of life to be tolerable than his own
Tartar—American—do not cavil at nor comprehend civilization, are[14]

Page f.7v

menagerie in S Africa
Bates[15] tells me June 21. That it was reported that the King of the Barotse, who lived near the Victoria Falls in the Zambesi, and who was driven out by the Caffres some (?20) years ago, Anyhow he was then a little before Sebituane's time;–had a menagerie. Hippopotami came to be fed from the river and he had many other animals, whether as pets or in confinement Bates does not know.

Page f.8

amazed at it.
237 despise ~~European~~ white men
383 a note. Excellent résumé of American character
388 The negroes are more improveable & though slaves look down on the Indians
Ulloa in Pinkerton XIV 315
He marvels that the descendants of the whole Peruvians shd be so case (quote this to shew the utter wreck of ancient efforts towards progress)
517 are remarkably slow but very persevering—mechanical aptitude
518 Debauch—do not gamble
520 Religious inaptitude
522 Little or no natural dread of death
523 Praises the Jesuits' pupils &

Page f.9

those taught among the Spanish at Quito
Buy Ramayana French transl.
Mahabharata
Zendavesta[16]
Vedas
Menu[17]
Shah Nameh (Heroic Hist. Persia)[18]
Mexican birds & beast. Hernandez the naturalist was sent at great cost to describe them. Works not pub: till after his death. Hernandez Plantarium Animalium & Mineralium Mexicanorum Historia fol. Rome 1651.[19]
Edgar Quinet. Génie des Religion[20]

---

14 Continued on page f.8.
15 Henry Walter Bates.
16 Zoroastrians.
17 Mánava-Dherma-Sástra, Or the Institutes of Menu.
18 Ferdowsi, *Persian Book of Kings*.
19 Francisco Hernandez, *Rerum Medicarum Novae Hispaniae Thesaurus, seu Plantarium, Animalium, Mineralium Mexicanorum Historia* (1651).
20 French historian, 1803–1875.

Page f.10

Fur seals way of leading them–half-domesticated Sir J. Simpson Voyage round world I 223.[21]
~~Hottentot Popes~~
Réaumur histoire des insectes[22] whose talent for observation seems almost to have amounted to genius
Coquettishness even in Butterflies Quatrefages 53[23]
~~Superstition among gregarious animals, vanity Character of the animal generally does not consist of number of independent factors any of which may be separately eliminated~~
~~Virtues exacted by gregariousness—French duellists, respect to law, superstition. Dull individuals have their thought reinforced by another's activities—Jackals never bite each other (Speer)~~

Page f.11

Monarchism in Thibet (Buddha & Rom. Caths)
Bannister British colonization & coloured tribes[24]
Howitt colonization & xtianity[25]
A.e.
Sir Charles Nicholson & conversation with Burton
The black Australian children are teachable & quite amiable. A year or two after puberty their sexual passions become ungovernable & they leave civilization for the bush. The black fellows are always gentlemen In our constitutional country the classes are so sharply divided that a man ~~has no practice in behaving himself~~ are subservient to those about them. In barbarous despotism a slave speaks, as a man, to a king. Excellent mimics.
Société Ethnologique Paris going on in 1839
Society for the Protection of Aborigines U.S.
Schomburgk Brazil (try)[26]

Page f.12

[Cortés Edifiantes]
p. 618 Meadows[27] says he has remarked and is sure "animals dogs for instance do not understand consonants" He wd construct a vocabulary for speaking to dogs mainly of e as in meet a as in father u as in bull also of the intermediate sounds a as in fate & o as in lord. For easy pronunciation by man he would systematically combine

Page f.13

them with the weak consonants h, w, m, l these "woomah" "celoo" wuh! surprise
~~A white feather in a pigeon punished with death from a hawk as much as indolence or any vice~~
Free will & predestination. Bird in a cage.
Civilized life does not eliminate short sightedness few children due to later marriages Children from selected unions. Keep down population. neuters polygamy & polyandry. Girls pass: nature early astray & lost
Antiquated implements are usually complicated—adaptation rather than complexity the scheme of nature.
Aristotle Morals in Eudeme II, 6-12[28]
Free will, (Will is a first cause, a creation FG)

---

[21] Sir George Simpson, *Narrative of a Journey Round the World: During the Years 1841 and 1842* (1847).
[22] Réaumur, René-Antoine Ferchault de, (1683–1757), *Mémoires pour servir à l'histoire des insectes.*
[23] Armand Quatrefages, *Metamorphoses of Man and the Lower Animals* (1864).
[24] Saxe Bannister (1790–1877), *British Colonization and Coloured Tribes* (London: William Ball, 1838).
[25] William Howitt (1792–1879), *Colonization and Christianity: A Popular History of the Treatment of the Natives by the Europeans in all their Colonies* (London: Longman, 1838).
[26] Sir Robert H. Schomburgk, *A Description of British Guiana, Geographical and Statistical* (1840).
[27] Thomas Taylor Meadows, *The Chinese and their rebellions* (1856).
[28] Eudemian Ethics.

Interest in descendants. Jews.

Page f.13b

Sir George Grey Australia.[29] "The savage is naturally of a thoughtful restless disposition". In G. Grey appendix abo. report
Their labour neither constant nor steady Eyre
Missionary Intelligencer look through, shew number of sporadic attempts none (?) have produced ~~borne~~ permanent effect
As a consequence of civilised inroads nature art & knowledge whatever may have been its previous development at once declines — what the white men bring is far superior. The natives are abashed & discouraged. Colonial Intelligencer
Sir F. Head's dispatches in the affairs of the Indians just before or in ~~about~~ 1847[30]
Dr. Thompson report (? also book) of New Zealand[31]
Women white children — vanity leads them to the whites
Aborigines 1, 71. New South Wales reply to a select committee of enquiry—Gunning. "the love of a wandering independent life prevents their settling anywhere to work, except for a short time, when compelled by want of food or clothing & encouraged by kind treatment"[32]
Meadows[33] 591. "savages are "fan fuk puhting" (as the Chinese call them) "hither & hither, not fixed" They are unstable and unpersevering"

Page f.13c

Shame – Sin is sense of dishonor
Chadwick address on Public Health Social Science Assoc. Glasgow 1860
p. 576[34] "The primary qualities of those whom we are accustomed to call Anglo Saxons including the lowland Scotch are great bodily strength, applied under the command of a steady persevering will, mental self-contentedness, impassability to external irrelevant impressions which carries them through the continued [repetition] of toilsome labour "steady as time"

Page f.14

G if [existing] & [knowable] by conscience, must be in accordance with conscience [from Duty &]
Shame – Sin is sense of dishonour

Page f.15r

Horse shying from skittishness is affectation
Horse buck [jumping]
Unsoundness of English horses they can be bred for Can breed what tempers you like
"intellect" Geoffroy St. Hilaire[35] Comptes Rendu III 29

---

[29] Possibly *Journals of Two Expeditions of Discovery in North-West and Western Australia* by George Grey 2 vols. (1841).

[30] See *Copies or Extracts of Despatches From Sir F. B. Head Bart. KCH on the Subject of Canada*: House of Commons, 1839.

[31] Possibly Dr. Arthur Saunders Thompson, 'Observations on the Stature, Bodily Weight, Magnitude of Chest, and Physical Strength of the New Zealand Race of Men' JRGS 23 (1853). Saunders also wrote a book, *The Story of New Zealand* (1859).

[32] *1846 New South Wales. Aborigines. Replies to a Circular Letter, addressed to the Clergy, of all Denominations, By Order of The Select Committee on the Condition of the Aborigines. Ordered, By the Council, To be Printed 31st October 1846.* Sydney: W.W. Davies, Government Printing Office. Galton is citing from the contribution by Reverend Robert Cartwright, Minister of the Church of England, Gunning, 6th April, 1846.

[33] Thomas Taylor Meadows, see above.

[34] Address of Edwin Chadwick, Esq., C.B., as vice president of the Public Health Section to the General Meeting of the National Association for the Promotion of Social Science, Glasgow, September 29th, 1860.

[35] Étienne Geoffroy Saint-Hilaire (1772–1844).

also Cours d'histoire naturelle des animaux
Proudhon de la Justice II, 279 [...][36] "Victrix causa diis placuit"[37]
vae victis[38]

Page f.16

A2
Lucretius II, 55
"For even as children are flurried and dread all things in the thick darkness, thus we in the daylight fear at times things not a whit more to be dreaded than those at which children shudder in the dark & fancy sure to be. This terror therefore & darkness of mind must be dispelled not by the rays of the sun & flittering shafts of day, but by the aspect and law of nature."
Sin — injustice to society or untruth to one's own nature. (F.G.)

Page f.17

Hereditary. G. H. Lewes II 378 quotes Prosper Lucas Traité de l'hérédité naturelle 1847–50.
Lectures on the Physiology of Breeding Orton 1855[39]
(turf annals Longman 1844)[40]
Traité de Generation Gibou (Girou) de Buzareingues 1828[41]
Moreux — La Physiologie morbide 1859[42]
Westminster Review July 1856 Hered: influence[43]
See also Herbert Spencer Principles of Psychology
See Parent [...] Da Gama Machado[44]
Australian Ullathorne[45] has written? where
Whitecross sketches & characters Saunders & Otley[46]
Butler Earp — Australia Life in the Bush[47]
North American Indians utter indifference to melody chivalrous, vivid sense of honour
Indians of California Bryant[48] Central Asiatic nations. stability of their character no spirit of curiosity or speculation p. 59. Hindus a very good passage quote

Page f.18

Read Adam Smith's moral philosophy
Chenevix on national character[49] in library of Athenaeum quote in my appendix his passage in the influence of individual geniuses in moulding the destiny of nations vol. 2 p 41–43 & after
Read his chapter on morality of different nations vol 1, 137 see some pages in
Xtians 240 million Buddhists 200 million } Give to [Bowman] 987 [...]
Glacier in (not [...]) like

---

36 De la justice la révolution et dans l'église
37 The victorious cause pleased the gods.
38 Woe to the vanquished.
39 Reginald Orton, *On the physiology of breeding. Two lectures, delivered to the Newcastle Farmers' Club* (1855).
40 Possibly John Orton, *Turf Annals of York and Doncaster to 1843* (1844).
41 Louis François Charles Girou de Buzareingues (1773–1856), *De la génération* Paris, 1828.
42 Jacques-Joseph Moreau (1804–1884), *La psychologie morbide dans ses rapports avec la philosophic de l'histoire.*
43 Lewes 1856. See *Wellesley Index to Victorian Periodicals: 1824–1900* Vol. 3: 625, for confirmation of the attribution to Lewes. This was a review of the three books on heredity and breeding by Prosper Lucas, Reginald Orton and Charles Girou, all of which are noted above by Galton.
44 *Théorie de resemblances* (1831).
45 William Bernard Ullathorne (1806–1889). Possibly *The Horrors of Transportation Briefly Unfolded to the People* (1838).
46 James William Whitecross, *Sketches and Characters; Or, The Natural History of the Human Intellects* (1853).
47 Possibly George Butler Earp, *What we did in Australia: being the practical experience of three clerks in the stock-yard and at the gold fields* (1853).
48 Edwin Bryant, *What I Saw in California* (1848).
49 Richard Chenevix, *An Essay Upon National Character* (1832).

moving waterfall
Types of [wonder]
Break up into new forms
Infinity [any sheer fall if ...]
Divide by billion x ... [not] rapid change enough
We part of bigger system than of others &c.&c. & their conscious relation between man & G. average of life happy we take our truth &

Page f.19

come in occasionally for something splendid
Macaulay says when the Puritans put down bear baiting their zeal was less inspired by pity for the bear than hostility to a popular amusement.
Jerome III 316
[...]

Page f.20

Individuality. Man perishes like a nation or family does breaks up & reconstitutes in new combinations
Electro biology individuality merged in another
A child is ignorant of the feelings of grown people yet consistent in itself —
x=0 is as true for trees as it is for men
individuality of trees — made up of cells each a life of its own
over [Progress] objectively valueless subjectively [hurtful]
Parental affection shewn by children to dolls
meddling of women in others concerns
Conscience may [impose] ... & Sabbath

Page f.21r

Bees ants patriotism loyalty
Disinterestedness. a person forgets himself [and] too readily & throws himself into others people's ideas & schemes &c.&c. see Mr. Douglas letter about Africans & Hindoos
Council meetings. forget all other matters for the time temporary individuality of the meeting. ditto of states, families, schools regiments
With entire ignorance of the purpose for which we are called into life & not withstanding the conviction of the disinterestedness & lost opportunities of nature we ~~half believe~~ also see a decided tendency to progress of type & of complexity & mutual harmonies in this world (of which alone we have any knowledge, that only during some 6000 years & a most partial one) we have therefore reason rather than the contrary to ~~believe the~~ confirm our strong natural hope (Anglo Saxon) that there is a purpose & that we are more than mere sports of nature
Colebrook Vedantic papers[50]
Wilson Rig Veda[51] — Vishnu Puranas
Ramohun[52] Rig Veda
Rodwell's Koran[53]
Bornouf[54]
Textbook Physical Geography published by Blackwood & son see Times advertisement Jan 13 1865 p. 12
Advanced textbook Phys Geogr. Page
Introductory ditto Page

[50] Henry Thomas Colebrook, *Transactions of the Royal Asiatic Society* (1827).
[51] Horace Hayman Wilson (1786–1860).
[52] Rammohan Roy.
[53] John Medows Rodwell (1808–1900).
[54] Eugène Burnouf (1801–1852) the French scholar of Persian.

Elements Modern Geogr. Mackay

- Vedas. 1200—1400 BC. Wilson
- Zoroaster. 589. Penny Cycl: see Bryant's analysis ancient mythology
- Daniel. 530
- Confucius. 55
- Abraham. 1900
- Moses. 1500
- 315 Buddha. ? 1000
- 252 Mahomet
- 111 Brahmaic. see Colebrook Asiatic Research VIII 369 -476 reprinted in his miscellaneous essays p 9–114
- 120 Christians

Page f.21v

Identity — what does the Llama of Thibet think of his identity see Penny Cycl V, 532 top left column.

[...]

Pouchet 64 Table by Linnaeus in Syst Nat

- Homo Americanus. Pertinax, contentus, liber
  Regitur consuetudine

Homo Europaeus. Levis, argutus, inventor
Regitur ritibus

Homo Asiaticus. Severus, fastuosus, avarus
Regitur opinionibus

Homo Afer. Vafer, segnis, negligens
Regitur arbitrio

Page f.22

Congregational worship

Apparitions are far stronger evidence than for example the return of X or of the spirits in the streets of Jerusalem at the time of the crucifixion Sight seers spiritualists might be bred Now is at most imperfect — want knowledge & cannot get it. Confucius

Page f.23r

Barbaric ancestry as surely as if well nurtured dogs is shewn by their sniffing at dirty messes.

Motive of duty against the grain— labour of love

Hodgson mortality among clergymen expectations of life ... 3d Jan 1865 C & E Layton[55]

Stanford's classified catalogue of educational works (Geograph)

American Indians indifference to death Ulloa (Pinkerton) XIV 522, Good passage to quote I have marked it

That great suicide of J to save mankind. We (encourage) allow suicide to save others or for the good of others not for one's own relief.

Page f.23v

Mr [Atkins] enclosed to Benjamin Handley 5 Park St. Westminster

Clergy on committees

Sawyer's opinion of s. of widows

Convocation — prayer at beginning

[55] Rev. John Hodgson, *Observations on the Mortality of the Clergy of England and Wales* (London: C & E Layton, 1865).

King Charles [extinct] Buccleuch (Southampton) Grafton (Northumberland) St Albans Richmond 4 extant 2 not so.[56]

Page f.24r

- Character. Dread of death
- American Indian. Is written out (Robertson, mainly). Ulloa Pink XIV 522
- Esqimaux. Hall when deserted to die was uncomplaining
- Hindoos
- Arabs
- Chinese. Huc & Gabet[57] China Vol II Chap 1 p 41-3
- Negroes
- Papuans
- English. Brodie Bacon

Page f.24v

(Spottiswoode's list of books)
Burnouf Introduction du Buddhism Indien
37 Proselytism 159 disinterested death 198 facility with wh poor were admitted by Brahmins
Spiegel Zoroaster
Wilson Rig Veda
Ramohun Rig Veda (as well as Wedant) Colebrook's Essays
Menu
Oxford Essays Max Müller

Page f.25r

Valhalla — Westminster Abbey & St. Paul's
Smiles self help — Campbell Judges & Ld Chancellors
Dr Guy's papers Statistical Journal
xvii 20. life of clergy & medical men (many the latter are exposed to contagious maladies p. 22)
xx. Summary of previous enquiries 65
Table of duration of life in the 3 learned professions 70
Conclusions — — 71
xxii mean age of sovereigns Gentry clergy [...]
p. 135 (? a right reference — not in index)
Summary of previous papers 337
Duration of life as affected by social occupation 355.
There are times when a man needs the support of sympathy — What he believes to be sympathy does as well. The auth: says there is no G. on that assumption the feeling of a devout man would be due to his belief in sympathy not to the reality of it. But is there any extraneous influence? Is there none than can be accounted for thus? Devout people say yes. Eminent men believe it, but they more than any others require such sympathy, for they are misunderstood by their associates.
Walford Men of the Time 1862 Routledge
2 Van Eyck's & sister Margaret picture of
Virgin & child in our National Gallery.

Page f.25v

See "Grandeur of the Law" by Foss.[58]

[56] Possibly a reference to the respective Earldoms.
[57] Évariste Régis Huc and Joseph Gabet *Travels in Tartary, Thibet, and China During the years 1844-5-6* (1850).
[58] Edward Foss, *The Grandeur of the Law* (1843).

Page f.26r

Campbell's lives of the Chancellors
Chancellors with distinguished relatives
Vol 1. Turketel 33. grandson of Alfred
Roger Pauper 58. nephew of Roger the Great but a job
Thomas a Becket 61. Father was sheriff of London, notwithstanding his poverty.
Longchamp 106. very eminent & a foreigner accused of inducting his relations into power
Godfrey Giffard 155 same family as predecessor—he did well, both as chancellor & as bishop
Robert Burnell 162 very able, accused of nepotism
Langton 173. related to Cardinal of the Magna Charta.
Stratfords 231. 2 brothers both chancellors
VIII 425 [Houz]
More 504. Son of Sir R. More, a judge
II Sir Nicholas Bacon see 112 in high praise of him
Lord Bacon 112 Son of above, had also a talented mother (Noble houses derived from the bar 82 in number p. 260)
Finch 542 Father a good lawyer.
Littleton 577 of an ancient family of the robe — father also a judge
III Whitelock 18 strongly connected with the law father & great uncle were judges. His son Wm rose to great eminence at the bar (p. 83) & his eldest son was distinguished as a soldier.
Herbert 109. His sons distinguished 109. Eldest rose to a high command in the army 2d the great naval officer who was created Lord Torrington 3d son became Chief Justice of KB followed James II into exile & was made keeper of great seat in partibus
Clarendon 110. Uncle chief Justice of K. B. daughter 169 Duchess of York (& mother of Queen Mary & Queen Anne) 199 with agreeable manner though plain. 270 3 sons who gained some distinction
Bridgman 272 Son of a Bishop
Shaftesbury 288. grandson of a clerk of the council (see for the profligacy of this ancestor of our Ld S)
Finch 300 Father was 1st cousin to the previous Finch. Also recorder of London & speaker of House of Commons — 428. "His second son almost recalled his father in the brilliancy & success of his professional career". ~~all the learning of the bar will not set up a man~~ in "a natural facility of speech such as the family of the Finches is eminent by" Made Solicitor Genl & Earl Aylesford He completed the list of eminent men that have made the name of Finch so conspicuous in our legal annals.

Page f.26v

[Image of page f.26v omitted.]

Page f.27r

Lucas (Prosper) Traité de 'hérédité naturelle[59]
I. 247–254 Longevity 249. susceptibility to opium coffee calomel
477. drunkenness – gambling, meat
478 sexual passion, women & men. 400. crime to property. 486 monomania to steal
496. crimes of violence
579 La génie veritable est toujours isolé, disent d'une commune voix Spurzheim Virey Lordach Burdach etc. il ne se réveille point dans sa postérité (references are given)
587. Eloquence & other qualities political as
583. Sculpture – painters — music
607 Dancing 605 Summary
II. Melancholy Johnson (see Burton?)
251 immunity from infection syphilis
252 Lord Althorp's reply to Mr. Baring about longevity. Short lives are hereditary 260 the man 81 [age] fath. 102 had beaten him for not saluting his grandfather 123[60] 341. Hindoos think son is a

[59] Prosper Lucas, *Traité philosophique et physiologique de l'hérédité naturelle* (1847).
[60] The 81 year old was beaten by the 102 year old father for not saluting the 123 year old grandfather.

perpetuation of father's life Inst: Manou[61] IX 518 also X, 58 & more on other authorities. Institution of castes due to hereditary belief in.

Page f.27v

Combe Phrenology[62]
ruling passions glory, wealth, intellect poetry power purity asceticism combativeness
172 on personal identity — girl with double consciousness
174 a lunatic thinks himself Napoleon & feels poignant remorse for the massacre at Jaffa.
241. Gall says but doubtingly, that in all animals where both the male & female concur in rearing the young, marriage for life exists — where the female is unaided, marriage is temporary. Vimont[63] mentions
many instances of social animals.
Van Tromp[64] Marcus Aurelius

Lucas (Prosper) continued from last page
371. Moral matter consists of sensation—sentiments intelligence—movement. They are so many forms of activity of one & the same principle.
372 sensations. Touch—varied sensitiveness—one calls not what another calls cold. Damp & fine weather. East wind. Hatred. of light. 382. Smell & taste. 385 Fetidness & sweetness are purely matters of individual taste. Strength of scent in some dogs. Gourmandizers
390 Hatred of flesh. 397. Table of congenital blindness
very important (399 quotes Huzzard sur l' amélioration des chevaux en France 1808).[65] 465 colour blindness 420 hearing. 434 Sentiments 461 (quote Pichard Manuel des Harras)[66] Horses 461

Page f.28r

read Broc Essai sur les races humaines.[67] George Leroy perfectibilité des animaux[68]

Chenevix 1, 137. morality of different nations
virtue submission with Xtian war with Mussulmans
(blood feuds a duty) hospitality. Trade cheating, pride, magnanimity vulgarity fashion nations lament degeneracy & magnify the deeds of old time, shews the standard becomes more lofty—man improves—laws of honor Paley's definition quoted 158 as "a system of rules constructed by people of fashion and calculated to facilitate their intercourse with one another & for no other purpose". Atone for an outrage by fighting a duel. A debt of honor has precedence over all other debts. Trade rules. Code of honor among thieves. (Code of honor allows cheating in selling horses because cheats are associated with—must do so in self-defence) Chivalry. Importance assigned to human life. Politeness among Chinese. Socialism Rules of war & Divine rights of Kings titles of kings. Suicide of Ch[...] Suttees. (Sacrifices, p. 181 human & others) Infanticide p. 183 (Which [shd be] the highest motive maternal satisfaction is the well being of the infant, is the good of the state?) (Cruelty (is quite an instinct) 181 onwards) Priest craft. Everything is fair in love. Monastic life. Men of taste Superior races improving away inferior races = Theophrastus (La Bruyère)[69] Brutality [...]

Page f.28v

---

61 Menu, or Manu.
62 George Combe (1788–1858), *Elements of Phrenology* (1824).
63 Joseph Vimont (1795–1857).
64 Presumably the Dutch admiral Maarten Tromp.
65 Jean-Baptiste Huzard, *Instruction sur l'amélioration des chevaux en France.*
66 Pichard, *Manuel des haras ou système de régénération des races de chevaux* (1812).
67 Pierre Paul Broc (1782–1848), *Essai sur les races humaines considérées sous les rapports anatomique et philosophique* (1836).
68 Charles George (1723–1789), *Lettres philosophiques sur l'intelligence et la perfectibilité des animaux* (1802).
69 Jean de La Bruyère (1645–1696), *The Characters of Theophrastus* (1688).

R. B. Walker
10 Melbourne Grove
West Brompton
Caesar Smith's Dictionary I 554 [V Column]
"His moral character as far as the connexion of the sexes goes, was as low as that of the rest of the Romans of the age. His intrigues with the most distinguished Roman ladies were notorious & he was equally lavish of his favour in the provinces". 552 Cleopatra had a son by Caesar who Caesarion, p. 556 was executed by order of Augustus. He married p. 537 [...] Cornelia, Pompeia & Calphurnia. Children Julia by Calphurnia[70] married [to] [Pompey] besides Caesarion.

Page f.29v

Read Renan Langue Sémitiques[71]
P. de Rémusat Revue deux Mondes 1854 Races Humaines.[72]
Population of Europe at time of black death in 1347 has been estimated "at 105,000,000,— a high estimate". Cornhill Mag 1865 Black Death.
Mitchell of Oxford is said by Brodrick[73] to have made out that all the 1st classmen of Oxford belong to a very limited number of families.
Relationships. Isaac Taylor's family. The Gregories
Burns & Father. Carlyle & father & brother of Dante.
Milton's father wrote a sonnet & composed some music
Ptolemies married sisters for ... generations. So did the Persian Kings.
(A see next page)
Guizot Hist. of Civilization[74]
At the beginning of the 10th centy barbarians had arrived at the extreme term (see p. 60) say 30 generations ago. Then came feudalism between the 10th & 13th centy p 80 instilling the right of personal resistance—a terrible unsocial right, which appeals to force & war, which is the destruction of society itself "the very nature of feudal society, was repugnant to order & legality" see p 421
426. Instances of the coldness of wandering
Savages Bruce Levaillant Iroquois Malthus
Robertson

Page f.30

Is there an instinct of expiation?
Tacitus XXVI "To devote both day & night to deep drinking is a disgrace to no man" "They yield to gambling with such ardour, that when they have lost every thing else they place their own liberty on the hazard of the dice"
Insert, education like fencing ... navvy work
monstrosities in dogs
individual variation
Gipsies
[...] Miss Kent
George Henry Lewes
Shew that selfishness of [...]
I don't depend on statistics alone but can't quote private cases without breach of confidence
Distinguished families
Hannibal. Philip of Macedon incl: Ptolemy. Charlemagne. Macaulay. Coleridge. Sheridan. Pollock. Wordsworth Stephen. Venn. Stanleys. Thompson (Math). Gregory [...] Palgrave. Kingsley. ?Dawson Turner. Lushingtons Butlers. Hooker. Darwin. Brontë. Merivale Wordsworth. Pitt (see Eothen). Herschel (Caroline) Nieburgh Napier Burns. Kennedy (see A last page)

---

70 Actually, by Cornelia.
71 Ernest Renan, *Histoire générale et système comparé des langues sémitiques* 1855.
72 Paul de Rémusat (1831–1897), 'Des Races Humaines' *Revue deux Mondes* (1854).
73 George Charles Brodrick (1831–1903), Warden of Merton.
74 François Guizot, *General History of Civilization in Europe* (1828).

Page f.32v

[...]
Some model of rocks in white plaster
[Altaussee] salt mine.

Page f.35v

Page – la Plata
Southey tale of Paraguay
Lettres Edifiantes (not much) VIII 177 Jesuits commenced their missions 1608—expelled 1768
Charlevoix seems the authority. 1756. Paris[75]
Chrétineau-Joly Histoire de la Compagnie de Jesus[76]
Ulloa Don Antonio
Tableau des Peuplades formées par les Jesuites Azara[77] (he discusses with animosity to the fathers)
Dobrizhoffer de Abiponibus (Azara thought it a compilation)[78] (Jesuit's pageants included wild beasts of all kinds)
Charlevoix
D'Arbiguy
Voltaire (where?)
Robertson letters
Sir Woodbine Parish[79] library
N. Duran. 1658. Relation de insignes progrez de la religion chrestienne[80]

Page f.36v

5 Lincoln Asylum 1848
6. Littlemore Asylum 1849
7 Statist Maladies Mentales Hübertz 1853[81] He was a Danish commissioner to report on Danish madmen in 1847. They were nearly 4000. He groups the returns but the returns do not enter into detail.
Report on Cretinism to Carlo Alberto Turin 1841
a Goo! (Blue?) book (an Italian) nothing about
religion apparently[82]
Journal Statis: Soc: IV, 278 Tuke[83] is quoted to say that 1 Quaker in 300 is insane.
XIX Lunatics at Sierra Leone.[84] Religious excitement 729 patients in 11 years.
VII, 310. Thurnam. relative liability of the 2 sexes[85] He shows an error of Esquirol[86] who takes the

[75] Pierre François Xavier de Charlevoix (1682–1761), *Histoire du Paraguay* 4 vols, Paris: Ganeau & al. (1756).
[76] Jacques Crétineau-Joly, *Histoire de la Compagnie de Jesus* (1846–1848).
[77] Félix Manuel de Azara (1746–1821).
[78] Martin Dobrizhoffer (1717–1791), *Historia de Abiponibus, equestri bellicosaque Paraquariae natione, locupletata copiosis barbarorum gentium, urbium, fluminum, ferarum, amphibiorum, insectorum, serpentium praecipuorum, piscium, avium, arborum, plantarum aliarumque ejusdem provinciae proprietatum observationibus* (1784).
[79] See Register.
[80] Nicola Mastrilli (1568–1653) also known as Nicolas Duran, *Relation des insignes progrez de la religion chrestienne faits au Paraguay et dans les vastes régions de Guair et d'Uruaig nouvellement découvertes par les Pères de la Compagnie de Jésus, és années 1626 et 1627* (1638).
[81] Hübertz, J. R., *Statistique des maladies mentales en Danemark.* Paris: Imprimerie de L. Martinet, 1853.
[82] *Relazione della Commissione di S.M. il Re di Sardegna 'Studiare il cretinismo'* (Torino: Stamperia Reale, 1848). Jacob Marc d'Espine did the statistical analysis.
[83] Possibly James Hack Tuke, Treasurer of the Retreat Asylum near York.
[84] Robert Clarke, "Short Notes of the Prevailing Diseases in the Colony of Sierra Leone" *Journal of the Statistical Society of London*, 1856.
[85] John Thurnam, "On the Relative Liability of the Two Sexes to Insanity" *Journal of the Statistical Society of London* Vol. 7, No. 1 (Apr., 1844).
[86] Jean-Étienne Dominique Esquirol (1772–1840).

case, existing & not occurring.

Page f.37v

Abendberg retreat for Cretins — [Paff]
Causes of Idiocy Howe (a report to Governor of Massachusetts [sic]) Edinburgh 1858.[87]
page x "Nature outraged in the persons of the parents, exacts her penalty from the persons of their children". p VIII 574 idiots were examined.
p 6. quotes Séguin[88] the type of an idiot is one who "knows nothing, can do nothing, cannot even desire to do anything".
The book ascribes much to interference of parents but is vague & American.
Causes of Insanity Jarvis
A popular lecture in Massachusetts.[89]
Burrows in Insanity 1828.[90]
p. 28. Cooper "The morbid tendency of Cooper

Page f.37r

to mental derangement was always counteracted by the consolatory influence of true religion. In the first paroxysm of his disease prayer to God afforded a temporary solace to his distractions and later in life the sense of his obligations & duty to his creator assisted his uplifted hand in the act of suicide.["]
There is a deal about religion in reference to insanity — & cases are given but the writing seems unsound—too much theory.
Tracts into Vol. 1 Alienation mentale — not much
Tours asylum.
Tracts Vol. 6 Insanity & Lunatic Asylum
1 Esquirol. Illusion chez Aliénés[91]
2 Renaudin Aliénés Bas Rhin. Paris 1840.[92]
number of mad much the same in all religions p. 29.
3 Statistics of the Retreat New York 1841.[93]
4 Stafford lunatic asylum 1844[94]

Page f.38v

Aliénation Mentale Aubanel et Thore Paris 1841.[95]
27. Causes no mention of SN.
69. Madness hereditary from 1: 2.89 Esquirol to
1:15.30 in Palermo.
p 75 Causes. Dévotion Exaltée 5 in 87. there are 2 cases

---

87 S. G. Howe, *On the causes of idiocy: being the supplement to a report by Dr. S.G. Howe and the other commissioners appointed by the Governor of Massachusetts to inquire into the condition of the idiots of the Commonwealth, dated February 26, 1848.* Maclachlan & Stewart, 1858.

88 Édouard Séguin (1812–1880).

89 Edward Jarvis, "Causes of Insanity—An Address Delivered before the Norfolk (Mass.) District Medical Society, May 14th, 1851, and Communicated for the Boston Medical and Surgical Journal" *Boston Medical and Surgical Journal* Nov 12, 1851.

90 George Man Burrows (1771–1846), *Commentaries on the causes, forms, symptoms, and treatment, moral and medical, of insanity.* London, T. and G. Underwood, 1828.

91 Étienne Esquirol (1772–1840), *Des illusions chez les aliénés; Question médico-légale sur l'isolement des aliénés* Crochard, Paris, 1832.

92 Louis-François-Émile Renaudin (1808–1865), *Notice statistique sur les aliénés du departement du Bas-Rhin.* Paris, 1840.

93 *The Statistics of the Retreat, (near York, England,) consisting of a report and tables, exhibiting the experience of that Institution for the Insane, from its establishment in 1796 to 1840.* York, England, 1841.

94 Staffordshire County Lunatic Asylum.

95 *Recherches statistiques sur l'aliénation mentale faites à l'hospice de Bicêtre.* Par H. Aubanel et A. M. Thore. Paris, 1841.

pf 6. Amour contrarié & [...]. Misére of 18
& Chagrin domestiques of 26, all the rest are less in that year 1839, in the 7 years previous [...] was only 1 in 272.
Hood Statistics of insanity Bethlem Hospital from 1846. 1855[96]
p. 30. Religious persuasion of patients admitted as curable. "Scarcely anything is more difficult than to determine the real influence of religion on insanity". he quotes Pritchard (? who)[97] largely
The Prussian states in the Rhine appear the only locality where a comparison between Romanists & Protestants maybe fairly instituted. He takes Düsseldorf Aix la Chapelle, Treves, Cologne & Coblenz,

Page f.38r

and finds mad Romanists & mad Lutherans 11:10
A statistical paper by Dr. R. Hübertz on the state of mental diseases in Denmark (see (?in the) Journal Physiological Medicine July 1853 p 441) notes that the proportion of idiots & insane varies considerable with the sect viz. in Suez 1000 inhabitants Romanists 3.34, Jews 5.85 Calvinists 9.16 The paper itself throws doubt on these Bethlem returns [which] are "altogether valueless" on the extent of the influence of religion as affecting the chances of recovery in death" p. 34 "this we hope will not be the case in future ["] p 37. Married & unmarried. Bethlem results do not confirm Dr. Pritchard & Esquirol, which give an enormous preponderance to unmarried page 63. Perverted religion 37 in 435 cases among men 11 in 545 among women. These wd probably be yet higher if more was known of the history of the Bethlem patients.

Back Cover

July 1864[98]

96 William Charles Hood (1824–1870), *Statistics of insanity: being a decennial report of Bethlem Hospital, from 1846 to 1855 inclusive* (London: David Batten, 1856).

97 James Cowles Prichard, *A Treatise on Insanity and Other Disorders Affecting the Mind* (1837).

98 Most likely the date that the notebook was initiated.

Appendix B

Ethnological Inquiries on the Innate Character and Intelligence of Different Races

By Francis Galton, F.R.S. &c.[1]

The questions solely refer to individuals who have been reared since childhood in European or American schools, families, asylums or missionary establishments. By this restriction, it is hoped to eliminate all peculiarities that are due to the abiding influence of early education, and to the manners and customs of their own people.

The *standard* to be kept in mind in answering these questions, is the *average* Anglo-Saxon character; paying strict regard to the influence of sex, age, education, and social position. Where there is no *decided* divergence from this standard, it will be best to reply—'ordinary'.

1. Signature, title and full address of the sender of the information.
2. Name or initials, sex and age of individual whose character is described.
3. His (or her) country and race. State specifically if his race is known to be pure, if not describe the admixture.
4. Age at which he was removed from his parents and people, also particulars showing the extent to which he has since been separated from their influence.
5. What language, or languages, does he commonly speak? Does he retain the use of his native tongue?
6. State any circumstances that may or may not justify his being considered a good typical specimen of his race.
7. Is he capable of steady and sustained hard labour; or, is he restless and irregular in his habits?
8. Is he capable of filling responsible situations? Does he show coolness of temper when in difficulty? (It is said that Hindoos are incapable of steering large ships, that is, of acting as quarter-masters; while in British vessels that duty is commonly performed by native Christians of the Philippines.)
9. Is he docile or obstinate?
10. Children of many races are fully as quick, and even more precocious than European children, but they mostly cease to make progress after the season of manhood. Their moral character changes for the worse at the same time. State if this has been observed in the present instance.

[1] F. Galton 1864a. This circular attempts to gather data for a transracial adoption study design, of the kind reported in Scarr and Weinberg 1983. It was certainly distributed, for Charles Andersson notes receiving a copy and circulating it on 1865-05-31, see Tredoux 2023c, but no replies or tabulations have been discovered yet. They may languish in the very large pile of unsorted data for *Hereditary Genius* in GALTON/2/4/1. Most of the circular was republished in K. Pearson 1924, 352-3 with some errors. The transcription here is based on the original document.

11. Has he any special aptitudes, or the reverse, such as in mimicry, sense of the ludicrous, taste in colours, music, poetry, dancing, calculating power, keenness of sight or hearing, quickness of observation, manual dexterity, horsemanship, ability to tend cattle?
12. Is he naturally polished and self-composed in manner or rude and awkward?
13. Is he modest and self-reliant, or servile and cringing? Is he vain?
14. Is he solitary or sociable; morose or cheerful?
15. Is the passion of sexual affection strongly developed in him, or the reverse?
16. Is he fond of children, and are children fond of him?
17. Does he cherish malice for long periods, or does he forgive frankly?
18. Is he liable to outbursts of rage?
19. Did he for long show uneasiness at the restrictions of civilised life, or did he readily accept them; such as keeping regular hours, acting on a steady system, wearing shoes and other clothing ?
20. Children of savages, who have been reared in missionary families, have been known to throw off their clothes, and quit the house in a momentary rage, and to go back to their people, among whom they were afterwards found in apparently contented barbarism. State authentic instances of this, if you know of any, with full particulars.
21. Has he a strong natural sense of right and wrong, and a sensitive conscience?
22. Does he exhibit to his religious teachers any strong conviction of an original sinfulness in his nature, or the reverse?
23. Is he much influenced by ceremonial observances, such as those of the Roman Catholic Church?
24. Is he a willing keeper of the Sabbath?
25. Has he any strong religious instinct; is he inclined to quiet devotion?
26. Is he ascetic, self-mortifying and self-denying, or the contrary?
27. Is he inclined to be unduly credulous or unduly sceptical?
28. Is he active or impassive in social duties?
29. Is he much governed by superstitious feelings, such as charms or omens of good or ill luck?
30. Has he any tendency to be sanctimonious and hypocritical?
31. Is he honest, truthful and open, or cunning and intriguing?
32. Is he grateful or ungrateful?
33. Does he, in conversation, make frequent use of abstract terms? Does he adequately understand their meaning when he employs them?
34. Are there any other marked peculiarities in his character or intellect?

Please address replies to
FRANCIS GALTON,
42 Rutland Gate,
London.

Appendix C

Hereditary Talent and Character, 1865

C.1 Part I [June 1865]

HEREDITARY TALENT AND CHARACTER.

BY FRANCIS GALTON.

PART I.[1]

The power of man over animal life, in producing whatever varieties of form he pleases, is enormously great. It would seem as though the physical structure of future generations was almost as plastic as clay, under the control of the breeder's will. It is my desire to show, more pointedly than—so far as I am aware—has been attempted before, that mental qualities are equally under control.

A remarkable misapprehension appears to be current as to the fact of the transmission of talent by inheritance. It is commonly asserted that the children of eminent men are stupid; that, where great power of intellect seems to have been inherited, it has descended through the mother's side; and that one son commonly runs away with the talent of a whole family. My own inquiries have led me to a diametrically opposite conclusion. I find that talent is transmitted by inheritance in a very remarkable degree; that the mother has by no means the monopoly of its transmission; and that whole families of persons of talent are more common than those in which one member only is possessed of it. I justify my conclusions by the statistics I now proceed to adduce, which I believe are amply sufficient to command conviction. They are only a part of much material I have collected, for a future volume on this subject; all of which points in the same direction. I should be very grateful to any of my readers for information that may help me in my further inquiries.

In investigating the hereditary transmission of talent, we must ever bear in mind our ignorance of the laws which govern the inheritance even of physical features. We know to a certainty that the latter exist, though we do not thoroughly understand their action. The breeders of our domestic animals have discovered many rules by experience, and act upon them to a nicety. But we have not advanced, even to this limited extent, in respect to the human race. It has been nobody's business to study them; and the study is difficult, for many reasons. Thus, only two generations are likely to be born during the life of any observer; clothing conceals the shape; and each individual rarely marries more than once. Nevertheless, all analogy assures us that the physical features of man are equally transmissible with those of brutes. The resemblances between parent and offspring, as they appear to a casual observer, are just as close in one case as in the other; and, therefore, as a nearer scrutiny has established strict laws of hereditary transmission in brutes, we have every reason for believing that the same could also be discovered in the case of man.

[1] F. Galton 1865b.

So far as I am aware, no animals have ever been bred for general intelligence. Special aptitudes are thoroughly controlled by the breeder. He breeds dogs that point, retrieve, that fondle, or that bite; but no one has ever yet attempted to breed for high general intellect, irrespective of all other qualities. It would be a most interesting subject for an attempt. We hear constantly of prodigies of dogs, whose very intelligence makes them of little value as slaves. When they are wanted, they are apt to be absent on their own errands. They are too critical of their master's conduct. For instance, an intelligent dog shows marked contempt for an unsuccessful sportsman. He will follow nobody along a road that leads on a well-known tedious errand. He does not readily forgive a man who wounds his self-esteem. He is often a dexterous thief and a sad hypocrite. For these reasons an over-intelligent dog is not an object of particular desire, and therefore, I suppose, no one has ever thought of encouraging a breed of wise dogs. But it would be a most interesting occupation for a country philosopher to pick up the cleverest dogs he could hear of, and mate them together, generation after generation—breeding purely for intellectual power, and disregarding shape, size, and every other quality.

As no experiment of this description has ever been made, I cannot appeal to its success. I can only say that the general resemblances in mental qualities between parents and offspring, in man and brute, are every whit as near as the resemblance of their physical features; and I must leave the existence of actual laws in the former case to be a matter of inference from the analogy of the latter. Resemblance frequently fails where we might have expected it to hold; but we may fairly ascribe the failure to the influence of conditions that we do not yet comprehend. So long as we have a plenitude of evidence in favour of the hypothesis of the hereditary descent of talent, we need not be disconcerted when negative evidence is brought against us. We must reply that just the same argument might have been urged against the transmission of the physical features of our domestic animals; yet our breeders have discovered certain rules, and make their living by acting upon them. They know, with accurate prevision, when particular types of animals are mated together, what will be the character of the offspring. They can say that such and such qualities will be reproduced to a certainty. That others are doubtful; for they may appear in some of the descendants and not in the rest. Lastly, that there are yet other qualities, excessive in one parent and defective in the other, that will be counterbalanced and be transmitted to the offspring in a moderate proportion.

I maintain by analogy that this prevision could be equally attained in respect to the mental qualities, though I cannot prove it. All I can show is that talent and peculiarities of character are found in the children, when they have existed in either of the parents, to an extent beyond all question greater than in the children of ordinary persons. It is a fact, neither to be denied nor to be considered of importance, that the children of men of genius are frequently of mediocre intellect. The qualities of each individual are due to the combined influence of his two parents; and the remarkable qualities of the one may have been neutralized in the offspring, by the opposite or defective qualities of the other. It is natural that contrast of qualities, in the parents' dispositions, should occur as frequently as harmony; for one of the many foundations of friendship and of the marriage union is a difference of character; each individual seeking thereby to supplement the qualities in which he feels his own nature to be deficient. We have also good reason to believe that every special talent or character depends on a variety of obscure conditions, the analysis of which has never yet been seriously attempted. It is easy to conceive that the entire character might be considerably altered, owing to the modification of any one of these conditions.

As a first step in my investigation, I sought a biographical work, of manageable size, that should contain the lives of the chief men of genius whom the world is known to have produced. I ultimately selected that of Sir Thomas Phillips, in his well-known work of reference, 'The Million of Facts'[2] because it is compiled with evident discrimination, and without the slightest regard

[2] Sir Richard Phillips (1767–1840) *A Million of Facts* (London: Darton 1853) which went through several

to the question on which I was engaged. It is, moreover, prefaced,—'It has been attempted to record, in brief, only the ORIGINAL MINDS, who founded or originated. Biography in general is filled with mere imitators, or with men noted only for chance of birth, or necessary position in society'. I do not mean to say that Sir Thomas Phillips's selection is the best that could have been made, for he was a somewhat crotchety writer. It did not, however, much matter whose biography I adopted, so long as it had been written in the abovementioned spirit, and so long as I determined to abide steadfastly within its limits, without yielding to the temptation of supplying obvious omissions, in a way favourable to any provisional theory.

According to this select biography, I find that 605 notabilities lived between the years 1453 and 1853. And among these are no less than 102 relationships, or 1 in 6, according to the following list:- [see Figure C.1, p. 418 for the full table with totals]

- ☐ J. Adams, Pres. U.S.A.; son Samuel also patriot; nephew, J. Quincey, president.
- ☐ W. Belsham, historian; brother of T. Belsham, Unitarian minister.
- ☐ J. Bernoulli, father of James and uncle of John, all mathematicians.
- ☐ Breughel, father and two sons, painters.
- ☐ Buxtorff, father and son, Hebraists.
- ☐ Caracci [Carracci]. An. and Ag. brothers, Lud. cousin, painter.
- ☐ Cartwright, reformer; brother, mechanist.
- ☐ Cassini, grandfather, father, and son, all mathematicians.
- ☐ Cooper, Privy Councillor to Cromwell; grandson, literary.
- ☐ De Witt, two brothers, patriots.
- ☐ Elizabeth, Queen, daughter of Henry VIII. and granddaughter of Sir T. Bullen.
- ☐ Fontana, two brothers, natural philosophers.
- ☐ Forster, father and son, naturalists (Cook's voyages).
- ☐ Gronovius, sons and grandsons, six in all, learned critics.
- ☐ Gustavus Adolphus, father of Christina and grandson of Gustavus Vasa.
- ☐ Herschel, father and son, astronomers.
- ☐ Hunter, two brothers, anatomists.
- ☐ Jussieu, uncle and nephew, botanists.
- ☐ Medici, grandfather, father, and son, and Catherine.
- ☐ Orleans, Égalité, and son Louis Philippe.
- ☐ Ostade, two brothers, painters.
- ☐ Perrault, four brothers, all writers.
- ☐ Penn, admiral; son, Quaker writer.
- ☐ Phillibert, Prince of Orange; cousin William, whose son was Maurice. His grandson was our William III.
- ☐ Pitt, father and son, statesmen.
- ☐ Scaliger, classical critic; son also.
- ☐ Sforzas, father and son.
- ☐ Shaftesbury, statesman; grandson, author.
- ☐ Sheridan, father and son.
- ☐ Staël, Madam, daughter of Necker, financier.
- ☐ Stephens, family of six, critics and editors.
- ☐ Teniers, father and son, painters.
- ☐ Tytler, historian and poet; son, Lord Woodhouselee.
- ☐ Vandervelde, father and son, painters.

editions stretching back to 1832 at least. Phillips was a publisher, bookseller and author in his own right. 'His publications included vast numbers of elementary classbooks and cheap manuals, issued under a variety of pseudonyms. ... Several of these works have passed through from one hundred to five hundred editions.' See 'Phillips, Richard (1767–1840)' in Lee et al. 1901, Vol. 45, 210-1. Galton appears to have confused him with the portrait painter Thomas Phillips (1770–1845).

- ☐ Vanderwerf, two brothers, famous for small history.
- ☐ Valnoo, two brothers, and nephew, painters.[3]
- ☐ Walpole, Sir Robert, statesman; Sir Horace, author.
- ☐ Van Tromp, father and son, admirals.
- ☐ Villiers, statesman; grandson, the reprobate poet.
- ☐ Vossius, father, son, and other relatives, all writers.
- ☐ Warton, editor of Pope; son, poet.

It will be observed that the number is swelled by four large families, such as those of Gronovius and Stephens, of six members each, and of the Medici and the House of Orange, of four members each. The two first might be objected to, as hardly worthy of the distinguished place they occupy. But we must adhere to our biography; there are many more relationships that could very fairly have been added, as a set-off against these names. Such are two more Vanderveldes, and the family of Richelieu; besides others, like Hallam the historian, and Watt the mechanic, whose sons died early, full of the highest promise. Even if sixteen names were struck out of our list, the proportion of the relationship would remain as $\frac{86}{605}$, or 1 in 7. And these are almost wholly referable to transmission of talent through the male line; for eminent mothers do not find a place in mere biographical lists. The overwhelming force of a statistical fact like this renders counter-arguments of no substantial effect.

Next, let us examine a biographical list of much greater extension. I have selected for this purpose an excellent brief dictionary by Mr C. Hone.[4] It is not yet published, but part of its proof sheets have been obligingly lent to me. The entire work appears to contain some 19,000 names; it is, therefore, more than thirty times as extensive as the list we have hitherto been considering. I have selected one part only of this long series of names for examination, namely, those that begin with the letter M. There are 1141 names that remain under this letter, after eliminating those of sovereigns, and also of all persons who died before A.D. 1453. Out of these, 103, or 1 in 11, are either fathers and sons, or brothers; and I am by no means sure that I have succeeded in hunting out all the relationships that might be found to exist among them.

It will be remarked that the proportion of distinguished relationships becomes smaller, as we relax the restrictions of our selection; and it is reasonable that it should be so, for we then include in our lists the names of men who have been inducted into history through other conditions than the possession of eminent talent.

Again, if we examine into the relationships of the notabilities of the present day, we obtain even larger proportions. Walford's 'Men of the Time' contains an account of the distinguished men of England, the Continent, and America, who are now alive.[5] Under the letter A there are 85 names of men, and no less than 25 of these, or 1 in 3½, have relatives also in the list; 12 of them are brothers, and 11 fathers and sons.

- ☐ Abbott, Rev. Jacob (U.S.A.), author on religious and moral subjects.
- ☐ Abbott, Rev. John, younger *brother* of above, author on religious and moral subjects.
- ☐ A'Beckett, Sir William, author, Solicitor-Gen. of New South Wales, and *brother* of late Gilbert Abbott A'Beckett.
- ☐ Adam Jean Victor, painter, *son* of an eminent engraver.
- ☐ Adams, American minister, *son* of John Quincey Adams.
- ☐ Ainsworth, William Francis, editor of 'Journal of Natural and Geographical Science,' 'Explorations in Asia Minor and Kurdistan.'[6]

[3] Van Loo, Jean-Baptiste (1684–1745). The error is copied from Sir Richard Phillips' *A Million of Facts*.

[4] Rev. Charles Hole *A Brief Biographical Dictionary* (London: Macmillan, 1865). Reprinted with additions in 1866. Hole (1823–1906) was a Trinity College man, where he was admitted in 1842, taking a BA in 1846, as the 30th Wrangler. He was the Curate of Shanklin, Isle of Wight, from 1858 to 1868.

[5] By Edward Walford (1823–1897), see Walford 1865b. There were numerous other editions, but the first involving Walford seems to have been in 1862.

[6] Galton must mean W. F. Ainsworth *Travels and Researches in Asia Minor, Mesopotamia, Chaldea, and*

- ☐ Ainsworth, William Harrison, novelist, *cousin* of above.
- ☐ Aïvazooski, Gabriel, Armenian, born in the Crimea, Professor of European and Oriental languages, and member of Historical Institute of France.
- ☐ Aïvazooski, Ivan, a marine painter, *brother* of above.
- ☐ Albemarle, Earl of (*brother* Keppel).
- ☐ Albert, Prince (*brother*).
- ☐ Aldis, Sir Charles, medical.
- ☐ Aldis, Charles J.B. medical, *son* of above.
- ☐ Alexander, James Waddell, American divine (*son* of a Professor).
- ☐ Alexander, Joseph Addison, Professor of Ancient languages, and of Biblical and Ecclesiastical history, *brother* of the above.
- ☐ Alison, Sir Archibald, historian, *son* of author of 'Essays on Taste;' his mother belonged to 'a family which has for two centuries been eminent in mathematics and the exact sciences.'
- ☐ Ampère, member of French Academy, and Professor in College of France (literary), *son* of the celebrated physicist of the same name.
- ☐ Arago, Étienne, journalist and theatrical writer, *brother* of the celebrated philosopher.
- ☐ Argyropopulo [Argyropoulos], statesman, *son* of grand interpreter to the Porte.
- ☐ Aristarchi, ecclesiastic and statesman, *son* of grand interpreter to the Porte.
- ☐ Arnold, Matthew, *son* of late Dr Arnold.
- ☐ Arwidson [Adolf Ivar Arwidsson], Librarian R. Library, Stockholm, author, *son* of a person who held a high position in the Church.
- ☐ Ashburton, Lord, *son* of Rt Hon. Alexander Baring.
- ☐ Azeglio, Massimo, statesman and painter.
- ☐ Azeglio, Marquis, *nephew* of above, diplomatist and painter.

So if we examine the biographies of artists. In Bryan's large 'Dictionary of Painters,'[7] the letter A contains 391 names of men, of whom 65 are near relatives, or 1 in 6: 33 of them are fathers and sons, 30 are brothers. In Fétis 'Biographie Universelle des Musiciens'[8] the letter A contains 515 names, of which 50 are near relations, or 1 in 10. Two-third are fathers and sons, one-third are brothers.

It is justly to be urged, in limitation of the enormous effect of hereditary influence, implied by the above figures, that when a parent has achieved great eminence, his son will be placed in a more favourable position for advancement, than if he had been the son of an ordinary person. Social position is an especially important aid to success in statesmanship and generalship; for it is notorious that neither the Legislature nor the army afford, in their highest ranks, an open arena to the ablest intellects. The sons of the favoured classes are introduced early in life to both these fields of trial, with every encouragement to support them. Those of the lower classes are delayed and discouraged in their start; and when they are near the coveted goal, they find themselves aged. They are too late: they are not beaten by the superior merit of their contemporaries, but by time; as was once touchingly remarked by Sir De Lacy Evans.[9]

In order to test the value of hereditary influence with greater precision, we should therefore extract from our biographical list the names (they are 330) of those that have achieved distinction in the more open fields of science and literature. There is no favour here beyond the advantage of a good education. Whatever spur may be given by the desire to maintain the family fame, and whatever opportunities are afforded by abundant leisure, are more than

Armenia (1842).

7 Michael Bryan (1757–1821) *Dictionary of painters and engravers* (London: Bohn, 1849). There were numerous editions.

8 François-Joseph Fétis (1784–1871) *Biographie universelle des musiciens et bibliographie générale de la musique* (Paris, 1837). There were many other editions.

9 General Sir George de Lacy Evans (1787–1870), the Anglo-Irish soldier.

neutralised by those influences which commonly lead the heirs of fortune to idleness and dilettantism.

Recurring to our list, we find fifty-one literary men who have distinguished relations. Therefore, no less than $\frac{51}{605}$, or one distinguished man in every twelve, has a father, son, or brother, distinguished in literature. To take a round number at a venture, we may be sure that there have been far more than a million students educated in Europe during the last four centuries, being an average of only 2,500 in each a year. According to our list, about 330 of these, or only 1 in 3,000, achieved eminent distinction: yet of those who did so, 1 in 12 was related to a distinguished man. Keeping to literature alone, it is 51 to 330 = 1 to 6½, that a very distinguished literary man has a very distinguished literary relative, and it is (leaving out the Gronovius and Stephenses) 20 to 330 = 1 to 16, and 12 to 330 = 1 to 28, that the relationship is father and son, or brother and brother, respectively.

The Law is, by far, the most open to fair competition of all the professions; and of all offices in the law there is none that is more surely the reward of the most distinguished intellectual capacity than that of the Lord Chancellor. It therefore becomes an exceedingly interesting question to learn what have been the relationships of our Lord Chancellors. Are they to any notable degree the children, or the parents, or the brothers of very eminent men? Lord Campbell's 'Lives of the Chancellors'[10] forms a valuable biographical dictionary for the purpose of this investigation. I have taken it just as it stands; including, as Lord Campbell does, certain Lord Keepers and Commissioners of the Great Seal, as of equal rank with the Chancellors. I may further mention, that many expressions in Lord Campbell's works show that he was a disbeliever in hereditary influence.

Now what are the facts? Since Henry VIII's time, when Chancellors ceased to be ecclesiastics, and were capable of marrying, we have had thirty-nine Chancellors, &c. whose lives have been written by Lord Campbell, or whom the following had eminent relationships:—

- ☐ Sir Nicholas Bacon, Lord Keeper: son, Lord Chancellor Bacon.
- ☐ Coventry: son of a very learned judge of the Common Pleas.
- ☐ Bacon: father as above.
- ☐ Littleton: son of a judge.
- ☐ Whitelock: son of a judge, father of two sons, one of great eminence as a lawyer, the other as a soldier.
- ☐ Herbert: three sons. One had high command in army; the second, the great naval officer, created Lord Torrington; the third, Chief Justice of Queen's Bench.
- ☐ Finch, son of Speaker of House of Commons, and first cousin to the Lord Chancellor Finch of previous years, had a son who 'almost rivalled his father,' and who was made Solicitor-General and Earl of Aylesford.
- ☐ Macclesfield: son, President of Royal Society.
- ☐ Talbot: father was bishop, consecutively, of Oxford, Salisbury, and Durham; had sons, of one of whom there were great hopes, but he died young; the other 'succeeded to his father's virtues'.
- ☐ Hardwick[11] had five sons, all very distinguished. One, a man of letters; second, Lord Chancellor Yorke; third, an ambassador; fourth, 'talented as the others;' fifth, Bishop of Ely.
- ☐ Northington: father was 'one of the most accomplished men of his day.'
- ☐ Pratt: father was Chief Justice of King's Bench; his son was distinguished for public service.

[10] Baron John Campbell (1779–1861) *The lives of the Lord Chancellors and Keepers of the Great Seal of England from the earliest times till the reign of King George IV* (London: John Murray, 1857). 4th edition in 10 volumes.

[11] Philip Yorke, 1st Earl of Hardwicke.

- ☐ Yorke: father was Lord Chancellor Hardwicke. (*See* above.)
- ☐ Bathurst: father was the Lord Bathurst of Queen Anne's time; his son was the Lord Bathurst who filled high office under George III and IV.
- ☐ Erskine: his brothers were nearly as eminent. The whole family was most talented.
- ☐ Eldon: brother was the famous Lord Stowell, Judge of Admiralty.

Thus out of the 39 Chancellors 16 had kinsmen of eminence. 13 of them—viz. Sir Nicholas Bacon, Lord Bacon, Coventry, Littleton, Whitelock, Herbert, Finch, Hardwick, Pratt, Yorke, Bathurst, Erskine, and Eldon—had kinsmen of great eminence. In other words, 13 out of 39—that is, 1 in every 3—are remarkable instances of hereditary influence.

It is astonishing to remark the number of the Chancellors, who rose from mediocre social positions, showing how talent makes its way at the Bar, and how utterly insufficient are favouritism and special opportunities to win the great legal prize of the Chancellorship. It is not possible accurately, and it is hardly worth while roughly, to calculate the numerical value of hereditary influence in obtaining the Chancellorship. It is sufficient to say that it is enormous. We must not only reckon the number of students actually at the Chancery bar, and say that the Lord Chancellor was the foremost man among them, but we must reckon the immense number of schools in England, in any one of which, if a boy shows real marks of eminence, he is pretty sure to be patronised and passed on to a better place of education; whence by exhibitions, and subsequently by University scholarships and fellowships, he may become educated as a lawyer. I believe, from these reasons, that the chances of the son of a Lord Chancellor to be himself also a Chancellor, supposing he enters the law, to be more than a thousandfold greater than if he were the son of equally rich but otherwise undistinguished parents. It does not appear an accident that, out of 54 Lord Chancellors or Lord Keepers, two—viz. Sir Nicholas Bacon and Lord Hardwick—should have had sons who were also Chancellors, when we bear in mind the very eminent legal relationships of Herbert, Finch, Eldon, and the rest.

The intellectual force of English boys has, up to almost the present date, been steadily directed to classical education. Classics form the basis of instruction at our grammar schools, so that every boy who possesses signal classical aptitudes has a chance of showing them. Those who are successful obtain exhibitions and other help, and ultimately find their way to the great arena of competition of University life.

The senior classic at Cambridge is not only the foremost of the 300 youths who take their degrees in the same year, but he is the foremost of perhaps a tenth part of the classical intellect of his generation, throughout all England. No industry, without eminent natural talent to back it, could possibly raise a youth into that position.

The institution of the class list at Cambridge dates from 1824; so there have been 41 senior classics up to the present year. Wherever two names had been bracketed together, I selected the one that stood best in other examinations, and then extracted the following names from the list of them, as instances of hereditary influence:—

1827. Kennedy: father was a classic of eminence; two brothers, *see* below; another brother, almost equally distinguished in classics.

1828. Selwyn: brother M.P. for Cambridge, an eminent lawyer.

1830. Wordsworth: nephew to the poet, brother of an almost equally distinguished classic, son of the Master of Trinity.

1831. Kennedy (*see* above).

1832. Lushington: brother (*see* below); nephew to the Right Hon. Sir Stephen Lushington. The family has numerous other members of eminent talent.

1834. Kennedy (*see* above).
1835. Goulbourn: father, Chancellor of Exchequer, nephew of Serjeant Goulbourn, cousin to Dr Goulbourn, Head Master of Rugby, the well-known preacher.
1835. Vaughan: many relationships like those of Goulbourn, including the Judge, the Professor at Oxford, and Mr Hawkins (see below).
1842. Denman: father was the eminent Chief Justice Lord Denman.
1846. Lushington: brother (*see* above).
1854. Hawkins: *see* Vaughan.
1855. Butler: son of Senior Wrangler of 1794; three brothers, of whom two held University Scholarships in Oxford, and the other was a double first-class man at Cambridge.

12 of the 41, or about 1 in 3½, show these influences in a more or less marked degree; 7 of them, or 1 in 6, viz. 3 Kennedy, 1 Wordsworth, 2 Lushington, and 1 Butler, very much so.

The data we have been considering are summed up in the following table:—

[See Figure C.3, p. 420]

Everywhere is the enormous power of hereditary influence forced on our attention. If we take a list of the most brilliant standard writers of the last few years, we shall find a large share of the number have distinguished relationships. It would be difficult to set off, against the following instances, the same number of names of men of equal eminence, whose immediate relatives were undistinguished. Brontë (Jane Eyre and her two sisters); Bulwer (and his brother the ambassador); Disraeli (father, author of 'Curiosities of Literature'); Hallam (son, the subject of 'In Memoriam'); Kingsley (two brothers eminent novelists, two others no less talented); Lord Macaulay (son of Zachary Macaulay); Miss Martineau (and her brother); Merivale, Herman and Charles (brothers); Dean Stanley (father the bishop, and popular writer on birds); Thackeray (daughter, authoress of 'Elizabeth'); Tennyson (brother also a poet); Mrs. Trollope (son, Anthony).

As we cannot doubt that the transmission of talent is as much through the side of the mother as through that of the father, how vastly would the offspring be improved, supposing distinguished women to be commonly married to distinguished men, generation after generation, their qualities being in harmony and not in contrast, according to rules, of which we are now ignorant, but which a study of the subject would be sure to evolve!

It has been said by Bacon that 'great men have no continuance.' I, however, find that very great men are certainly not averse to the other sex, for some such have been noted for their illicit intercourses, and, I believe, for a corresponding amount of illegitimate issue. Great lawyers are especially to be blamed in this, even more than poets, artists, or great commanders. It seems natural to believe that a person who is not married, or who, if married, does not happen to have children, should feel himself more vacant to the attractions of a public or a literary career than if he had the domestic cares and interests of a family to attend to. Thus, if we take a list of the leaders in science of the present day, the small number of them who have families is very remarkable. Perhaps the best selection of names we can make, is from those who have filled the annual scientific office of President of the British Association. We will take the list of the commoners simply, lest it should be objected, though unjustly, that some of the noblemen who have occupied the chair were not wholly indebted to their scientific attainments for that high position. Out of twenty-two individuals, about one-third have children; one-third are or have been married and have no children; and one-third have never been married. Among the children of those who have had families, the names of Frank Buckland and Alexander Herschel are already well-known to the public.

There has been a popular belief that men of great intellectual eminence, are usually of feeble constitution, and of a dry and cold disposition. There may be such instances, but I believe the general rule to be exactly the opposite. Such men, so far as by observation and reading extend, are usually more manly and genial than the average, and by the aid of these very qualities, they obtain a recognised ascendancy. It is a great and common mistake to suppose that high intellectual powers are commonly associated with puny frames and small physical strength. Men of remarkable eminence are almost always men of vast powers of work. Those among them that have fallen into sedentary ways will frequently astonish their friends by their physical feats, when they happen to be in the mood of a vacation ramble. The Alpine Club contains a remarkable number of men of fair literary and scientific distinction; and these are among the strongest and most daring of the climbers. I believe, from my own recollections of the thews and energies of my contemporaries and friends of many years at Cambridge, that the first half-dozen class-men in classics or mathematics would have beaten, out of all proportion, the last half-dozen class-men in any trial of physical strength or endurance. Most notabilities have been great eaters and excellent digesters, on literally the same principle that the furnace which can raise more steam than is usual for one of its size must burn more freely and well than is common. Most great men are vigorous animals, with exuberant powers, and an extreme devotion to a cause. There is no reason to suppose that, in breeding for the highest order of intellect, we should produce a sterile or a feeble race.

Many forms of civilization have been peculiarly unfavourable to the hereditary transmission of rare talent. None of them were more prejudicial to it than that of the Middle Ages, where almost every youth of genius was attracted into the Church, and enrolled in the ranks of a celibate clergy.

Another great hindrance to it is a costly tone of society, like that of our own, where it becomes a folly for a rising man to encumber himself with domestic expenses, which custom exacts, and which are larger than his resources are able to meet. Here also genius is celibate, at least during the best period of manhood.

A spirit of caste is also bad, which compels a man of genius to select his wife from a narrow neighbourhood, or from the members of a few families.

But a spirit of clique is not bad. I understand that in Germany it is very much the custom for professors to marry the daughters of other professors, and I have some reason to believe, but am anxious for further information before I can feel sure of it, that the enormous intellectual digestion of German literary men, which far exceeds that of the corresponding class of our own countrymen, may, in some considerable degree, be traceable to this practice.

So far as beauty is concerned, the custom of many countries, of the nobility purchasing the handsomest girls they could find for their wives, has laid the foundation of a higher type of features among the ruling classes. It is not so very long ago in England that it was thought quite natural that the strongest lance at the tournament should win the fairest or the noblest lady. The lady was the prize to be tilted for. She rarely objected to the arrangement, because her vanity was gratified by the éclat of the proceeding. Now history is justly charged with a tendency to repeat itself. We may, therefore, reasonably look forward to the possibility, I do not venture to say the probability, of a recurrence of some such practice of competition. What an extraordinary effect might be produced on our race, if its object was to unite in marriage those who possessed the finest and most suitable natures, mental, moral, and physical!

Let us, then, give reins to our fancy, and imagine a Utopia—or a Laputa, if you will—in which a system of competitive examination for girls, as well as for youths, had been so developed as to embrace every important quality of mind and body, and where a considerable sum was yearly allotted to the endowment of such marriages as promised to yield children who would grow into eminent servants of the State. We may picture to ourselves an annual ceremony in that Utopia or Laputa, in which the Senior Trustee of the Endowment Fund would address ten deeply-blushing young men, all of twenty-five years old, in the following terms:—

> "Gentlemen, I have to announce the results of a public examination, conducted on established principles; which show that you occupy the foremost places in your year, in respect to those qualities of talent, character, and bodily vigour which are proved, on the whole, to do most honour and best service to our race. An examination has also been conducted on established principles among all the young ladies of this country who are now of the age of twenty-one, and I need hardly remind you, that this examination takes note of grace, beauty, health, good temper, accomplished housewifery, and disengaged affections, in addition to noble qualities of heart and brain. By a careful investigation of the marks you have severally obtained, and a comparison of them, always on established principles, with those obtained by the most distinguished among the young ladies, we have been enabled to select ten of their names with especial reference to your individual qualities. It appears that marriages between you and these ten ladies, according to the list I hold in my hand, would offer the probability of unusual happiness to yourselves, and, what is of paramount interest to the State, would probably result in an extraordinarily talented issue. Under these circumstances, if any or all of these marriages should be agreed upon, the Sovereign herself will give away the brides, at a high and solemn festival, six months hence, in Westminster Abbey. We, on our part, are prepared, in each case, to assign 5000*l.* as a wedding-present, and to defray the cost of maintaining and educating your children, out of the ample funds entrusted to our disposal by the State."

If a twentieth part of the cost and pains were spent in measures for the improvement of the human race that is spent on the improvement of the breed of horses and cattle, what a galaxy of genius might we not create! We might introduce prophets and high priests of civilization into the world, as surely as we can propagate idiots by mating *crétins.* Men and women of the present day are, to those we might hope to bring into existence, what the pariah dogs of the streets of an Eastern town are to our own highly-bred varieties.

The feeble nations of the world are necessarily giving way before the nobler varieties of mankind; and even the best of these, so far as we know them, seem unequal to their work. The average culture of mankind is become so much higher than it was, and the branches of knowledge and history so various and extended, that few are capable even of comprehending the exigencies of our modern civilization; much less of fulfilling them. We are living in a sort of intellectual anarchy, for the want of master minds. The general intellectual capacity of our leaders requires to be raised, and also to be differentiated. We want abler commanders, statesmen, thinkers, inventors, and artists. The natural qualifications of our race are no greater than they used to be in semi-barbarous times, though the conditions amid which we are born are vastly more complex than of old. The foremost minds of the present day seem to stagger and halt under an intellectual load too heavy for their powers.

To be continued.

C.2 Part II [August 1865]

SECOND PAPER.[12]

[12] F. Galton 1865c. [F.G.] Note.—I take this opportunity of correcting a small *erratum* in my last paper. The name of the author of the forthcoming Brief Biographical Dictionary from which I quoted should have been the Rev. Charles Hole, not Hone.

I have shown, in my previous paper, that intellectual capacity is so largely transmitted by descent that, out of every hundred sons of men distinguished in the open professions, no less than eight are found to have rivalled their fathers in eminence. It must be recollected that success of this kind implies the simultaneous inheritance of many points of character, in addition to mere intellectual capacity. A man must inherit good health, a love of mental work, a strong purpose, and considerable ambition, in order to achieve successes of the high order of which we are speaking. The deficiency of any one of these qualities would certainly be injurious, and probably be fatal to his chance of obtaining great distinction. But more than this: the proportion we have arrived at takes no account whatever of one-half of the hereditary influences that form the nature of the child. My particular method of inquiry did not admit of regard being paid to the influences transmitted by the mother, whether they had strengthened or weakened those transmitted by the father. Lastly, though the talent and character of both of the parents might, in any particular case, be of a remarkably noble order, and thoroughly congenial, yet they would necessarily have such mongrel antecedents that it would be absurd to expect their children to invariably equal them in their natural endowments. The law of atavism prevents it. When we estimate at its true importance this accumulation of impediments in the way of the son of a distinguished father rivalling his parent—the mother being selected, as it were, at haphazard—we cannot but feel amazed at the number of instances in which a successful rivalship has occurred. Eight per cent is as large a proportion as could have been expected on the most stringent hypothesis of hereditary transmission. No one, I think, can doubt, from the facts and analogies I have brought forward, that, if talented men were mated with talented women, of the same mental and physical characters as themselves, generation after generation, we might produce a highly-bred human race, with no more tendency to revert to meaner ancestral types than is shown by our long-established breeds of race-horses and fox-hounds.

It may be said that, even granting the validity of my arguments, it would be impossible to carry their indications into practical effect. For instance, if we divided the rising generation into two castes, A and B, of which A was selected for natural gifts, and B was the refuse, then, supposing marriage was confined within the pale of the caste to which each individual belonged, it might be objected that we should simply differentiate our race—that we should create a good and a bad caste, but we should not improve the race as a whole. I reply that this is by no means the necessary result. There remains another very important law to be brought into play. Any agency, however indirect, that would somewhat hasten the marriages in caste A, and retard those in caste B, would result in a larger proportion of children being born to A than to B, and would end by wholly eliminating B, and replacing it by A.

Let us take a definite case, in order to give precision to our ideas. We will suppose the population to be, in the first instance, stationary; A and B to be equal in numbers; and the children of each married pair who survive to maturity to be rather more than 2½ in the case of A, and rather less than 1½ in the case of B. This no extravagant hypothesis. Half the population of the British Isles are born of mothers under the age of thirty years.

The result in the first generation would be that the total population would be unchanged, but that only one-third part of it would consist of the children of B. In the second generation, the descendants of B would be reduced to two-ninths of their original numbers, but the total population would begin to increase, owing to the greater preponderance of the prolific caste A. At this point the law of natural selection would powerfully assist in the substitution of caste A for caste B, by pressing heavily on the minority of weakly and incapable men.

The customs that affect the direction and date of marriages are already numerous. In many families, marriages between cousins are discouraged and checked. Marriages, in other respects appropriate, are very commonly deferred, through prudential considerations. If it was generally felt that intermarriages between A and B were as unadvisable as they are supposed to be between cousins, and that marriages in A ought to be hastened, on the ground of prudential considerations, while those in B ought to be discouraged and retarded, then, I believe, we should have agencies amply sufficient to eliminate B in a few generations.

I hence conclude that the improvement of the breed of mankind is no insuperable difficulty. If everybody were to agree on the improvement of the race of man being a matter of the very utmost importance, and if the theory of the hereditary transmission of qualities in men was as thoroughly understood as it is in the case of our domestic animals, I see no absurdity in supposing that, in some way or other, the improvement would be carried into effect.

It remains for me in the present article to show that hereditary influence is as clearly marked in mental aptitudes as in general intellectual power. I will then enter into some of the considerations which my views on hereditary talent and character naturally suggest.

I will first quote a few of those cases in which characteristics have been inherited that clearly depend on peculiarities of organization. Prosper Lucas was among our earliest encyclopædists on this subject.[13] It is distinctly shown by him, and agreed to by others, such as Mr. G. Lewes,[14] that predisposition to any form of disease, or any malformation, may become an inheritance. Thus disease of the heart is hereditary; so are tubercles in the lungs; so also are diseases of the brain, of the liver, and of the kidney; so are diseases of the eye and of the ear. General maladies are equally inheritable, as gout and madness. Longevity on the one hand, and premature deaths on the other, go by descent. If we consider a class of peculiarities, more recondite in their origin than these, we shall still find the law of inheritance to hold good. A morbid susceptibility to contagious disease, or to the poisonous effects of opium, or of calomel, and an aversion to the taste of meat, are all found to be inherited. So is a craving for drink, or for gambling, strong sexual passion, a proclivity to pauperism, to crimes of violence, and to crimes of fraud.

There are certain marked types of character, justly associated with marked types of feature and of temperament. We hold, axiomatically, that the latter are inherited (the case being too notorious, and too consistent with the analogy afforded by brute animals, to render argument necessary), and we therefore infer the same of the former. For instance, the face of the combatant is square, coarse, and heavily jawed. It differs from that of the ascetic, the voluptuary, the dreamer, and the charlatan.

Still more strongly marked than these, are the typical features and characters of different races of men. The Mongolians, Jews, Negroes, Gipsies, and American Indians; severally propagate their kinds; and each kind differs in character and intellect, as well as in colour and shape, from the other four. They, and a vast number of other races, form a class of instances worthy of close investigation, in which peculiarities of character are invariably transmitted from the parents to the offspring.

In founding argument on the innate character of different races, it is necessary to bear in mind the exceeding docility of man. His mental habits in mature life are the creatures of social discipline, as well as of inborn aptitudes, and it is impossible to ascertain what is due to the latter alone, except by observing several individuals of the same race, reared under various influences, and noting the peculiarities of character that invariably assert themselves. But, even when we have imposed these restrictions to check a hasty and imaginative conclusion, we find there remain abundant data to prove an astonishing diversity in the natural characteristics of different races. It will be sufficient for our purpose if we fix our attention upon the peculiarities of one or two of them.

The race of the American Indians is spread over an enormous area, and through every

[13] Prosper Lucas *Traité philosophique et physiologique de l'hérédité naturelle* (1847).

[14] G. H. Lewes (unsigned) 'Hereditary Influence, Animal and Human'. *Westminster Review* 66 (July 1856): 75–90. See *Wellesley Index to Victorian Periodicals: 1824–1900* Vol. 3: 625, for confirmation of the attribution to Lewes.

climate; for it reaches from the frozen regions of the North, through the equator down to the inclement regions of the South. It exists in thousands of disconnected communities, speaking nearly as many different languages. It has been subjected to a strange variety of political influences, such as its own despotisms in Peru, Mexico, Natchez, and Bogotá, and its numerous republics, large and small. Members of the race have been conquered and ruled by military adventures from Spain and Portugal; others have been subjugated to Jesuitical rule; numerous settlements have been made by strangers on its soil; and, finally, the north of the continent has been colonized by European races. Excellent observers have watched the American Indians under all these influences, and their almost unanimous conclusion is as follows:—

The race is divided into many varieties, but it has fundamentally the same character throughout the whole of America. The men, and in a less degree the women, are naturally cold, melancholic, patient, and taciturn. A father, mother, and their children, are said to live together in a hut, like persons assembled by accident, not tied by affection. The youths treat their parents with neglect, and often with such harshness and insolence as to horrify Europeans who have witnessed their conduct. The mothers have been seen to commit infanticide without the slightest discomposure, and numerous savage tribes have died out in consequence of this practice. The American Indians are eminently non-gregarious. They nourish a sullen reserve, and show little sympathy with each other, even when in great distress. The Spaniards had to enforce the common duties of humanity by positive laws. They are strangely taciturn. When not engaged in action they will sit whole days in one posture without opening their lips, and wrapped up in their narrow thoughts. They usually march in Indian file, that is to say, in a long line, at some distance from each other, without exchanging a word. They keep the same profound silence in rowing a canoe, unless they happen to be excited by some extraneous cause. On the other hand, their patriotism and local attachments are strong, and they have an astonishing sense of personal dignity. The nature of the American Indians appears to contain the minimum of affectionate and social qualities compatible with the continuance of their race.

Here, then, is a well-marked type of character, that formerly prevailed over a large part of the globe, with which other equally marked types of character in other regions are strongly contrasted. Take, for instance, the typical West African Negro. He is more unlike the Red man in his mind than in his body. Their characters are almost opposite, one to the other. The Red man has great patience, great reticence, great dignity, and no passion; the Negro has strong impulsive passions, and neither patience, reticence, nor dignity. He is warm-hearted, loving towards his master's children, and idolised by the children in return. He is eminently gregarious, for he is always jabbering, quarrelling, tom-tom-ing, or dancing. He is remarkably domestic, and he is endowed with such constitutional vigour, and is so prolific, that his race is irrepressible.

The Hindu, the Arab, the Mongol, the Teuton, and very many more, have each of them their peculiar characters. We have not space to analyse them on this occasion; but, whatever they are, they are transmitted, generation after generation, as truly as their physical forms.

What is true for the entire race is equally true for its varieties. If we were to select persons who were born with a type of character that we desired to intensify,—suppose it was one that approached to some ideal standard of perfection—and if we compelled marriage within the limits of the society so selected, generation after generation; there can be no doubt that the offspring would ultimately be born with the qualities we sought, as surely as if we had been breeding for physical features, and not for intellect or disposition.

Our natural constitution seems to bear as direct and stringent a relation to that of our forefathers as any other physical effect does to its cause. Our bodies, minds, and capabilities of development have been derived from them. Everything we possess at our birth is a heritage from our ancestors.

Can we hand anything down to our children, that we have fairly won by our own independent exertions? Will our children be born with more virtuous dispositions, if we ourselves have acquired virtuous habits? Or are we no more than passive transmitters of a nature we have received, and which we have no power to modify? There are but a few instances in which habit even[15] seems to be inherited. The chief among them are such as those of dogs being born excellent pointers; of the attachment to man shown by dogs; and of the fear of man, rapidly learnt and established among the birds of newly-discovered islands. But all of these admit of being accounted for on other grounds than the hereditary transmission of habits. Pointing is, in some faint degree, a natural disposition of all dogs. Breeders have gradually improved upon it, and created the race we now possess. There is nothing to show that the reason why dogs are born staunch pointers is that their parents had been broken into acquiring an artificial habit. So as regards the fondness of dogs for man. It is inherent to a great extent in the genus. The dingo, or wild dog of Australia, is attached to the man who has caught him when a puppy, and clings to him even although he is turned adrift to hunt for his own living. This quality in dogs is made more intense by the custom of selection. The savage dogs are lost or killed; the tame ones are kept and bred from. Lastly, as regards the birds. As soon as any of their flock has learned to fear, I presume that its frightened movements on the approach of man form a language that is rapidly and unerringly understood by the rest, old or young; and that, after a few repetitions of the signal, man becomes an object of well-remembered mistrust. Moreover, just as natural selection has been shown to encourage love of man in domestic dogs, so it tends to encourage fear of man in all wild animals—the tamer varieties perishing owing to their misplaced confidence, and the wilder ones continuing their breed.

If we examine the question from the opposite side, a list of life-long habits in the parents might be adduced which leave no perceptible trace on their descendants. I cannot ascertain that the son of an old soldier learns his drill more quickly than the son of an artizan. I am assured that the sons of fishermen, whose ancestors have pursued the same calling time out of mind, are just as sea-sick as the sons of landsmen when they first go to sea. I cannot discover that the castes of India show signs of being naturally endowed with special aptitudes. If the habits of an individual are transmitted to his descendants, it is, as Darwin says, in a very small degree, and is hardly, if at all, traceable.

We shall therefore take an approximately correct view of the origin of our life, if we consider our own embryos to have sprung immediately from those embryos whence our parents were developed, and these from the embryos of *their* parents, and so on for ever. We should in this way look on the nature of mankind, and perhaps on that of the whole animated creation, as one continuous system, ever pushing out new branches in all directions, that variously interlace, and that bud into separate lives at every point of interlacement.

This simile does not at all express the popular notion of life. Most persons seem to have a vague idea that a new element, specially fashioned in heaven, and not transmitted by simple descent, is introduced into the body of every newly-born infant. Such a notion is unfitted to stand upon any scientific basis with which we are acquainted. It is impossible it should be true, unless there exists some property or quality in man that is not transmissible by descent. But the terms *talent* and *character* are exhaustive: they include the whole of man's spiritual nature so far as we are able to understand it. No other class of qualities is known to exist, that we might suppose to have been interpolated from on high. Moreover, the idea is improbable from *à priori* considerations, because there is no other instance in which creative power operates under our own observation at the present day, except it may be in the freedom in action of our own wills. Wherever else we turn our eyes, we see nothing but law and order, and effect following cause.

But though, when we look back to our ancestors, the embryos of our progenitors may be conceived to have been developed, in each generation, immediately from the one that preceded

[15] Likely a typo for 'ever'.

it, yet we cannot take so restricted a view when we look forward. The interval that separates the full-grown animal from its embryo is too important to be disregarded. It is in this interval that Darwin's law of natural selection comes into play; and those conditions are entered into, which affect, we know not how, the 'individual variation' of the offspring. I mean those that cause dissimilarity among brothers and sisters who are born successively, while twins, produced simultaneously, are often almost identical. If it were possible that embryos should descend directly from embryos, there might be developments in every direction, and the world would be filled with monstrosities. But this is not the order of nature. It is her fiat that the natural tendencies of animals should never disaccord long and widely with the conditions under which they are placed. Every animal before it is of an age to bear offspring, has to undergo frequent stern examinations before the board of nature, under the laws of natural selection; where to be 'plucked' is not necessarily disgrace, but is certainly death. Never let it be forgotten that man, as a reasonable being, has the privilege of not being helpless under the tyranny of uncongenial requirements, but that he can, and that he does, modify the subjects in which nature examines him, and that he has considerable power in settling beforehand the relative importance in the examination that shall be assigned to each separate subject.

It becomes a question of great interest how far moral monstrosities admit of being bred. Is there any obvious law that assigns a limit to the propagation of supremely vicious or supremely virtuous natures? In strength, agility, and other physical qualities, Darwin's law of natural selection acts with unimpassioned, merciless severity. The weakly die in the battle for life; the stronger and more capable individuals are alone permitted to survive, and to bequeath their constitutional vigour to future generations. Is there any corresponding rule in respect to moral character? I believe there is, and I have already hinted at it when speaking of the American Indians. I am prepared to maintain that its action, by insuring a certain fundamental unity in the quality of the affections, enables men and the higher order of animals to sympathise in some degree with each other, and also, that this law forms the broad basis of our religious sentiments.

Animal life, in all but the very lowest classes, depends on at least one, and, more commonly, on all of the four following principles:—There must be affection, and it must be of four kinds: sexual, parental, filial, and social. The absolute deficiency of any one of these would be a serious hindrance, if not a bar to the continuance of any race. Those who possessed all of them, in the strongest measure, would, speaking generally, have an advantage in the struggle for existence. Without sexual affection, there would be no marriages, and no children; without parental affection, the children would be abandoned; without filial affection, they would stray and perish; and, without the social, each individual would be single-handed against rivals who were capable of banding themselves into tribes. Affection for others as well as self, is therefore a necessary part of animal character. Disinterestedness is as essential to a brute's well-being as selfishness. No animal lives for itself alone, but also, at least occasionally, for its parent, its mate, its offspring, or its fellow. Companionship is frequently more grateful to an animal than abundant food. The safety of her young is considered by many a mother as a paramount object to her own. The passion for a mate is equally strong. The gregarious bird posts itself during its turn of duty as watchman on a tree, by the side of the feeding flock. Its zeal to serve the common cause exceeds its care to attend to its own interests. Extreme selfishness is not a common vice. Narrow thoughts of self by no means absorb the minds of ordinary men; they occupy a secondary position in the thoughts of the more noble and generous of our race. A large part of an Englishman's life is devoted to others, or to the furtherance of general ideas, and not to directly personal ends. The Jesuit toils for his order, not for himself. Many plan for that which they can never live to see. At the hour of death they are still planning. An incompleted will, which might work unfairness among those who would succeed to the property of a dying man, harasses his mind. Personal obligations of all sorts press as heavily as in the fullness of health, although the touch of death is known to be on the point of cancelling them. It is so with animals. A dog's thoughts are towards his master, even when he suffers the extremest pain. His mind is largely filled at all times with sentiments of affection. But disinterested feelings are more necessary to man than to any other animal, because of the long period of his dependent childhood, and also because of his great social needs, due to his physical helplessness. Darwin's law of natural selection would therefore be expected to develop these sentiments among men, even among the lowest barbarians, to a greater degree than among animals.

I believe that our religious sentiments spring primarily from these four sources. The institution of celibacy is an open acknowledgment that the theistic and human affections are more or less convertible; I mean that by starving the one class the other becomes more intense and absorbing. In savages, the theistic sentiment is chiefly, if not wholly, absent. I would refer my readers, who may hesitate in accepting this assertion, to the recently published work of my friend Sir John Lubbock, 'Prehistoric Times,'[16] p. 467-472, where the reports of travellers on the religion of savages are very ably and fairly collated. The theistic sentiment is secondary, not primary. It becomes developed within us under the influence of reflection and reason. All evidence tends to show that man is directed to the contemplation and love of God by instincts that he shares with the whole animal world, and that primarily appeal to the love of his neighbour.

Moral monsters are born among Englishmen, even at the present day; and, when they are betrayed by their acts, the law puts them out of the way, by the prison or the gallows, and so prevents them from continuing their breed. Townley, the murderer, is an instance in point.[17] He behaved with decorum and propriety; he was perfectly well-conducted to the gaol officials, and he corresponded with his mother in a style that was certainly flippant, but was not generally considered to be insane. However, with all this reasonableness of disposition, he could not be brought to see that he had done anything particularly wrong in murdering the girl that was disinclined to marry him. He was thoroughly consistent in his disregard for life, because, when his own existence became wearisome, he ended it with perfect coolness, by jumping from an upper staircase. It is a notable fact that a man without a conscience, like Townley, should be able to mix in English society for years, just like other people.

How enormous is the compass of the scale of human character, which reaches from dispositions like those we have just described, to that of a Socrates! How various are the intermediate types of character that commonly fall under everybody's notice, and how differently are the principles of virtue measured out to different natures! We can clearly observe the extreme diversity of character in children. Some are naturally generous and open, others mean and tricky; some are warm and loving, others cold and heartless; some are meek and patient, others obstinate and self-asserting; some few have the tempers of angels, and at least as many have the tempers of devils. In the same way, as I showed in my previous paper, that by selecting men and women of rare and similar talent, and mating them together, generation after generation, an extraordinarily gifted race might be developed, so a yet more rigid selection, having regard to their moral nature, would, I believe, result in a no less marked improvement of their natural disposition.

Let us consider an instance in which different social influences have modified the inborn dispositions of a nation. The North American people has been bred from the most restless and combative class of Europe. Whenever, during the last ten or twelve generations, a political or religious party has suffered defeat, its prominent members, whether they were the best, or only the noisiest, have been apt to emigrate to America, as a refuge from persecution. Men fled to America for conscience' sake, and for that of unappreciated patriotism. Every scheming knave, and every brutal ruffian, who feared the arm of the law, also turned his eyes in the

[16] Lubbock 1865.

[17] George Victor Townley (1838–1865) stabbed his former fiancee, Elizabeth Caroline Goodwin, to death in 1863.

same direction. Peasants and artisans, whose spirit rebelled against the tyranny of society and the monotony of their daily life, and men of a higher position, who chafed under conventional restraints, all yearned towards America. Thus the dispositions of the parents of the American people have been exceedingly varied, and usually extreme, either for good or for evil. But in one respect they almost universally agreed. Every head of an emigrant family brought with him a restless character, and a spirit apt to rebel. If we estimate the moral nature of Americans from their present social state, we shall find it to be just what we might have expected from such a parentage. They are enterprising, defiant, and touchy; impatient of authority; furious politicians; very tolerant of fraud and violence; possessing much high and generous spirit, and some true religious feeling, but strongly addicted to cant.

We have seen that the law of natural selection develops disinterested affection of a varied character even in animals and barbarian man. Is the same law different in its requirements when acting on civilized man? It is no doubt more favourable on the whole to civilized progress, but we must not expect to find as yet many marked signs of its action. As a matter of history, our Anglo-Saxon civilization is only skin-deep. It is but eight hundred years, or twenty-six generations, since the Conquest, and the ancestors of the large majority of Englishmen were the merest boors at a much later date than that. It is said that among the heads of the noble houses of England there can barely be found one that has a right to claim the sixteen quarterings—that is to say, whose great-great-grandparents were, all of them (sixteen in number), entitled to carry arms. Generally the nobility of a family is represented by only a few slender rills among a multiplicity of non-noble sources.

The most notable quality that the requirements of civilization have hitherto bred in us, living as we do in a rigorous climate and on a naturally barren soil, is the instinct of continuous steady labour. This is alone possessed by civilized races, and it is possessed in a far greater degree by the feeblest individuals among them than by the most able-bodied savages. Unless a man can work hard and regularly in England, he becomes an outcast. If he only works by fits and starts he has not a chance of competition with steady workmen. An artizan who has variable impulses, and wayward moods, is almost sure to end in intemperance and ruin. In short, men who are born with wild and irregular dispositions, even though they contain much that is truly noble, are alien to the spirit of a civilized country, and they and their breed are eliminated from it by the law of selection. On the other hand, a wild, untameable restlessness is innate with savages. I have collected numerous instances where children of a low race have been separated at an early age from their parents, and reared as part of a settler's family, quite apart from their own people. Yet, after years of civilized ways, in some fit of passion, or under some craving, like that of a bird about to emigrate, they have abandoned their home, flung away their dress, and sought their countrymen in the bush, among whom they have subsequently been found living in contented barbarism, without a vestige of their gentle nurture. This is eminently the case with the Australians, and I have heard of many others in South Africa. There are also numerous instances in England where the restless nature of gipsy half-blood asserts itself with irresistible force.

Another difference, which may either be due to natural selection or to original difference of race, is the fact that savages seem incapable of progress after the first few years of their life. The average children of all races are much on a par. Occasionally, those of the lower races are more precocious than the Anglo-Saxons; as a brute beast of a few weeks old is certainly more apt and forward than a child of the same age. But, as the years go by, the higher races continue to progress, while the lower ones gradually stop. They remain children in mind, with the passions of grown men. Eminent genius commonly asserts itself in tender years, but it continues long to develop. The highest minds in the highest race seem to have been those who had the longest boyhood. It is not those who were little men in early youth who have succeeded. Here I may remark that, in the great mortality that besets the children of our poor, those who are members of precocious families, and who are therefore able to help in earning wages at a very early age, have a marked advantage over their competitors. They, on the whole, live, and breed their like, while the others die. But, if this sort of precocity be unfavourable to a race—if it be generally followed up by an early arrest of development, and by a premature old age—then modern industrial civilization, in encouraging precocious varieties of men, deteriorates the breed.

Besides these three points of difference—endurance of steady labour, tameness of disposition, and prolonged development—I know of none that very markedly distinguishes the nature of the lower classes of civilized man from that of barbarians. In the excitement of a pillaged town the English soldier is just as brutal as the savage. Gentle manners seem, under those circumstances, to have been a mere gloss thrown by education over a barbarous nature. One of the effects of civilization is to diminish the rigour of the application of the law of natural selection. It preserves weakly lives, that would have perished in barbarous lands. The sickly children of a wealthy family have a better chance of living and rearing offspring than the stalwart children of a poor one. As with the body, so with the mind. Poverty is more adverse to early marriages than is natural bad temper, or inferiority of intellect. In civilized society, money interposes her ægis between the law of natural selection and very many of its rightful victims. Scrofula and madness are naturalised among us by wealth; short-sightedness is becoming so. There seems no limit to the morbific tendencies of body or mind that might accumulate in a land where the law of primogeniture was general, and where riches were more esteemed than personal qualities. Neither is there any known limit to the intellectual and moral grandeur of nature that might be introduced into aristocratical families, if their representatives, who have such rare privilege in winning wives that please them best, should invariably, generation after generation, marry with a view of transmitting those noble qualities to their descendants. Inferior blood in the representative of a family might be eliminated from it in a few generations. The share that a man retains in the constitution of his remote descendants is inconceivably small. The father transmits, on an average, one-half of his nature, the grandfather one-fourth, the great-grandfather one-eighth; the share decreasing step by step, in a geometrical ratio, with great rapidity. Thus the man who claims descent from a Norman baron, who accompanied William the Conqueror twenty-six generations ago, has so minute a share of that baron's influence in his constitution, that, if he weighs fourteen stone, the part of him which may be ascribed to the baron (supposing, of course, there have been no additional lines of relationship) is only one-fiftieth of a grain in weight—an amount ludicrously disproportioned to the value popularly ascribed to ancient descent. As a stroke of policy, I question if the head of a great family, or a prince, would not give more strength to his position, by marrying a wife who would bear him talented sons than one who would merely bring him the support of high family connexions.

With the few but not insignificant exceptions we have specified above, we are still barbarians in our nature, and we show it in a thousand ways. The children who dabble and dig in the dirt have inherited the instincts of untold generations of barbarian forefathers, who dug with their nails for a large fraction of their lives. Our ancestors were grubbing by the hour, each day, to get at the roots they chiefly lived upon. They had to grub out pitfalls for their games, holes for their palisades and hut-poles, hiding-places, and ovens. Man became a digging animal by nature; and so we see the delicately-reared children of our era very ready to revert to primeval habits. Instinct breaks out in them, just as it does in the silk-haired, boudoir-nurtured spaniel, with a ribbon round its neck, that runs away from the endearments of its mistress, to sniff and revel in some road-side mess of carrion.

It is a common theme of moralists of many creeds, that man is born with an imperfect nature. He has lofty aspirations, but there is a weakness in his disposition that incapacitates him from carrying his nobler purposes into effect. He sees that some particular course of action is his duty, and should be his delight; but his inclinations are fickle and base, and do not conform to his better judgment. The whole moral nature of man is tainted with sin, which prevents him from doing the things he knows to be right.

I venture to offer an explanation of this apparent anomaly, which seems perfectly satisfactory from a scientific point of view. It is neither more nor less than that the development of our nature, under Darwin's law of natural selection, has not yet overtaken the development of our religious civilization. Man was barbarous but yesterday, and therefore it is not to be expected that the natural aptitudes of his race should already have become moulded into accordance with his very recent advance. We men of the present centuries are like animals suddenly transplanted among new conditions of climate and of food: our instincts fail us under the altered circumstances.

My theory is confirmed by the fact that the members of old civilizations are far less sensible than those newly converted from barbarism of their nature being inadequate to their moral needs. The conscience of a Negro is aghast at his own wild, impulsive nature, and is easily stirred by a preacher, but it is scarcely possible to ruffle the self-complacency of a steady-going Chinaman.

The sense of original sin would show, according to my theory, not that man was fallen from a high estate, but that he was rapidly rising from a low one. It would therefore confirm the conclusion that has been arrived at by every independent line of ethnological research—that our forefathers were utter savages from the beginning; and, that, after myriads of years of barbarism, our race has but very recently grown to be civilized and religious.

Art.	Lit. & science.	More distant.	Brothers.	Father & son.	Number.	NOTABLE PERSONS.
–	–	1	–	2	3	J. Adams, Pres. U.S.A.; son Samuel also patriot; nephew, J. Quincey, president.
–	2	–	2	–	2	W. Belsham, historian; brother of T. Belsham, Unitarian minister.
–	3	–	–	2	3	J. Bernouilli, father of James and uncle of John, all mathematicians.
3	–	–	–	2	3	Breughel, father and two sons, painters.
–	2	–	–	2	2	Buxtorff, father and son, Hebraists.
3	–	1	2	–	3	Caracci, An. and Ag. brothers, Lud. cousin, painter.
–	1	–	2	–	2	Cartwright, reformer; brother, mechanist.
–	3	–	–	3	3	Casini, grandfather, father, and son, all mathematicians.
–	1	2	–	–	2	Cooper, Privy Councillor to Cromwell; grandson, literary.
–	–	–	2	–	2	De Witt, two brothers, patriots.
–	–	–	–	3	3	Elizabeth, queen, daughter of Henry VIII. and granddaughter of Sir T. Bullen.
–	2	–	2	–	2	Fontana, two brothers, natural philosophers.
–	2	–	–	2	2	Forster, father and son, naturalists (Cook's voyages),
–	6	–	–	6	6	Gronovius, sons and grandsons, six in all, learned critics.
–	–	1	–	2	3	Gustavus Adolphus, father of Christina and grandson of Gustavus Vasa.
–	2	–	–	2	2	Herschel, father and son, astronomers.
–	2	–	2	–	2	Hunter, two brothers, anatomists.
–	2	2	–	–	2	Jussieu, uncle and nephew, botanists.
–	–	–	–	4	4	Medici, grandfather, father, and son, and Catherine.

Figure C.1: Notable persons in families, part 1.

–	–	–	–	2	2	Orleans, Egalité, and son Louis Phillippe.
–	–	–	2	–	2	Ostade, two brothers, painters.
–	4	–	4	–	4	Perrault, four brothers, all writers.
–	1	–	–	2	2	Penn, admiral ; son, Quaker writer.
–	–	2	–	2	4	Phillibert, Prince of Orange ; cousin William, whose son was Maurice. His grandson was our William III.
–	–	–	–	2	2	Pitt, father and son, statesmen.
–	2	–	–	2	2	Scaliger, classical critic ; son also.
–	–	–	–	2	2	Sforzas, father and son.
–	1	2	–	–	2	Shaftesbury, statesman ; grandson, author.
–	1	–	–	2	2	Sheridan, father and son.
–	1	–	–	2	2	Staël, Madam, daughter of Necker, financier.
–	6	–	–	6	6	Stephens, family of six, critics and editors.
2	–	–	–	2	2	Teniers, father and son, painters.
–	–	–	–	2	2	Tytler, historian and poet ; son, Lord Woodhouselee.
2	–	–	–	2	2	Vanderwelde, father and son, painters.
–	2	–	2	–	2	Vanderwurf, two brothers, famous for small history.
–	–	1	2	–	3	Valnoo, two brothers, and nephew, painters.
–	1	–	–	2	2	Walpole, Sir Robert, statesman ; Sir Horace, author.
–	–	–	–	2	2	Van Tromp, father and son, admirals.
–	1	2	–	–	2	Villiers, statesman ; grandson, the reprobate poet.
–	2	–	–	2	2	Vossius, father, son, and other relatives, all writers.
–	2	–	–	2	2	Warton, editor of Pope ; son, poet.
9	52	14	22	66	102	

Figure C.2: Notable persons in families, part 2.

Number of cases.		Occurrence of near male relationship.	Percentages.	
			Distinguished father has a distinguished son.	Distinguished man has a distinguished brother.
605	All the men of "original minds" (Sir T. Phillips) and of every profession between 1453 and 1853	1 in 6 cases.	6 times in 100 cases.	2 times in 100 cases.
85	Living notabilities (Walford's "Men of the Times," letter A)	1 in 3½ cases.	7 ,, ,,	7 ,, ,,
391	Painters of all dates (Bryan's Dicty. A)	1 in 6 cases.	5 ,, ,,	4 ,, ,,
515	Musicians (Fétis Dicty. A)	1 in 10 cases.	6 ,, ,,	3 ,, ,,
54	Lord Chancellors (Lord Campbell) .	1 in 3 cases.	16 ,, ,,	4 ,, ,,
41	Senior Classics of Cambridge . . .	1 in 4 cases.	Too recent	10 ,, ,,
	Averages	1 in 6 cases.	8 in 100 cases.	5 in 100 cases.

Figure C.3: Eminence runs in families.

Appendix D

Men of Science Respondents

Those marked * responded to the survey. Last name, First names; Birth–Death; Field; Year of accession to FRS where applicable; Occupation [Father's occupation].

1. Abel, Frederick A.; 1827–1902; Chemistry; FRS 1860.
2. Acland, Henry W.; 1815–1900; Medicine; FRS 1847.
3. Adams, James C.; 1819–1892; Astronomy; FRS 1849.
4. Adams, William G.; 1836–1915; Physics; FRS 1872.
5. Airy, George B.; 1801–1892; Astronomy; FRS 1836.
6. *Alderson, James; 1794–1882; Medicine; FRS 1841; Consulting Physician, St. Mary's Hospital, London [Physician].
7. *Allman, George James; 1812–1898; Biology; FRS 1854; Prof. of Natural History, Edinburgh.
8. *Andrews, Thomas; 1813–1885; Chemistry; FRS 1849; Prof. of Chemistry, Queen's College, Belfast [Linen Merchant].
9. Apjohn, James; 1796–1886; Geography; FRS 1853.
10. *Ansted, David Thomas; 1814–1880; Geology; FRS 1849; Prof. of Geology, King's College, London.
11. Argyll, Duke of; 1823–1900; Geology; FRS 1851.
12. *Armstrong, William; 1810–1900; Engineering; FRS 1846; Engineering entrepreneur [Merchant].
13. Arnott, James M.; 1794–1885; Medicine; FRS 1843.
14. *Babington, Charles; 1808–1895; Botany; FRS 1851; Prof. of Botany, Cambridge [Physician].
15. *Back, George; 1796–1878; Geography; FRS 1847; Admiral, Royal Navy.
16. *Balfour, Thomas Graham; 1813–1891; Medicine; FRS 1858; Deputy Inspector General, Army Medical Department [Merchant].
17. *Balfour, John Hutton; 1808–1884; Botany; FRS 1856; Prof. of Botany, Glasgow [Army Surgeon].
18. *Ball, John; 1818–1889; Botany; FRS 1868; MP for County Carlow, Ireland and Under-Secretary of State for the Colonies [Irish Judge].
19. *Bastian, Henry Charlton; 1837–1915; Medicine; FRS 1863; Prof. of Clinical Medicine, UCL.
20. *Bateman, John Frederick; 1810–1889; Engineering; FRS 1860; Civil (water-supply) engineer.
21. *Bateman, James; 1811–1897; Botany; FRS 1838; Iron and engineering entrepreneur and banker [Landowner].
22. Beale, Lionel S.; 1828–1906; Medicine; FRS 1857.
23. Beck, Thoms S.; 1814–1877; Medicine; FRS 1851.
24. Bell, Thomas; 1792–1880; Zoology; FRS 1828.

25. Bennett, John J.; 1801–1876; Mathematics; FRS 1841.
26. *Bentham, George; 1800–1884; Botany; FRS 1862; Independent means (heir to his uncle Jeremy Bentham's fortune) [Naval Architect].
27. Booth, James; 1806–1878; Mathematics; FRS 1846.
28. *Bowman, William; 1816–1892; Physiology; FRS 1841; Ophthalmologist in private practice and Prof. of Physiology, King's College [Banker].
29. Brodie, Benjamin C.; 1817–1880; Chemistry; FRS 1849.
30. *Brooke, Charles; 1804–1879; Medicine; FRS 1847; Surgeon, Metropolitan Free Hospital [Mineralogist].
31. *Buckton, George; 1818–1905; Zoology; FRS 1857; Research assistant at Royal College of Chemistry.
32. *Burdon-Sanderson, John; 1828–1905; Physiology; FRS 1867; Prof. of Physiology, UCL.
33. Burrows, George; 1801–1887; Medicine; FRS 1847.
34. Busk, George; 1807–1886; Biology; FRS 1850.
35. *Carpenter, William Benjamin; 1813–1885; Physiology; FRS 1844; Registrar, University of London.
36. *Carrington, Richard Christopher; 1826–1875; Astronomy; FRS 1860; Brewer [Brewer].
37. *Carter, Henry John; 1813–1895; Biology; FRS 1859; Surgeon-Major, Army of the East-India Company.
38. *Cayley, Arthur; 1821–1895; Mathematics; FRS 1852; Prof. of Math, Cambridge [Russia Merchant].
39. Challis, James; 1803–1882; Astronomy; FRS 1848.
40. Clarke, Jacob A.L.; 1817–1880; Medicine; FRS 1854.
41. Clifton, Robert B.; 1836–1921; Physics; FRS 1868.
42. *Cobbold, Thomas Spencer; 1828–1886; Biology; FRS 1864; Independent means, Prof. of Botany, Royal Veterinary College [Clergy].
43. *Currey, Frederick; 1819–1881; Biology; FRS 1858; Barrister, conveyancer and equity draughtsman [Parliamentary Clerk].
44. *Darwin, Charles Robert; 1809–1882; Biology; FRS 1839; Independent means [Physician].
45. Davidson, Thomas; 1817–1885; Biology; FRS 1857.
46. *Debus, Henry; 1824–1916; Chemistry; FRS 1861; Prof. of Chemistry, Royal Naval College.
47. *De La Rue, Warren; 1815–1889; Astronomy; FRS 1850; Stationer [Paper Manufacturer].
48. Devonshire, Duke of; 1808–1891; Biology; FRS 1829.
49. Duncan, Peter M.; 1821–1891; Biology; FRS 1868.
50. *Egerton, Philip Grey; 1806–1881; Biology; FRS 1831; M.P. and Baronet [Clergy].
51. Enniskillen, Earl of; 1807–1886; Zoology; FRS 1829.
52. *Evans, John; 1823–1908; Archaeology; FRS 1864; Partner in John Dickinson & Co., stationers [Teacher].
53. *Evans, Frederick John; 1815–1885; Geography; FRS 1862; Captain, Royal Navy [Master, Royal Navy].
54. *Fairbairn, William; 1789–1874; Engineering; FRS 1850; Ship and bridge builder, Fairbairn and Lillie Engine Makers [Farm Bailiff].
55. Farr, William; 1807–1883; Statistics; FRS 1855.
56. *Farre, Arthur; 1811–1887; Medicine; FRS 1839; Private practice and Prof. of Obstetric Medicine, King's College, London [Physician].
57. Fergusson, James; 1808–1886; Archaeology; FRS 1863.
58. *Fergusson, William; 1808–1877; Medicine; FRS 1848; Private practice and Prof. of Surgery, King's College Hospital.

59. *Flower, William Henry; 1831–1899; Biology; FRS 1869; Director of Natural History Museum [Author, Brewer].
60. *Forbes, David; 1828–1876; Geology; FRS 1858; Partner in the Birmingham nickel-smelters Evans & Askin.
61. *Foster, George Carey; 1835–1919; Chemistry; FRS 1869; Prof. of Natural Philosophy, UCL.
62. Foster, Michael; 1836–1907; Physiology; FRS 1872; Professor of Physiology, Cambridge [Physician].
63. *Fox, Wilson; 1831–1887; Medicine; FRS 1870; Prof. of Clinical Medicine, UCL [Manufacturer].
64. Frankland, Edward; 1825–1899; Chemistry; FRS 1853.
65. Galton, Francis; 1822–1911; Anthropology; FRS 1860; Private means.
66. *Galton, Douglas Strutt; 1822–1899; Engineering; FRS 1859; Sanitary engineer [Independent].
67. Gassiot, John P.; 1797–1877; Physics; FRS 1840.
68. Geikie, Archibald; 1835–1925; Geology; FRS 1865.
69. Gilbert, Joseph H.; 1817–1901; Chemistry; FRS 1860.
70. *Gladstone, John Hall; 1827–1902; Chemistry; FRS 1853; Prof. of Chemistry, Queen's College, Belfast [Draper].
71. Glaisher, James; 1809–1903; Astronomy; FRS 1849; Royal Observatory, Greenwich.
72. Godwin-Austen, Robert; 1808–1884; Geology; FRS 1849.
73. Grant, Robert; 1814–1892; Astronomy; FRS 1865.
74. *Gray, John Edward; 1800–1875; Zoology; FRS 1832; Keeper of Zoology, British Museum [Pharmacist].
75. *Greg, Robert Hyde; 1795–1875; Statistics; Industrialist [Mill owner].
76. *Grove, William; 1811–1896; Physics; FRS 1840; Judge, Prof. of Experimental Philosophy, Royal Institution [Magistrate].
77. *Günther, Albert Charles; 1830–1914; Zoology; FRS 1867; Keeper of Zoology, British Museum [Bursar of Estates].
78. *Guy, William Augustus; 1810–1885; Statistics; FRS 1866; Prof. of Forensic Medicine, King's College, London [Physician].
79. *Harcourt, Augustus; 1834–1919; Chemistry; FRS 1868; Tutor, Christ Church, Oxford [Admiral].
80. Harkness, Robert; 1816–1878; Geology; FRS 1856.
81. *Haughton, Samuel; 1821–1897; Geology; FRS 1858; Prof. of Geology, Trinity College, Dublin [Flour Merchant, Reformer].
82. Hawkshaw, John; 1811–1891; Engineering; FRS 1855.
83. *Henslow, John Stevens; 1796–1861; Botany; FRS 1818; Prof. of Botany, Cambridge [Solicitor].
84. *Heywood, James; 1810–1897; Statistics; FRS 1839; Barrister of independent means and M.P..
85. *Hill, Rowland; 1795–1879; Statistics; FRS 1857; Director of the London and Brighton Railway Company and Secretary to the Postmaster General [Teacher].
86. Hind, John Russell; 1823–1895; Astronomy; FRS 1863.
87. *Hirst, Thomas; 1830–1892; Mathematics; FRS 1861; Director of Naval Studies, Royal Naval College [Woolstapler].
88. *Holland, Henry; 1788–1873; Medicine; FRS 1815; Private practice, Physician Ordinary and Extraordinary to Royalty [Physician].
89. *Hooker, Joseph Dalton; 1817–1911; Botany; FRS 1847; Director of Kew Gardens [Botanist].

90. Huggins, William; 1824–1910; Astronomy; FRS 1865.
91. *Humphry, George; 1820–1896; Medicine; FRS 1859; Prof. of Human Anatomy, Cambridge [Lawyer].
92. *Huxley, Thomas Henry; 1825–1895; Zoology; FRS 1851; Hunterian Prof., Royal College of Surgeons [Teacher].
93. James, Henry; 1803–1877; Geology; FRS 1848.
94. Jardine, William; 1800–1874; Biology; FRS 1860.
95. *Jeffreys, John Gwyn; 1809–1885; Zoology; FRS 1840; Barrister [Solicitor].
96. *Jenkin, Fleeming; 1833–1885; Engineering; FRS 1865; Prof. of Engineering, Edinburgh [Captain Royal Navy].
97. Jenner, William; 1815–1898; Medicine; FRS 1864.
98. *Jevons, William Stanley; 1835–1882; Statistics; FRS 1872; Prof. of Political Economy, Owens College, Manchester [Iron Merchant].
99. *Jones, Thomas Rupert; 1819–1911; Geology; FRS 1872; Prof. at Royal Military College, Sandhurst [Silk Merchant].
100. Joule, James P.; 1818–1889; Physics; FRS 1850.
101. Kane, Robert J.; 1807–1890; Chemistry; FRS 1849.
102. Kiernan, Francis; 1800–1890; Physiology; FRS 1834.
103. *Lassell, William; 1799–1880; Astronomy; FRS 1849; Brewer.
104. *Lawes, John Bennet; 1814–1900; Chemistry; FRS 1854; Fertilizer entrepreneur, founder of Rothamsted [Landowner].
105. Lee, Robert; 1793–1877; Physiology; FRS 1830.
106. Lefevre, John G.; 1797–1879; Medicine; FRS 1820.
107. Lloyd, Humphrey; 1800–1881; Astronomy; FRS 1836.
108. Lockyer, Joseph N.; 1836–1920; Astronomy; FRS 1869.
109. *Lubbock, John; 1834–1913; Anthropology; FRS 1858; Banker [Banker].
110. Lyell, Charles; 1797–1875; Geology; FRS 1826.
111. *Main, Robert; 1808–1878; Astronomy; FRS 1860; Assistant to Astronomer Royal, Oxford.
112. *Marshall, John; 1818–1891; Medicine; FRS 1857; Prof. of Clinical Surgery at UCL [Solicitor].
113. *Maskelyne, Nevil Story; 1823–1911; Geology; FRS 1870; Prof. of Mineralogy, Oxford [Anthony Mervyn Reeve Story-Maskelyne, FRS].
114. *Maxwell, James Clerk; 1831–1879; Physics; FRS 1861; Prof. of Experimental Physics, Cambridge [Landowner].
115. *Merrifield, Charles; 1827–1884; Engineering; FRS 1863; Principal of the Royal School of Architecture.
116. *Miers, John; 1789–1879; Botany; FRS 1843; Engineering Entrepreneur in Chile and Brazil [Jeweller, Mineralogist].
117. *Miller, William Hallowes; 1801–1880; Geology; FRS 1838; Prof of Mineralogy, Cambridge.
118. Milnes, Richard Monckton; 1809–1885; Statistics; FRS 1868.
119. *Mivart, St-George Jackson; 1827–1900; Medicine; FRS 1869; Chair in Zoology, St Mary's Hospital medical school [Hotel Owner].
120. *Newmarch, William; 1820–1882; Statistics; FRS 1861; Banker.
121. *Newton, Alfred; 1829–1907; Biology; FRS 1870; Prof. of Zoology and Comparative Anatomy, Cambridge [Statesman].
122. Olding, William; 1829–1921; Chemistry; FRS 1859.
123. *Osborn, Sherard; 1829–1875; Geography; FRS 1870; Rear-admiral [Army].
124. *Owen, Richard; 1804–1892; Medicine; FRS 1834; Superintendent of Natural History,

British Museum [West India Merchant].

125. *Paget, James; 1814–1899; Medicine; FRS 1851; Surgeon at St. Barts, London [Brewer, Shipowner].
126. *Parker, William Kitchen; 1823–1890; Medicine; FRS 1865; Prof. of Comparative Anatomy, Royal College of Surgeons [Farmer].
127. *Parkes, Edmund Alexander; 1819–1876; Medicine; FRS 1861; Prof. of Hygiene, Army Medical College [Landowner].
128. *Parsons, Laurence; 1840–1908; Astronomy; FRS 1867; Landowner (Earl of Rosse) [Astronomer, Landowner].
129. Pavy, Frederick W.; 1829–1911; Medicine; FRS 1863.
130. Percy, John; 1817–1889; Chemistry; FRS 1847.
131. Pettigrew, James B.; 1834–1908; Physiology; FRS 1868.
132. *Phillips, John; 1800–1879; Geology; FRS 1834; Reader of Geology at Oxford [Excise Officer].
133. *Playfair, Lyon; 1818–1898; Chemistry; FRS 1848; Prof. of Chemistry and MP, Edinburgh University [Bengal Hospital Inspector].
134. *Pole, William; 1814–1900; Engineering; FRS 1861; Prof. of Civil Engineering, UCL.
135. *Prestwich, Joseph; 1812–1896; Geology; FRS 1853; Inherited and ran "Joseph Prestwich & Co." the wine-merchant in Mark-lane [London Wine Merchant].
136. Price, Bartholomew; 1818–1898; Mathematics; FRS 1852.
137. Pritchard, Charles; 1808–1893; Astronomy; FRS 1840.
138. *Ramsay, Andrew Crombie; 1814–1891; Geology; FRS 1849; Prof. of Geology, UCL and Director-General of the Geological Survey [Manufacturing Chemist].
139. Rawlinson, Henry C.; 1810–1895; Archaeology; FRS 1850.
140. Rennie, John; 1794–1874; Engineering; FRS 1823.
141. Richards, George H.; 1820–1896; Engineering; FRS 1866.
142. Richardson, Benjamin W.; 1823–1896; Medicine; FRS 1867.
143. Robinson, Thomas; 1792–1882; Astronomy; FRS 1856.
144. Rolleston, George; 1829–1881; Medicine; FRS 1862.
145. *Roscoe, Henry Enfield; 1833–1915; Chemistry; FRS 1863; Prof. of Chemistry, Owens College, Manchester [Judge].
146. Russell, John S.; 1808–1882; Engineering; FRS 1849.
147. *Sabine, Edward; 1788–1883; Astronomy; FRS 1818; General.
148. *Salisbury, Robert; 1830–1903; Physics; FRS 1869; Statesman [Marquess of Salisbury].
149. Salmon, George; 1819–1904; Mathematics; FRS 1863.
150. *Sclater, Philip Lutley; 1829–1913; Biology; FRS 1861; Barrister [Landowner].
151. *Scott, Robert Henry; 1833–1916; Geography; FRS 1870; Meteorologist.
152. Sharpey, William; 1802–1886; Physiology; FRS 1839.
153. Sibson, Francis; 1814–1876; Medicine; FRS 1849.
154. *Siemens, Charles William; 1823–1883; Engineering; FRS 1862; Industrial engineer and entrepreneur [Farmer].
155. Simon, John; 1816–1904; Medicine; FRS 1845.
156. Smith, Henry J.; 1826–1883; Mathematics; FRS 1861.
157. *Smith, Archibald; 1813–1872; Mathematics; FRS 1856; Prof. of Math, Cambridge [Author, Geologist].
158. Smith, Edward; 1818–1874; Medicine; FRS 1860.
159. *Smyth, Warington Wilkinson; 1817–1890; Geology; FRS 1858; Professor of Mineralogy, School of Mines [Admiral].
160. *Smyth, Charles Piazzi; 1819–1900; Astronomy; FRS 1857; Astronomer Royal of Scotland [Admiral].

161. Smythe, William; 1816–1887; Engineering; FRS 1864.
162. *Spencer, Herbert; 1820–1903; Philosophy; —; Civil Engineer and Writer [Schoolmaster].
163. *Spottiswoode, William; 1825–1883; Mathematics; FRS 1853; Printer [Printer].
164. Stanhope, Earl; 1805–1873; History; FRS 1827.
165. *Stenhouse, John; 1809–1880; Chemistry; FRS 1848; Proprietor of a chemistry enterprise in London [Calico Printer].
166. *Stewart, Balfour; 1828–1887; Physics; FRS 1862; Director of Kew Observatory [Tea Merchant].
167. *Stokes, William; 1804–1878; Medicine; FRS 1861; Prof. of Medicine, Trinity College, Dublin [Professor of Medicine].
168. Stokes, George G.; 1819–1903; Physics; FRS 1851.
169. *Strachey, Richard; 1817–1908; Geography; FRS 1854; Lieut.-General, Royal Bengal Engineers [India Civil Service].
170. Strange, Alexander; 1818–1876; Geography; FRS 1868.
171. Strzelecki, Paul E-de; 1796–1873; Geography; FRS 1853.
172. Sylvester, James J.; 1814–1897; Mathematics; FRS 1839.
173. Talbot, Christopher R.; M.P.; 1803–1890; FRS 1831.
174. Talbot, William Fox; 1800–1877; Mathematics; FRS 1831.
175. *Thomson, Allen; 1809–1884; Biology; FRS 1848; Prof. of Anatomy, Glasgow [Physician].
176. Thomson, Thomas; 1817–1878; Botany; FRS 1855.
177. Thomson, William; 1824–1907; Physics; FRS 1851.
178. Todhunter, Isaac; 1820–1884; Mathematics; FRS 1862.
179. Tylor, Edward B.; 1832–1917; Anthropology; FRS 1871.
180. Tyndall, John; 1820–1893; Physics; FRS 1852.
181. Vignoles, Charles H.; 1793–1875; Engineering; FRS 1855.
182. *Voelcker, John; 1822–1884; Chemistry; FRS 1870; Prof. of Chemistry, Royal Agricultural College [German Merchant].
183. Walden, Arthur; 1824–1878; Zoology; FRS 1871.
184. *Watson, Thomas; 1792–1882; Medicine; FRS 1859; Private practice and Prof. Medicine at King's College, London.
185. Waugh, Andrew S.; 1810–1878; Engineering; FRS 1858.
186. Webster, Thomas; 1810–1875; Physics; FRS 1847.
187. Wheatstone, Charles; 1802–1875; Physics; FRS 1836.
188. *Williamson, William Crawford; 1816–1895; Biology; FRS 1854; Prof. of Natural History, Owens College, Manchester [Gardener].
189. *Williamson, Alexander William; 1824–1904; Chemistry; FRS 1855; Prof. of Chemistry, UCL [India House Clerk].
190. Willis, Robert; 1800–1875; Mathematics; FRS 1830.
191. *Wood, William Page; 1801–1881; Statistics; FRS 1836; Lord Chancellor, as Lord Hatherley [Political Reformer].
192. *Yorke, Phillip James; 1799–1874; Chemistry; FRS 1849; Lieut.-Colonel in the Scots Fusiliers [Prebendary of Ely].

Appendix E

The Law of Errors

In his use of the Law of Errors, Galton was working with the 19th Century understanding of the distribution parameters, which used a mean of μ and a 'modulus of dispersion' c:

$$f(x) = \frac{1}{c\sqrt{\pi}} exp[-(x-\mu)^2/c^2] \tag{E.1}$$

whereas the Normal distribution more familiar today is parameterized by the mean μ and variance σ^2, where $2\sigma^2 = c^2$.

$$f(x) = \frac{1}{\sigma\sqrt{2\pi}} exp[-\frac{1}{2}(x-\mu)^2/\sigma^2] \tag{E.2}$$

The proof that the sum of two random variables obeying the Law of Errors also obeyed the Law of Errors, which Galton had shown directly via the Quincunx, was already well-known.[1] The proof of this is not difficult but either requires some scaffolding using characteristic functions, or a proof by convolution with some tedious algebra.[2] The key result here is that the variance of the sum of independent random variables is the sum of the variances. Or, the squared modulus of dispersion of the sum is the sum of the squared moduli of dispersion of the components. This is actually more generally true, regardless of the distributions involved, though Galton would not have known that. If we have two distributions A and B, with $A \perp\!\!\!\perp B$

$$Var(A+B) = Var(A) + Var(B). \tag{E.3}$$

Equivalently $c^2 = a^2 + b^2$ where c is the modulus of $A+B$, and a, b are the moduli of A, B respectively. Another well-known result is that, in our terms

$$Var(kA) = k^2 Var(A) \tag{E.4}$$

with the equivalent relation for the squared modulus. It follows that when A, B come from the same distribution with variance σ^2 (as in the mid-parent below)

$$\begin{aligned} Var[\frac{1}{2}(A+B)] &= (\frac{1}{2})^2[Var(A) + Var(B)] \\ &= (\frac{1}{2})^2[2\sigma^2] \\ &= \frac{1}{2}\sigma^2 \end{aligned} \tag{E.5}$$

Galton also needed a straightforward result involving the product of exponential functions.

$$exp[-\frac{(x-\mu)^2}{a^2}]exp[-\frac{(x-\mu)^2}{b^2}] = exp[-(\frac{1}{a^2} + \frac{1}{b^2})(x-\mu)^2] \tag{E.6}$$

[1] Airy 1861.

[2] For a modern treatment using Normal variables, see Casella and Berger 2002.

We will consider the simplest case first, a population with 'simple descent' and no natural selection or differential fecundity.

1. Start with a population with some squared modulus of dispersion, say c_1^2.
2. A reverted population derived from the above. Given a coefficient of reversion r this would, using Equation E.4, p. 427, give squared modulus

$$c_2^2 = r^2 c_1^2$$

3. The reverted population above, now with 'family dispersion' of say v, giving

$$c_4^2 = c_2^2 + v^2$$

If the first population is the same as the last population, not having evolved, we must have equal dispersion, so that

$$c_1^2 = c_4^2 = c_2^2 + v^2 = r^2 c_1^2 + v^2 \implies c_1^2 = v^2/(1 - r^2)$$

Galton pointed out (without giving formulae) that as successive generations ensued, reversion and family variation would accumulate. Then as long as r and v are constant, and $n >= 1$, the $n+1$ th generation subject to *both* these influences (the even-numbered cases above) would have a modulus m_{n+1} given by a recurrence relation.[3]

$$\begin{aligned} m_{n+1}^2 &= r^2 m_n^2 + v^2 \\ &= \{(r^2)^n + ... + (r^2)^2 + (r^2)^1 + 1\}v^2 \\ &= v^2 \{\frac{1 - (r^2)^{n+1}}{1 - r^2}\} \end{aligned} \tag{E.7}$$

This approaches an equilibrium of $\frac{v^2}{(1-r^2)}$ as $n \to \infty$.

In the more complex case, five successive population transformations are modelled. Each one produces another distribution obeying the Law of Errors, but the 'moduli of dispersion' are related to each other as given below.

1. Start with some original population with squared modulus of dispersion, say c_0^2.
2. Offspring from two parents, drawn from the original population. Galton reasoned that, using the mid-parent value $(A + B)/2$ for the offspring (see Equation E.5, p. 427), this would have squared modulus

$$c_1^2 = \frac{1}{2} c_0^2$$

3. A reverted population derived from the above. Given a coefficient of reversion r this would, using Equation E.4, p. 427, give squared modulus

$$c_2^2 = r^2 c_1^2$$

4. Also different fertility at the extremes, so that giants and midgets produce fewer offspring (this is similar to the natural selection step below). The frequencies of offspring would be proportional to the product of a dispersion with modulus of say f and the above dispersion of c_2. A little algebra of exponential functions (see Equation E.6, p. 427) shows that this must give a squared modulus of

$$c_3^2 = (\frac{1}{c_2^2} + \frac{1}{f^2})^{-1} = \frac{c_2^2 f^2}{c_2^2 + f^2}$$

[3] See Bulmer 2003, 218, who argues that it is not necessarily true in practice that r and v are constant over generations.

5. The preceding, now with 'family dispersion' of say v, giving squared modulus

$$c_4^2 = c_3^2 + v^2$$

6. Additional stabilizing (centripetal) natural selection. Supposing the selection function to have a modulus of s, and again supposing that frequencies of offspring would be proportional to the product of that selection function and the above frequency with dispersion c_4,[4] similar algebra gives

$$c_5^2 = \left(\frac{1}{c_4^2} + \frac{1}{s^2}\right)^{-1} = \frac{c_4^2 s^2}{c_4^2 + s^2}$$

Now if the first population is the same as the last population, $c_5 = c_0$. Then if values are known for the other constants involved, the size of the required 'reversion to mediocrity' r can be calculated, as in the simpler case above, from the given equations.

4 See Manly 1985, 9 where this formulation is also used.

Appendix F

Assortative Mating

Recall that in Galton's era, the 'probable error' γ, rather than our standard deviation σ, was the usual measure of dispersion. However, since heights are normally distributed one is a multiple of the other ($0.67449\,\sigma = \gamma$). Consider the case of his height data from *English Men of Science*, where he analyzed *dissimilarity* in parental heights of his scientists, for which he had 62 complete cases.[1] In modern terms, with more familiar algebra, we have

$$\sigma^2_{m-f} = \sigma^2_m + \sigma^2_f - 2\,\sigma_{m,f}.$$

If there is no systematic covariation between the heights of fathers (m) and mothers (f), either to match or to contrast, we expect that the term for it is approximately zero, i.e. $-2\,\sigma_{m,f} \approx 0$, and can be discarded, so that

$$\sigma^2_{m-f} \approx \sigma^2_m + \sigma^2_f$$

From his data, Galton could calculate σ^2_{m-f}, expressed in terms of 'probable errors'. He found that the second equation is approximately true.

Taking as given Galton's rounded 'probable errors' for heights of fathers (1.7) and mothers (1.9) and the difference of the heights (2.3), we can calculate the standard deviations by dividing by 0.67449, and then

$$r = \frac{\sigma_{m,f}}{\sigma_m \sigma_f}$$

The estimated covariance is

$$\frac{1}{2}\{\sigma^2_m + \sigma^2_f - \sigma^2_{m-f}\} = 1.329855$$

and the correlation is then

$$r = 0.1873065 \approx 0.19$$

which is pretty weak but not zero. See also K. Pearson 1924, 149, who finds a similar correlation of about 0.19 for this difference in heights.

Taking the case of hair color, one must take into account the prevalence of 'brown' in the population. The theory to test is whether or not the hair colors of fathers and mothers are *independent*. Galton's protégé and future biographer Karl Pearson would invent a test for this in 1901 which would be adopted broadly. The *observed* frequencies with which hair colors go together may be compared against their *expected* frequencies on the assumption of independence. If the discrepancy is too large, then the assumption of independence can be rejected, a procedure known as Pearson's 'χ^2 Test'. But this depends on an approximation that is inaccurate for cases like this where some of the combinations have low counts.

Instead an exact permutation test due to Ronald Fisher—invented in 1922—can be used, provided that some of the categories are merged to make the problem computationally feasible.

[1] F. Galton 1874c, 31-32. This presentation has been substantially corrected from the first edition.

In Table F.1 we have combined 'black', 'dark' and 'dark brown' as 'darkish'; 'brown' and 'light brown' as 'brownish'; and 'light brown', 'light' and 'fair' as 'fairish'. Fisher's exact test yields p=0.035. In other words, if the hair colors are independent, data at least as extreme as this would only have been observed at most 3.5 percent of the time. Naively one might conclude that the colors are *not* independent, but this actually depends on the grouping used above. Other groupings do not work, and the grouping that did work had to be searched for, so that we had multiple bites at the cherry. This data set does not decide the issue.

Table F.1: Hair color inheritance.

	Father	darkish	brownish	fairish
Mother	darkish	11	5	5
	brownish	9	23	5
	fairish	5	5	5

Appendix G

The Mathematics of Natural Inheritance (1889)

When first working out the details of regression, Galton was considering the bivariate normal case, an inevitability given his fondness for the 'law of error' that he had renamed the 'normal'. Modern proofs have the benefit of the simplifications that abstraction can bring, but the proofs are a lot messier in the original formulation.

Since he feared that his mathematics skills were rusty, Galton sought the help of J. D. Hamilton Dickson—a Fifth Wrangler who was then a mathematics instructor at Cambridge—and set two problems for him. Both of these were phrased in the archaic style then common on the Cambridge Tripos.[1]

Here we rehydrate the problems and the proofs as given by Dickson, which will be unfamiliar to most modern readers. It will be seen that simply stating the problems entailed the solutions given, which Galton would have had no trouble finding himself if he took the time to do so.

Recall that Galton is regressing the height of offspring on the height of the 'mid-parent'. This is the average of the father and the 1.8 scaled-mother (to compensate for sexual dimorphism in stature). But the problems as posed to Hamilton Dickson made no mention of this.

G.1 Problem 1

> A point P is capable of moving along a straight line P'OP, making an angle $\tan^{-1}(2/3)$ with the axis of y, which is drawn through O the mean position of P; the probable error of the projection of P on Oy is 1.22 inch: another point p, whose mean position at any time is P, is capable of moving from P parallel to the axis of x (rectangular co-ordinates) with a probable error of 1 .50 inch. To discuss the "surface of frequency" of p.

The hardest part of Dickson's solution is really the problem statement sent to him by Galton, because its language is now archaic, though it posed no problem to Dickson, who was after all a Fifth Wrangler.

Here y is the child and x is the mid-parent, and P corresponds to the height deviation of the child y. Galton describes the regression line through the origin O on which P must lie, somewhat confusingly labelling this line P'OP. The slope of this line is given as 2/3, which is equivalent to an angle to the y axis of $\tan^{-1}(2/3)$. Galton also states the actual value of the conditional variance of y given x, which we would write as ${\sigma^2}_{y|x}$.

However, in that era the 'probable error' was used rather than the variance. This is the interval $(-e, e)$ containing 50% of the area under the normal curve, so any value one may observe is just as likely to be within $(-e, e)$ as it is without, in $(\infty, -e)$ or (e, ∞). It is easy to get e from the Quartile function Q (the Inverse Cumulative Normal, which Galton called the Ogive), $e = \frac{1}{2}[Q(0.75) - Q(0.25)]$. Galton could easily estimate this by ranking his observations and doing the arithmetic. Modern usage favours σ rather than e, where 2/3 of the area under the normal curve lies between $(-\sigma, \sigma)$. However the probable error e is just a multiple of the

[1] F. Galton 1886b, 1889c.

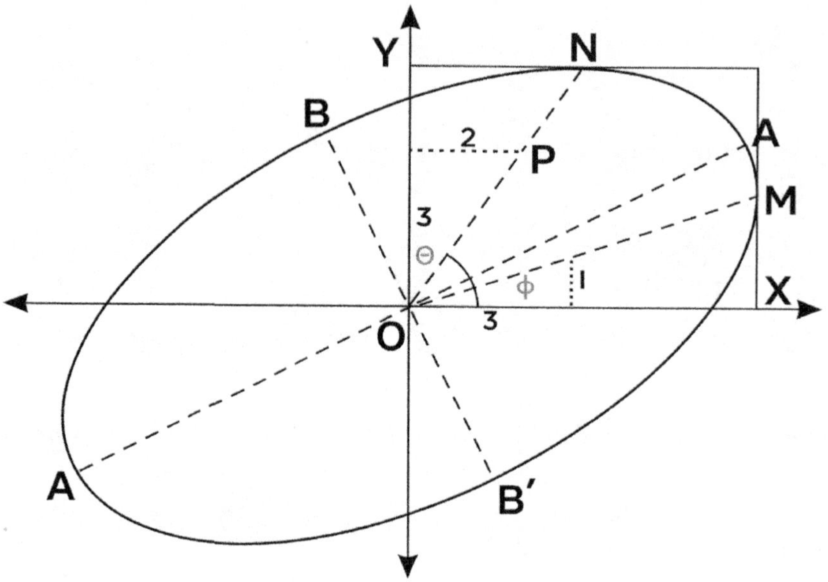

Figure G.1: Ellipse with regression lines intersecting the tangents.

standard deviation $e \approxeq 0.6745\,\sigma$. The standard deviation is plotted for the child height data in Figure G.2, p. 435, and the probable error in Figure G.3, p. 435.

Since modern readers think naturally in terms of the Normal Distribution parameterized by the standard deviation σ and variance σ^2, the discussion below uses those terms, but can easily be rewritten using the probable error as the constant multiplier will cancel out.[2] The following table shows how to translate Dickson and Galton's notation (which, tediously, is not the same as Dickson's) into the scheme used here.

$$\begin{aligned} a^2 &= \sigma_X^2 = p \\ b^2 &= \sigma_{Y|X}^2 = f \\ c^2 &= \sigma_Y^2 = p \text{ (confusingly)} \\ e^2 &= \sigma_{X|Y}^2 = c^2 \\ \tan(\phi) &= \beta_{X|Y} = w \\ \tan(\theta) &= \beta_{Y|X} = w\prime \end{aligned}$$

Using the data, the estimates $\sigma_X^2 = 1.22$ and $\sigma_{Y|X}^2 = 1.50$ were provided by Galton. Dickson first worked out the problem using these specific estimates, then generalized it, but it is easier to just work out the general form. The Scot was easily able to obtain the joint density defining the 'surface of frequency' asked for by Galton, using the definition of the conditional density $f_{X|Y}$. Note that the problem is stated in terms of height deviations from the mean height, so the means (medians) of both X and Y are 0. Also, it is given from the regression line that the mean of $X|Y$ is a multiple of Y, so that $\mu_{X|Y} = \beta_{X|Y}Y$. Thus, ignoring the constants of integration:

$$\begin{aligned} f(x,y) &\propto f(y)f_{X|Y}(x) \\ &\propto \exp[-(\frac{y}{\sigma_Y})^2]\exp[-(\frac{x-\mu_{X|Y}}{\sigma_{X|Y}})^2] \end{aligned}$$

[2] See Bulmer 2003 for a discussion of some of this material in terms of probable errors.

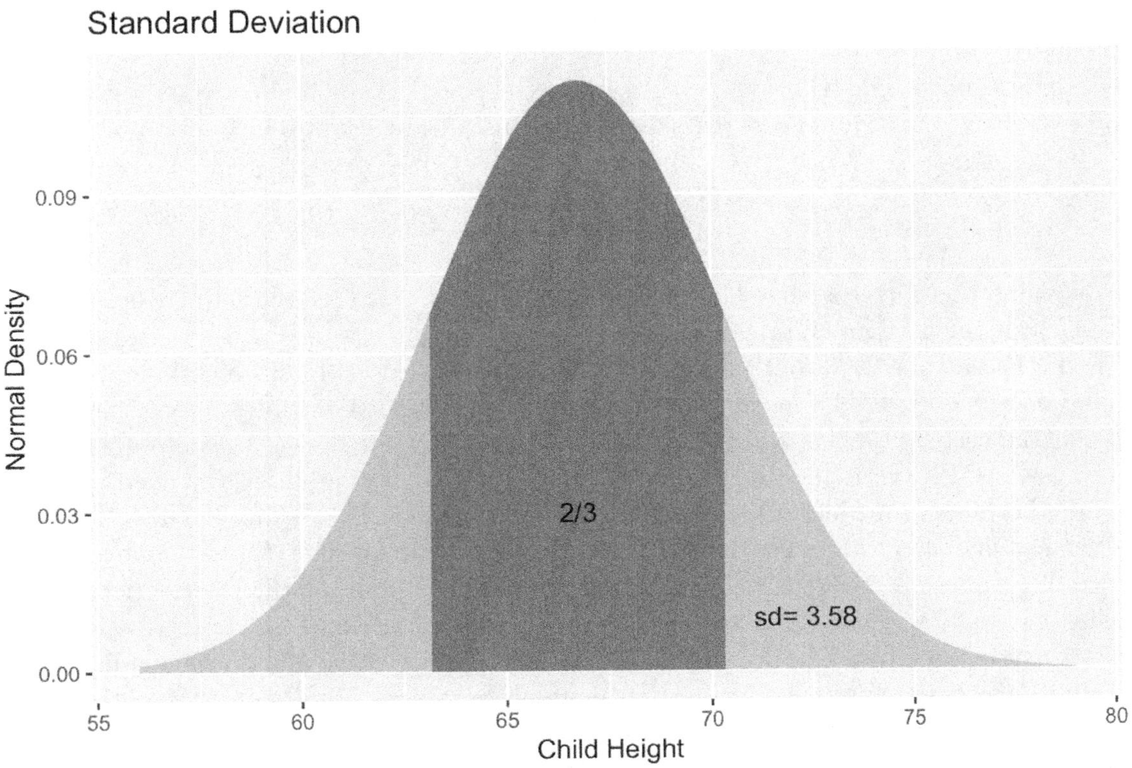

Figure G.2: The standard deviation corresponds to 2/3 the area under the normal curve

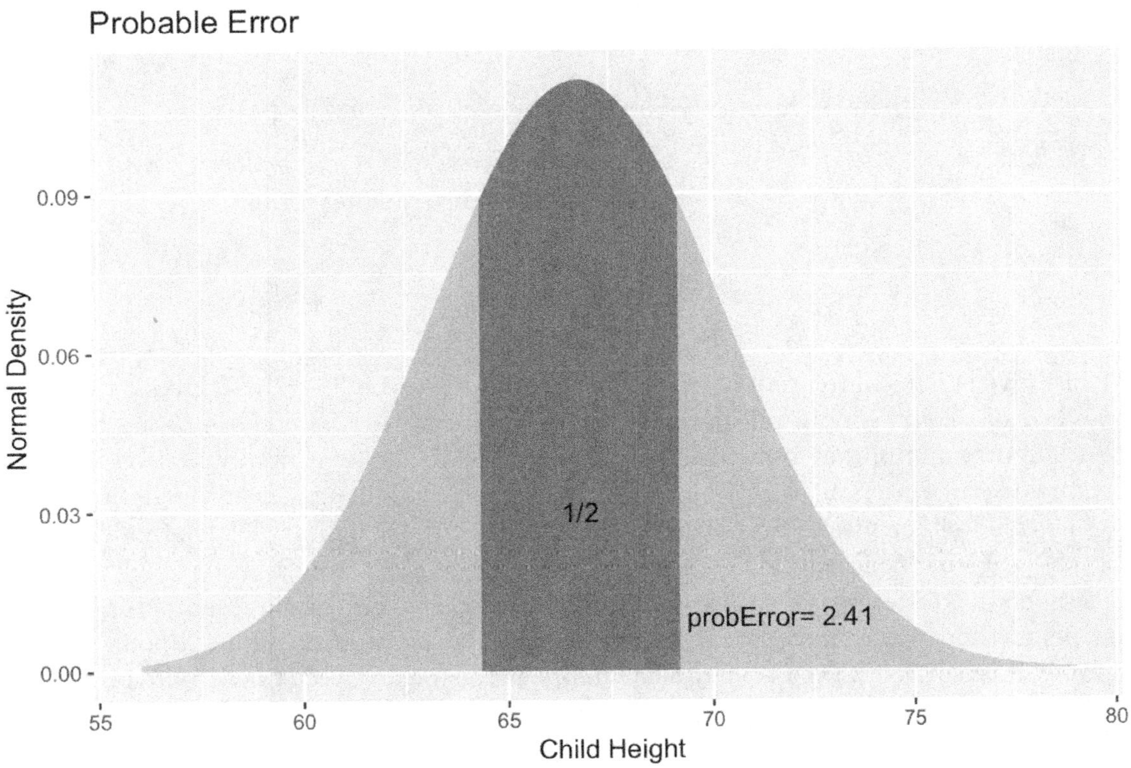

Figure G.3: The probable error corresponds to 1/2 the area under the normal curve, $e \approxeq 0.6745\sigma$.

$$= \exp[-(\frac{y}{\sigma_Y})^2 - (\frac{x - \beta_{X|Y}y}{\sigma_{X|Y}})^2].$$

Since exp is a monotonic function, $p = q \Longleftrightarrow exp(p) = exp(q)$ so points of equal density can be found by setting the argument of exp equal to a constant, which is the equation of an ellipse:

$$(\frac{y}{\sigma_Y})^2 + (\frac{x - \beta_{X|Y}y}{\sigma_{X|Y}})^2 = C$$

This explains the patterns that Galton had found in his 'map' of the data, where equal frequencies are joined by an elliptical contour.

Next, Dickson observed that the line OPN in Figure G.1, p. 434, on which $Y \propto X$, intersects the ellipse at its horizontal tangent point N. Likewise the line OM on which $X \propto Y$ intersects the ellipse at the vertical tangent point M. This is because each line bisects the ellipse, horizontally in the case of OPN, vertically in the case of OM. Or in other words, for any fixed point y, the mean of the (horizontal line of) x values lies on OPN. Similarly for any fixed x, the mean of the (vertical line of) y values lies on OM. As these lines approach the ellipse they bisect smaller and smaller segments until the segment consists of a single point at the intersections.

This gave Dickson two facts to use for each point N and M. They both obey the ellipse equation, and since there is a tangent at each point the derivative of a corresponding function (the ellipse restricted to some domain on which one variable is a function of the other) is zero. He simply differentiated the ellipse equation, without fussing about the domain restriction needed. Here x is a function of y.

$$(\frac{y}{\sigma_Y})^2 + (\frac{x - \beta_{X|Y}y}{\sigma_{X|Y}})^2 = C \tag{G.1}$$

$$\Longrightarrow \frac{2y}{\sigma_Y^2} + 2(\frac{x - \beta_{X|Y}y}{\sigma_{X|Y}^2})(\frac{dx}{dy} - \beta_{X|Y}) = 0 \tag{G.2}$$

$$\Longrightarrow \frac{y}{\sigma_Y^2} + (\frac{x - \beta_{X|Y}y}{\sigma_{X|Y}^2})(-\beta_{X|Y}) = 0 \text{ since dx/dy} = 0 \tag{G.3}$$

$$\Longrightarrow y = (\frac{\beta_{X|Y}\sigma_Y^2}{\sigma_{X|Y}^2 + \beta_{X|Y}^2\sigma_Y^2})x \text{ after rearranging terms} \tag{G.4}$$

$$\Longrightarrow y = \beta_{Y|X}x \text{ where } \beta_{Y|X} = (\frac{\sigma_Y^2}{\sigma_{X|Y}^2 + \beta_{X|Y}^2\sigma_Y^2})\beta_{X|Y} \tag{G.5}$$

Thus, given that Galton had already estimated $\beta_{Y|X}$ to be 2/3, the regression of children on parents, one could simply calculate the regression of parents on children using the formula for $\beta_{X|Y}$ above, and obtain 1/3. This completely explained the at-first surprising result that the two regressions are not equal, which Galton had obtained directly from the data. To obtain $\beta_{Y|X}$ one could either fit a line by eye or use least squares, though neither method was part of Dickson's problem or his solution.

Dickson's next step was to work out an expression for $\sigma_{Y|X}$ in terms of $\sigma_{X|Y}$. We will follow his method. He worked directly from the joint density above. Fixing a point (x_1, y_1) on the probability surface he considered what happened as one moved away from this point in a direction parallel to the y axis. This is to take a vertical slice through the 3-dimensional figure to get a 2-dimensional normal curve $f_{Y|X}$. Below we can ignore constant terms K_1, K_2, K_3.

$$f_{Y|X}(y) = f[-\frac{q}{2}] \text{ where}$$

$$
\begin{aligned}
q &= [\frac{y_1 + y}{\sigma_Y}]^2 + [\frac{x_1 - \beta_{X|Y}(y_1 + y)}{\sigma_{X|Y}}]^2 \\
&= (\frac{1}{\sigma_Y^2} + \frac{\beta_{X|Y}^2}{\sigma_{X|Y}^2})y^2 + 2y(\frac{y_1}{\sigma_Y^2} + \frac{y_1\beta_{X|Y}^2}{\sigma_{X|Y}^2} - \frac{x_1\beta_{X|Y}}{\sigma_{X|Y}^2}) + K_1 \\
&= (\frac{1}{\sigma_Y^2} + \frac{\beta_{X|Y}^2}{\sigma_{X|Y}^2})y^2 + 2y[\frac{y_1}{\sigma_Y^2} + \frac{(-y_1\beta_{X|Y} + x_1)(-\beta_{X|Y})}{\sigma_{X|Y}^2}] + K_2 \\
&= (\frac{1}{\sigma_Y^2} + \frac{\beta_{X|Y}^2}{\sigma_{X|Y}^2})y^2 + K_3 \\
&\text{since } \frac{y_1}{\sigma_Y^2} + \frac{(-y_1\beta_{X|Y} + x_1)(-\beta_{X|Y})}{\sigma_{X|Y}^2} = 0 \text{ from G.3} \\
&= \frac{y^2}{(\frac{1}{\sigma_Y^2} + \frac{\beta_{X|Y}^2}{\sigma_{X|Y}^2})^{-1}} + K_3
\end{aligned}
$$

Therefore

$$\sigma_{Y|X}^2 = (\frac{1}{\sigma_Y^2} + \frac{\beta_{X|Y}^2}{\sigma_{X|Y}^2})^{-1}.$$

In the same way one can show that

$$\sigma_{X|Y}^2 = (\frac{1}{\sigma_X^2} + \frac{\beta_{Y|X}^2}{\sigma_{Y|X}^2})^{-1}.$$

Rearranging the first expression gives a useful ratio

$$\sigma_{Y|X}^2 = \frac{\sigma_Y^2\sigma_{X|Y}^2}{\sigma_{X|Y}^2 + \sigma_Y^2\beta_{X|Y}^2} \tag{G.6}$$

$$= [\frac{\sigma_Y^2}{\sigma_{X|Y}^2 + \sigma_Y^2\beta_{X|Y}^2}]\sigma_{X|Y}^2 \tag{G.7}$$

$$= \frac{\beta_{Y|X}}{\beta_{X|Y}}\sigma_{X|Y}^2 \text{ by G.5} \tag{G.8}$$

Plugging in the values that Galton obtained from the data we see that

$$\sigma_{X|Y}^2 = (\frac{1}{1.22} + \frac{4/9}{1.50})^{-1} \approxeq 1.50^2\frac{1}{2}$$

The reason why $\beta_{X|Y} = \frac{1}{2}\beta_{Y|X}$ in Galton's data is due to the construction of the mid-parents X as the weighted sum of the father and mother: $X = X_F + 1.08X_M$. This reduces the variance (probable error) of X since, using the laws of variance

$$
\begin{aligned}
\sigma_{X|Y}^2 &= Var(\frac{1}{2}(X_F + 1.08X_M)|Y]) \\
&= \frac{1}{4}[Var(X_F|Y) + Var(1.08X_M|Y)] \\
&\approxeq \frac{1}{4}[Var(X_F|Y) + Var(X_F|Y)] \\
&= \frac{1}{4}2Var(X_F|Y)
\end{aligned}
$$

$$\approx \frac{1}{2}\sigma^2_{Y|X} \text{ from the data}$$
$$= \frac{1}{2}\sigma^2_{Y|X}$$
$$\Longrightarrow \frac{\beta_{X|Y}}{\beta_{Y|X}} = \frac{\sigma^2_{X|Y}}{\sigma^2_{Y|X}} \approx \frac{1}{2} = \frac{1/3}{2/3} \text{ as estimated.}$$

Dickson then uses the regression equation for Y in terms of X, to which the variance operator may be applied, to neatly reduce the variance ratio to an unconditional form.[3]

$$\begin{aligned}
Y &= \beta_{Y|X}X + \text{error} \\
\Longrightarrow \text{Var}(Y) &= \text{Var}(\beta_{Y|X}X) + \text{Var}(\text{error}) \\
\Longrightarrow \quad \sigma^2_Y &= \beta^2_{Y|X}\sigma^2_X + \sigma^2_{Y|X} \\
\Longrightarrow \quad \beta_{X|Y} &= \frac{\sigma^2_X}{\sigma^2_Y}\beta_{Y|X} \text{ by G.5 and the above} \\
\Longrightarrow \quad \frac{\beta_{Y|X}}{\beta_{X|Y}} &= \frac{\sigma^2_Y}{\sigma^2_X} = \frac{\sigma^2_{Y|X}}{\sigma^2_{X|Y}} \text{ by G.8 and above}
\end{aligned}$$

The reader should note that the ratio is required for the simplification since in general $\sigma_X \neq \sigma_{X|Y}$ and $\sigma_Y \neq \sigma_{Y|X}$.

Galton's acolyte Karl Pearson would later generalize regression to any random variable, showing that it does not depend on the normal distribution. Suppose that the expectations below exist, then

$$\begin{aligned}
Y - \mu_Y &= \beta(X - \mu_X) + e \\
\Longrightarrow \quad (Y - \mu_Y)(X - \mu_X) &= \beta(X - \mu_X)(X - \mu_X) + e(X - \mu_X) \\
\Longrightarrow E(Y - \mu_Y)(X - \mu_X) &= E\beta(X - \mu_X)^2 + E(e)E(X - \mu_X)) \\
\Longrightarrow \quad \text{Cov}(X, Y) &= \beta\,\text{Var}(X) \\
\Longrightarrow \quad \beta &= \text{Cov}(X, Y)/\text{Var}(X).
\end{aligned}$$

G.2 Problem 2

> An index q moves under some restraint up and down a bar AQB, its mean position for any given position of the bar being Q; the bar, always carrying the index with it, moves under some restraint up and down a fixed frame YMY', the mean position of Q being M: the movements of the index relatively to the bar and of the bar relatively to the frame being quite independent. For any given observed position of q, required the most probable position of Q (which cannot be observed); it being known that the probable error of q relatively to Q in all positions is b, and that of Q relatively to M is c. The ordinary law of error is to be assumed.

Simply put, the final position of the point on the bar is the addition of two independent influences: the point moving up and down the bar, and the bar itself moving. The problem actually derives from the pairs of brothers that Galton was analyzing, whose variability in height was tricky to estimate from his limited data. He worked primarily from his 'special' data set of heights for brothers, but then repeated all the work using his 'R.F.F.' data set.

Dickson wrote the problem more clearly as $Z = X + Y$ where X is the deviation of the point and Y is the deviation of the bar and both are normally distributed around zero. Note

[3] F. Galton 1889c, 223 introduces a typo in formula 11 that was not present in F. Galton 1886d. The correct value is $\frac{a^2}{c^2}$ not $\frac{a^2}{b^2}$. In both sources 'plane of xz' should read 'plane of yz' since Dickson is calculating σ^2_Y.

that X and Y are independent. If we know the value of Z, say Z=a, then the problem is to find the most probable value of X corresponding to that. Dickson provided a terse solution, but the following explanation may help. If the variables are rescaled to x/σ_X and y/σ_Y then their distribution is wholly symmetrical around $(0,0)$ and the likelihood is maximized when points are closest to the origin. To minimize the distance from the origin is the same as to minimize

$$\frac{x^2}{\sigma_X^2} + \frac{y^2}{\sigma_Y^2} \tag{G.9}$$

which is also square of the joint probable error of (x, y). Using the given constraint $X + Y = a$ to substitute y or x and setting derivatives to 0 gives the solutions

$$x = (\frac{\sigma_X^2}{\sigma_X^2 + \sigma_Y^2})a \qquad y = (\frac{\sigma_Y^2}{\sigma_X^2 + \sigma_Y^2})a$$

To calculate the (square of the) probable error of either X or Y, one can plug this estimate into G.9 above and simplify it. Alternatively one can work it out from the regression

$$\begin{aligned}
X &= \beta Z + e \\
\Longrightarrow \mathrm{Var}(X) &= \mathrm{Var}(\beta Z) + \mathrm{Var}(e) \\
\Longrightarrow \qquad \sigma_X^2 &= \beta^2(\sigma_X^2 + \sigma_Y^2) + \sigma_{X|Z}^2 \\
\Longrightarrow \qquad \sigma_{X|Z}^2 &= \sigma_X^2 - \frac{(\sigma_X^2)^2}{\sigma_X^2 + \sigma_Y^2} \\
&= \frac{\sigma_X^2 \sigma_Y^2}{\sigma_X^2 + \sigma_Y^2} \qquad \text{(G.10)}
\end{aligned}$$

This suggests an easier method to estimate X from $Z = a$ *if we already know that X must be a linear function of Z*. This is the regression equation $X = \beta Z + e$ and we can start there by estimating β in the usual way:

$$\begin{aligned}
\beta &= \frac{\mathrm{Cov}(X, Z)}{\mathrm{Var}(X + Y)} \\
&= \frac{\mathrm{Cov}(X, X + Y)}{\mathrm{Var}(X)} \\
&= \frac{\mathrm{Cov}(X, X) + \mathrm{Cov}(X, Y)}{\mathrm{Var}(X) + \mathrm{Var}(Y)} \\
&= \frac{\sigma_X^2}{\sigma_X^2 + \sigma_Y^2} \quad \text{since } \mathrm{Cov}(X, Y) = 0
\end{aligned}$$

Thus $E(X) = \beta\, a = a\, \sigma_X^2/(\sigma_X^2 + \sigma_Y^2)$. One may calculate the variance as in G.10.

Alternatively, one can use a Bayesian argument to deduce X from Z=a. Dickson would have called this 'inverse probability' but he does not mention it.

$$\begin{aligned}
P(X|Z = a) &= P(Z|X)P(X)/P(Z = a) \\
&\propto P(Z|X)P(X)
\end{aligned}$$

In terms of normal distributions we can write this as follows

$$\begin{aligned}
f_{X|Z=a}(x, z) &\propto f_{Z|X}(x|y) f_X(x) \\
&= f_{X,X+Y}(x, z)
\end{aligned}$$

Since we know by the independence of X and Y that the joint distribution $f_{X,Y} = f_X f_Y$ we can use a transformation of coordinates $g(x, y) = (x, x + y)$ and the fact that $\mu_X = 0, \mu_Y = 0$ to get

$$\begin{aligned} f_{X,X+Y}(x,z) &= f_X(x) f_Y(z-x)1 \\ &\propto exp[-\frac{1}{2\sigma_X^2}x^2]\, exp[-\frac{1}{2\sigma_Y^2}(z-x)^2] \\ &\text{completing the square and factoring constants,} \\ &\propto exp[-\frac{1}{2}(\frac{\sigma_X^2+\sigma_Y^2}{\sigma_X^2\sigma_Y^2})(x - a\frac{\sigma_X^2}{\sigma_X^2+\sigma_Y^2})^2] \\ &\text{which is } N(a\frac{\sigma_X^2}{\sigma_X^2+\sigma_Y^2}, \frac{\sigma_X^2\sigma_Y^2}{\sigma_X^2+\sigma_Y^2}) \end{aligned}$$

This gives the same estimate of X and simultaneously confirms that the variance of the estimate is

$$\sigma_X^2\sigma_Y^2/(\sigma_X^2+\sigma_Y^2).$$

G.3 Fraternal Variance

Estimating the within-family variance of fraternal heights, which he needed to understand the regression of brother on brother, brought out a great deal of ingenuity from Galton.[4] As explained above, he did not have enough data to directly estimate this variance, given that each family was pretty small. He devised four different ways of finding an estimate and averaged the results to get a final best estimate, anticipating a procedure now called 'model averaging'.

1. This was method IV but it is easiest to explain first. From his complete set, Galton repeatedly drew random pairs of brothers, call these $Y_{1,j}$ and $Y_{k,j}$ in order of age, and calculated differences in height, $|Y_{1,j} - Y_{k,j}|$. Here $k > 1$. He could easily calculate the variance of this set of differences. Then

$$\begin{aligned} \text{Var}|Y_{1,j} - Y_{k,j}| &= \text{Var}(Y_{1,j}) + \text{Var}(Y_{k,j}) \\ &= 2\,\text{Var}(Y_{1,j}) \text{ by assumption} \\ \implies \quad \text{Var}(Y_{1,j}) &= \frac{1}{2}\text{Var}|Y_{1,j} - Y_{k,j}| \end{aligned}$$

   and we have one method of estimating the variability. For his data, Galton obtained $\sigma_{Y_{1,j}} = 1.10$.
2. For method II, Galton could directly get the regression coefficient of brother-on-brother using all his data, and estimated it to be close (once again) to 2/3. If we decompose Z_{ij}, the height of the i-th brother in the j-th family as the sum of two influences, a family-specific influence X_i and an idiosyncratic component Y_{ij}, both normally distributed with mean 0, then for the eldest $Z_{1j} = X_i + Y_{1j}$. Note that Y_{ij} is supposed to be independent of all other influences. Galton knew that $\sigma_{X_j}^2 + \sigma_{Y_{2j}}^2 = \sigma_{Z_{2j}}^2 = 1.7^2$ directly from his data, but he wanted to determine the component $\sigma_{Y_{2j}}^2$. From the regression, $E(Z_{1j}|Z_{1j} = a) = \frac{2}{3}a$.

$$\begin{aligned} \frac{2}{3}a &= E(Z_{1j}|Z_{2j} = a) \\ &= E(X_j + Y_{1j}|Z_{2j} = a) \\ &= E(X_j|Z_{2j} = a) + E(Y_{1j}|Z_{2j} = a) \end{aligned}$$

[4] The exposition here owes a lot to Stigler 1986, 293-4.

$$
\begin{aligned}
&= E(X_j|Z_{2j}=a)+E(Y_{1j})\\
&= E(X_j|Z_{2j}=a)\\
&= \frac{\sigma^2_{X_j}}{\sigma^2_{X_j}+\sigma^2_{Y_{2j}}}a \text{ G.10}\\
\implies \sigma^2_{X_j} &= \frac{2}{3}(\sigma^2_{X_j}+\sigma^2_{Y_{2j}})
\end{aligned}
$$

He could then substitute $\sigma^2_{X_j}=1.7^2-\sigma^2_{Y_{2j}}$ into the above equation.

$$
\begin{aligned}
1.7^2-\sigma^2_{Y_{2j}} &= \frac{2}{3}(1.7^2-\sigma^2_{Y_{2j}}+\sigma^2_{Y_{2j}})\\
\implies \qquad \sigma_{Y_{2j}} &= \sqrt{\frac{1}{3}1.7^2}\approxeq 0.98
\end{aligned}
$$

3. Method III is similar to II but uses the expression for the *variance* of the unobserved component X_j, rather than the regression coefficient. From the data Galton could calculate $\text{Var}(Z_{1j}|Z_{2j})$ directly, and it is about 1.24^2 in each case. However we can also derive a formula for it.

$$
\begin{aligned}
1.24^2 &= \text{Var}(Z_{1j}|Z_{2j})\\
&= \text{Var}(X_j+Y_{1j}|Z_{2j})\\
&= \text{Var}(X_j|Z_{2j})+\text{Var}(Y_{1j}|Z_{2j})\\
&= \frac{\sigma^2_{X_j}\sigma^2_{Z_{2j}}}{\sigma^2_{X_j}+\sigma^2_{Z_{2j}}}+\sigma^2_{Y_{1j}} \text{ by G.10}\\
&= \frac{\sigma^2_{Y_{1j}}(1.7^2-\sigma^2_{Y_{1j}})}{1.7^2}+\sigma^2_{Y_{1j}}\\
\implies \qquad 0 &= [\sigma^2_{Y_{1j}}]^2-2(1.7^2)[\sigma^2_{Y_{1j}}]+1.24^2 1.7^2
\end{aligned}
$$

Galton correctly gave one solution to this quadratic equation $\sigma^2_{Y_{1j}}=1.7(1.7-\sqrt{(}1.7^2-1.24^2))$ but *calculated the value incorrectly*. It gives $\sigma_{Y_{1j}}\approxeq 0.96$ and not his 1.10.[5] But he has a more serious problem in that $\sigma^2_{Y_{1j}}=1.7(1.7+\sqrt{(}1.7^2-1.24^2))$ is *also* a solution to the quadratic, yielding $\sigma_{Y_{1j}}=2.206122\approx 2.21$ and he gives no reason to exclude it. It is well-defined.[6]

4. Method I involves a different decomposition. One may write $Y_{ij}=\text{mean}(Y_j)+(Y_{ij}-\text{mean}(Y_j))$. Then Galton assumes that their covariance is small, writing

$$
\text{Var}(Y_{ij})=\text{Var}(\text{mean}(Y_j))+\text{Var}(Y_{ij}-\text{mean}(Y_j))
$$

Now he could estimate the second summand directly from the data by pooling differences from families of the same size n_j. But this was too laborious, so he only worked this out for families of size 4, 5, 6 and 7. Call this d_j. This gives

$$
\begin{aligned}
\sigma^2_{Y_{ij}} &= \text{Var}(\sum_{i=1}^{n_j}Y_{ij})+d_j\\
&= \frac{1}{n_j}\sigma^2_{Y_{ij}}+d_j
\end{aligned}
$$

[5] F. Galton 1889c, 128 gives 0.98, which is closer, but the table on the very next page 129 restores the incorrect 1.10 from F. Galton 1886d!

[6] This was not noticed by Stigler 1986.

$$\implies \sigma^2_{Y_{ij}} = \frac{n_j}{n_j - 1} d_j^2$$

Plugging in the four estimates $d_4, ..., d_7$ (which he does not disclose) yielded 1.01, 1.01, 1.20, and 1.80, for a mean of $\sigma^2_{Y_{ij}} = 1.075 \approxeq 1.07$. Though he added this average to the pot, it would arguably have been better to include the four estimates individually.

Galton averaged the estimates above $(1.10, 0.98, 1.10, 1.07)$ finding $1.0625 \approxeq 1.06$, his estimate for the 'Special' data set. If we correct his method III estimate to 0.96, include the additional solution from there of 2.21 and separately include $(1.01, 1.01, 1.20, 1.80)$ from method I, we get (1.10, 0.98, 0.96, 2.21, 1.01, 1.01, 1.20, 1.80) with mean 1.28, which is not far off 1.06.

Appendix H

The Ancestral Law

Galton based his law of ancestral inheritance on a causal theory he derived from first principles.[1] This supposes that a deviation in a character expressed by a person is the result of a combination of *specific* deviations, one from each ancestor. These ancestral contributions are a *latent* source, the parental are *patent*. Each contribute a half. He worked in terms of mid-parents.

We will use different notation to capture Galton's ideas more precisely. In fact he never stated it very precisely himself, leading to considerable confusion in later years.[2] Call the trait peculiarity of the i-th generation d_i, where i indexes through 0 (the person), 1 (mid-parent), 2 (mid-grandparent), and so on.

$$d_0 = \sum_1^{\infty} \text{ancestral deviation}_i \tag{H.1}$$

He supposed that with each generation a person inherits half of their peculiarity directly from their mid-parent and the other half from more distant ancestors, via the stirp.

$$\text{ancestral deviation}_i = (\frac{1}{2})^i \, d_i \tag{H.2}$$

This gives the form of the law he used in practice to make predictions. If one knows all the values of the d_i this is straightforward.

$$d_0 = \sum_1^{\infty} (\frac{1}{2})^i \, d_i \tag{H.3}$$

In practice one knows a few generations of d_i at most. He solved this problem by proportionally imputing the value of the remaining unknown terms using the known population frequency of the trait. An approximation to be sure, but one that he was comfortable with.

Galton was aware that his halves were open to question. So he tried to cross-check this. One way was to allow the fraction between generations to take on different values

$$\text{ancestral deviation}_i = \beta_i d_i \tag{H.4}$$

Then he equated the simple regression prediction $d_0 = \frac{2}{3}d_1$ with the sum of the ancestral deviations. (He provided no particular justification for this). By imposing restrictions on the form of β_i he could solve for the introduced parameter(s). But he certainly did not think of these as a series of unrelated constants, they had to arise from the generational process and so would be related systematically to each other. He considered two straightforward alternatives of this kind for β_i.

[1] F. Galton 1885f, 1886c, 1889c, 1897b.

[2] See especially Provine 2001.

(i). The first was a simple constant, β. This corresponds to the idea that the operation of heredity between generations is usually the same, which is what he had assumed by setting the fraction to half.

$$\frac{2}{3}d_1 = \sum_1^{\infty} \beta\, d_i \tag{H.5}$$

This can be solved by finding values for the d_i. Here he used regression from offspring to mid-parents, $d_{i+1} = \frac{1}{3}d_i$. From the recursion he obtained:

$$\frac{2}{3}d_1 = d_1 \sum_1^{\infty} \beta\, (\frac{1}{3})^i \tag{H.6}$$

Assuming that $d_1 \neq 0$ and using the fact that $\sum_i^{\infty} r^i = 1/(1-r)$ when $|r| < 1$, this gives

$$\frac{2}{3} = \frac{3}{2}\beta \tag{H.7}$$

and thus $\beta = \frac{4}{9} \approx \frac{1}{2}$.

(ii). The second value Galton tried for β_i was an exponentially decreasing value, which we may write as β^i. We must now solve:

$$\frac{2}{3}d_1 = d_1 \sum_1^{\infty} (\frac{1}{3}\beta)^i \tag{H.8}$$

Using the same method as above, provided that $|\beta/3| < 1$ we get

$$\begin{aligned} \frac{2}{3} &= \frac{3\beta}{3-\beta} \\ \Longrightarrow \beta &= \frac{6}{11} \approx \frac{1}{2} \end{aligned}$$

It was no more than a rough sanity check on the halving proportions used for the specific ancestral components. It was not a derivation of the law per se. It still uses assumption Equation H.1, p. 443, which must be taken as given and is really the essence of what Galton was doing. His use of the arbitrary regression 2/3 and its converse 1/3 is perhaps problematic.[3] A simple constant may be an inspired simplification though, since modern evidence suggests that most traits are about ½ heritable, if pretensions to precision are dropped.[4]

At no stage did Galton ever think of this in terms of generalized multiple regression, which had not even been invented at the time that he first formulated his law.

F. Galton 1897b returned to the justification he obviously felt was still lacking for his halving theory, offering some additional arguments. He invoked recent observations of binary subdivision in germ cells, but he did not know this has nothing to do with ancestral contributions, and is a consequence of the finite size of the genome that the child can inherit, so each parent halves its own contribution. His contention that halving is required to make all ancestral contributions sum to one begs the question whether they should contribute at all, or as the simple sum he proposed (the assumption in Equation H.1, p. 443, that he had never justified).

—oOo—

[3] Stigler 1986.
[4] Polderman et al. 2015.

There are more complicated scenarios when applying the ancestral law, with halving, in cases where partial information only is known. It may be helpful to work through these to see how his predictions can be replicated. The simple scenario is predicting a son from a father's total contributions:

$$\frac{1}{4}[d_1] + 2\frac{1}{16}[\frac{1}{3}d_1] + 4\frac{1}{64}[\frac{1}{9}d_1] + ... \tag{H.9}$$

The infinite geometric series above converges and can easily be summed with a little manipulation, to give $0.30\, d_1$, which is Galton's estimate of the total inherited contribution to the son through the father. Combining this with the mother, assumed for simplicity to also have d_1, 0.60 of the peculiarity of the child comes from the two lines of parents, and the rest from other sources.

If we know say h for only a grandparent, and neither parent, then—since we must estimate the son of the grandparent to be $(1/3)h$, and note that we are not using the mid-parent—similar reasoning produces

$$\frac{1}{4}[\frac{1}{3}h] + \frac{1}{16}[h] + 2\frac{1}{64}[\frac{1}{3}h] + 4\frac{1}{128}[\frac{1}{9}h] + ... \tag{H.10}$$

which sums to $0.158h \approxeq 0.16h$. Taken together, the 4 grandparental lines contribute $4 \times 0.16 = 0.64$ to the peculiarity of the child.

Consider now the eye colour calculations in *Natural Inheritance.* Three cases are shown below for predicting light-eyed offspring (y_l) in a family with n members: the first where light-eyed parents (p_l) are known, the second where light-eyed (g_l) and hazel-eyed (g_h) grandparents are known, and the third where both are known. Each formula has a final residual term from the table which in general is distributed between predicted light and dark eyes proportionally, i.e. 2/3 to light-eyes below.

$$\begin{aligned} y_l/n &= 0.30\, p_l + \frac{2}{3}0.4 \\ y_l/n &= 0.16\, g_l + 0.10\, g_h + \frac{2}{3}0.36 \\ y_l/n &= 0.25\, p_l + 0.08\, g_l + 0.05\, g_h + \frac{2}{3}0.18 \end{aligned} \tag{H.11}$$

(Here the coefficient 0.30 was derived as above.) Thus in a family of $n = 12$, with $p_l = 2$ light-eyed parents, $g_l = 3$ light-eyed grandparents, and one hazel-eyed grandparent, $g_h = 1$, the third formula gives

$$y_l/12 = 0.25\,(2) + 0.08\,(3) + 0.05\,(1) + \frac{2}{3}0.18$$

By comparing these predicted values against his actual data Galton satisfied himself that the correspondence was close.

——oOo——

Karl Pearson doubted Galton's causal theory, which had the flimsiest of justifications and is, as we now know, incorrect. Instead he interpreted 'Galton's Ancestral Law' in terms of multiple regression to see if he could recover its results without a causal theory. The applied mathematician was in any event an early positivist, philosophically opposed to causal theories and ghostly mechanisms. The thing to do, and the only thing to do, was *predict* characters of offspring based on those of ancestors. (Again, this was not Galton's own point of view.)

Since there were many ancestors this led Pearson first to formulate a completely general technique of multiple regression, building on prior work by F. Y. Edgeworth.[5] In general form, for a character deviation d_i from the mean in generation i this may be written as

$$E[d_0|d_1, d_2, d_3, ...] = \beta_1 d_1 + \beta_2 d_2 + \beta_3 d_3 + ...$$

Galton had deduced from his causal argument and repeated substitution that the β_i form a geometric series. Unfortunately in the multiple regression context, this meant, as Pearson showed, that ancestors beyond the immediate parents contribute nothing at all to descendants. The trouble was the i-th generation coefficient of $1/3^i$ that Galton had used. But the paper of 1897 on Basset hounds impressed Pearson because of the concordance between the predictions and observations found there.[6]

The following year, Pearson was relieved to conclude that Galton was on the right track after all—provided one assumes a coefficient of $0.6(1/2^i)$ between the descendants and their i-th generation ancestors, rather than Galton's troublesome $1/3^i$. That gives coefficients of 0.3, 0.15, 0.075, ...[7]

Still, the mathematician was bothered by the idea of a magic constant like 0.6, regardless of the trait involved. So he generalized the law, replacing 0.6 with a parameter γ which could be estimated (given an additional constraint) from the data obtained for the particular trait involved. He would revisit the ancestral law in later publications to handle situations like discrete characters. But the discussion is already very far removed in intent and execution from anything that Galton had contemplated, so we will not follow him further. See Bulmer 2003 for a more complete exposition of this history, which may appeal to the specialist.

——oOo——

For inexplicable reasons, in his discussion of Galton's 'ancestral law' coefficients Pearson refers to *correlations*, which Galton had not yet invented, Similarly Bulmer talks of correlation coefficients in this context, more than once.[8] Perhaps Pearson is thinking of correlations then when he doubts that Galton's coefficients can be multiplied as they indeed are. In general correlations are not necessarily transitive, but these are just regression coefficients and are not measures of association.

Suppose $X = \beta_Y Y + e_Y$ and that $Y = \beta_Z Z + e_Z$ where $E(e_Y) = 0 = E(e_Z)$. Then

$$\begin{aligned} X &= \beta_Y(\beta_Z Z + e_Z) + e_Y \\ &= (\beta_Y \beta_Z)Z + (\beta_Y e_Z + e_Y) \end{aligned}$$

We may write this as $X = \beta Z + e$ since $E(\beta_Y e_Z + e_Y) = 0$.

5 K. Pearson 1896; Stigler 1986.
6 F. Galton 1897a.
7 K. Pearson 1898. This inordinately ponderous and abhorrently-written paper fully justifies the characterization of Pearson's peculiar abilities in Fisher 1956.
8 K. Pearson 1930a, 39, Provine 2001, Bulmer 2003.

Appendix I

Regression of Artistic Ability

From his Record of Family Faculties responses Galton figured that roughly 30% of the population may be considered definitely artistic, and 70% not. From his tables for the Normal Distribution, this threshold is a 'probable error grade' of 0.78 (or about 0.53 standard deviations). To simplify he chose to reason about the average of these definitely musical people, that is the average of all those with grade exceeding $P + 0.78$, which he roughly calculated as the threshold $P + 1.71$.

If regression to the mean held for artistic ability as it did for stature, with a coefficient of 2/3, he expected the mean of the children of that mid-parent to be $P + (2/3)1.71 = P + 1.14$. Among these children, any with grade in excess of $P + 0.78$ would definitely be musical. This may be written as $(P + 1.4) - 0.36$. Treating this as a normal distribution with mean $P + 1.4$, if one knows the probable error of the children, then the corresponding percentage of the artistic among them can be looked up.

Here Galton assumed for simplicity that there would be the same relationship as there was for stature, so that a probable error of 1 among the parents reduces to $1.5/1.7 \approx 0.88$. One should find about 60% of the children above $(P + 1.4) - 0.36$, and 40% below.[1] From his observed data Galton saw 'very happy agreement' with 64% children counted artistic and 36% not.

He made a similar calculation for the non-artistic parents and again found 'a very fair coincidence' of expected and observed percentages. Thus the data was, in his view, at least consistent with regression to the mean for artistic faculty.

[1] In standard deviations, $\text{pnorm}(-0.36 * 0.6744, 0, 1.5/1.7) \approx 0.392$.

Sources

Principal primary sources are listed here. Other sources are listed individually in the text.

1. The Collected Works of Francis Galton
   `http://galton.org`.
2. GALTON. Francis Galton papers, Special Collections, University College, London, UK.
3. PEARSON. Karl Pearson papers, Special Collections, University College, London, UK.
4. APS. American Philosophical Society, Philadelphia, USA.
5. DARWIN. The Darwin Correspondence Project.
   `https://www.darwinproject.ac.uk/`.
6. ANDERSSON. Andersson Papers, photocopy in the South African Library, Cape Town, South Africa. See also the microfilm in the University of Witwatersrand library, Gauteng, South Africa. The originals reside with the Andersson family, Dolobran, Parktown, Gauteng, South Africa.
7. LIVINGSTONE. The David Livingstone papers, at `https://livingstoneonline.org`.
8. Wellcome Collection, London, UK.
9. Alfred Russel Wallace Correspondence.
   `https://epsilon.ac.uk`.
10. Smithsonian Institution, Washington D.C., USA.
11. British Library Manuscripts, London, UK.
12. Bateson and Airy Papers, Cambridge University Library, UK.
13. Houghton Papers, Wren Library, Trinity College, Cambridge, UK.
14. Royal Geographical Society, London, UK.
15. Royal Anthropological Institute, London, UK.
16. Huntington Library, San Marino, California, USA.
17. Special Collections, Keele University, UK.
18. National Archives, Kew, London, UK.
19. Darwin Papers, University of Virginia, Virginia. USA.

FG: Francis Galton.
FVG: Frances Violetta Galton.
STG: Samuel Tertius Galton.
EG: Elizabeth Galton.

Glossary

Bentham, George 1800–1884. The botanist nephew of Jeremy Bentham. A gifted linguist, he had known his uncle's associate John Stuart Mill as a young man. Before he took up botany, he had interests in logic, publishing *Outline of a New System of Logic* in 1827, which introduced predicate quantifiers (All X is Y), but was mostly ignored, only to be rediscovered later and feature in a priority debate in the *Athenaeum* in 1850-1 (*Athenaeum*, No. 1208, December 1850, 1351. See also F. Galton 1908, 174–175, which slightly misremembers the piece). Galton had befriended Bentham by 1856, when he was no longer concerned with symbolic logic. He served as president of the Linnaean Society several times. 51

Brandram, Eleonora Anne 1830–1896. The daughter of Francis Holles Brandram, Justice of the Peace for Kent and Sussex, by his wife Maria. The Galtons first met 'Miss Brandram' in Switzerland in September 1862. When they looked her up at Hastings in 1866, she agreed to accompany them to Europe the following year as a companion. In 1875 she became the second wife, and then the widow, of the archaeologist John Ferguson McLennan (1827–1881) but remained a close friend of the Galtons. McLennan was a bitter critic of one of Galton's friends, Henry Sumner Maine. 'Maine, knowing that I was well acquainted with him, begged me to do what I could to moderate his controversial tone; I tried in vain' (F. Galton 1908, 67). 'She was a great friend to both of us; a companion and kind nurse to my wife when she was ill, an excellent walking companion to myself, and always ready to be of service. She helped me much in revising some of my earlier writings, especially the last edition of my *Art of Travel*' (F. Galton 1908, 192). After the death of her husband she edited his *Studies in Ancient History* (1896). She was travelling with the Galtons in France later that year when a fall near Nice, while crossing a stream, led, in Galton's opinion, to her death at home sometime later from internal injuries. 33, 35

Crackanthorpe, Montague 1832–1913. Son of Christopher Cookson of Wellington. Educated at St John's College, Oxford, taking a double first in classics and mathematics and winning a fellowship. Practiced successfully at the bar, Q.C. in 1875. Took the name Crackanthorpe after inheriting the Newbiggin estate in Westmoreland in 1888. Stood as a Liberal for Parliament twice, becoming a unionist. Lived at 65 Rutland Gate, and so was a neighbour of Galton, who lived at number 42. Wrote widely in the periodical press, covering subjects like population control, which led to a book in 1907, *Population and Progress*. On June 25th, 1908, he hosted a meeting of the Eugenics Education Society at his house where Galton spoke on eugenics, which led to widely increased membership when reported. Succeeded Sir James Crichton Browne as president of the society, at Galton's desire.
335

Darwin, George Howard 1845–1912. Physicist and mathematician. The fifth child of Charles Robert Darwin the naturalist, born at Down House. Studied at Trinity College, Cambridge, under the renowned mathematics coach Edward Routh. Second Wrangler in the Tripos of 1868 and winner of the second Smith's Prize. He was a late developer in mathematics. He was awarded a fellowship at Trinity in 1868 but soon chose to study law instead. He passed the bar in 1874 but did not practice, returning to a second fellowship at Trinity College instead. After 1875 he concentrated on astrophysics, quickly becoming distinguished in the field. He was inducted into the prestigious Royal Society in 1879. Unlike his father he received a K.C.B., awarded in 1905. He collaborated with Galton on statistical questions, such as the properties of the normal distribution, in the 1870s. His brother Francis, in his memoir of George, found 'many records of visits to Francis Galton with whom he was united alike by kinship and affection' but does not give clear dates (F. Darwin 1916, xvi). George Darwin married an American, Maud Du Puy and had 5 children, 4 of whom survived to adulthood. 318

Farrar, Frederic William 1831.viii.07–1903.iii.23. A schoolmaster and ordained reforming clergyman, rising to Dean of Canterbury. He was the son of a missionary and was educated at Trinity College, Cambridge (1850–1856), where, despite enrolling only as a sizar, he soon became an Apostle and was made a Fellow. He taught at Harrow, under Galton's brother in law H. M. Butler, and was the head of Marlborough School when Galton applied to him for statistics of schoolboys. He was a Darwinian philologist, and both Galton and Darwin helped to elect him to the Royal Society in 1866. Later his influence aided Galton's efforts to bury Darwin in Westminster Abbey. Farrar's books, of which there were at least a dozen, included the boarding school novel *Eric, or Little by Little* (1858)—overtly about the perils of impure thoughts, covertly about the deadly effects of masturbation—which went through at least 36 editions and was famously despised many namesakes of 'Eric', including George Orwell (Eric Blair). 12, 68, 158

Farrer, Thomas Henry 1819.vi.24–1899.x.11. A civil servant who worked with Galton on the report to the Board of Trade that became known as the 'Galton Report'; at that time Farrer was a secretary of the Board. He was educated at Eton and Oxford. 29

Galton, Erasmus 1815–1909. The younger of two older brothers of Francis. Joined the navy as a midshipman at age 13, serving on *H.M.S. Southampton* and various other ships. 'Erasmus, then a boy of twelve or thirteen, entered the navy, and showed himself to us in his uniform, with the dagger or 'dirk' that was part of it. I, a child of five or so, fingered it with awe, and with my little head full of Greeks and Trojans looked upon him as a hero, like Achilles, and can perfectly recall my sense of increased security from knowing that England could henceforth avail herself of his puissant arm and terrible weapon'. Erasmus left the service, immediately after passing out of the Royal Navy College, in 1836. His sister Bessy attributed this to the lack of influence of their father in securing promotion, but other theories were advanced (see below). He farmed at Loxton, near Weston-super-Mare. In 1850 he ascended Mont Blanc, after previously climbing Mount Etna. He joined the 2nd Warwickshire Militia, serving as a Captain at the time of the Crimean War. He was at Aldershot in 1856 when Francis gave lectures there on the Arts of Campaigning. Though he rose to Major in 1858, he retired from the militia in 1862. Afterward he served as Magistrate for Somerset. His sister remembered him as 'fond of reading, especially scientific works'. He also published in *The Field* and the *Journal of the Royal Agricultural Society* (e.g. on his device for taming savage bulls, 1843; on canals, 1845; on shaking bogs, 1846). Never married, he had religious beliefs. At his death, the estate was worth over £83,000.

'DEATH OF MR. ERASMUS GALTON. [*Leamington Spa Courier*, Friday 26 February 1909, 7.] Mr. Erasmus Galton, who has for very many years resided at the Regent Hotel during the winter months, died Tuesday at the advanced age of 93. The 27th of last month he slipped and fell on the pavement on the Parade, fracturing his thigh, and confinement to bed undoubtedly accelerated complications which caused his death. Mr. Galton entered the Navy in 1828, joining *H.M.S. Southampton*, a 52 gun frigate, flagship

of Admiral Sir Edward Owen, and served in her the East India Station till 1832, when she was ordered to join the English and French fleets in the Downs, and from thence to sail to the blockade of Antwerp. While in the Downs she was struck by lightning, which did much damage and set the ship on fire in about 30 places. Owing to the assistance rendered by other ships of the two fleets, the fires were speedily extinguished. From 1833 to 1836 Mr. Galton served on *H.M.S. Favourite*, an 18 gun corvette, on the Mediterranean Station. In 1836 he left the Navy, and turned his attention to farming, and settled down at Loxton, near Weston-super-Mare, the property belonging to his father, which, on the death of the latter in 1844, came into his own possession. In 1852 he joined the 2nd Warwickshire Militia, and served in it till retired as Major in 1862. During the Crimean War, in 1855, and again at the time of the Indian Mutiny, the regiment was embodied, and did service at Aldershot and in Ireland. Mr. Galton was the second son of the late Mr. Samuel Tertius Galton, of Duddeston, Warwickshire, and F. A. Violetta Darwin, daughter of Dr. Erasmus Darwin, a grandfather of Charles Darwin, the naturalist. Mr. Galton was the nephew of Captain Barclay, the pedestrian, celebrated for having first performed the feat of walking 1,000 miles in 1,000 hours. largely inherited the pedestrian powers and love of walking of his relative, and ascended Mont Blanc in what was then record time. Mr. Galton was J.P. and D.L. for Somersetshire, and lord of the manor of Loxton.
An Appreciation. In the 'twenties of last century two happy little boys mounted on scampering ponies were a familiar sight in the then rural outskirts of Birmingham. They were Darwin and Erasmus Galton, so named after their distinguished grandfather, Dr. Erasmus Darwin, F.R.S., physician, poet, and philosopher, grandfather also to Charles Darwin, the great naturalist. Darwin Galton died not long since, having left his mark as Warwickshire country gentleman, and now Erasmus, whose bent figure was familiar in Leamington, has followed him, at the age of 93. Erasmus joined the Navy a boy of 14, and had various experiences in it, especially when detached in charge of one of the ship's boats for days together, engaged in the suppression of piracy in the Malay waters. The discipline of the Navy made a profound impression on his character, although he left the Service before he had even gained a lieutenancy. The cause was this. He and a brother officer, while walking in Malta, were attacked from behind by two ruffians, one of whom dashed a heavy stone down upon the head of Erasmus, felling him to the ground. His companion saved the situation. The offenders were captured and afterwards severely punished; Erasmus was carried unconscious to the hospital. There he had to be left, the ship being under immediate orders to sail. The effects of the blow made further service in the Navy impossible to him, and also modified his character as such injuries often do, making him for a long time particularly sensitive and touchy. He then studied farming in Scotland, and afterwards settled on property in Somersetshire, belonging to his father, called Loxton, which subsequently became his own, and which he much improved by judicious forestry and by additions. He was a strong walker, and one of the earlier climbers of Mont Blanc. It is amusing now to read his grotesquely exaggerated ideas of the difficulty of such ascents, due to the rarity of the air and to mountain dangers, which use has long since rendered familiar to Alpine climbers. He joined the Warwickshire Militia in the fifties, and being at Aldershot when the camp was first established, he took a fancy to the simplicity of the accommodation provided there at that time, and built himself a stone house at Loxton modelled on the design of an Aldershot wooden hut. It was and continued to be comically deficient in the ordinary comforts of an English home; there were not even means for boiling a potato, the operation, when required, being performed at a somewhat distant cottage. But stoicism was a luxury to him. He always rose very early, even at 2 or 3 a.m., occupied himself methodically at his accounts and current business, read much, and led a very original life which suited his disposition. He collected

numerous nautical and other appliances for dealing with heavy weights, stone or timber, carefully labelled and arranged in a most methodical way, all kept scrupulously clean and effective, and most it in frequent use. He was an extremely kind but a punctilious landlord. The children of his tenants loved him. and he was seen at his best with those children clustering about him, his pockets filled with goodies for them. His movements during the year were almost as regular those of the celestial bodies; he visited Loxton, Ryde, and Leamington each in their due season, living always in a hotel in the two latter towns, for he hated the encumbrance of a house. His life was guided by particularly strict sense of duty, perhaps more strict in trifles than consistent with taste, and he never spared himself. He had a strongly religious bent of mind, while abhorring dogma. He habitually faced death without fear, and has now happily passed away with little suffering'.
'WILLS AND BEQUESTS. [*Leamington Spa Courier* 1909-05-07, 4.] Mr. Erasmus Galton, Loxton, Somerset, and of Leamington, D.L., J.P. for Somerset, formerly of the Royal Navy, and afterwards major of the 2nd Warwickshire Militia, Lord of the Manor of Loxton, who died at the Regent Hotel, Leamington, on the 23rd February last, aged 93 years, second son of the late Mr. Samuel Tertius Galton of Duddeston, Warwickshire, and cousin of Charles Darwin, the naturalist, left estate of the gross value of £84,523 11s 8d, of which the net personalty has been sworn at £60,849 10s 4d. Probate of his will, dated 29th November, 1906, with codicil of the 1st December last, has been granted to his nephew, Mr. Edward Galton Wheler, J.P., of Claverdon Leys, Warwick, his cousin Miss Hilda Galton, of Leamington, and Mr. Henry Field, solicitor, of Leamington. The testator left £500 to his cousin, Miss Hilda Galton, of Leamington; £150 to Mr. Henry Field, solicitor, of Leamington; a life annuity of £300 to his brother, Francis Galton; £300 for distribution equally between the children of Mrs. Cameron Galton, late Shelsey, Worcester; £300 equally between the children of Mrs. Lucy Amelia Biggs, formerly the wife of the Rev. Hesketh Biggs; £300 each to Mr. Herman Galton and Miss Dora Galton; £200 to Mrs. Emily Dodd, widow of former Rector of Loxton; £200 to Mrs. Margaret Ashwin, of Christon Parish, Loxton; £100 to Mrs. Barker, 5, Queen's Gate, Plymouth, daughter of the Rev. Wm. Littlehales, formerly Vicar of Compton Bishop, Somerset. The testator also directed that his remains should be cremated, and that no mourning of any kind should be used at his funeral or worn by any of his relatives, and if he should die at Loxton, each of his cottagers who should carry his temporary coffin should receive £1. He also left £100 to the Birmingham Hospital, £100 to the Warneford Hospital, Leamington; £100 to the Weston-super-Mare Hospital; £100 to the South Hants Infirmary, Southampton; £300, a further £10 for each year of service since January, 1878, his wearing apparel, and a life annuity of £50 to his valet, John James Morrish; a life annuity of £30 to Sarah Ann Morrish, wife of his said valet; £10 to each of the five children of his said valet; £200 to George Southwood, his headman at Loxton; £25 to William Henry Stark, labourer at Loxton; £15 to each of his cottage tenants at Loxton who has signed his or her agreement and who has been his tenant for ten complete years; £5 to each such tenants similarly qualified of five years' tenancy; 6s. per week for life to his cottager at Loxton, Eliza Sevier; to Eliza Sevier, Francis Scott, and Mrs. Joseph Every, cottagers at Loxton, each the use for life at a nominal rental of 2s. 6d. per annum of the cottages now occupied by them, and provided that they shall live in such cottages, and shall not take lodgers and shall not sell either beer, cider, or other intoxicating liquor, but his successor in the Loxton estate has power to buy out their rights at fifteen years' purchase on the proper annual value of these cottages as entered in his estate book; 5s. to each child belonging to Loxton Parish who was present at the last Government school inspection for that parish previous to his decease, whether under the School Board or any other Education authority, this sum being paid to the credit of a Post Office Savings Bank

account in the name of each child at Axbridge Post Office; £10 to the nurse attending him at his decease and in his last illness. He directed that all wages for which he was liable at his decease, whether at Loxton or elsewhere, are to be continued out of his estate for the six months following his decease, provided that the recipient so long live and be not dismissed for misconduct. The residue of his property he left to his nephew, Mr. Edward Galton Wheler, of Claverdon Leys, and failing him he had left his residuary estate to his great-nephew, John Guy Baron Lethbridge, of Tregan, Cornwall'. 43

Gifi, Albert Galton's Swiss manservant for some 40 years. Possibly Rudolf Albert Gifi, who died intestate at 19 Montpelier Row, Knightsbridge, on the 18th of April 1917, survived by his wife Maria Gifi (*Kensington News and West London Times*, Friday 31 October 1919). 358

Grove, Justice William Robert 1811–1896. Active in the Royal Society, which he was responsible for reforming on meritocratic lines. An expert on patent law, he was educated at Oxford, and had early interests in chemistry and electricity. Galton knew him from the Royal Society, and spent a number of vacations shooting with him. Grove contributed a piece on 'Nabloos and the Samaritans' to Galton's series *Vacation Tourists and Notes of Travel in 1861*, F. Galton 1862. Galton was also friendly with his daughter, Mrs. Anna Hills. 29

Hallam, Henry Fitzmaurice 1824–1850. A close friend of Galton from Cambridge, where they were in the Historical Society together. They were part of a circle including Henry Sumner Maine, Charles Astor Bristed and Franklin Lushington. 'Harry' was a son of Henry Hallam the historian, and a younger brother of the poet Arthur Henry Hallam (1811–1833), the lamented subject of Tennyson's *In Memoriam* (1850). He shared his brother's knack for inspiring worshipful devotion, and was a brilliant but uneven and idiosyncratic scholar, gaining a Chancellor's Medal. After failing to be selected as a Fellow at Trinity, he read for the bar instead, joining the Midlands circuit. He died on October 25th 1850 of fever while travelling with his family in Italy, at Siena. The autopsy suggested that Hallam, who was always delicate, could not have expected to live much longer. He was buried inside St Andrew's Church ('obscure and solitary') in Clevedon, North Somerset, along with his siblings. 36

Hawkins, Francis Vaughan 1833–1908. A barrister and founding member of the Alpine Club. He was educated at Harrow and Cambridge, where he was Senior Classic in 1854, getting a record score, and 41st Wrangler. Galton traces the Hawkins family in *Hereditary Genius* (F. Galton 1869b). Hawkins published a notable book, *A Concise Treatise on the Construction of Wills* (1872). His account of an attempt to climb the Matterhorn, accompanied by Prof. John Tyndall, appeared in Galton's own *Vacation Tourists and Notes of Travel in 1860* (F. Galton 1862). He never married. 31

Hollond, Mrs. Robert 1822–1884. Born Ellen Julia Teed in Madras, but raised at Stanmore in Middlesex. She married the wealthy Whig M.P. Robert Hollond (1808–1877) in 1840, but was chiefly active in Paris, where she was an Orleanist bluestocking. Galton met the Hollonds through his wife Louisa, whose Butler family was well-connected with the broader intelligentsia. 31

Hopkins, William 1793–1866. Galton's (second) mathematics coach at Cambridge. Hopkins was one of the two most illustrious freelance coaches in Victorian Cambridge, approached only (in later years) by his own pupil Edward J. Routh. He was Seventh Wrangler himself in 1827, having entered college at the advanced age of 30 after an earlier career as a farmer. He only took pupils in their second year; these included Stokes, Cayley, Sylvester and Thomson. His (second) marriage as a student meant that he could not be elected a fellow, but he earned considerably more as the leading private tutor than most full professors. 'I never had a tutor whom I reverenced and loved so entirely as

Hopkins' (F. Galton 1908, 65). He was part of the committee that recommended Galton as the second General Secretary to the British Association in 1863; Hopkins was the other secretary. 'The office is a very pleasant and gentlemanly one', he wrote to his former pupil, 'I should rejoice if you could be installed as my partner' (Hopkins to Galton, 1863-08-03, GALTON/3/2/2/23). Galton accepted the position, but noticed sad changes, as Hopkins' 'health had by then declined and his fine intellect begun to fail'. Galton's diary for February 1865 already noted 'letters about Hopkins failure of intellect'. Hopkins died in Northumberland House Lunatic Asylum, Green Lanes, Stoke Newington (Craik 2008, 107). 41

Jephson, Dr. Henry 1798–1878. An immensely successful doctor who practiced in Leamington, tending to the well-heeled, including Samuel Tertius Galton. He had risen rapidly, starting as an apothecary's apprentice in 1823, and taking time off to complete a medical degree at Glasgow in 1826-8, before returning to his burgeoning practice at Leamington in 1828. At times he earned over £20,000 a year, a very large sum for the times, allowing him to build a mansion, 'Beech Lawn', in Leamington (since demolished). In 1848 he went blind and was forced to retire to the consolations of well-connected philanthropy—though he continued to treat some patients after this, including Francis Galton himself in 1866-7. His success was founded on the inspiration of his personality and his command of patients. His treatments consisted mostly of 'plain' diets—especially avoiding stimulants, but not excluding essentials like sherry and game—outdoor exercise and taking the sulphurous waters of the Spa, internally and externally. One of his stand-by medicines was an emetic preparation called 'the broom'. 33

Johnson, Alice Neville Vowe 1869–1938. Present at Flinders Petrie's camp in Egypt when Galton visited him there in January 1900. She appeared to already be known to Galton before that. Later she qualified as a doctor in Brussels. She lived with Alice Abadam (1856–1940) for many years. She also corresponded with and visited Galton at Rutland Gate, see GALTON/3/3/10/8.

'Lambeth School Guardians on Wednesday appointed a lady doctor as medical officer for their schools at West Norwood. The schools accommodate about 500 children. The newly appointed officer is Dr. Alice N. V. Johnson. She is thirty-four years of age, and has held appointments previously in the Royal Infirmary, Edinburgh, the Children's Hospital in Great Ormond-street, St. Pierre Hospital, Brussels, and the Children's Hospital, Vienna' (*Lake's Falmouth Packet and Cornwall Advertiser*, 1904-01-09, 5).

'Dr. Alice Neville Vowe Johnson, who died on December 14 at Clifton Road, Wimbledon, was for twenty years medical officer of the "Bird in Bush" Mothers and Infants Welfare Centre in the Old Kent Road. Born in London in 1869, the daughter of Edward Johnson, she was educated at Katherine Lodge. London, at Les Ruches, Fontainebleau, and at the London School of Medicine for Women. She qualified as L.S.A. in 1893 and L.R.C.P. &. S Ed. in 1901; she also obtained the M.D. of the University of Brussels with high distinction, and the F.R.C.S.L in 1903; and held the Cambridge D P H. Dr. Alice Johnson's first appointment was that of assistant medical officer to the Joint Counties Asylum, Carmarthen; she then became medical officer and examiner under the L.C.C. and the Board of Education. She was also for a time clinical assistant to out-patients at the Elizabeth Garrett Anderson Hospital and medical officer to the Central Foundation School Unit. During the war she worked as honorary medical officer at the Chelsea Red Cross Hospital. She had been a member of the British Medical Association for forty-four years, and in the intervals of her work for the welfare of children she enjoyed music and Alpine climbing' (*British Medical Journal*, Vol. 2, No. 4069, Dec. 31, 1938). 360

Lloyd, Francis 1846?–1880. Philosopher, author of a tract, *A Scientific View of Mr Francis Galton's Theories of Heredity*, controverting Galton—see Lloyd 1876. Claimed association with the universities of Halle and Athens. Died at Smyrna in 1880-04-17 'in great suffering'. Also the author of *Prussia's representative man* (1875), about B.H.W. Kleist. Both of these works were produced in collaboration with William Newton (1838?–?). In 1877 he had written to Charles Darwin for financial help, and Darwin responded with a cheque for £10 but thought a subscription would not succeed. (See also `http://levantineheritage.com/pdf/Smyrna_BMD_19th_century_British_Newspapers.pdf`.) 115

Lushington, Sir Franklin 1823–1901. Galton had known Franklin Lushington, whose brothers were part of Tennyson's circle, at Trinity College, Cambridge. Henry Fitzmaurice Hallam was a friend in common, and Lushington, along with Main, would help to edit and publish Hallam's literary remains after Harry's premature death in 1850. Franklin was Senior Classic in 1846—like his elder brother Henry had been in 1832—and gained a fellowship at Trinity. He was appointed to the Supreme Court of the Ionian Islands, then administered by Britain, and, though he had an 'intense friendship' on Corfu with the artist and poet Edward Lear, married the daughter of a vicar, Kate Maria Morgan, on January 21, 1862. In 1869 he was appointed the Thames police court magistrate, later becoming the Chief Magistrate. The Lushingtons had two mature sons, George Henry Fitzjames (1872–?) and Franklin (1875–1926), and two daughters. Franklin was the 4th son of Edmond Henry Lushington. His relatives, the twins Vernon (1832–1912) and Godfrey (1832–1907), were sons of Stephen Lushington (1782–1873) and were among the subjects of Galton's twin investigations. 36, 127

Murray, Rear-Admiral Henry Anthony 1810–1865. The third and youngest son of George, fifth Earl of Dunmore, and Lady Susan Hamilton. A lifelong bachelor, Murray was an officer in the Royal Navy who was actively involved in the RGS in the 1850s and early 1860s. He joined the navy in 1823, obtaining his commission in 1831, and rising to Commander in 1838. After retiring in 1842, Murray travelled widely. He published a travelogue describing his trip through the Americas, *Lands of the Slave and the Free* (1855). He cultivated young male companions and explorers, hosting them at well-attended Monday evening gatherings at his quarters for 'single gentlemen' in D4, the Albany—attendees included Francis Galton, Richard Burton, Bishop Wilberforce, John Petherick, Mansfield Parkyns, Samuel Baker and Tom Hughes. As was the usual practice, in 1862 he was awarded the rank of Rear-Admiral in retirement. 'Death of Admiral the Hon. Henry Murray.—At his residence the Albany, expired the other day, the Hon. Henry Anthony Murray, youngest son of the fifth Earl of Dunmore by Lady Susan Hamilton, daughter of the ninth Duke of Hamilton and Brandon. The gallant admiral entered the navy in 1823, and, after passing his examination, obtained his first commission, in 1831, from which date to 1834, he served in the *Revenge*—Captain, the Hon Donald Hugh Mackay—off Lisbon and in the *Endymion*, Captain Sir Samuel Roberts. In his next vessel, the *Tyne*, Captain Lord Ingestre, he was employed for three years in the Mediterranean. His next appointment was in 1838, as senior lieutenant to the *Modeste*, on the home station. He acquired commander's rank in the same year, and from December 1840, until paid off in the spring of 1842, he served in that capacity again in the Mediterranean, on board the *Wasp*, a small 16 gun vessel. In November following he was advanced to the rank of captain, and was appointed a rear-admiral on the reserved list in 1862. The deceased was a Knight Grand Cross of the Bavarian Order of St. Michael of Merit' (*Kentish Chronicle*, Saturday 25 February 1865, 6). 36

Petrie, William Matthew Flinders 1853–1942. Pioneering archaeologist and Egyptologist. Largely self-taught. After disproving Piazzi Smyth's mystical speculations about the pyramids, he acquired a backer in the form of Amelia Ann Blanford Edwards (1831–1892, spinster, novelist and travel writer) and the Egypt Exploration Fund. He soon fell out with the latter and worked on his own, with financing from a connection of Edwards, the

Manchester textile manufacturer Jesse Haworth (1835–1921). Along with Pitt-Rivers, he introduced systematic excavation, preserving the context of finds. He also excavated in Palestine. Petrie appears to have met Galton when he submitted details of his mental imagery to the scientist. Several of the 'number-forms' illustrating *Human Faculty*, F. Galton 1883d, were from Petrie. A few years later Petrie obtained a grant from the British Association, with the assistance of Galton, for photographing portraits from Egyptian sources to show racial variation. A selection of these were published in *Racial Photographs from the Egyptian Monuments* (1887). Later he would also send many examples of ancient skulls and skeletons to Karl Pearson for biometric analysis. He was appointed Professor of Egyptology at UCL in 1892, after Amelia Edwards endowed a chair for him in her will. In January of 1900, Galton stayed with Petrie for a week, near Luxor in Egypt, to witness excavations in the desert. Soon after this Galton forwarded Petrie's name for election to the prestigious Royal Society (he was admitted in 1902). Petrie was a eugenicist in the Galtonian mould and two (out of his over one hundred) books deal with dysgenic and eugenic themes, *Janus in Modern Life* (1907) and *The Revolutions of Civilization* (1911). The Petrie family was included in Galton and Schuster's *Noteworthy Families* (1906). He was knighted in 1923. 167, 359

Rentoul, Robert Reid 1856–1925. Medical practitioner, born in Ireland and educated at Edinburgh, Dublin, Belfast and London (*Br. Med. J* 1925; 2). Son of John Lawrence and Dorcas Rentoul. He practiced in Liverpool, and was an active member of the British Medical Association, which he deluged with caustic comments and motions on contentious topics. He has been described as a eugenicist, though his beliefs about racial degeneration long preceded that movement and he was not influential in its bodies, which considered him an extremist. He forcibly advocated various form of sterilization to prevent, as he considered it, genetic degeneration. Rentoul was the author of, among other works, *Race Culture or Race Suicide: a Plea for the Unborn* (Walter Scott, 1906). 331

Spottiswoode, William 1825–1883. Applied mathematician and businessman. He was educated at Eton (but expelled for making explosives), Harrow and Oxford, where he studied mathematics. Though he was born in London, he came from an old Scottish family. He knew Galton from the Alpine club, the Geographical Society, the Ethnological Society, the Royal Society, and the British Association (he did not care for the Athenaeum Club). He took over the running of the family printing business, to save it from bankruptcy, but continued his own research into applied mathematics and experimental physics, when he was not writing up his extensive vacation travels, enlivened by his knack for foreign languages. An account of one such journey, written by his brother George, who had accompanied him, appeared in Galton's *Vacation Tourists and Notes of Travel in 1860* (F. Galton 1861). He was a member of the X-Club of Darwinians. Galton and Spottiswoode were Honorary Secretaries at the Geographical Society at the same time, and both resigned their positions at the time of Galton's clash with Norton Shaw in 1863. Spottiswoode was elected the president of the Royal Society in 1878. He died in office, succumbing to typhoid fever in 1883, and was buried in Westminster Abbey. 14, 36, 62

Tyndall, John 1820–1893. An Anglo-Irish experimental physicist from County Carlow, based at the Royal Institution in London. He was educated at the University of Marburg in Germany. Tyndall was a frequent walking companion of Galton—they were both members of the Alpine Club. They may have first met in the Alps in 1857 when Tyndall was climbing the Jardin in the company of T.H. Huxley: 'We met Mr. Galton, the African traveller, returning from an attempt upon the Jardin; and learning that his guides had lost their way in the fog, we deemed it prudent to return' (John Tyndall *The glaciers of the Alps, & Mountaineering in 1861* (London, 1861) [Everyman, 1906]: 59). Tyndall's emphasis on visual demonstrations of physical phenomena, in his lectures at the Royal

Institution, may have influenced Galton's use of analogue demonstrations of statistical concepts like the Quincunx. Galton is known to have attended those lectures. 'My late but passionate love for mountaineering was one cause that subsequently brought me into frequent contact with Professor Tyndall ... who was then at his very best physically and mentally. He, I, and Vaughan Hawkins ... an eminent classic in his Harrow and Cambridge days and of first rank in mountaineering, made a tour together in Cornwall. We chose our way on Tyndall's principle, that it is easy to find difficult places to climb elsewhere than in the high mountains. Certainly he was skilful at discovering them. One of his freaks sent my heart into my mouth. It was at a gully, strewn deeply with loose stones that led over a sea cliff. Down he dashed, the stones were all set in motion like an avalanche, but somehow he extricated himself in time and got clear to one side of them. At another place an isolated needle or cone of rock was separated from the shore by a narrow strait through which the sea swirled, but which could be leapt at low water. We leapt it, and clambered up, he declaring that it was as difficult a bit of rock-work as he had ever been on. We reached the top and got back successfully, jump and all, to the mainland, where I was glad to feel in safety. The Irish dash in Tyndall's blood gave a charm to all he did' (F. Galton 1908, 192–192). Tyndall and Galton were allies in the debate over the role of religion in science—see their contributions to the 'statistical efficacy of prayer' controversy in 1871. (Various pieces to do with this, including some by Tyndall and Galton, were collected by the Congregational Publishing Society in *The Prayer Gauge Debate* (Boston, 1876).) A lifelong struggle with insomnia led to Tyndall's death from a self-administered chloral overdose in 1893. 31

Wallace, Alfred Russel 1823–1913. Co-discoverer with Darwin of evolution by natural selection. Galton corresponded extensively with him in later years, after Darwin's death. Wallace reviewed *Hereditary Genius* in very favourable terms (Wallace 1870). 71

Watson, Rev. Henry William 1827–1903. A mathematician and cleric, who collaborated with Galton on branching processes and other mathematical problems. He lived at Berkswell in Warwickshire, near the Coventry border. At Cambridge he had been a member of the Apostles, and was ranked as Second Wrangler in 1850, also becoming a Smith's Prizeman. After the usual fellowship at Trinity he had taught at the City of London School, King's College and Harrow. Like many of the Cambridge graduates of his era he took holy orders after marrying, and in 1865 was offered the leisurely Rectorship of Berkswell, where he established his family (two daughters and a son) and produced a string of textbooks. Watson's daughter Maud Edith Eleanor was the first Wimbledon Lady's Tennis Champion in 1884. 37, 53

Welby, Victoria Lady 1837–1912. Self-taught philosopher. Daughter of the Hon. Charles and Lady Emmeline Stuart-Wortley, christened Victoria Alexandrina Maria Louisa. Queen Victoria agreed to become her Godmother. After the death of her father Charles—a hunting accident in 1844—she travelled continuously through Europe, Mexico, Panama and the Levant with her mother. She was entirely self-schooled, through reading, association and experience. Upon the death of her travel-narrating poetaster mother Emmeline—from dysentery, somewhere between Antioch and Beirut—she was left an orphan at 18, and had to be rescued by the British Consul at Aleppo. Once back in England she was sheltered by a succession of relatives. Her fortunes improved when she was appointed Maid of Honour to Queen Victoria in 1861. She married Sir William Welby-Gregory in 1863. By 1881 she was in print, publishing *Links and Clues* as 'Vita'. Over the years she assembled a salon of scholarly friends and a library of over 1,500 volumes and some 1,000 pamphlets, annotating them with copious notes and indexes. A stream of articles emerged in the 1890s, honing in on the subject of meaning, which she came to call 'Significs'. These resulted in some books a decade later (Welby 1903, 1911). She befriended Francis Galton

List of Figures

Bibliography

Airy, George (1861). *On the Algebraical and Numerical Theory of Errors of Observations and the Combination of Observations.* London: Macmillan, 1861.

Allen, Grant (1883). "Inquiries into Human Faculty." In: *Academy* (July 14, 1883). A review of Inquiries into Human Faculty, pp. 30–31.

Allingham, H and E. Baumer Williams (1911). *Letters to William Allingham.* London: Longmans, 1911.

Anonymous (1869a). "Hereditary Genius." In: *Spectator* 2161 (Nov. 27, 1869), pp. 1391–1392.

— (1869b). "Hereditary Genius." In: *Globe* (Dec. 3, 1869), p. 3.

— (1869c). "Hereditary Genius." In: *Saturday Review* (Dec. 25, 1869), pp. 833–834.

— (1869d). "Hereditary Genius." In: *Scotsman* (Dec. 28, 1869), p. 6.

— (1869e). "Untitled review of Hereditary Genius." In: *Daily News* (Dec. 16, 1869).

— (1870a). "Hereditary Genius." In: *Times* (Jan. 7, 1870), p. 10.

— (1870b). "Hereditary Genius." In: *Chambers' Journal* 47.315 (Jan. 8, 1870). Fourth series, pp. 118–122.

— (1870c). "Hereditary Genius." In: *British Quarterly Review* 51 (Jan. 1870), pp. 242–4.

— (1870d). "Hereditary Genius." In: *Westminster Review* CLXXXIII (Jan. 1870), pp. 144–145.

— (1875a). "English Men of Science." In: *Scotsman* (May 13, 1875), p. 3.

— (1875b). "English Men of Science." In: *Morning Post* (1875), p. 3.

— (1875c). "English Men of Science." In: *The Lancet* (Jan. 23, 1875). A review of English Men of Science, pp. 129–130.

— (1875d). "Hereditary Scientific Capacity." In: *The Examiner* (Feb. 27, 1875). A review of English Men of Science, pp. 248–249.

— (1883a). "[Review of Inquiries into Human Faculty]." In: *The Spectator* (Aug. 11, 1883), pp. 1029–1030.

— (1883b). "[Review of Inquiries into Human Faculty]." In: *The Continent* (Sept. 5, 1883), pp. 348–349.

— (1883c). "[Review of Inquiries into Human Faculty]." In: *Daily News* (Oct. 18, 1883).

— (1883d). "[Review of Inquiries into Human Faculty]." In: *Guardian* (July 4, 1883), p. 1001.

— (1883e). "Contemporary Literature." In: *Westminster Review* (July 1883). Review of Inquiries into Human Faculty, p. 88.

— (1883f). "Contemporary Literature." In: *British Quarterly Review* (July 1883). Review of Inquiries into Human Faculty, pp. 112–114.

— (1883g). "Galton on Human Faculty." In: *The Critic* (July 18, 1883). A review of Inquiries into Human Faculty, p. 325.

— (1883h). "Galton's Development of the Hunan Faculty." In: *The Nation* (June 14, 1883). Review of Inquiries into Human Faculty, pp. 512–513.

— (1883i). "Galton's Human Faculty and its Development." In: *American Naturalist* (Dec. 1883). A review of Inquiries into Human Faculty, pp. 1265–1266.

— (1883j). "Galton's Inquiries into Human Faculty." In: *Saturday Review* (May 26, 1883). Review of Inquiries into Human Faculty, pp. 668–689.

Anonymous (1883k). "Human Faculty and its Development." In: *English Mechanic and World of Science* (July 29, 1883). Review of Inquiries into Human Faculty, pp. 376–377.

— (1883l). "Literature." In: *Scotsman* (June 2, 1883). Review of Inquiries into Human Faculty, p. 11.

— (1883m). "Literature." In: *Glasgow Herald* (Sept. 4, 1883). Review of Inquiries into Human Faculty, p. 5.

— (1883n). "Mr. Galton on Human Faculty." In: *Pall Mall Gazette* (June 18, 1883). Review of Inquiries into Human Faculty, pp. 4–5.

— (1883o). "New Books." In: *Mind* (July 1883). Review of Inquiries into Human Faculty, p. 453.

— (1883p). "Reviews and Notices of Books." In: *Lancet* (Oct. 13, 1883). Review of Inquiries into Human Faculty, pp. 636–637.

— (1883q). "The Development of Human Faculty." In: *St James Gazette* (May 12, 1883). Review of Inquiries into Human Faculty, pp. 6–7.

— (1888). "A Morning with the Anthropometric Detectives." In: *Pall Mall Gazette* (Nov. 16, 1888), pp. 1–2.

— (1889a). "Galton's 'Natural Inheritance'." In: *The Critic* 279 (May 4, 1889), p. 218.

— (1889b). "Head growth in students at the University of Cambridge." In: *Nature* 40 (Aug. 1, 1889). Signed 'F. M. T.', pp. 317–318.

— (1889c). "Head growth in students at the University of Cambridge." In: *Nature* 40 (Oct. 17, 1889). Signed 'H. J. P. Trinity College', pp. 593–594.

— (1892). "Telegraphing to Mars." In: *Spectator* 2161 (Aug. 13, 1892), pp. 218–219.

— (1895). "Before Mr. Loveland Loveland." In: *Times* (Sept. 18, 1895), 5a.

Ansell, Charles (1874). *On the Rate of Mortality at Early Periods of Life.* London: National Life Insurance Society, 1874.

Bailey, Liberty Hyde (1892). "Cross-breeding and hybridizing." In: *The Rural Library Series* 1 (1892), pp. 1–44.

— (1895). *Plant Breeding.* London: Macmillan, 1895.

Bainton, George, ed. (1891). *The Art of Authorship. Literary Reminiscences, Methods of Work, and Advice to Young Beginners Personally Contributed by Leading Authors of the Day.* New York: Appleton, 1891.

Ball, W. W. Rouse (1922). *Mathematical Recreations and Essays.* 10th. London: Macmillan, 1922.

Banks, Charlotte (1983). "Professor Sir Cyril Burt: Selected Reminiscences." In: *Association of Educational Psychologists Journal* 6.1 (1983), pp. 21–42.

Barkow, Jerome, Leda Cosmides, and John Tooby (1992). *The Adapted Mind. Evolutionary Psychology and the Generation of Culture.* London: Oxford University Press, 1992.

Bateson, Beatrice, ed. (1928). *William Bateson, F.R.S. Naturalist. His Essays and Addresses Together with a Short Account of His Life.* Cambridge: Cambridge University Press, 1928.

Bateson, William (1894). *Materials for the Study of Variation with Especial Regard to Discontinuity in the Origin of Species.* Cambridge: Cambridge University Press, 1894.

— (1902). *Mendel's Principles of Heredity. A Defence.* Cambridge: Cambridge University Press, Apr. 1902.

— (1905). "Heredity in the Physiology of Nations. A review of The Principles of Heredity by G. Archdall Reid." In: *William Bateson, F.R.S. Naturalist. His Essays and Addresses Together with a Short Account of His Life.* Ed. by Beatrice Bateson. Cambridge: Cambridge University Press, 1905.

— (1909). *Mendel's Principles of Heredity.* Edition of 1913, reissued by Dover in 2010. Cambridge: Cambridge University Press, 1909.

— (1912). "Biological Fact and the Structure of Society. The Herbert Spencer Lecture." In: *William Bateson, F.R.S. Naturalist. His Essays and Addresses Together with a Short Account of His Life*. Ed. by Beatrice Bateson. Cambridge: Cambridge University Press, 1912.

— (1913). *Problems of Genetics*. New Haven, CT: Yale University Press, 1913.

— (1919). "Common-sense in Racial Problems. The Galton Lecture." In: *William Bateson, F.R.S. Naturalist. His Essays and Addresses Together with a Short Account of His Life*. Ed. by Beatrice Bateson. Cambridge: Cambridge University Press, 1919.

Beavan, Colin (2001). *Fingerprints. The Origins of Crime Detection and the Murder Case that Launched Forensic Science*. New York: Hyperion, 2001.

Beck, Lester F. (1933). "The role of speed in intelligence." In: *Psychological Bulletin* 30.2 (1933), pp. 169–178.

Bennett, R. Henry, ed. (1982). *Natural Selection, Heredity and Eugenics. Including selected correspondence of R. A. Fisher with Leonard Darwin and others*. Oxford: Oxford University Press, 1982.

Binet, Alfred and Théophile Simon (1916). *The Development of Intelligence in Children*. Translated by Elizabeth Kite. Vineland, NJ: Publications of the training school at Vineland, 1916.

Boswell, James (1887). *The Life of Samuel Johnson, LL.D.* Ed. by G. B. Hill. Originally published in 1791. Oxford: Oxford at the Clarendon Press, 1887.

Bouchard, Thomas J. and P. Propping, eds. (1993). *Twins as a Tool of Behavior Genetics*. New York: John Wiley, 1993.

Box, Joan Fisher (1978). *R. A. Fisher, the Life of a Scientist*. The author was a daughter of Fisher. New York: John Wiley, 1978.

Brand, Christopher (1996). *The g Factor. General Intelligence and its Implications*. Chichester: John Wiley, 1996.

Bulmer, Michael (2003). *Francis Galton. Pioneer of Heredity and Biometry*. Baltimore, Maryland: Johns Hopkins University Press, 2003.

— (2012). "Francis Galton and the Twin Method." In: *International Journal of Epidemiology* 41 (2012), pp. 913–917.

Burbridge, David (1994). "Galton's 100: an exploration of Francis Galton's imagery studies." In: *British Journal for the History of Science* 27 (1994), pp. 443–463.

Burt, Austin and Robert Trivers (2006). *Genes in Conflict. The Biology of Selfish Genetic Elements*. London: Harvard University Press, 2006.

Burt, Cyril (1909). "Experimental Tests of General Intelligence." In: *British Journal of Psychology* 3 (1909), pp. 94–177.

— (1952). "Autobiography." In: *History of Psychology in Autobiography*. Ed. by E. G. Boring. Vol. 4. Clark University Press, 1952.

— (1962). "Francis Galton and His Contributions to Psychology." In: *British Journal of Statistical Psychology* 15.1 (1962), pp. 1–49.

Candolle, Alphonse de (1873). *Histoire des sciences et des savants depuis deux siècles*. History of science and scholars for two centuries. Geneva: H. Georg, 1873.

Caroe, Gwendy (1985). *The Royal Institution. An Informal History*. London: John Murray, 1985.

Casella, George and Robert L. Berger (2002). *Statistical Inference*. Pacific Grove, CA: Duxbury, 2002.

Cattell, James McKeen (1890). "Mental Tests and Measurements." In: *Mind* 15 (July 1890), pp. 373–381.

— (1924). "The Interpretation of Intelligence Tests." In: *The Scientific Monthly* 18.5 (May 1924), pp. 508–516.

Cattell, James McKeen (1929). "Psychology in America." In: *Science* 70.1815 (Oct. 11, 1929), pp. 335–347.
Chambers, Robert (1844). *Vestiges of the Natural History of Creation.* London: Churchill, 1844.
Chapman, David L. (1994). *Sandow the Magnificent. Eugen Sandow and the Beginning of Body Building.* Chicago: Illinois University Press, 1994.
Clark, Stuart (2007). *The Sun Kings: the unexpected tragedy of Richard Carrington and the tale of how modern astronomy began.* Princeton, NJ: Princeton University Press, 2007.
Cock, Alan G. and Donald Forsdyke (2008). *Treasure your Exceptions. The Science and Life of William Bateson.* New York: Springer, 2008.
Cole, Simon (2001). *Suspect Identities. A History of Fingerprinting and Criminal Identification.* Cambridge, MA: Harvard University Press, 2001.
Colp, Ralph Jr. (2008). *Darwin's Illness.* Gainesville, Florida: University Press of Florida, 2008.
Cotton, James S. (1875). "English Men of Science." In: *Academy* 7 (Jan. 23, 1875). Review, pp. 92–93.
Cowan, Ruth Leah Schwartz (1969). "Sir Francis Galton and the Study of Heredity in the Nineteenth Century." PhD. Johns Hopkins, 1969.
Crackanthorpe, Montague (1907). *Population and Progress.* London: Chapman and Hall, 1907.
— (1911). "Sir Francis Galton." In: *Times* (Jan. 25, 1911). Obituary remarks, p. 10.
Craik, A. D. D. (2008). *Mr. Hopkins' Men.* London: Springer Verlag, 2008.
Cunningham, Joseph Thomas (1899). "Professor Weldon's Evidence of the Operation of Natural Selection." In: *Natural Science* 14 (1899), pp. 38–45.
Dalby, William Bartlett (1885). *Lectures on Diseases and Injuries of the Ear: Delivered at St. George's Hospital.* Third edition. London: Churchill, 1885.
Darlington, Cyril (1953). *The Facts of Life.* London: George Allen & Unwin, 1953.
— (1959). *Darwin's Place in History.* London: Oxford, 1959.
— (1964). *Genetics and Man.* London: George Allen and Unwin, 1964.
Darwin, Charles (1859). *On the Origin of Species by Means of Natural Selection, or the Preservation of Favoured Races in the Struggle for Life.* London: John Murray, 1859.
— (1868). *Variation of Plants and Animals Under Domestication.* First edition. 2 volumes. London: John Murray, 1868.
— (1871a). "Pangenesis." In: *Nature* 3 (Apr. 27, 1871), pp. 502–503.
— (1871b). *The Descent of Man and Selection in Relation to Sex.* Franklin Library edition, 1980. London: John Murray, 1871.
— (1872). *The Expression of the Emotions in Man and Animals.* London: John Murray, 1872.
— (1875). *Variation of Plants and Animals Under Domestication.* Second edition. 2 volumes. London: John Murray, 1875.
Darwin, Francis, ed. (1916). *Scientific Papers by Sir George Howard Darwin.* Vol. 5, Supplementary. 1916.
Darwin, George Howard (1873). "On Beneficial Restrictions to Liberty of Marriage." In: *Contemporary Review* (Aug. 1873), pp. 412–426.
— (1875). "Marriages between first cousins in England and their effects." In: *Journal of the Statistical Society of London* 38 (June 1875), pp. 153–184.
Dawes, Robyn (1979). "The Robust Beauty of Improper Linear Models in Decision Making." In: *American Psychologist* 34.7 (July 1979), pp. 571–82.
Dawkins, Richard (1982). *The Extended Phenotype. The Gene as the Unit of Selection.* London: Oxford, 1982.
Deary, Ian J. (2000). *Looking Down on Human Intelligence. From Psychometrics to the Brain.* London: Oxford University Press, 2000.

Dewey, John (1889). "Galton's Statistical Methods." In: *Quarterly Publications of the American Statistical Association* 7 (Sept. 1889), pp. 331–4.
Doran, John (1870). "Hereditary Genius." In: *Athenaeum* 2203 (Jan. 15, 1870), pp. 85–87.
Drower, ed. (2004). *Letters from the desert. The Correspondence of Flinders and Hilda Petrie.* Oxford: Aris and Phillips, 2004.
Du Cane, Edward Frederick (1876). "Address on Repression of Crime." In: *Transactions of the National Association for the Promotion of Social Science. Brighton Meeting, 1875.* Ed. by Charles Wager Ryalls. London: Longmans, Green, and Co., 1876.
Edgeworth, Francis Ysidro (1889). "Natural Inheritance [Review]." In: *Nature* 39 (Apr. 25, 1889), pp. 603–604.
Elderton, Ethel M. and Karl Pearson (1910). *A First Study of the Influence of Parental Alcoholism on the Physique and Ability of the Offspring.* Eugenics Laboratory Memoirs. London: Dulau, 1910.
Elderton, William Palin and Ethel M. Elderton (1909). *A Primer of Statistics.* Preface by Francis Galton. London: Black, 1909.
Ellegård, Alvar (1958). *Darwin and the General Reader. The reception of Darwin's Theory of Evolution in the British Periodical Press, 1859–1872.* Göteborg: Göteborg, 1958.
— (1971). "The Readership of the Periodical Press in Mid-Victorian Britain: II. Directory." In: *Victorian Periodicals Newsletter* 13 (1971), pp. 3–22.
Eysenck, Hans J. (1998). *Intelligence. A new look.* New Brunswick, New Jersey: Transaction Publishers, 1998.
Fancher, Raymond (1985). *The Intelligence Men.* New York: Norton, 1985.
Farrar, Frederic William (1870). "Hereditary Genius." In: *Fraser's Magazine* 5115 (Aug. 1870), pp. 251–265.
Faulds, Henry (1879). "Remarks on pottery shards." In: *Transactions of the Asiatic Society of Japan* 8 (Nov. 11, 1879), p. 91.
— (1880). "On the Skin-furrows of the Hand." In: *Nature* 22 (Oct. 28, 1880), p. 605.
— (1885). *Nine Years in Nipon. Sketches of Japanese life and manners.* London: A. Gardner, 1885.
— (1894). "On the Identification of Habitual Criminals by Finger-Prints." In: *Nature* (Oct. 4, 1894), p. 548.
— (1905a). "[Letter]." In: *Birmingham Post* (May 16, 1905).
— (1905b). *A Guide to Finger-Print Identification.* Hanley: Wood, Mitchell and Co., 1905.
— (1912). *Dactylography, or the study of finger-prints.* Halifax: Milner, 1912.
— (1915). *How the English fingerprint method arose.* Hanley: Wood, Mitchell, and co., 1915.
— (1917a). *The Hidden Hand. A contribution to the history of finger prints.* 1917.
— (1917b). "The Permanence of Finger-Print Patterns." In: *Nature* 18 (Jan. 18, 1917), pp. 388–9.
— (1920). *The Hidden Hand. Vestigia nulla retrorsum.* Hanley: Wood, Mitchell and Co., 1920.
— (1923). *A Manual of Practical Dactylography.* London: Police Review Publishing, 1923.
— (1926). *Was Sir E. R. Henry the Originator of the Finger Print System?* Hanley: Webberley, 1926.
Fechner, Gustav (1860). *Elemente der Psychophysik.* Leipzig: Breitkopf and Härtel, 1860.
Filon, L. N. G et al. (1934). *Speeches Delivered at a Dinner Held in University College, London in Honour of Professor Karl Pearson, 23 April 1934.* Privately printed. Cambridge: University Press, 1934.
Fisher, R. A. (1911). "Mendelism and Biometry." In: *Natural Selection, Heredity and Eugenics. Including selected correspondence of R. A. Fisher with Leonard Darwin and others.* Ed. by R. Henry Bennett. Oxford: Oxford University Press, 1911.
— (1930). *The Genetical Theory of Natural Selection.* London: Oxford University Press, 1930.

Fisher, R. A. (1936). "Has Mendel's work been rediscovered?" In: *Annals of Science* 1 (1936), pp. 115–137.
— (1956). *Statistical Methods and Scientific Inference.* New York: Hafner, 1956.
Fletcher, Ronald (1991). *Science, Ideology and the Media. The Cyril Burt Scandal.* New Brunswick, New Jersey: Transaction Publishers, 1991.
Focke, Wilhelm Olbers (1881). *Pflanzen-mischlinge. Ein Beitrag zur Biologie der Gewächse.* Berlin: Gebrüder Bornträger, 1881.
Forrest, Derek William (1974). *Francis Galton. The Life and Work of a Victorian Genius.* London: Paul Elek, 1974.
Franklin, Allan et al., eds. (2008). *Ending the Mendel-Fisher Controversy.* Pittsburgh: University of Pittsburgh Press, 2008.
Freeman, Richard Broke (1977). *The Works of Charles Darwin: An Annotated Bibliographical Handlist.* 2nd ed. Simultaneously published by Archon Books, Hamden, CT. Folkestone: Dawson, 1977.
Frisby, Craig L. and William T. O'Donohue, eds. (2018). *Cultural Competence in Applied Psychology.* New York: Springer, 2018.
Galton, Elizabeth Anne (2003). *Elizabeth Anne Galton (1808–1906). A Well-connected Gentlewoman.* Ed. by Andrew Moilliet. Diary and reminiscences of Bessy. Northwich, Cheshire: Leonie Press, 2003.
Galton, Francis (1849). *The Telotype. A Printing Electric Telegraph.* London: J. Ridgway, 1849.
— (1850). *The Telotype. A Printing Electric Telegraph.* Post-dated. London: J. Weale, June 1850.
— (1853). *The Narrative of an Explorer in Tropical South Africa.* First edition. London: John Murray, 1853.
— (1855). *The Art of Travel; or, Shifts and Contrivances Available in Wild Countries.* First edition. London: John Murray, 1855.
— (1860). *The Art of Travel; or, Shifts and Contrivances Available in Wild Countries.* Third edition. London: John Murray, 1860.
— ed. (1861). *Vacation tourists and Notes of Travel in 1860.* London: Macmillan, 1861.
— ed. (1862). *Vacation Tourists and Notes of Travel in 1861.* London: Macmillan, Apr. 26, 1862.
— (1863a). *Meteorographica. or Methods of Mapping the Weather.* London: Macmillan, Aug. 8, 1863.
— (1863b). *The First Steps Towards the Domestication of Animals.* Privately printed. London: Spottiswoode & Co., Dec. 1863.
— (1864a). *Ethnological Inquiries on the Innate Character and Intelligence of Different Races.* Pamphlet. London: Privately Printed, 1864.
— (1864b). *The Knapsack Guide for Travellers in Switzerland.* First edition. Not credited. London: John Murray, 1864.
— (1865a). *A Handbook for Travellers in Switzerland.* Not credited. London: John Murray, 1865.
— (1865b). "Hereditary Talent and Character: Part I." In: *Macmillan's Magazine* 12 (68 June 1865), pp. 157–166.
— (1865c). "Hereditary Talent and Character: Second Paper." In: *Macmillan's Magazine* 12 (70 Aug. 1865), pp. 318–327.
— (1865d). "On spectacles for divers, and on the vision of amphibious animals." In: *Report of the British Association for the Advancement of Science* 35 (1865), pp. 10–11.
— (1865e). "On stereoscopic maps, taken from models of mountainous countries." In: *Journal of the Royal Geographical Society* 35 (Mar. 13, 1865), pp. 99–104.

— (1866a). “[Report on the Meteor as seen at Boulogne].” In: *Report of the British Association for the Advancement of Science* 36 (Aug. 1866), p. 130.
— (1866b). “Meteor.” In: *Times* (June 21, 1866), 7f.
— (1866c). “On an error in the usual method of obtaining meteorological statistics of the ocean.” In: *Report of the British Association for the Advancement of Science* 36 (Aug. 1866), pp. 16–7.
— (1866d). “On the conversion of wind-charts into passage charts.” In: *Philosophical Magazine* 32 (Nov. 1866), pp. 345–9.
— (1866e). “On the conversion of wind-charts into passage charts.” In: *Report of the British Association for the Advancement of Science* 36 (Aug. 24, 1866), pp. 17–20.
— (1867a). *The Art of Travel; or, Shifts and Contrivances Available in Wild Countries.* Fourth edition. London: John Murray, 1867.
— (1867b). *The Knapsack Guide for Travellers in Switzerland.* Second edition. Not credited. London: John Murray, 1867.
— (1868). “Hereditary Genius.” In: *Notes and Queries on China and Japan* 2.8 (Aug. 1868), p. 120.
— (1869a). *Hereditary Genius. An Inquiry into its Laws and Consequences.* First edition. London: Macmillan, 1869.
— (1869b). “Hereditary Genius. The Judges of England between 1660 and 1865.” In: *Macmillan’s Magazine* 19 (Mar. 1869), pp. 424–431.
— (1870a). “Barometric Predictions of Weather.” In: *Nature* 2 (Oct. 20, 1870), pp. 501–503.
— (1870b). “Description of the pantagraph designed by Mr. Galton.” In: *Report of the Meteorological Committee of the Royal Society for 1870* (1870), pp. 31–36.
— (1870c). “Geographical Prizes.” In: *Nature* 2 (July 14, 1870), p. 210.
— (1870d). “On the reported current in the Suez Canal.” In: *Nature* 2 (July 7, 1870), pp. 31–3.
— (1871a). “Experiments in pangenesis, by breeding from rabbits of a pure variety, into whose circulation blood taken from other varieties had previously been largely transfused.” In: *Proceedings of the Royal Society* 19.126 (Mar. 30, 1871), pp. 393–410.
— (1871b). “On the teaching of geography.” In: *Journal of the Society of Arts* 19.126 (June 16, 1871), p. 601.
— (1871c). “Pangenesis.” In: *Nature* 4 (May 4, 1871), pp. 5–6.
— (1872a). “On blood-relationship.” In: *Proceedings of the Royal Society* 20 (June 13, 1872), pp. 394–402.
— (1872b). “Statistical Inquiries into the Efficacy of Prayer.” In: *Fortnightly Review* 12 (Aug. 1872), pp. 125–35.
— (1872c). *The Art of Travel; or, Shifts and Contrivances Available in Wild Countries.* Fifth edition. London: John Murray, 1872.
— (1872d). “The Efficacy of Prayer.” In: *Spectator* 24 (Aug. 24, 1872), p. 1073.
— (1873a). “Africa for the Chinese.” In: *Times* (June 5, 1873).
— (1873b). “Hereditary Improvement.” In: *Fraser’s Magazine* 7 (Jan. 1873), pp. 116–30.
— (1873c). “On the Causes which Operate to Create Scientific Men.” In: *Fortnightly Review* 13 (Mar. 1, 1873), pp. 345–51.
— (1873d). “Problem 4001: On the extinction of surnames.” In: *Educational Times* 26 (Apr. 1873), p. 17.
— (1873e). “Proposal to apply for anthropological statistics from schools.” In: *Journal of the Anthropological Institute* 3 (Nov. 11, 1873), pp. 308–11.
— (1874a). “[Prefatory remarks to H. W. Watson, ’On the probability of the extinction of families’].” In: *Journal of the Anthropological Institute* 4 (May 12, 1874), pp. 138–9.

Galton, Francis (1874b). "Appendix C: Statistics." In: *Notes and Queries on Anthropology. For the Use of Travellers and Residents in Uncivilized Lands.* Ed. by John Beddoe. First edition. London: Edward Stanford, 1874.

— (1874c). *English Men of Science.* London: Macmillan, 1874.

— (1874d). "Notes on the Marlborough School statistics." In: *Journal of the Anthropological Institute* 4 (May 12, 1874), pp. 130–5.

— (1874e). "On a Proposed Statistical Scale." In: *Nature* 9 (Mar. 5, 1874), pp. 342–343.

— (1874f). "On men of science, their nature and their nurture." In: *Proceedings of the Royal Institution* 7 (Feb. 27, 1874), pp. 227–236.

— (1875a). "A Theory of Heredity." In: *Journal of the Anthropological Institute* 5 (Nov. 9, 1875), pp. 329–348.

— (1875b). "Discussion of 'Ultra Centenarian Longevity' by G. D. Gibb." In: *Journal of the Anthropological Institute* 5 (Mar. 9, 1875), pp. 98–9.

— (1875c). "Discussion on Cameron's Examination of the Southern Half of Lake Tanganyika." In: *Proceedings of the Royal Geographical Society* 19 (Mar. 8, 1875), p. 262.

— (1875d). "On certain verses wrongly ascribed to Rogers." In: *Notes and Queries* (Feb. 13, 1875), pp. 122–124.

— (1875e). "Rate of Bicycles." In: *Field* (June 19, 1875), p. 621.

— (1875f). "Review of Heredity; a Psychological Study of its Phenomena, Laws, Causes, and Consequences by Th. Ribot." In: *Academy* (Jan. 30, 1875), pp. 118–119.

— (1875g). "Short notes on Heredity &c. in Twins." In: *Journal of the Anthropological Institute* 5 (Nov. 23, 1875), pp. 324–9.

— (1875h). "Statistics by intercomparison with remarks on the Law of Frequency of Error." In: *Philosophical Magazine* 49 (Jan. 1875), pp. 33–46.

— (1875i). "The history of twins, as a criterion of the relative powers of nature and nurture." In: *Journal of the Anthropological Institute* 5 (Nov. 23, 1875), pp. 391–406.

— (1875j). "The history of twins, as a criterion of the relative powers of nature and nurture." In: *Fraser's Magazine* 12 (Nov. 1875), pp. 566–576.

— (1876a). "Apparatus for the Statistical Treatment of Large Numbers of Seeds; Apparatus Affording Physical Illustration of the Action of the Law of Error or of Dispersion; Working Model of a Trace Computer; Whistles for Producing Shrill Notes, Within and Beyond the Limits of Ordinary Audition; Sun Signals, for the Use of Travellers; Models Illustrating Two Methods of Verifying Sextants Employed at the Kew Observatory; Spectacles for Divers, for Use in Water; Models to Illustrate the Arts of Camp Life; Stereoscopic Maps." In: *Catalogue of the special loan collection of scientific apparatus at the South Kensington Museum 1876.* Second edition. London: Eyre and Spottiswoode, 1876, pp. 10–11, 15–16, 151, 642–3, 674–5.

— (1876b). *Galton's Whistles for determining the upper limits of audible sound in different individuals.* London: H. and W. Brown, 1876.

— (1877a). *Address to the Anthropological Department of the British Association, Plymouth, 1877.* London: William Clowes and Sons, 1877.

— (1877b). "Considerations adverse to the maintenance of Section F." In: *Journal of the Statistical Society* 40 (Sept. 1877), pp. 468–473.

— (1877c). "On means of combining various data in maps and diagrams." In: *Conferences held in connection with the special loan collection of scientific apparatus, 1876, Vol. 2: Chemistry, Biology, etc.* London: Chapman and Hall, 1877, pp. 312–5.

— (1877d). "Typical laws of heredity." In: *Proceedings of the Royal Institution* 8 (Feb. 9, 1877), pp. 282–301.

— (1878a). *Composite portraits made by combining those of many different persons into a single figure.* With Additions. London: William Harrison and Sons, 1878.

— (1878b). “Giddiness.” In: *Nature* 18 (Oct. 24, 1878), p. 669.
— (1878c). “On the advancement of geographical teaching.” In: *Nature* 18 (July 25, 1878), p. 337.
— (1879a). *Generic Images.* With additions. London: William Clowes and Sons, 1879.
— (1879b). *Psychometric Experiments.* London: William Clowes and Sons, 1879.
— (1879c). “The geometric mean, in vital and social statistics.” In: *Proceedings of the Royal Society* 29 (Nov. 20, 1879), pp. 365–7.
— (1880a). “Jewish Physiognomy.” In: *Notes and Queries* 6th series, I (Mar. 13, 1880), p. 226.
— (1880b). “Statistics of Mental Imagery.” In: *Mind* (July 1880), pp. 301–18.
— (1880c). *Visualised Numerals.* London: Harrison and Sons, 1880.
— (1880d). “Visualised numerals.” In: *Nature* 21 (Jan. 15, 1880), pp. 252–6.
— (1880e). “Visualised numerals.” In: *Nature* 21 (Feb. 5, 1880), p. 323.
— (1880f). “Visualised numerals.” In: *Nature* 21 (Mar. 25, 1880), pp. 494–5.
— (1881a). “On the construction of isochronic passage charts.” In: *Proceedings of the Royal Geographical Society* 3 (Sept. 1, 1881), pp. 657–8.
— (1881b). “On the construction of isochronic postal charts.” In: *Report of the British Association for the Advancement of Science* 51 (Sept. 1881), pp. 740–1.
— (1881c). “Remarks concerning “On the Laws affecting the relations between civilized and savage life, as bearing on the dealings of colonists with aborigines” by Bartle Frere.” In: *Journal Anthropological Society* (July 28, 1881).
— (1882a). “On apparatus for testing the delicacy of the muscular and other senses in different persons.” In: *Journal of the Anthropological Institute* 12 (Nov. 14, 1882), pp. 469–477.
— (1882b). “Photographic chronicles from childhood to age.” In: *Fortnightly Review* 31 (Jan. 1882), pp. 26–31.
— (1882c). “The anthropometric laboratory.” In: *Fortnightly Review* 31 (Mar. 1882), pp. 332–8.
— (1882d). “The Late Mr. Darwin: a Suggestion.” In: *Pall Mall Gazette* (Apr. 27, 1882).
— (1883a). “Final Report of the Anthropometric Committee.” In: *Report of the British Association* (Sept. 1883), pp. 253–306.
— (1883b). “Hydrogen Whistles.” In: *Nature* 27 (Mar. 22, 1883), pp. 491–2.
— (1883c). “Hydrogen Whistles.” In: *Nature* 28 (May 17, 1883), p. 54.
— (1883d). *Inquiries into Human Faculty and its Development.* London: Macmillan, Apr. 24, 1883.
— (1883e). *Life History Album.* First edition. London: Macmillan, Dec. 1883.
— (1883f). “Medical Family Registers.” In: *Fortnightly Review* 34 (Aug. 1883), pp. 244–250.
— (1883g). “Obituary of William Spottiswoode.” In: *Proceedings of the Royal Geographical Society* 5.8 (Aug. 1883), pp. 489–491.
— (1883h). *Outfit for an anthropometric laboratory.* Mar. 1883.
— (1883i). *Record of Family Faculties.* London: Macmillan, Dec. 1883.
— (1884a). *Anthropometric Laboratory; Arranged by Francis Galton, F.R.S., for the Determination of Height, Weight, Span, Breathing Power, Strength of Pull And' Squeeze, Quickness of Blow, Hearing, Seeing, Colour-Sense, and Other Personal Data.* London: William Clowes for the International Health Exhibition, May 1884.
— (1884b). “Extracts from Family Records.” In: *Nature* 29 (Jan. 10, 1884), p. 244.
— (1884c). “Family records.” In: *The Times* (Jan. 9, 1884), 10b.
— (1884d). “Free-will—observations and inferences.” In: *Mind* 9 (July 1884), pp. 406–13.
— (1884e). “Measurement of character.” In: *Fortnightly Review* 36 (Aug. 1884), pp. 179–185.
— (1884f). *On the Anthropometric Laboratory at the late International Health Exhibition.* London: William Harrison and Sons, Nov. 1884.

Galton, Francis (1884g). "Some results of the Anthropometric Laboratory." In: *Journal of the Anthropological Institute* 14 (Nov. 25, 1884), pp. 275–87.
— (1884h). "The weights of British noblemen during the last three generations." In: *Nature* 29 (Jan. 17, 1884), pp. 266–268.
— (1885a). "[Presidential address, Section H, Anthropology]." In: *Report of the British Association for the Advancement of Science* 55 (1885), pp. 1206–1214.
— (1885b). "Anthropometric percentiles." In: *Nature* 31 (Jan. 8, 1885), pp. 223–5.
— (1885c). "Exhibition of composite photographs of skulls by Francis Galton." In: *Journal of the Anthropological Institute* 15 (Nov. 24, 1885), pp. 390–1.
— (1885d). "Opening remarks by the President and discussion on 'Experiments in testing the character of school children' by Sophie Bryant." In: *Journal of the Anthropological Institute* 15 (Nov. 10, 1885), pp. 336–8, 350.
— (1885e). "Photographic composites." In: *Photographic News* 29 (Apr. 17, 1885), pp. 234–45.
— (1885f). "Regression towards mediocrity in hereditary stature." In: *Journal of the Anthropological Institute* 15 (June 9, 1885), pp. 246–263.
— (1885g). "The measure of fidget." In: *Nature* 32 (June 25, 1885), pp. 174–175.
— (1886a). "[Royal Society Award Speech]." In: *Times* (Dec. 1, 1886).
— (1886b). "Family likeness in stature." In: *Proceedings of the Royal Society* 40 (Jan. 21, 1886), pp. 42–63.
— (1886c). "Family-likeness in eye-colour." In: *Proceedings of the Royal Society* 40 (May 27, 1886), pp. 402–416.
— (1886d). "Hereditary Stature [President's address]." In: *Journal of the Anthropological Institute* 15 (1886), pp. 488–499.
— (1886e). "On recent designs for anthropometric instruments." In: *Journal of the Anthropological Institute* 16 (Feb. 9, 1886), pp. 2–8.
— (1886f). "President's address [Hereditary Stature]." In: *Journal of the Anthropological Institute* 15 (Jan. 26, 1886), pp. 488–499.
— (1887a). "Good and bad temper in English families." In: *Fortnightly Review* 42 (July 1887), pp. 21–30.
— (1887b). *List of Anthropometric Apparatus.* Cambridge: Cambridge Scientific Instrument Company, 1887.
— (1887c). "Photography and silhouettes." In: *Photographic News* 31 (July 8, 1887), pp. 429–30, 462.
— (1887d). "Supplementary notes on 'Prehension in idiots'." In: *Mind* 12 (Jan. 1887), pp. 79–82.
— (1887e). "Thoughts without Words [Letters]." In: *Open Court* 1 (1887), pp. 441–4, 472–4, 498–50.
— (1888a). "[Discussion: 'On an Ancient British Settlement excavated near Rushmore, Salisbury' by Pitt Rivers]." In: *Journal of the Royal Anthropological Institute* 17 (June 28, 1888), pp. 199–200.
— (1888b). "Anniversary Address." In: *Journal of the Royal Anthropological institute* 17 (Jan. 24, 1888), pp. 346–54.
— (1888c). "Co-relations and their measurement, chiefly from anthropometric data." In: *Proceedings of the Royal Society* 45 (Dec. 13, 1888), pp. 135–145.
— (1888d). "On head growth in students at the University of Cambridge." In: *Journal of the Anthropological Institute* 18 (Apr. 24, 1888), pp. 155–156.
— (1888e). "Personal identification and description." In: *Proceedings of the Royal Institution* 12 (May 1888), pp. 346–60.
— (1888f). "Personal identification and description." In: *Journal of the Royal Anthropological Institute* 18 (1888), pp. 177–91.

— (1888g). "Remarks on Replies by Teachers to Questions Respecting Mental Fatigue." In: *Journal of the Anthropological Institute* 18 (Apr. 10, 1888), pp. 157–68.
— (1889a). "Head growth in students at the University of Cambridge." In: *Nature* 40 (Aug. 1, 1889), p. 318.
— (1889b). "Head measures at Cambridge." In: *Nature* 40 (Oct. 31, 1889), p. 643.
— (1889c). *Natural Inheritance*. London: Macmillan, 1889.
— (1889d). "On the advisability of assigning marks for bodily efficiency in the examination of candidates for public services." In: *Report of the British Association for the Advancement of Science* 59 (Sept. 1889), pp. 471–3.
— (1889e). "On the principle and methods of assigning marks for bodily efficiency." In: *Nature* 40 (Oct. 24, 1889), pp. 631–2.
— (1889f). "On the principle and methods of assigning marks for bodily efficiency." In: *Report of the British Association for the Advancement of Science* 59 (Sept. 1889), pp. 474–8.
— (1889g). "On the principle and methods of assigning marks for bodily efficiency (part 2)." In: *Nature* 40 (Oct. 31, 1889), pp. 649–52.
— (1889h). "President's Address." In: *Journal of the Royal Anthropological Institute* 18 (Jan. 22, 1889), pp. 401–19.
— (1889i). "Table of Observations." In: *Journal of the Royal Anthropological Institute* 18 (Jan. 22, 1889). Four hundred observations from the Anthropometric Laboratory of 1884, pp. 420–30.
— (1890a). "A Plea for Physical Examination." In: *Education* 1 (May 1890), pp. 86–7.
— (1890b). "Bodies and Brains in Examinations." In: *Spectator* 64 (Apr. 5, 1890), pp. 472–3.
— (1890c). "Cambridge anthropometry." In: *Nature* 41 (Mar. 13, 1890), p. 454.
— (1890d). "Kinship and correlation." In: *North American Review* 150 (Apr. 1890), pp. 419–431.
— (1890e). "Marks for Physical Efficiency." In: *Times* (Mar. 28, 1890), 4f.
— (1890f). "Physical tests in competitive examinations." In: *Journal of the Society of Arts* 39 (Nov. 28, 1890), pp. 19–27.
— (1890g). "Remarks on James McKeen Cattell's 'Mental Tests and Measurements'." In: *Mind* 15 (July 1890), p. 381.
— (1890h). "The patterns in thumb and finger marks." In: *Proceedings of the Royal Society* 48 (Nov. 27, 1890). abstract, pp. 455–7.
— (1890i). "The patterns in thumb and finger marks." In: *Philosophical Transactions of the Royal Society B* 182 (Nov. 27, 1890), pp. 1–23.
— (1890j). "Why do we measure mankind?" In: *Lippincott's Monthly Magazine* 45 (Feb. 1890), pp. 236–41.
— (1891a). "Identification by finger tips." In: *Nineteenth Century* 30 (Aug. 1891), pp. 303–11.
— (1891b). "Methods of indexing finger-marks." In: *Philosophical Transactions of the Royal Society B* 49 (May 28, 1891), pp. 540–48.
— (1892a). *Finger Prints*. London: Macmillan, 1892.
— (1892b). *Hereditary Genius. An Inquiry into its Laws and Consequences*. Second edition. London: Macmillan, 1892.
— (1892c). "Quincunx, Trace Computer of Vapour Tension, and Drill Pantagraph." In: *Katalog Mathematische und Mathematisch-Physikalischer Modelle, Apparate und Instrumente Deutsche Mathematiker-Vereinigung* III.6 (1892), pp. 154–5, 164–8, 232–4.
— (1892d). "Sun Signals to Mars." In: *Times* (Aug. 6, 1892), 7f.
— (1893a). *Decipherment of Blurred Finger Prints. Supplementary Chapter to 'Finger Prints'*. London: Macmillan, 1893.
— (1893b). "Identification." In: *Times* (July 7, 1893), 4e.
— (1893c). "Measure of the imagination." In: *Nature* 47 (1893), pp. 319–21.

Galton, Francis (1893d). "The just-perceptible difference." In: *Proceedings of the Royal Institution* 14 (Jan. 27, 1893), pp. 13–26.
— (1894a). "Arithmetic by smell." In: *Psychological Review* 1 (1894), pp. 61–2.
— (1894b). "Discontinuity in evolution." In: *Mind* 3 (July 1, 1894), pp. 362–72.
— (1894c). "Psychology of mental arithmeticians and blindfold chess-players." In: *Nature* 51 (Nov. 22, 1894). Review of Alfred Binet, Psychologie des Grands Calculateurs et Joueurs, pp. 73–4.
— (1894d). "The part of religion in human evolution." In: *National Review* 23 (Aug. 1894), pp. 755–63.
— (1895). *Fingerprint Directories*. London: Macmillan, 1895.
— (1896). "Intelligible signals between neighbouring stars." In: *Fortnightly Review* 60 (Nov. 1, 1896), pp. 657–64.
— (1897a). "Rate of Racial Change that accompanies Different Degrees of Severity in Selection." In: *Nature* 55.1435 (Apr. 29, 1897), p. 606.
— (1897b). "The average contribution of each several ancestors to the total heritage of the offspring." In: *Proceedings of the Royal Society* 61 (June 3, 1897), pp. 401–13.
— (1898a). "Corporal punishment." In: *Times* (Oct. 4, 1898), p. 606.
— (1898b). "Photographic measurement of horses and other animals." In: *Nature* 57 (Jan. 6, 1898), pp. 230–2.
— (1898c). "Photographic records of pedigree stock." In: *Live Stock Journal* (Sept. 30, 1898), p. 393.
— (1898d). "Photographic records of pedigree stock." In: *Report of the British Association for the Advancement of Science* (1898), pp. 597–603.
— (1898e). "Photographic records of pedigree stock." In: *Nature* 58 (Oct. 13, 1898), p. 584.
— (1899). "Strawberry cure for gout." In: *Nature* 60 (1899), p. 125.
— (1900a). "Analytical photography." In: *Photographic Journal* 25 (Dec. 31, 1900), pp. 135–8.
— (1900b). "Analytical portraiture." In: *Nature* 62 (Aug. 2, 1900), p. 320.
— (1900c). "Identification Offices in Egypt and India." In: *Nineteenth Century* 48 (1900), pp. 118–26.
— (1901a). "Biometrics." In: *Biometrika* 1 (1901), pp. 7–10.
— (1901b). "On the probability that the son of a very highly-gifted father will be no less gifted." In: *Nature* 65 (1901), p. 79.
— (1901c). "Souvenirs d'Egypte." In: *Bulletin de la Société Khédiviale de Géographie du Caire* V (1901), pp. 375–380.
— (1902). *Life History Album. Tables and charts for recording the development of body and mind from childhood upwards, with introductory remarks.* Second edition. London: Macmillan, 1902.
— (1904a). *A eugenics investigation. Index to achievements of near kinsfolk of some of the Fellows of the Royal Society.* London: R. Clay, 1904.
— (1904b). "Distribution of successes and of natural ability among the kinsfolk of Fellows of the Royal Society." In: *Nature* 70 (1904), pp. 354–6.
— (1904c). "Eugenics. Its definition, scope and aims." In: *American Journal of Sociology* 10 (1904), pp. 1–6.
— (1904d). "Eugenics. Its definition, scope and aims." In: *Nature* 70 (1904), p. 82.
— (1905a). "[Review of] Guide to Finger Print Identification by Henry Faulds." In: *Nature Supplement* 72 (1905), pp. iv–v.
— (1905b). "Education and Physique." In: *Nature* 72 (Oct. 19, 1905). Review of Demeny, pp. v–vi.
— (1905c). "Eugenics. Its definition, scope and aims." In: *Sociological Papers* 1 (1905), pp. 45–50, 78–9.

— (1905d). "Number of strokes of the brush in a picture." In: *Nature* 72 (June 29, 1905), p. 198.
— (1906). "Cutting a round cake on scientific principles." In: *Nature* 75 (Dec. 20, 1906), p. 173.
— (1907a). *Inquiries into Human Faculty.* Second edition. Three chapters from the first edition were suppressed. London: Everyman, 1907.
— (1907b). "One vote, one value." In: *Nature* 75 (Feb. 28, 1907), p. 414.
— (1907c). "The ballot box." In: *Nature* 75 (Mar. 28, 1907), pp. 509–510.
— (1907d). "Vox Populi." In: *Nature* 75 (Mar. 7, 1907), pp. 450–1.
— (1908). *Memories of My Life.* London: Methuen, 1908.
— (1909). "Foreword." In: *Eugenics Review* 1 (1909), pp. 1–2.
— (1910a). "Alcoholism and offspring." In: *Times* (39289 June 3, 1910), p. 6.
— (1910b). "Eugenic qualities of primary importance." In: *Eugenics Review* 1 (1910), pp. 74–6.
— (1910c). "Note on the effects of small and persistent influences." In: *Eugenics Review* 1 (1910), pp. 148–9.
Galton, Francis and F. A. Mahomed (1882). "An inquiry into the physiognomy of phthisis by the method of 'composite portraiture'." In: *Guy's Hospital Reports* 25 (Feb. 1882).
Galton, Francis and Edgar Schuster (1906). *Noteworthy Families (Modern Science). An index to kinships in near degrees between persons whose achievements are honourable, and have been publicly recorded.* London: John Murray, 1906.
Gökyi it, Emel Aileen (1994). "The Reception of Francis Galton's 'Hereditary Genius' in the Victorian Periodical Press." In: *Journal of the History of Biology* 27.2 (1994), pp. 215–240.
Goodhart, J. F. and W. H. A. Jacobson (1886). "In memoriam. Frederick Horatio Akbar Mahomed." In: *Guy's Hospital Reports* 43.10 (1886), pp. 1–10.
Gruber, Howard E. and Paul H. Barrett (1974). *Darwin on Man. A Psychological Study of Scientific Creativity, together with Darwin's Early and Unpublished Notebooks.* New York: E. P. Dutton, 1974.
Hahn, Carl Hugo (1984). *Diaries 1837–1860 Part II: 1846–1851.* Ed. by Brigitte Lau. Windhoek, South-West Africa: Archives Services Division of the Department of National Education, Namibia, 1984.
Haier, Richard (2017). *The Neuroscience of Intelligence.* London: Cambridge University Press, 2017.
Hamilton, William D. (1964a). "The genetical evolution of social behaviour. I." In: *Journal of Theoretical Biology* 7.1 (1964), pp. 1–16.
— (1964b). "The genetical evolution of social behaviour. II." In: *Journal of Theoretical Biology* 7.1 (1964), pp. 17–52.
Harris, George (1870). "Hereditary Genius." In: *Journal of Anthropology* 1.1 (July 1870), pp. 56–65.
Hartl, Daniel and Vítězslav Orel (1992). "What Did Gregor Mendel Think He Discovered?" In: *Genetics* 131 (1992), pp. 245–253.
Hasluck, Alexandra (1973). *Royal Engineer. A Life of Sir Edmund DuCane.* London: Angus and Robertson, 1973.
Hayward, John (2013). "Charles Darwin's Mitochondria." In: *Genetics* 194.1 (May 2013), pp. 21–25.
Hearnshaw, L. S. (1979). *Cyril Burt, Psychologist.* London: Hodder and Stoughton, 1979.
Herrnstein, Richard J. and Charles Murray (1994). *The Bell Curve. Intelligence and Class Structure in American Life.* New York: Free Press, 1994.
Herschel, William J. (1880). "Skin Furrows of the Hand." In: *Nature* 23 (Oct. 25, 1880), p. 76.
— (1894). "Finger-Prints." In: Nature (Nov. 22, 1894), pp. 77–8.

Herschel, William James (1916). *The Origin of Fingerprinting.* London: Oxford, 1916.
Hilts, Victor L. (1975). "A Guide to Francis Galton's *English Men of Science.*" In: *Transactions of the American Philosophical Society* New Series 65.5 (Aug. 1975), pp. 3–85.
Hogben, Lancelot (1933). *Nature and Nurture.* London: Allen and Unwin, 1933.
Hollander, Paul (1981). *Political Pilgrims. Travels of Western Intellectuals to the Soviet Union, China, and Cuba, 1928–1978.* Oxford: Oxford University Press, 1981.
Holmes, Samuel (1924). *A Bibliography of Eugenics.* Berkeley, CA: University of California Press, 1924.
Hughes, Alun (2013). "'On the cards': Collective investigation of disease and medical life histories in the nineteenth century." In: *International Journal of Epidemiology* 42 (2013), pp. 683–688.
Huxley, Julian (1932). *A scientist among the Soviets.* New York: Harper, 1932.
Huxley, Leonard, ed. (1902). *Life and Letters of Thomas H. Huxley.* Vol. 2. New York: Appleton, 1902.
Huxley, T. H. (1865). "Emancipation, Black and White." In: *Reader* 5 (1865), pp. 561–2.
— (1894). "Professor Tyndall." In: *Nineteenth Century* CIII (Jan. 1894), pp. 1–11.
Huxley, Thomas Henry (1893). *Darwiniana. Essays.* London: Macmillan, 1893.
— (1896). *Science and Education. Essays.* New York: Appleton, 1896.
— (1910). *Lectures and Lay Sermons.* London: Dent and Dutton, 1910.
Ioannidis, John P. A. (2005). "Why Most Published Research Findings Are False." In: *PLOS Medicine* 2.8 (Aug. 30, 2005).
Jackson, Benjamin Daydon (1906). *George Bentham.* London: J. M. Dent, 1906.
Jackson, Roland (2018). *The Ascent of John Tyndall.* Oxford: Oxford University Press, 2018.
Jacobs, Joseph (1883). "Review of Human Faculty." In: *Athenaeum* 2906 (July 7, 1883). Unsigned, pp. 20–21.
— (1885). "The Jewish Type and Galton's Composite Photographs." In: *The Photographic News* 29.1390 (Apr. 24, 1885), pp. 268–9.
— (1891). *Studies in Jewish Statistics, Social, Vital and Anthropometric.* London: D. Nutt, 1891.
Jenkin, Fleeming (1867). "The Origin of Species." In: *The North British Review* 46 (June 1867). Unsigned, pp. 149–171.
Jensen, Arthur R. (1980). *Bias in Mental Testing.* New York: Free Press, 1980.
— (1998). *The g Factor. The Science of Mental Ability.* Westport, Connecticut: Praeger, 1998.
Jevons, William Stanley (1871). "The Power of Numerical Discrimination." In: *Nature* 3 (Feb. 9, 1871), pp. 281–282.
— (1874). "English Men of Science." In: *Nature* 11 (Dec. 31, 1874), pp. 161–164.
Johnson, Ronald et al. (1985). "Galton's Data a Century Later." In: *The American psychologist* 40 (Sept. 1985), pp. 875–92.
Johnson, Samuel (2006). *Lives of the Poets.* Ed. by Roger Lonsdale. 4 volumes. Oxford: Oxford University Press, 2006.
Jones, M. C. and Arthur Pewsey (2009). "Sinh-arcsinh distributions." In: *Biometrika* 96.4 (2009), pp. 761–780.
Joynson, Robert B. (1989). *The Burt Affair.* London: Routledge, 1989.
Kendall, David G. (1966). "Branching Processes Since 1873." In: *Journal of the London Mathematical Society* s1.41.1 (1966), pp. 385–406.
— (1975). "The Genealogy of Genealogy Branching Processes before (and after) 1873." In: *Bulletin of the London Mathematical Society* 7.3 (1975), pp. 225–253.
Kevles, Daniel (1985). *In the Name of Eugenics. Genetics and the Uses of Human heredity.* New York: Knopf, 1985.
Kidd, Benjamin, ed. (1894). *Social Evolution.* London: Macmillan, 1894.

Koga, Yukiyoshi and G. M. Morant (1923). "On the degree of association between reaction times in the case of different senses." In: *Biometrika* 15 (Dec. 1, 1923), pp. 346–372.
Kottler, Malcolm J. (2006). "Hugo de Vries and the rediscovery of Mendel's laws." In: *Annals of Science* 36.5 (2006), pp. 517–538.
Lambourne, Gerald (1984). *The Fingerprint Story.* London: Chambers, 1984.
Laski, Harold (1912). "A Mendelian View of Racial Heredity." In: *Biometrika* 8.3 (Jan. 1912), pp. 424–430.
Laski, Harold et al. (1913). *On the Correlation of Fertility With Social Value: a Cooperative Study.* Eugenics Laboratory Memoir Series: Volume 18. London: Dulau, 1913.
Lay, Wilfrid (1898). "Mental imagery: Experimentally and subjectively considered." In: *The Psychological Review: Monograph Supplements* 2.3 (1898), pp. 1–59.
Lee, Sidney et al., eds. (1901). *Dictionary of National Biography.* London: Smith, Elder & Co., 1901.
Lewes, George Henry (1856). "Hereditary Influence, Animal and Human." In: *Westminster Review* 66 (July 1856), pp. 75–90.
Lloyd, Francis (1876). *A Scientific View of Mr Francis Galton's Theories of Heredity.* Ed. by William Newton. London: Trübner, 1876.
Lorenz, Konrad (1970). *Studies in Animal and Human Behaviour.* 2 volumes. Cambridge, MA: Harvard University Press, 1970.
Lubbock, Sir John (1865). *Pre-historic Times: As Illustrated by Ancient Remains, and the Manners and Customs of Modern Savages.* First edition. London and Edinburgh: Williams and Norgate, 1865.
Mackenzie, Norman and Jeanne (1982). *The Diary of Beatrice Webb.* Cambridge, MA: Harvard University Press, 1982.
Magnello, Eileen (1993). "Karl Pearson: Evolutionary Biology and the emergence of a Modern Theory of Statistics (1884–1936)." D.Phil. Oxford University, 1993.
— (2004). "The Reception of Mendelism by the Biometricians and the Early Mendelians (1899–1909." In: *A Century of Mendelism in Human Genetics.* Ed. by Milo Keynes, A. W. F. Edwards, and Robert Peel. Boca Raton, FL: CRC Press, 2004, pp. 19–32.
Manly, Bryan (1985). *The Statistics of Natural Selection.* London: Chapman and Hall, 1985.
Marchant, James (1916). *Alfred Russel Wallace. Letters and Reminiscences.* 2 volumes. London: Cassell, 1916.
Marcy, Randolph (1859). *The Prairie Traveler.* New York: Harper, 1859.
McDaniel, M. A. (2005). "Big-brained people are smarter: a meta-analysis of the relationship between in vivo brain volume and intelligence." In: *Intelligence* 33 (2005), pp. 337–346.
McDougall, William (1930). "Autobiography." In: *History of Psychology in Autobiography.* Ed. by Carl Murchison. Vol. 1. Worcester, MA: Clark University Press, 1930.
Means, John O. (1876). *The Prayer-Gauge Debate.* Editor is not credited but is thought to be Means. Boston: Congregational Publishing Society, 1876.
Mendel, Johann Gregor (1866). "Versuche über Pflanzen-Hybriden." In: *Verhandlungen des naturforschenden Vereines in Brünn* 4 (1866), pp. 3–47.
— (1870). "Über einige aus künstlicher Befruchtung gewonnenen Hieracium-Bastarde." In: *Verhandlungen des naturforschenden Vereines in Brünn* 8 (1870), pp. 26–31.
Merrivale, Herman (1870). "Hereditary Genius." In: *Edinburgh Review* 132.269 (July 1870), pp. 100–125.
Mitchell, Frank D. (1907). "Mathematical Prodigies." In: *The American Journal of Psychology* 18.1 (Jan. 1907), pp. 61–143. DOI: 10.2307/1412172. URL: https://www.jstor.org/stable/1412172.
Mitchell, Kevin (2018). *Innate. How the Wiring of Our Brains Shapes Who We Are.* Princeton, NJ: Princeton University Press, 2018.

Mivart, St George Jackson (1874). "Primitive man: Tylor and Lubbock." In: *Quarterly Review* 137 (July 1874), pp. 40–77.
Morant, G. M. (1925). "A Study of Egyptian Craniology from Prehistoric to Roman Times." In: *Biometrika* 17.1 (June 1925), pp. 1–52.
Muller, Herman F. (1925). "Mental Traits and Heredity as Studied in a Case of Identical Twins Reared Apart." In: *Journal of Heredity* 16 (1925), pp. 433–448.
Museum, South Kensington (1877). *Catalogue of the special loan collection of scientific apparatus at the South Kensington Museum. 1876.* 2 volumes. London: G. E. Eyre and W. Spottiswoode, 1877.
Naroll, Raoul (1961). "Two Solutions to Galton's Problem." In: *Philosophy of Science* 28 (1 Jan. 1961), pp. 15–39.
Neale, Michael C. and Lon R. Cardon (1992). *Methodology for Genetic Studies of Twins and Families.* Dordrecht: Kluwer, 1992.
Neville-Rolfe, Sybil (1949). *Social Biology and Welfare.* London: Allen and Unwin, 1949.
Norton, Bernard and Egon Sharples Pearson (1976). "A Note on the Background to, and Refereeing of, R. A. Fisher's 1918 Paper 'On the Correlation between Relatives on the Supposition of Mendelian Inheritance'." In: *Notes and Records of the Royal Society of London* 31 (1976), pp. 151–162.
Oakeshott, Michael (1991). *Rationalism in Politics.* New and expanded edition, originally published by Methuen 1962. Indianapolis: Liberty Fund, 1991.
Olby, Robert C. (1966). *Origins of Mendelism.* New York: Schocken Books, 1966.
Ormsby, John (1869). "Hereditary Genius." In: *Pall Mall Gazette* (Feb. 18, 1869). Unsigned.
Paget, Stephen, ed. (1901). *Memoirs and Letters of Sir James Paget.* London: Longmans, 1901.
Pearson, Hesketh (1921). *Modern Men and Mummers.* London: Allen and Unwin, 1921.
— (1938). *Thinking it Over. The Reminiscences of Hesketh Pearson.* New York: Harper & Brothers, 1938.
— (1964). *Extraordinary People.* New York: Harper and Row, 1964.
Pearson, Karl (1893). "Contributions to the Mathematical Theory of Evolution." In: *Proceedings of the Royal Society of London* 54 (1893), pp. 329–333.
— (1894). "Contributions to the Mathematical Theory of Evolution." In: *Philosophical Transactions of the Royal Society of London. A* 185 (1894), pp. 71–110.
— (1896). "Regression, Heredity and Panmixia." In: *Philosophical Transactions of the Royal Society of London. A* 187 (1896), pp. 253–318.
— (1898). "On the Law of Ancestral Heredity." In: *Proceedings of the Royal Society of London* 62 (1898), pp. 386–412.
— (1900). *The Grammar of Science.* Second edition. Revised and enlarged. London: Adam and Charles Black, 1900.
— (1902). "On the Correlation of Intellectual Ability with the Size and Shape of the Head. (Preliminary Notice. A New Year's Greeting to Francis Galton, 1902)." In: *Proceedings of the Royal Society of London* 69 (1902), pp. 333–342.
— (1904). "On a generalised theory of alternative Inheritance, with special reference to Mendel's laws." In: *Philosophical Transactions of the Royal Society of London. Series A* 20 (1904), pp. 353–86.
— (1906). "On the Relationship of Intelligence to Size and Shape of Head, and to Other Physical and Mental Characters." In: *Biometrika* 5.1 (Oct. 1906), pp. 105–146.
— (1909). "On the ancestral gametic correlations of a mendelian population mating at random." In: *Proceedings of the Royal Society of London B* 81 (1909), pp. 225–229.
— (1911). "The Late Sir Francis Galton." In: *Nature* 88 (1911), p. 76.

— (1914). *The Life, Letters and Labours of Francis Galton. Volume I: Birth 1822 to Marriage 1853.* Cambridge: Cambridge University Press, 1914.
— (1924). *The Life, Letters and Labours of Francis Galton. Volume II: Researches of Middle Life.* Cambridge: Cambridge University Press, 1924.
— (1930a). *The Life, Letters and Labours of Francis Galton. Volume IIIa: Correlation, Personal Identification and Eugenics.* Cambridge: Cambridge University Press, 1930.
— (1930b). *The Life, Letters and Labours of Francis Galton. Volume IIIb: Characterisation, Especially by Letters; Index.* Cambridge: Cambridge University Press, 1930.
Petrie, Flinders (1887). *Racial Photographs from the Egyptian Monuments.* London: R. C. Murray, 1887.
— (1931). *Seventy Years in Archaeology.* London: Sampson Low, 1931.
Philip, Levy (1995). "Charles Spearman's contributions to test theory." In: *British Journal of Mathematical and Statistical Psychology* 48 (1995), pp. 221–235.
Pitt-Rivers, W. H. R., ed. (1901). *Reports of the Cambridge Anthropological Expedition to the Torres Straits. Physiology and Psychology.* Cambridge: Cambridge University Press, 1901.
Plomin, Robert (1994). *Genetics and Experience.* New York: Sage, 1994.
— (2018). *Blueprint.* New York: MIT Press, 2018.
Plomin, Robert et al. (2012). *Behavioral Genetics.* Sixth edition. New York: Worth Publishers, 2012.
Polderman, Tinca J. C. et al. (2015). "Meta-analysis of the heritability of human traits based on fifty years of twin studies." In: *Nature Genetics* 47 (May 18, 2015), p. 702.
Poppenoe, Paul (1922). "Twins Reared Apart." In: *Journal of Heredity* 13 (1922), pp. 142–144.
Popper, Karl (1959). *The Logic of Scientific Discovery.* London: Hutchinson, 1959.
Porter, Theodore (2004). *Karl Pearson. The Scientific Life in a Statistical Age.* Princeton, NJ: Princeton University Press, 2004.
Pritchard, Chris (2018). *A Common Family Weakness for Statistics. Essays on Francis Galton, George Darwin and the Normal Curve of Evolutionary Biology.* Leicester: Mathematical Association, 2018.
Pritchard, Christopher B. (2005). "The Normal Curve of Evolutionary Biology, 1869–1877, with Special Reference to the Support Given to Francis Galton by George Darwin." PhD. Open University, 2005.
— (2006). "Bagatelle as the inspiration for Galton's Quincunx." In: *BSHM Bulletin: Journal of the British Society for the History of Mathematics* 21.2 (2006), pp. 102–110.
Provine, William B. (2001). *The Origins of Theoretical Population Genetics. With a New Afterword.* Second edition. First published in 1971. Chicago: University of Chicago Press, 2001.
Raoul Naroll and Roy G. D'Andrade (1963). "Two Further Solutions to Galton's Problem." In: *American Anthropologist* 65 (5 1963), pp. 1053–1067.
Rende, Richard D., Robert Plomin, and Steven G. Vandenberg (1990). "Who Discovered the Twin Method?" In: *Behavior Genetics* 20.2 (1990), pp. 277–285.
Rentoul, Robert Reid (1906). *Race Culture or Race Suicide. a Plea for the Unborn.* London: Walter Scott, 1906.
Rhodes, Henry T. F. (1956). *Alphonse Bertillon. Father of Scientific Detection.* London: Harrap, 1956.
Robertson, Cunningham (1878). "The Late Mr. G. P. Bidder." In: *Spectator* (Dec. 28, 1878), pp. 1634–5.
Romanes, George J. (1883). "Human Faculty and its Development." In: *Nature* 28 (May 31, 1883). Review of Inquiries into Human Faculty, pp. 97–98.

Rushton, J. and C. Davison Ankney (1996). "Brain size and cognitive ability: Correlations with age, sex, social class, and race." In: *Psychonomic Bulletin and Review* 3.1 (1996), pp. 21–36.
Saleeby, C. W. (1910). "The Progress of Eugenics." In: *New Age* 7.1 (May 5, 1910). Supplement, pp. 3–4.
Scarr, Sandra and Richard A. Weinberg (1983). "The Minnesota Adoption Studies: Genetic Differences and Malleability." In: *Child Development* (1983), pp. 260–7.
Scheper-Hughes, Nancy (1979). *Saints, Scholars and Schizophrenics. Mental illness in Rural Ireland.* Berkeley, CA: University of California Press, 1979.
Schmidt, Frank L. and John E. Hunter (2004). *Methods of Meta Analysis. Correcting error and bias in research findings.* Second edition. New York: Sage, 2004.
Schmidt, Paul J. (1968). "Transfusion in America in the Eighteenth and Nineteenth Centuries." In: *N Engl J Med* 279.24 (Dec. 1968), pp. 1319–1320. DOI: 10.1056/NEJM196812122792406. URL: https://doi.org/10.1056/NEJM196812122792406.
Schuster, Edgar (1912). *Eugenics.* London: Collins, 1912.
Schwantes, Robert S. (1953). "Christianity versus Science: a conflict of Ideas in Meiji Japan." In: *Far Eastern Quarterly* 12.2 (1953), pp. 123–132.
Schwartz, Joel S. (2010). *Darwin's Disciple. George John Romanes, a Life in Letters.* Philadelphia: APS Lightning Rod Press, 2010.
Secrist, Horace (1933). *The triumph of mediocrity in business.* Evanston, IL: Bureau of Business Research, Northwestern University, 1933.
Segal, Nancy L. (1999). *Entwined Lives. Twins and what they tell us about human behavior.* New York, NY: Dutton, 1999.
— (2012). *Born Together–Reared Apart. The Landmark Minnesota Twin Study.* Cambridge, MA: Harvard University Press, 2012.
Shadwell, Arthur (1905). "Identification by Fingerprints." In: *Times Literary Supplement* (Oct. 13, 1905), p. 335.
Sharp, Stella Emily (1899). "Individual Psychology: A study in psychological method." In: *American Journal of Psychology* 10.3 (1899), pp. 329–391.
Smith, John Maynard and David Harper (2003). *Animal Signals.* Oxford: Oxford University Press, 2003.
Smyth, Charles Piazzi (1864). *Our Inheritance in the Great Pyramid.* London: Alexander Strahan, 1864.
Sokal, Michael Mark (1972). "The Education and Psychological Career of James McKeen Cattell, 1860–1904." PhD Dissertation. Case Western Reserve University, 1972.
— (1981). *James McKeen Cattell's Journal and Letters from Germany and England, 1880–1888.* Cambridge, MA: MIT Press, 1981.
Solomon, Yvonne (2023). *The Galton Family.* Quiet Waters, The Street, Ewelme, Wallingford. OX10 6HQ, England: Privately published, 2023.
Spaeth, Joseph (1860). "Studien über Zwillinge." In: *Zeitschrift der K. K. Gesellschaft der Ärzte zu Wien* 16.15 (1860), pp. 225–231.
Spearman, Charles (1904a). "'General Intelligence,' Objectively Determined and Measured." In: *The American Journal of Psychology* 15.2 (Apr. 1904), pp. 201–292.
— (1904b). "The Proof and Measurement of Association between Two Things." In: *The American Journal of Psychology* 15.1 (Jan. 1904), pp. 72–101.
— (1927). *The Abilities of Man. Their Nature and Measurement.* London: Macmillan, 1927.
Spottiswoode, William (1861). "On Typical Mountain Ranges: An Application of the Calculus of Probabilities to Physical Geography." In: *Journal of the Royal Geographical Society of London* 31 (1861). Read before the RGS on 1860-04-23, pp. 149–154.

Stern, Curt and Eva R. Sherwood, eds. (1966). *The Origin of Genetics. A Mendel Sourcebook.* San Francisco: W. H. Freeman, 1966.
Stigler, Stephen M. (1986). *The History of Statistics. The Measurement of Uncertainty before 1900.* Cambridge, MA: Harvard University Press, 1986.
— (1989). "Francis Galton's Account of the Invention of Correlation." In: *Statistical Science* 4.2 (1989), pp. 73–86.
— (1997). "Regression towards the mean, historically considered." In: *Statistical Methods in Medical Research* 6.2 (1997), pp. 103–114.
Stulp, Gert and Simons, Mirre J. P. and Grasman, Sara and Pollet, Thomas V. (2017). "Assortative mating for human height: A meta-analysis." In: *American Journal of Human Biology* 29.1 (2017). Jan-Feb.
Taylor, Thomas (1877). "Hand Marks Under the Microscope." In: *American Journal of Microscopy and Popular Science* 2 (1877), p. 89.
Thompson, Sir Henry (1872a). "On Prayer." In: *Contemporary Review* 20 (Oct. 1872). Unsigned, pp. 767–777.
— (1872b). "The Prayer for the Sick." In: *Contemporary Review* 20 (July 1872). Unsigned, communicated by John Tyndall, pp. 205–210.
Thomson, William (1889). *Popular Lectures and Addresses.* 3 volumes. London: Macmillan, 1889.
Tobias, Phillip Vallentine (1991). *Images of Humanity. The Selected Writings of Phillip V. Tobias.* Rivonia, South Africa: Ashanti Publishing, 1991.
Torrey, Edwin Fuller (1980). *Schizophrenia and Civilization.* New York: Jason Aronson, 1980.
Torrey, Edwin Fuller and Judy Miller (2001). *The Invisible Plague. The Rise of Mental Illness from 1750 to the Present.* Rutgers, New Brunswick: Rutgers University Press, 2001.
Tredoux, Gavan, ed. (2016). *The Book of Burtoniana.* 4 volumes. burtoniana.org, 2016.
— (2018). *Comrade Haldane is too busy to go on holiday. The Genius who Spied for Stalin.* New York City: Encounter, 2018.
— (2023a). *Francis Galton. Selected Papers.* galton.org, 2023.
— (2023b). *Francis Galton and Alphonse de Candolle. Notes and Correspondence.* galton.org, 2023.
— (2023c). *Francis Galton's Nature and Nurture. 1822–1865.* galton.org, 2023.
— (2023d). *Galton on Mars. The Discontinuous Variation Notebook.* galton.org, 2023.
— (2023e). *Sir Francis Sacheverel Darwin.* galton.org, 2023.
Tukey, John (1977). *Exploratory Data Analysis.* Reading, MA: Addison-Wesley, 1977.
Tylor, Edward Burnett (1874). "Nuts and Men." In: *Spectator* (June 6, 1874), p. 722.
Tyndall, John (1865a). "Prayer and Cholera." In: *Pall Mall Gazette* II.212 (Oct. 12, 1865), p. 4.
— (1865b). "The Constitution of the Universe." In: *Fortnightly Review* 3 (Dec. 1865), pp. 129–144.
Venn, John (1888). "Cambridge Anthropometry." In: *Journal of the Royal Anthropological Institute* 18 (Apr. 24, 1888), pp. 140–154.
— (1889). "Natural Inheritance." In: *Mind* 14.55 (July 1889), pp. 414–420.
— (1890). "Cambridge Anthropometry." In: *Nature* 41 (Mar. 13, 1890), pp. 450–454.
Verburgt, Lukas M. (2022). *John Venn. Unpublished Writings and Selected Correspondence.* London: Springer, 2022.
Walford, Edward (1865a). *Men of the Time. a biographical dictionary of eminent living characters, (including women).* London: Routledge, 1865.
— (1865b). *Men of the Time: A Biographical Dictionary of Eminent Living Characters (including women).* London: Routledge, 1865.

Walker, Margot (1997). "James Murie 1832–1925: The Aggressive Librarian." In: *The Linnean* 13.3 (1997), pp. 23–24.
Wallace, Alfred Russel (1870). "Hereditary Genius." In: *Nature* 1 (Mar. 17, 1870), pp. 501–503.
— (1890). "Human Selection." In: *Fortnightly Review* 38.285 (Sept. 1890), pp. 325–337.
— (1906). *My Life. A Record of Events and Opinions.* 2 volumes. New York: Dodd and Mead, 1906.
Wasserman, John (2012). "A history of Intelligence Assessment." In: *Contemporary Intellectual Assessment.* Ed. by Dawn M. Flanagan and Patti L. Harrison. Third edition. New York: Guilford, 2012, pp. 3–55.
Watson, H.W. (1873). "Solution to Problem 4001." In: *Educational Times* 1 (1873), p. 115.
Webb, Beatrice (1926). *My Apprenticeship.* London: Longmans, 1926.
Webb, Sidney (1907). *The Decline in the Birth Rate.* London: Fabian Society, 1907.
Welby, Victoria Lady (1903). *What is meaning? Studies in the Development of Significance.* London: Macmillan, 1903.
— (1911). *Significs and Language.* London: Macmillan, 1911.
Weldon, W. F. R. (1890). "The Variations Occurring in Certain Decapod Crustacea.– I. Crangon vulgaris." In: *Proceedings of the Royal Society of London* 47 (1890), pp. 445–453.
— (1892). "Certain Correlated Variations in Crangon vulgaris." In: *Proceedings of the Royal Society of London* 51 (1892), pp. 1–21.
— (1893). "On Certain Correlated Variations in Carcinus maenas." In: *Proceedings of the Royal Society of London* 54 (1893), pp. 318–329.
— (1895). "An Attempt to Measure the Death-rate Due to the Selective Destruction of Carcinus maenas with Respect to a Particular Dimension." In: *Proceedings of the Royal Society* 57 (1895), pp. 360–79.
— (1898). "Presidential address to the zoological section of the British Association." In: *Report of the British Association for the Advancement of Science* (1898), pp. 887–902.
— (1902a). "Mendel's Laws of Alternative Inheritance in Peas." In: *Biometrika* 1 (1902). Part 2, October 1901 to August 1902, pp. 228–254.
— (1902b). "On the Ambiguity of Mendel's Categories." In: *Biometrika* 2 (1902). November 1902 to November 1903, pp. 44–55.
Whymper, Edward (1865). "Camping Out." In: *Alpine Journal* 2 (Mar. 1865), pp. 1–11.
— (1871). *Scrambles amongst the Alps in the years 1860-69.* London: John Murray, 1871.
Wilson, Edward O. (2014). *Sociobiology. The New Synthesis.* Cambridge, MA: Harvard University Press, 2014.
Wilton, George Wilton (1938). *Fingerprints. History, Law and Romance.* London: W. Hodge and Co., 1938.
— (1962). *Fingerprints. Six years' strife with British government departments, a tale documentarily told of groundless opposition to civil list pensions for [the daughters] of Henry Faulds.* North Berwick: Tantallon Press, 1962.
Wissler, Clark (1901). "The Correlation of Mental and Physical Tests." In: *Psychological Review Monograph Supplements* 3.6 (1901). Based on the author's PhD dissertation at Columbia.
Witkowski, Jan A. (2008). "Charles Benedict Davenport, 1866–1944." In: *Davenport's Dream. 21st Century Reflections on Heredity and Eugenics.* Ed. by Jan A. Witkowski and John R. Inglis. Cold Spring Harbor, NY: Cold Spring Harbor Press, 2008, pp. 35–58.
Wolf, Theta H. (1973). *Alfred Binet.* Chicago: University of Chicago, 1973.
Wrangham, Richard (2019). *The Goodness Paradox. The Strange Relationship Between Virtue and Violence in Human Evolution.* New York: Pantheon, 2019.
Yule, George Udny (1902). "Mendel's Laws and their probable relations to intra-racial heredity." In: *New Phytologist* 1 (1902), pp. 193–207, 228–238.

Zenderland, Leila (1998). *Measuring Minds. Henry Herbert Goddard and the Origins of American Intelligence Testing.* Cambridge: Cambridge University Press, 1998.

Printed in Dunstable, United Kingdom

91343480R00279